MICROELECTRONICS

MICROELECTRONICS

edited by

Jerry C. Whitaker

CRC Press
Boca Raton London New York Washington, D.C.

Library of Congress Cataloging-in-Publication Data

Microelectronics / Jerry C. Whitaker, editor.
p. cm.
Includes bibliographical references.
ISBN 0-8493-0050-9 (alk. paper)
1. Microelectronics. I. Whitaker, Jerry C.
TK7874.M4587 1999
621.381—dc21 99-44874
CIP

The material in this book was taken from The Electronics Handbook, edited by Jerry D. Whitaker.

International Standard Book Number 0-8493-0050-9
Library of Congress Card Number 99-44874
Printed in the United States of America 1 2 3 4 5 6 7 8 9 0
Printed on acid-free paper

Preface

Microelectronics is intended for engineers and technicians involved in the design, production, installation, operation, and maintenance of microelectronic devices and systems. This publication provides the reader with an expansive understanding of the technologies shaping microelectronics and how these technologies affect the use of the devices. It details the aspects of the field integral to practicing electrical engineers, such as design, the interrelationship of memory to overall performance, microprocessor programmability and program development tools, the impact of multichip module technology, Field Programmable Gate Array Devices design tools, and testing environments. References are provided throughout the book to direct readers to more detailed information on important subjects.

The purpose of this book is to provide in a single volume a comprehensive reference for the practicing engineer in industry, government, and academia. *Microelectronics* is the most convenient and complete reference to microelectronics applications. It presents the fundamentals of microelectronics, a summary of the current state of electronics, and the innovative directions it is taking.

Contributors

Samuel O. Agbo
California Polytechnic University
San Luis Obispo, California

W.F. Ames
Georgia Institute of Technology
Atlanta, Georgia

Constantine N. Anagnostopoulos
Eastman Kodak Corporation
Rochester, New York

Carl Bentz
Intertec Publishing
Overland Park, Kansas

Bruce W. Bomar
The University of Tennessee Space Institute
Tullahoma, Tennessee

George Cain
Georgia Institute of Technology
Atlanta, Georgia

Jonathon A. Chambers
Imperial College of Science, Technology and Medicine
London, England

Tom Chen
Colorado State University
Fort Collins, Colorado

James G. Cottle
Hewlett-Packard
Palo Alto, California

Eugene D. Fabricius
California Polytechnic University
San Luis Obispo, California

Bradly K. Fawcett
Xilinx, Inc.
San Jose, California

Robert J. Feugate, Jr.
Northern Arizona University
Flagstaff, Arizona

Paul D. Franzon
North Carolina State University
Raleigh, North Carolina

Susan A. R. Garrod
Purdue University
West Lafayette, Indiana

Robert D. Greenberg
Federal Communications Commision
Washington, DC

Paul P. K. Lee
Eastman Kodak Corp.
Rochester, New York

Shih-Lien L. Lu
Oregon State University
Corvallis, Oregon

Victor Meeldijk
Diagnostic Retrieval Systems, Inc.
Oakland, New Jersey

Wayne Needham
Intel Corporation
Chandler, Arizona

James F. Shackelford
University of California
Davis, California

Sawasd Tantaratana
National Electronics and Computer Technology Center
Bangkok, Thailand

Jerry C. Whitaker
Technical Press
Beaverton, Oregon

Contents

Microelectronics

Conversion Factors, Standards, and Constants
Jerry C. Whitaker,

I

Microelectronics

1 Introduction

Susan A.R. Garrod
Purdue University

MICROELECTRONICS HAVE PLAYED A FUNDAMENTAL ROLE in shaping the entire electronics industry as well as related industries that rely on electronic components and subsystems. In an industry where changes happen frequently and dramatically, the constant themes that have persisted are miniaturization, increased speed, reduced power consumption, and reduced cost. These effects have resulted in an increased demand for microelectronics in all sectors of consumer, industrial, and military products. The advancements in manufacturing have enabled these devices to be produced in very high volumes which reduces the cost per device. In turn, the lower cost fuels future demand which pushes the industry for further miniaturization and higher volume manufacturing. *Business Week* reported that "cheap technology" has now "assumed a central role in economies around the world" [March 6, 1995:77].

The combination of reduced size, increased speed, and increased capacity of microelectronics was originally observed by Gordon E. Moore, Chairman of Intel, when during the 1960s he commented that the feature size of semiconductor transistors reduced by 10% per year. In fact, the reduction has been even more dramatic than that. The capacity of dynamic random access memory (DRAM) integrated circuits has quadrupled approximately every three years. The increased density of transistors contained in the microelectronic devices has resulted in a phenomenon of virtually "free" computing power. The result of having such free computing power available is that the potential to use that power is available to a greater cross section of society.

The societal demand for microelectronics has evolved from largely a military-driven demand to one that is now largely consumer-driven. Consequently, the device features have also been targeted at consumer needs, such as low power, low cost, and mass market applications, rather than military needs, such as meeting military specifications for reliability and packaging, specialized applications, and the resulting high cost of such devices. The performance of microelectronics is measured, thus, from the viewpoint of the technological aspects of the device, as well as from the viewpoint of end user effectiveness. The goal is to enable the end user of the devices to perform more complex tasks in a more efficient manner than what was previously possible.

This book focuses on the technological issues within specific microelectronic technologies and how they affect the push of technology that drives the next generation of microelectronics. The chapters describe the areas of integrated circuit manufacturing and design, digital logic circuits, memory, microprocessors, digital to analog (D/A) and analog to digital (A/D) converters, application specific integrated circuits (ASICs), digital filters, multichip module technology, software development tools, IC packaging, and on-chip testing.

The first chapter on integrated circuits provides further detailed information on the historical as well as projected advancements expected in the design and manufacturing of semiconductors. The data suggest that by the year 2007 the industry will manufacture 16-Gb DRAMs with a 1-GHz clock rate. The author discusses the engineering challenges presented by these design projections, such as optimal gate design, interconnection parasitic effects, and timing issues.

0-8493-0050-9/00/$0.00+$.50

The next chapter describes in further depth the IC design process for very large-scale integrated (VLSI) circuits. This includes a discussion of gate arrays, standard cells, programmable logic devices, and programmable logic arrays. The concerns of parasitic capacitance and transmission line delays in the VLSI design are addressed as well.

An overview of digital logic families is presented in the next chapter, and many comparisons are drawn to illustrate the benefits and drawbacks of the specific TTL, CMOS, and ECL families. Both silicon and gallium arsenide semiconductors are included in the discussion. Finally, the comparative discussion is extended to programmable logic devices such as programmable logic arrays, field programmable gate arrays, and complex PLDs.

Memory devices are the subject of Chapter 5. Memory devices are described from the standpoint of the functions of the devices as well as the interrelationship of memory to the overall performance and operation of a computing system, consisting of a microprocessor plus various forms of memory. The concept of a memory hierarchy is presented to explain the interrelated use of different memory technologies to optimize the overall performance of the computing system.

Chapter 6 provides an overview of microprocessors and related areas of concern. The focus of the chapter is on microprocessor architectures, historical developments, and a comparison of complex instruction set computer (CISC) and reduced instruction set computer (RISC) processors. In addition, the programmability of microprocessors, and the program development tools, are discussed as these affect the ability of the manufacturer to bring the product to market quickly.

The next chapter summarizes fundamental concepts concerning A/D and D/A conversion. Various A/D and D/A techniques are described with reference to the types of applications best suited for each conversion technique.

Chapter 8 provides an overview of ASICs. Full custom and semicustom ASIC design techniques are explained. Attention is given to describing the design process, its time and cost issues, and the choices of tools a designer has within the ASIC design environment.

Chapter 9 provides a detailed and quantitative overview of digital filters. The discussion focuses on finite impulse response (FIR) and infinite impulse response (IIR) digital filter design.

Chapter 10 describes multichip module technology and explains how it provides designers opportunities to improve the performance and reduce the cost of systems employing this technology. Although the multichip module technology is a packaging technology, the author stresses that it has a significant impact on the design of semiconductors as well as on the end user value realized from the systems.

Software development tools for field programmable gate array (FPGA) devices are described in the next chapter, where the growth in the use of FPGAs has driven the demand for these software tools. The FPGA design process is described in terms of the tools available for each step in the process.

An overview of integrated circuit packages is provided in the next chapter. The packaging described includes surface mount, chip-scale, bare die, through-hole, and module assemblies.

An appropriate topic for the final chapter is the testing of integrated circuits. On-chip testing techniques are described in terms of the differences of the test environment vs the operating environment and the problems encountered when testing integrated circuits.

Overall, this book strives to give the reader a broad understanding of the technologies shaping microelectronics and how these technologies affect the end uses of the devices. After reading the chapters, the reader should have a sound impression of the future directions of microelectronic technology developments and the likely areas of growth in applications of the devices.

2

Integrated Circuits

Tom Chen
Colorado State University

2.1 Introduction

Transistors and their fabrication into *very large scale integrated (VLSI)* circuits are the invention that has made modern computing possible. Since its inception, integrated circuits have been advancing rapidly from a few transistors on a small silicon die in the early 1960s to more than 4 million transistors integrated on to a single large silicon substrate. It has been predicted that by the year 2000, the number of transistors on a single silicon substrate could exceed 15 million. The dominant type of transistor used in today's integrated circuits is the metal-oxide-semiconductor (MOS) type transistor. The rapid technological advances in integrated circuit (IC) technology accelerated during and after the 1980s, and one of the most influential factors for such a rapid advance is the technology scaling, that is, the reduction in MOS transistor feature sizes. The MOS feature size is typically measured by the MOS transistor channel length. During the last 24 years the minimum feature size has evolved from 6 to 0.35 μm currently. By the year 2000, the minimum feature size will reach 0.18 μm. Table 2.1 shows the Semiconductor Industry Association's (SIA) technology roadmap through the year 2007. The smaller the transistors, more dense the integrated circuits in terms of the number of transistors packed on to a unit area of silicon substrate, and the faster the transistor can switch. Not only can we pack more transistors onto a unit silicon area, the chip size also increases with the projection that it will reach 800 mm^2 in the year 2000 for microprocessors and 500 mm^2 for dynamic random access memory (DRAM) chips. As the transistor gets smaller and silicon chip size gets bigger, the transistor's driving capability decreases and the interconnect parasitics (interconnect capacitance and resistance) increases. Consequently, the entire VLSI system has to be designed very carefully to meet the speed demands of the future. Common design issues include optimal gate design and transistor sizing, minimization of clock skew and proper timing budgeting, and realistic modeling of interconnect parasitics.

TABLE 2.1 SIA's Technology Roadmap

	1995	1998	2001	2004	2007
Feature size, μm	0.35	0.25	0.18	0.12	0.1
Gates/chip	800 K	2 M	5 M	10 M	20 M
Bits/chip for DRAM	64 M	256 M	1 G	4 G	16 G
Bits/chip for SRAM	16 M	64 M	256 M	1 G	4 G
Logic chip size, mm	400	600	800	1000	1250
Memory chip size, mm	200	320	500	700	1000
Wafer diameter, mm	200	200–400	200–400	200–400	200–400
No. of metal levels	4–5	5	5–6	6	6–7
Max. power, W/die	15	30	40	40–120	40–200
Power supply, V	2.2	2.2	1.5	1.5	1.5
Number of I/Os	750	1500	2000	3500	5000
On-chip clock rate, mHz	200	350	500	700	1000
Off-chip clock rate, mh	100	175	250	350	500

Source: Semiconductor Industry Association.

0-8493-0050-9/00/$0.00+$.50

2.2 High-Speed Design Techniques

A modern VLSI device typically consists of several *mega-cells*, such as memory blocks and data-path arithmetic blocks, and a lot of basic MOS logic gates, such as inverters and NAND/NOR gates. **Complementary MOS (CMOS)** is one of the most widely used logic families, mainly because of its low-power consumption and high-noise margin. Other logic families include NMOS and PMOS logic. Because of its popularity, only the CMOS logic will be discussed. Many approaches to high-speed design discussed here are equally applicable to other logic families.

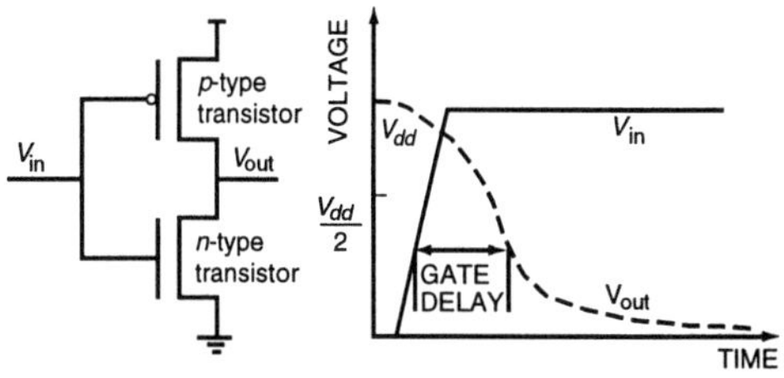

FIGURE 2.1 Gate delay in a single inverter.

Optimizing a VLSI device for high-speed operation can be carried out at the system level, as well as at the circuit and logic level. To achieve the maximum operating speed at the circuit and logic levels for a given technology, it is essential to properly set the size of each transistor in a logic gate to optimally drive the output load. If the output load is very large, a string of drivers with geometrically increasing sizes will be needed. The size of transistors in a logic gate is also determined by the impact of the transistors as a load to be driven by their preceding gates.

Optimization of Gate Level Design

To optimize the gate level design, let us look at the performance of a single CMOS inverter as shown in Fig. 2.1. Delay of a gate is typically defined as the time difference between input transition and output transition at 50% of supply voltage. The inverter gate delay can be analytically expressed as

$$T_d = C_l(A_n/\beta_n + A_p/\beta_p)/2$$

where C_l is the load capacitance of the inverter; β_n and β_p are the forward current gains of n-type and p-type transistors, respectively, and are proportional to the transistor's channel width and inversely proportional to the transistor's channel length; A_n and A_p are process related parameters for a given supply voltage and they are determined by

$$A_n = [2n/(1-n) + \ell n((2(1-n) - V_0)/V_0)][V_{dd}(1-n)]$$
$$A_n = [-2p/(1+p) + \ell n((2(1+p) - V_0)/V_0)][V_{dd}(1+p)]$$

where $n = V_{\text{th}n}/V_{dd}$ and $p = V_{\text{th}p}/V_{dd}$. $V_{\text{th}n}$ and $V_{\text{th}p}$ are gate threshold voltages for n-channel and p-channel transistors, respectively. This expression does not take the input signal slope into account. Otherwise, the expression would become more complicated. For more complex CMOS gates, an equivalent inverter structure is constructed to reflect the effective strength of their p-tree and n-tree in order to apply the inverter delay model.

In practice, CMOS gate delay is treated in a simple fashion. The delay of a logic gate can be divided into two parts: the intrinsic delay D_{ins}, and the load-related delay D_{load}. The gate intrinsic delay is determined by the internal characteristics of the gate including the implementing technology, the gate structure, and the transistor sizes. The load-related delay is a function of the total load capacitance at the gate's output. The total gate delay can be expressed as

$$T_d = D_{\text{ins}} + C_l^* S$$

where C_l is the total load capacitance and S is the factor for gate's driving strength. $C_l^* S$ represents the gate's load-related delay. In most CMOS circuits using leading-edge submicron technologies, the total delay of a gate can be dominated by the load-related delay. For an inverter in a modern submicron CMOS technology of around 0.5-μm feature size, D_{ins} can range from 0.08 to 0.12 ns and S can range from 0.00065 to 0.00085 ns/fF depending on specifics in the technology and the minimum transistor feature size. For other more complex gates such as NAND and NOR gates, D_{ins} and S will generally increase.

To optimize a VLSI circuit for its maximum operating speed, **critical paths** must be identified. A critical path in a circuit is a signal path with the longest time delay from a primary input to a primary output. The time delay on the critical path in a circuit determines the maximum operating speed of the circuit. The time delay of a critical path can be minimized by altering the size of the transistors on the critical path. Using the lumped resistor–capacitor (RC) delay model, the problem of transistor sizing can be formulated to an optimization problem with a convex relationship between the path delay and the sizes of the transistors on the path. This optimization problem is simple to solve. The solutions often have 20–30% deviation, however, compared to the **SPICE** simulation results. Realistic modeling of gate delay taking some second-order variables, such as input signal slope, into consideration has shown that the relationship between the path delay and the sizes of the transistors on the path is not convex. Such detailed analysis led to more sophisticated transistor sizing algorithms. One of these algorithms suggested using genetic methods to search for an optimal solution and has shown some promising results.

Clocks and Clock Schemes in High-Speed Circuit Design

Most of the modern electronic systems are **synchronous systems.** The clock is a central pace setter in a synchronous system to step the desired system operations through various stages of the computation. Latches are often used to facilitate catching the output data at the end of each clock cycle. Figure 2.2 shows the typical synchronous circuit with random logic clusters as computational blocks and latches as pace setting devices. When there exist feedbacks, as shown in Fig. 2.2, the circuit is referred to as *sequential circuit.*

A latch is also called a *register* or a *flip-flop*. The way a latch catches data depends on how it is triggered by the clock signal. Generally, there are level-triggered and edge-triggered latches, the former can be further subdivided according to the triggering polarity as positive or negative level or edge-triggered latches. The performance of a digital circuit is often determined by the maximum clock frequency the circuit can run. For a synchronous digital circuit to function properly, the longest delay through any combinational clusters must be less than the clock cycle period. Therefore, the following needs to be done for high-speed design:

- Partition the entire system so that the delays of all of the combinational clusters are as balanced as possible.
- Design the circuit of the combinational clusters so that the delay of critical paths in the circuit is minimized and less than the desired clock cycle period.
- Use a robust clock scheme to ensure that the entire system is free of race conditions and has minimal tolerable clock skew.

The first item listed is beyond the scope of this section. The second item was discussed in the preceding subsection.

Race conditions typically occur with the use of level triggered latches. Figure 2.3 shows a typical synchronous system based on level triggered latches. Because of delays on the clock distribution network, such as buffers and capacitive parasitics on the interconnect, the timing difference caused by such a distribution delay is often referred

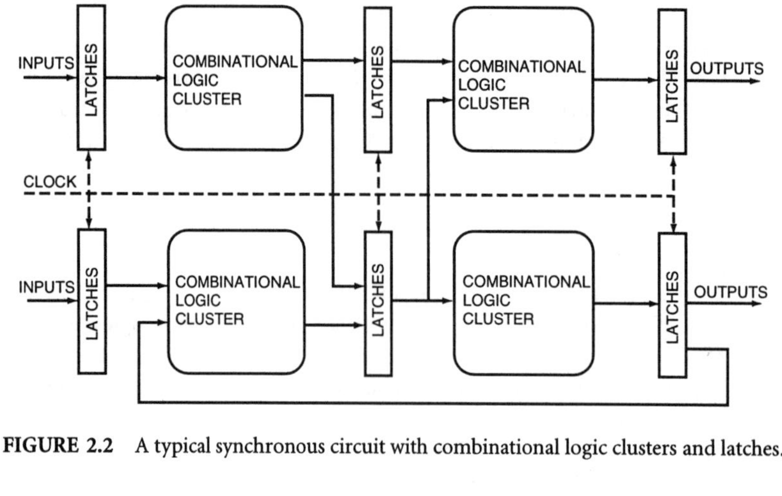

FIGURE 2.2 A typical synchronous circuit with combinational logic clusters and latches.

to as **clock skew** which is modeled by a delay element in Fig. 2.3. For the system to operate properly, at the positive edge each latch is supposed to capture its input data from the previous clock cycle. If the clock skew is severe shown as skewed clock clk'', however, it could be possible that the delay from $Q1$ to $D2$ becomes short enough that $D1$ is not only caught by the clk, but also caught by clk'. The solution to such a race condition caused by severe clock skew is as follows:

- Change the latch elements from level triggered to edge triggered or pseudoedge triggered latches such as latches using two-phase, nonoverlapping clocks.
- Resynthesize the system to balance to the critical path delay of different combinational clusters.
- Reduce the clock skew.

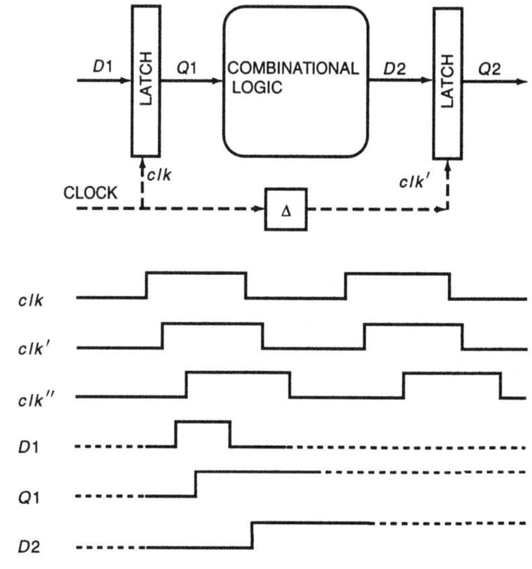

FIGURE 2.3 An example of race condition caused by a severe clock skew in the system.

Clock skew can also cause other types of circuit malfunction if it is not attended properly. In dynamic logic, severe clock skew may result in different functional blocks being in different stages of precharging or evaluating. Severe clock skew eats away a significant amount of precious cycle time in VLSI systems. Therefore, reducing clock skew is an important problem in high-speed circuit design.

As we mentioned previously, clock skew is mainly caused by the imbalance of the clock distribution network. Such an imbalance can be the result of distance differences from the nearest clock driver, different functional blocks driven by different clock drivers with different driving strengths, temperature difference on the same die, device characteristic differences on the same die due to process variation, etc. Two general approaches are often taken to minimize the clock skew. The first approach deals with the way the clock signal is distributed. The geometric shape of the clock distribution network is a very important attribute. Depending on the type of system operation, several popular distribution network topologies are illustrated in Fig. 2.4. Among these topologies, **H-tree** presents least amount of clock skew and, therefore, is widely used in high-performance systems.

The second approach deals with employing additional on-chip circuits to force the clock signals of two different functional blocks to be aligned, or to force the on-chip clock signal to be aligned with the global clock signal at the system level. The widely used circuits for this purpose include **phase-locked loop** (**PLL**) and **delay-locked loop** (**DLL**). Figure 2.5 shows a simple phase-locked loop. A simple PLL consists of four components: a digital phase detector, a charge pump, a low-pass filter, and a voltage controlled oscillator (VCO). The phase detector accepts reference clock, CLK_ref, and a skewed clock, CLK_out, and compares the phase difference of the two clocks to charge or discharge the charge pump. The low-pass filter is used to convert the phase difference between the reference frequency and the skewed frequency to a voltage level. This voltage is then fed into the VCO to reduce the difference of the reference and skewed clocks until they are locked to each other.

One of the most important design parameters for PLL is the output jitter. The output jitter is demonstrated by the random deviation of the output clock's phase from the reference clock signal. Significant peak-to-peak jitter will effectively reduce the clock period. The main contributor of the output jitter is the noise on the input of the VCO. Additional jitter can be induced by the noise on power supply rails that are common to high-speed VLSI circuits. Furthermore, acquisition time of PLL, in the several microsecond range, is often longer than desirable. This is mainly attributed to the response time of the VCO. In a typical scenario where clock skew is caused by the imbalance of the distribution network, the skewed clock often has the correct frequency. What needs to be corrected is the relative phase of the clock signals. Therefore, there is no need to have a VCO. Rather, a simple delay logic can be used to modify the clock signal's phase. This type of simplified phase correct circuit is referred to as a delay-locked loop. By replacing the VCO with a simple programmable delay line, DLL is simpler, yet exhibits less jitter than its PLL counterpart.

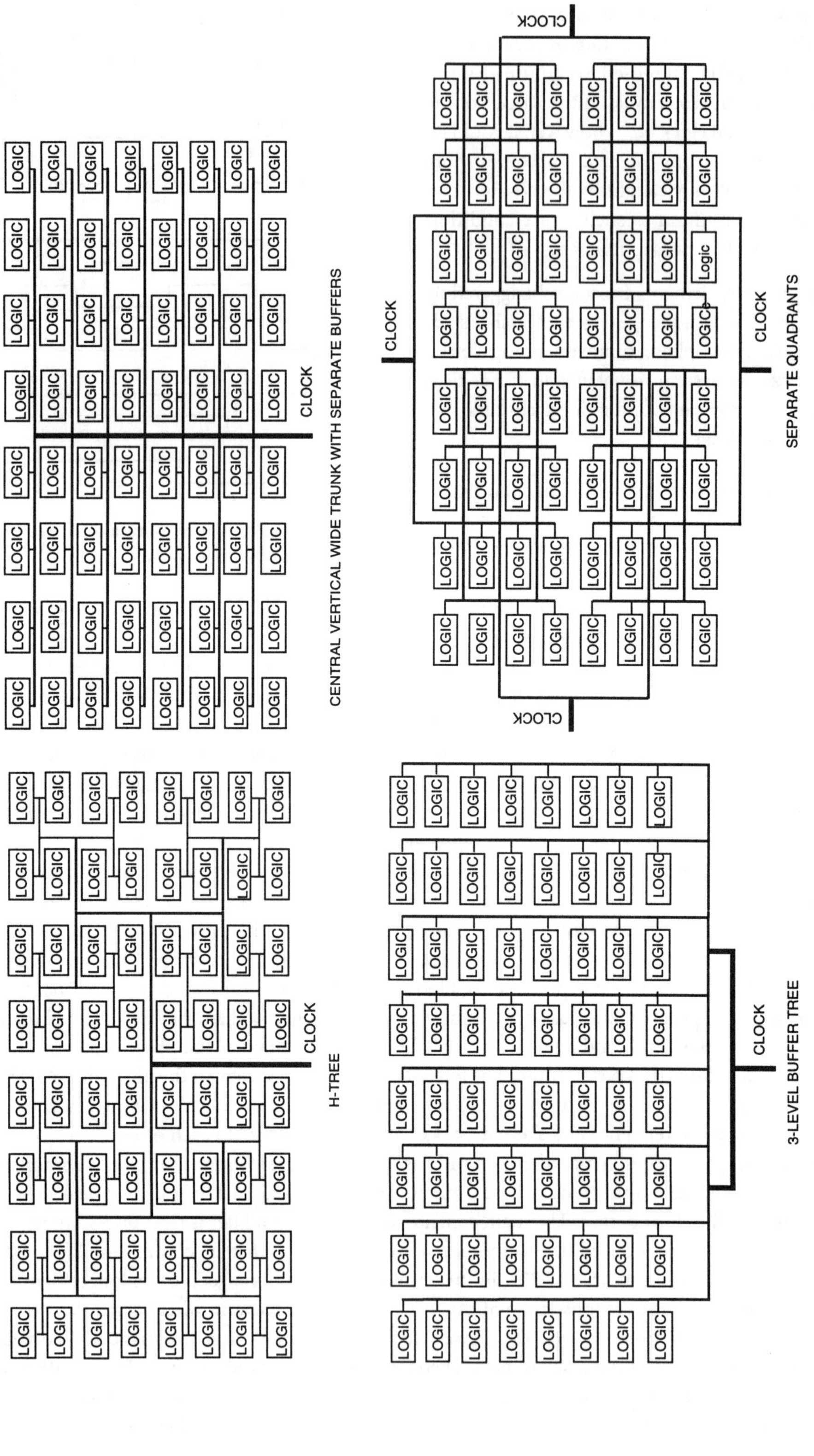

FIGURE 45.4 Various clock distribution structures.

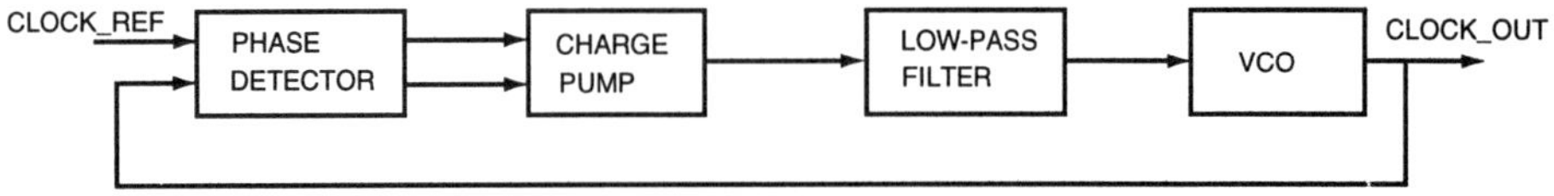

FIGURE 2.5 A simple phase-locked loop structure.

Asynchronous Circuits and Systems

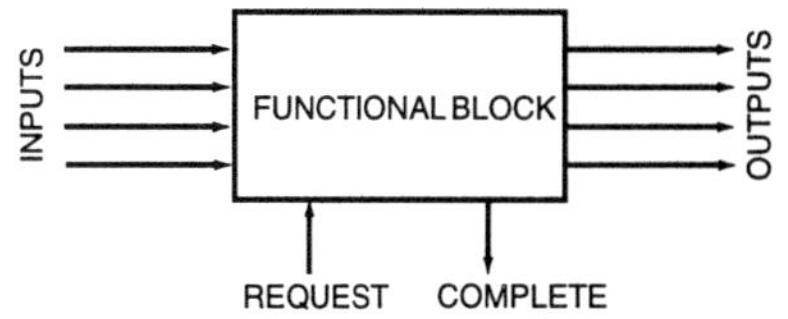

FIGURE 2.6 The single request and complete are two extra signals for a typical asynchronous circuit.

Clock distribution within large VLSI chips is becoming more and more of a problem for high-speed digital systems. Such a problem may be surmountable using state-of-the-art computer aided design (CAD) tools and on-chip PLL/DLL circuits. Asynchronous circuits have, nevertheless, gained a great deal of attention lately. An asynchronous circuit does not require an external clock to get it through the computation. Instead, it works on the principle of handshaking between functional blocks. Therefore, execution of a computational operation is totally dependent on the readiness of all of the input variables of the functional block. The biggest advantage of asynchronous circuits over synchronous circuits is that the correct behavior of asynchronous circuits is independent of the speed of their components or signal interconnect delays.

In a typical asynchronous circuit, functional blocks will have two more signals, *request* and *complete*, apart from their input and output signals as shown in Fig. 2.6. These two binary signals are necessary and sufficient for handshaking purposes. Even though asynchronous circuits are speed independent, the order of computation is still maintained by connecting the complete signal from one block to the request signal to another block. When the request signal is active for a functional block, indicating the computation of the preceding functional block is completed, the current functional block starts computation evaluation using its valid inputs from the preceding functional block. Once the evaluation is completed, the current functional block sets the complete signal to active to activate other functional blocks for computation. Figure 2.7 shows a schematic of such a communication protocol where block A and block B are connected in a pipeline.

To ensure that asynchronous circuits function correctly regardless of individual block speed, the *request* signal of a functional block should only be activated if the functional block has already completed the current computation. Otherwise, the current computation would be overwritten by incoming computation requests. To prevent this situation from happening, an interconnection block is required with an *acknowledge* signal from the current functional block to the preceding functional block. An active acknowledge signal indicates to the preceding function block that the current block is ready to accept new data from it. This two-way communication protocol with request and acknowledge is illustrated in Fig. 2.7. The interconnect circuit is unique to asynchronous circuits. It is often referred to as a **C-element**. Figure 2.8 shows a design of the C-element.

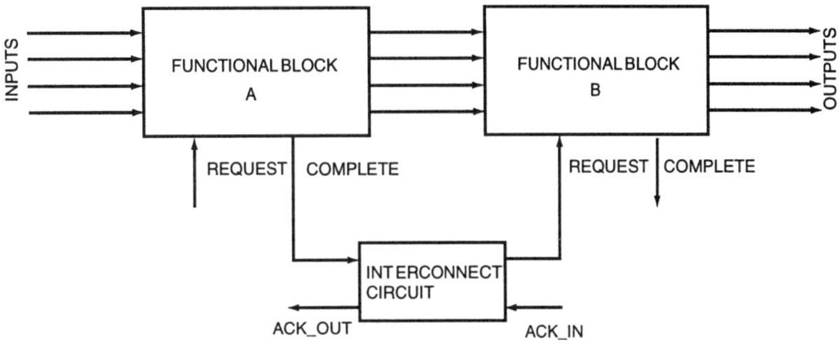

FIGURE 2.7 The schematic of a communication protocol in asynchromous systems.

In recent years, much effort has been spent on applying the asynchronous circuits to real-world applications. Several totally asynchronous designs of microprocessors have demonstrated their commercial feasibility. Several issues that still need to be addressed with regard to asynchronous circuits include acceptable amounts of silicon overhead, power efficiency, and performance as compared to their synchronous counterparts.

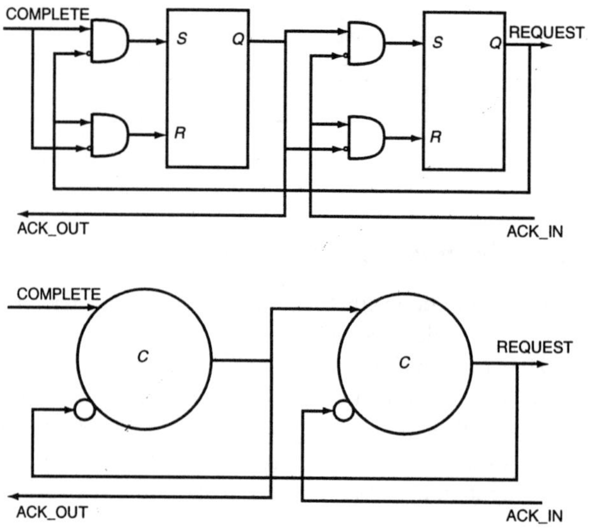

FIGURE 2.8 A schematic design of the C-element.

Interconnect Parasitics and Their Impact on High-Speed Design

On-chip interconnects present parasitic capacitance and resistance as loads to active circuits. Such parasitic loads had little impact on earlier ICs because the intrinsic gate delay dominated the total gate delay. With aggressive scaling in the VLSI process, the gate intrinsic delay decreases dramatically. The interconnect parasitics does not scale proportionally, however, and the wire resistance tends to increase, resulting in the delay caused by the interconnect load parasitics gradually becoming a dominant factor in the total gate delay. The problem is further exacerbated by the fact that when the operating speed reaches several hundred megahertz for modern digital systems, the traditional lumped RC model is no longer accurate. It has been suggested that such a lumped RC model should be modified to include a grounded resistor and an inductor. The RLC interconnect model includes nonequilibrium initial conditions and its response waveform may be nonmonotonic. Such a model may be more accurate because the existence of the inductance reduces the rate of increase in current and, therefore, increases the signal transition time. When the operating speed increases further such that the rise time of a signal is much less than the signal transmission time from point A to point B, a transmission line model should be used. On-chip interconnect, thus, is typically modeled as a microstrip. The characteristics of a transmission line are determined by its relative dielectric constant and magnetic permeability. Table 2.2 shows the signal transmission velocity in some common materials used in VLSI. As a rule of thumb, the transmission line phenomena become significant when

$$t_r < 2.5^* t_f$$

where t_r is the rise time of a signal and t_f is the signal transmission time, which is the interconnect length divided by the signal traveling velocity in the given material. The interconnect can be treated as a lumped RC network when

$$t_r > 5^* t_f$$

The signal rise time depends on the driver design and the transmission line's characteristic impedance Z_0. In MOS ICs, the load device at the receiving end of the transmission line can always be treated as an open circuit. Therefore, driver design is a very important aspect of high-speed circuit design. The ideal case is to have the driver's output impedance match the transmission line's characteristic impedance. Driving an unterminated transmission line (the MOS IC case) with its output impedance lower than the line's characteristic impedance, however, can increase driver's settling time due to excess ringing and, therefore, is definitely to be avoided. Excess ringing at the receiving end could also cause the load to switch undesirably. Assuming MOS transistor's threshold is 0.6–0.8 V, to ensure that no undesirable switching takes place, the output impedance of the drive should be at least a third of the charactertstic

TABLE 2.2 Transmission Line Velocity in Some Common Materials

	Velocity, cm/ns
Polymide	16–19
SiO_2	15
Epoxy glass (PCB)	13
Aluminium	10

impedance of the transmission line. When the output impedance is higher than the line's characteristic impedance, multiple wave trips of the signal may be required to switch the load. To ensure that only one wave trip is needed to switch the load, the output impedance of the driver should be within 60% of the characteristic impedance of the transmission line.

For a lossy transmission line due to parasitic resistance of on-chip interconnects, an exponential attenuating transfer function can be applied to the signal transfer at any point on the transmission line. The rate of the attenuation is proportional to the unit resistance of the interconnect. When operating frequency increases beyond a certain level, the on-chip transmission media will exhibit the *skin effect* in which the time-varying currents concentrate near the skin of the conductor. Therefore, the unit resistance of the transmission media increases dramatically.

2.3 VLSI Devices

Modern VLSI devices can be categorized into standard ICs and **application specific ICs (ASICs)**. Standard ICs are produced as generic products that can be used as building blocks to construct complex electronic systems. Standard IC parts are, typically, small in terms of their die size and have low logic complexity except memory chips. In general, standard ICs take aim at niche segments of the marketplace. We have seen some trend for some standard IC parts gradually becoming more and more application specific. For instance, DRAMs are being architecturally modified and optimized for certain specific application fields such as hand-held telecom systems, laptop computers, and graphics intensive processing systems.

ASICs are produced for only one or a few customers or applications. ASIC parts are, typically, large and much more complicated than their standard IC counterparts. ASICs have been one of the fastest growing segment of IC industry. To increase their application areas, many ASICs are gradually becoming **application specific standard products (ASSPs)**, such as graphics ICs, DRAM management ICs, and mass storage devices. These ASSP devices typically have a large volume for production. On the other hand, many ASIC devices are made using **programmable logic devices (PLDs)**, gate array devices, and **field programmable gate array (FPGA)** devices.

General purpose microprocessors are continuing to be the technology driver of the IC industry. Design complexity, design methodology, CAD capability, die size, etc., are often benchmarked against the state-of-the-art microprocessors. The latest release of Intel's Pentium and IBM/Motolora/Apple's PowerPC microprocessors are examples of latest VLSI design and manufacturing capabilities. The functionality and the performance of a VLSI device is often judged by the number of transistors in the chip and the maximum clock speed at which the chip is capable of operating. Table 2.3 lists some of the popular microprocessors and their characteristics. With several million transistors integrated on today's leading-edge microprocessors, many functions that were previously implemented in software or firmware, such as floating-point operations and advanced branch prediction and memory management schemes, are now implemented in silicon, thus, greatly improving the computation speed. It is believed that the microprocessors in 1997 will contain 20 million transistors and the leading-edge microprocessors in the year 2000 will contain 100 million transistors. If this prediction becomes a reality, the 100 million transistor microprocessor will probably offer today's mainframe computational power on one chip.

TABLE 2.3 Some Popular Microprocessors and Their Characteristics

	Power PC	486DX2	Pentium	R4000SC	R4400SC	SuperSPARC+	Alpha 21064	PA7100
Clock rate, MHz	66	66	66	100	150	50	150	99
SPECint92	>60	32.2	64.5	61.7	88 (est)	65.2	84.4	80
SPECfp92	>80	16.1	56.9	63.4	97 (est)	83	127.7	150.6
External cache	N/A	256 K	256 K	1 M	1 M	1 M	512 K	512 K
On-chip cache	32 K	8 K	16 K	16 K	32 K	36 K	16 K	None
Die size, mm	121	81	294	184	184	256	234	196
Transistors, millions	2.8	1.2	3.1	1.35	2.2	3.1	1.7	0.85
Process feature size, μm	0.65	0.8	0.8	0.8	0.6	0.7	0.75	0.75

Source: Microprocessor Report.

General purpose **digital signal processors** (**DSPs**) follow the same trend as microprocessors with more and more on-chip floating-point mathematics capabilities. DSPs are widely used in areas such as digital mobile radio, fax machines, medical imaging, disk drives, and modems. The high-end DSPs mainly aim at image processing applications where processing speed and bandwidth are very important. MVP, the latest DSP chip from Texas Instruments packs 4 million transistors on a single chip and is capable of performing many video signal processing tasks in real-time.

TABLE 2.4 Typical and Expected Features of Various DRAM Densities

Density, bit	Min. feature, μm	Cell size, μm	Capacitor structure
1 Mb	1	30	Planar
4 Mb	0.8	10	3-D
16 Mb	0.5	3.6	3-D
64 Mb	0.35	1.2	3-D
256 Mb	0.25	0.4	3-D
1 Gb	0.15	0.13	3-D

Source: Semiconductor International.

Memory is still believed to be the IC process driver with its processing technology one generation ahead of logic process. Using its advanced process, we have seen the DRAM's capacity quadrupling every 3 years in the past. The single-chip, 64-megabyte DRAM is already widely available. Whether such a trend will continue at its present pace is debatable. However, 256-megabyte and 1-Gb DRAM chips have been reported, and it is predicted that these devices will be widely available by the turn of the century. Table 2.4 shows the typical and expected features of various DRAM densities.

One of the fastest growing sectors in memory ICs in the past has been the flash memory sector. Flash memories are expected to replace most erasable programmable read only memory (EPROMs) and high-density EEPROMs, some DRAMs and a large portion of SRAMs now used with battery backup for portable computers. The reason flash memories can compete with various other data storage devices can be summarized in Table 2.5, which shows clearly some advantages of using flash memories in some applications. Market share of the flash memory will certainly grow as new application fields are found.

The techniques used for flash memories can also be used for programmable logic devices. Other types of programming techniques include fuse/antifuse, EPROM-based, and SRAM-based programming. Field programmable gate arrays primarily use the SRAMs and antifuse-based techniques. The major market force driving PLD usage continues to be the time-to-market factor. The leading-edge PLD/FPGA devices typically have between 6,000 and 10,000 usable gates for configuring a custom system.

TABLE 2.5 Comparison of Different Data Storage Techniques

	Flash memory	SRAM	DRAM	Hard disk drive
Access time	80 ns	25 ns	60 ns	1–10 μs
Erase operation	Unnecessary	Unnecessary	Unnecessary	Unnecessary
Current draw (operation), mA	30	30	100	500
Current draw (standby), μA	Unnecessary	50	250	Unnecessary
Impact resistance	High	High	High	Low
Relative per-bit cost	12	30	10	1

Source: JEE/ICE.

Defining Terms

Application-specific integrated circuit (ASIC): Device designed specifically for a particular application.

Application-specific standard product (ASSP): Device designed specifically for one area of applications, such as graphics and video processing.

Asynchronous system: A system in which the progress of a computation is driven by the readiness of all the necessary input variables for the computation through a handshaking protocol. Therefore, no central clock is needed.

C-element: A circuit used in an asynchronous as an interconnect circuit. The function of this circuit is to facilitate the handshaking communication protocol between two functional blocks.

Clock skew: A phase difference between two clock signals at different part of a chip/system due to imbalance of the distribution media and the distribution network.

Complementary metal-oxide silicon (CMOS): It is a very popular integrated circuit type in use today.
Critical path: A signal path from a primary input pin to a primary output pin with the longest delay time in a logic block.
Delay-locked loop (DLL): It is similar to PLL except that it has better jitter suppression capability.
Digital signal processor (DSP): A processing device specialized in popular math routines used by signal processing algorithms.
Field programmable gate array (FPGA): A popular device which can be tailored to a particular application by loading a customizing program on to the chip.
H-tree: A popular clock distribution tree topologically that resembles the H shape. It introduces the least amount of clock skew compared to other distribution topologies.
Phase-locked loop (PLL): A circuit that can detect the phase difference of two signals and reduce the difference in the presence of the phase difference.
Programmable logic devices (PLD): A class of IC products which are easy to customize for a particular application.
SPICE: A popular circuit level simulation program to perform detailed analysis of circuit behavior.
Synchronous system: A system in which a computation is divided into unit periods defined by a central clock signal. Signal transfer within the system typically occurred at the transition edge of the clock signal.

References

Bakoglu, H.B. 1991. *Circuits, Interconnections, and Packaging for VLSI.* Addison–Wesley, Reading, MA.
Dill, D.L. 1989. *Trace Theory for Automatic Hierarchical Verification of Speed-Independent Circuits.* MIT Press, Cambridge, MA.
Gardner, F.M. 1979. *Phaselock Techniques*, 2nd ed. Wiley, New York.
Jeong, D. et al. 1987. Design of PLL-based clock generation circuits. *IEEE J. Solid-State Circuits* SC-22(2): 255–261.
Johnson, M. and Hudson, E. 1988. A variable delay line PLL for CPU-coprocessor synchronization. *IEEE J. Solid-State Circuits* (Oct.):1218–1223.
Meng, T.H. 1991. *Synchronization Design for Digital Systems.* Kluwer Academic, Norwell, MA.
Rosenstark, S. 1994. *Transmission Lines in Computer Engineering.* McGraw–Hill, New York.
Sapatnekar, S., Rao, V., and Vaidya, P. 1992. A convex optimization approach to transistor sizing for CMOS circuits. *Proc. ICCAD*, pp. 482–485.
Wang, X. and Chen, T. 1995. Performance and area optimization of VLSI systems using genetic algorithms. *Int. J. of VLSI Design* 3(1):43–51.
Weste, N. and Eshraghian, K. 1993. *Principle of CMOS VLSI Design: A Systems Perspective*, 2nd ed. Addision–Wesley, Reading, MA.

Further Information

For general information on the VLSI design process and various design issues, consult several excellent reference books, two of which are listed in the reference section, including Mead and Conway's *Introduction to VLSI Systems*, Glasser and Dobberpuhl's *The Design and Analysis of VLSI Circuits*, and Geiger's *VLSI Design Techniques for Analog and Digital Circuits. IEEE Journal of Solid-State Circuits* provides an excellent source for the latest development of novel and high-performance VLSI devices.

Some of the latest applications of PLLs and DLLs can be found in the *Proceedings of International Solid-State Circuit Conference*, the *Symposium on VLSI Circuits*, and the *Custom Integrated Circuit Conference.* The article by Tanoi et al. "A 250–622 MHz Deskew and Jitter-Suppressed Clock Buffer Using a Frequency and Delay-Locked Two-Loop Architecture" in 1995 *Sym. on VLSI Circuits* is a good example of applying DLL to the clock skew problem.

For information on modeling of VLSI interconnects and their transmission line treatment, consult the *Proceedings of Design Automation Conference* and *International Conference on Computer-Aided Design. IEEE Transactions on CAD* is also an excellent source of information on the subject.

3

Integrated Circuit Design

Samuel O. Agbo
California Polytechnic State University

Eugene D. Fabricius
California Polytechnic State University

3.1 Introduction

Integrated circuits (ICs) are classified according to their levels of complexity: small-scale integration (SSI), medium-scale integration (MSI), large-scale integration (LSI) and very large-scale integration (VLSI). They are also classified according to the technology employed for their fabrication [bipolar, N metal oxide semiconductor (NMOS), complementary metal oxide semiconductor (CMOS), etc.]. The design of integrated circuits needs to be addressed at the SSI, MSI, LSI, and VLSI levels. Digital SSI and MSI typically consist of gates and combinations of gates. Design of digital SSI and MSI is presented in Sec. 3.3, and consists largely of the design of standard gates. These standard gates are designed to have large noise margins, large fan out, and large load current capability, in order to maximize their versatility.

In principle, the basic gates are sufficient for the design of any digital integrated circuit, no matter how complex. In practice, modifications are necessary in the basic gates and MSI circuits like flip-flops, registers, adders, etc., when such circuits are to be employed in LSI or VLSI design. For example, circuits to be interconnected on the same chip can be designed with lower noise margins, reduced load driving capability, and smaller logic swing. The resulting benefits are lower power consumption, greater circuit density, and improved reliability. On the other hand, several methodologies have emerged in LSI and VLSI design that are not based on interconnections or modification of SSI and MSI circuits. Both approaches to LSI and VLSI design are presented in the following sections.

3.2 An Overview of the IC Design Process

The effort required for the design of an integrated circuit depends on the complexity of the circuit. The requirement may range from several days effort for a single designer to several months work for a team of designers.

0-8493-0050-9/00/$0.00+$.50

Custom design of complex integrated circuits is the most demanding. For example, the Motorola 68000 Microprocessor required 52 man-years of design. By contrast, **semicustom design** of LSI and VLSI that utilize pre-existing designs, such as standard cells and gate arrays, requires less design effort.

IC design is performed at many different levels and Fig. 3.1 is a nonunique depiction of these levels. Level 1 presents the design in terms of subsystems (standard cells, gate arrays, custom subcircuits, etc.) and their interconnections. Design of the system **layout** begins with the floor plan of level 3. It does not involve the layout of individual transistors and devices, but is concerned with the geometric arrangement and interconnection of the subsystems. Level 4 involves the circuit design of the subsystems. Levels 2 and 5 involve system and subcircuit simulations, respectively, which may lead to modifications in levels 1 and/or 4.

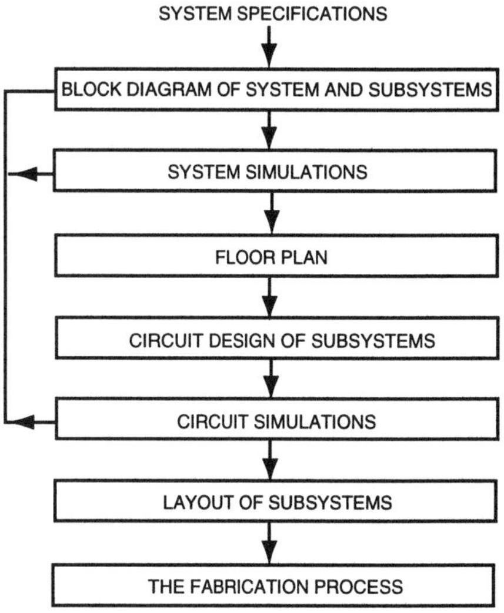

FIGURE 3.1 Different levels in IC design.

Discussion here will focus primarily on the system design of level 1 and the subsystem circuit design of level 4. Lumped under the fabrication process of level 7 are many tasks, such as **mask** generation, process simulation, wafer fabrication, testing, etc. Broadly speaking, floor plan generation is a part of layout. For large ICs, layout design is often relevant to system and circuit design.

3.3 General Considerations in IC Design

Layout or actual arrangement of components on an IC is a design process involving design tradeoffs. Circuit design will often influence layout and vice versa, especially in LSI and VLSI. Thus, it is helpful to outline some general considerations in IC design:

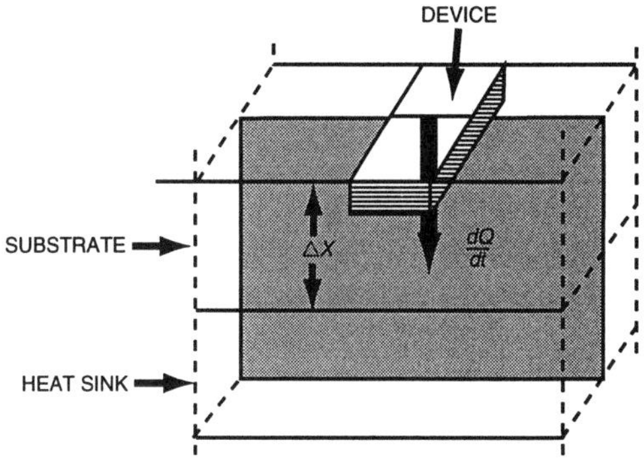

FIGURE 3.2 Heat flow in an IC device.

- Circuit components: Chip area is crucial in IC design. The usual IC components are transistors, resistors, and capacitors. Inductors are uneconomical in area requirements and are hardly ever used in ICs except in certain microwave applications. Transistors require small chip areas and are heavily utilized in ICs. Resistor area requirements increase with resistor values. IC resistors generally range between 50 Ω and 100 kΩ. Capacitors are area intensive and tend to be limited to 100 pF.
- Isolation regions: Usually, different components would be placed in different isolation regions. The number of isolation regions, however, should be minimized in the interest of chip area economy by placing more than one component in an isolation region whenever possible. For example, several resistors could share one isolation region.
- Design rules: Geometric design rules that specify minimum dimensions for features and separation between features must be followed for a given IC process.
- Power dissipation: Chip layout should allow for adequate power dissipation and avoid overheating or development of hot spots on the chip. In low-power circuits such as CMOS ICs, device size is determined by lithographic constraints. In circuits with appreciable power dissipation, device size is determined by thermal constraints and may be much larger than lithographic constraints would allow.

Device size determination from power-density considerations is illustrated with the aid of Fig. 3.2. The device resides close to the surface of the substrate. Thus, heat flow is approximately one dimensional, from the device to the substrate, although it is actually a three-dimensional flow. Assuming an infinite heat sink of ambient temperature T_A, substrate thickness ΔX, thermal conductivity α, device of surface area A, and temperature $T_A + \Delta T$, the rate of heat flow towards the heat sink, dQ/dt, is given by

$$\frac{dQ}{dt} = \alpha A\left(\frac{\Delta T}{\Delta X}\right) \tag{3.1}$$

The power density or power dissipation per unit area of the device is

$$\frac{P}{A} = \alpha\left(\frac{\Delta T}{\Delta X}\right) \tag{3.2}$$

Device Scaling

The trend in IC design, especially VLSI, is progressively toward smaller sizes. Scaling techniques permit the design of circuits that can be shrunk as technological developments allow progressively smaller sizes. Two basic approaches to scaling are full scaling and constant voltage scaling.

- Full scaling: All device dimensions, both surface and vertical, and all voltages are reduced by the same scaling factor S.
- Constant voltage (CV) scaling: All device dimensions, both surface and vertical, are reduced by the same scaling factor S, but voltages are not scaled. Voltages are maintained at levels compatible with transistor–transistor logic (TTL) supply voltages and logic levels.

Scaling of device dimensions has implications for other device parameters. Full scaling tends to maintain constant electric field strength and, hence, parameters that are no worse off, as device dimensions are reduced, but does not ensure TTL voltage compatibility. Table 3.1 compares effect on device parameters of the two scaling approaches. A compromise that is often adopted is to use full scaling for all internal circuits and maintain TTL voltage compatibility at the chip input/output (I/O) pins.

Although many scaling relationships are common to MOS field effect transistors (MOSFETs) and bipolar ICs [Keyes 1975], the scaling relationships of Table 3.1 more strictly applies to MOSFETs. Bipolar doping levels, unlike MOSFETs, are not limited by oxide breakdown. Thus, in principle, miniaturization can be carried further in bipo-

TABLE 3.1 Effect of Full Scaling and Constant Voltage Scaling on IC Device Parameters

Parameter	Full Scaling	CV Scaling
Channel length, L	$1/S$	$1/S$
Channel width, W	$1/S$	$1/S$
Oxide thickness, t_{OX}	$1/S$	$1/S$
Supply voltage, V_{DD}	$1/S$	1
Threshold voltage, V_{TD}	$1/S$	1
Oxide capacitances, C_{OX}, C_{SW}, C_{FOX}	S	S
Gate capacitance, $C_g = C_{OX}WL$	$1/S$	$1/S$
Transconductances, K_N, K_p	S	S
Current, I_D	$1/S$	$1/S$
Power dissipation per dense, P	$1/S^2$	S
Power dissipation per unit area, P/A	1	S^3
Packing density	S^2	S^2
Propagation delay, t_p	$1/S$	$1/S^2$
Power-delay product, Pt_p	$1/S^3$	$1/S$

TABLE 3.2 MOS Implementation System (MOSIS) NMOS Design Rules

Mask Level	Feature	Size (times λ)
N^+ Diffusion	diffusion width	2
	diffusion spacing	3
Implant mask	implant-gate overlap	2
	implant to gate spacing	1.5
Buried contact mask	contact to active device	2
	overlap with diffusion	1
	contact to poly spacing	2
	contact to diffusion spacing	2
Poly mask	poly width	2
	poly spacing	2
	poly to diffusion spacing	1
	gate extension beyond diffusion	2
	diffusion to poly edge	2
Contact mask	contact width	2
	contact-diffusion overlap	1
	contact-poly overlap	1
	contact to contact spacing	2
	contact to channel	2
	contact-metal overlap	1
Metal mask	metal width	3
	metal spacing	3

lar processing technology. However, bipolar scaling is more complex. One reason is that the junction voltages required to turn on a bipolar junction transistor (BJT) does not scale down with dimensions.

Geometric Design Rules

Design rules specify minimum device dimensions, minimum separation between features and maximum misalignment of features on an IC. Such rules tend to be process and equipment dependent. For example, a design rule for a 2-μm process may not be appropriate for a 0.5-μm process. Design rules should protect against fatal errors such as short circuits due to excessive misalignment of features, or open circuits due to too narrow a metal or polysilicon conductive path.

Generalized design rules that are portable between processes and scalable to facilitate adaptation to shrinking minimum geometries as processes evolve are desirable. Other advantages of generalized design rules include increased design efficiency due to fewer levels and fewer rules, automatic translation to final layout, layout-rule and electrical-rule checking, simulation, verification, etc. The Mead–Conway approach [1980] to generalized design rules is to define a scalable and process-dependent parameter, lambda (λ), as the maximum misalignment of a feature from its intended position on a wafer or half the maximum misalignment of two features on different mask layers. Table 3.2 shows a version of the Mead-Conway scalable design rules for NMOS [Fabricius 1990].

CMOS ICs utilize both NMOS and PMOS devices. Starting with a p-substrate, the NMOS would be fabricated on the p-substrate and the PMOS in an n-well, and vice versa. With the addition of an n-well, p-well or twin tub process, CMOS fabrication is similar to that for NMOS, although more complex. Table 3.3 [Fabricius 1990] shows the Mead–Conway scalable CMOS design rules. The dimensions are given in multiples of λ

TABLE 3.3 MOSIS Portable CMOS Design Rules

Mask Level	Feature	Size (times λ)
n-well and p-well	well width	6
	well to well spacing	6
n^+, p^+ active diffusion or implant	active width	3
	active to active spacing	3
	source/drain to well edge	6
	substrate/well contact	3
	active to well edge	3
Poly mask	poly width or spacing	2
	gate overlap of active	2
	active overlap of gate	2
	field poly to active	1
p-select, n-select	select-space (overlap) to (of) channel	3
	select-space (overlap) to (of) active	2
	select-space (overlap) to (of) contact	1
Simpler contact to poly	contact size	2 × 2
	active overlap of contact	2
	contact to contact spacing	2
	contact to gate spacing	2
Denser contact to poly	contact size	2 × 2
	poly overlap of contact	1
	contact spacing on same poly	2
	contact spacing on different poly	5
	contact to non-contact poly	4
	space to active short run	2
	space to active long run	3
Simpler contact to active	contact size	2 × 2
	active overlap of contact	2
	contact to contact spacing	2
	contact to gate spacing	2
Denser contact to active	contact size	2 × 2
	active overlap of contact	1
	contact spacing on same active	2
	contact spacing on different active	6
	contact to different active	5
	contact to gate spacing	2
	contact to field poly short run	2
	contact to field poly long run	3
Metal 1	width	3
	metal 1 to metal 1	3
	overlap of contact to poly	1
	overlap of contact to active	1
via	size	2 × 2
	via-to-via separation	2
	metal 1/via overlap	1
	space to poly or active edge	2
	via to contact spacing	2
Metal 2	width	3
	metal 2 to metal 2 spacing	4
	metal overlap of via	1
Overglass	bonding pad (with metal 2 undercut)	100 × 100 μm
	probe pad	75 × 75 μm
	pad to glass edge	6 μm

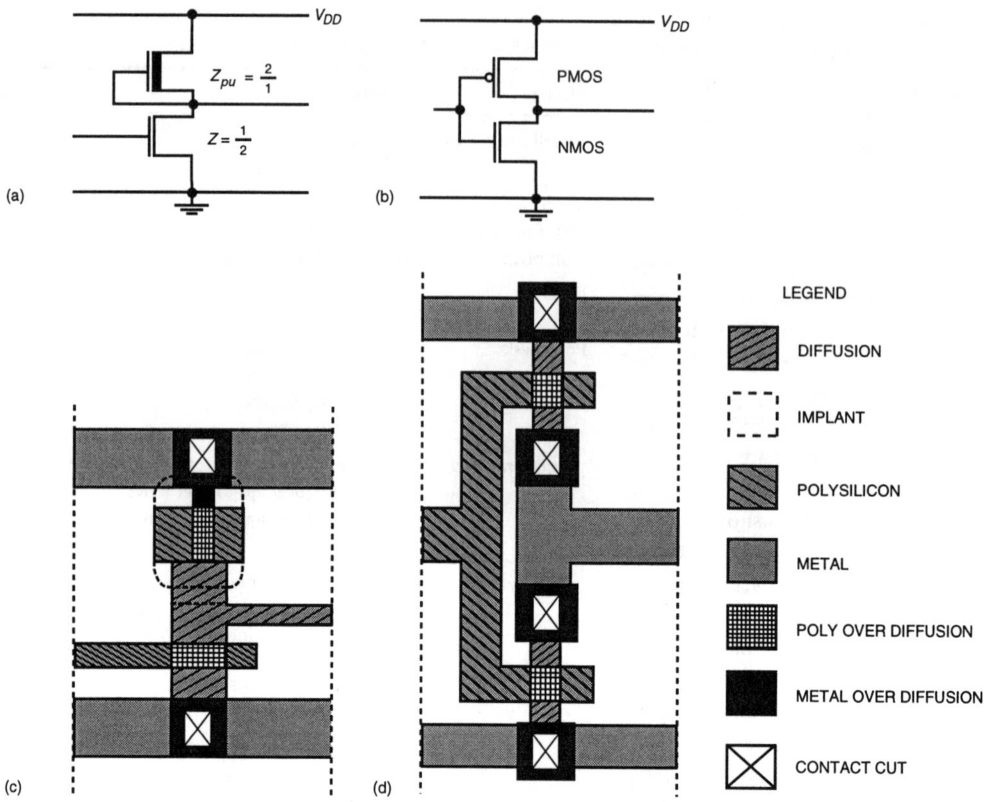

FIGURE 3.3 Layout design for NMOS inverter with depletion load and a CMOS inverter: (a) NMOS inverter, (b) CMOS inverter, (c) NMOS inverter layout, (d) CMOS inverter layout.

and the rules are specified by the MOS Implementation System (MOSIS) of the University of Southern California Information Sciences Institute. Further clarification of the rules may be obtained from MOSIS manuals [USCISI 1984, 1988].

Figure 3.3 illustrates the layout of an NMOS inverter with depletion load and a CMOS inverter, employing the NMOS and CMOS scalable design rules just discussed. Figures 3.3(a) and 3.3(b) show the circuit diagrams for the NMOS and CMOS, respectively, with the aspect ratios Z defined as the ratio of the length to the width of a transistor gate. Figures 3.3(c) and 3.3(d) give the layouts for the NMOS and CMOS, respectively.

A simplified design rule for a bipolar *npn* comparable to those previously discussed for NMOS and CMOS is presented in Table 3.4. As before, the minimum size of a feature in the layout is denoted by lambda (λ). The following six masks are required for the fabrication: n^+ buried-layer diffusion, p^+ isolation diffusion, p base region diffusion, n^+ emitter and collector diffusions, contact windows, and metalization.

TABLE 3.4 Simplified Design Rules for *npn* Transistor

Mask Level	Feature	Size (times λ)
Isolation mask	width of isolation wall	1
	wall edge to buried layer spacing	2.5
Base mask	base to isolation region spacing	2.5
Emitter	area	2×2
	emitter to base diffusion spacing	1
	multiple emitter spacing	
Collector contact	area	1×5
	n^+ to base diffusion	1
Contact windows	base contact	1×2
	emitter contact	1×1
	collector contact	1×2
	base contact to emitter spacing	1
Metallization	width	1.5
	metal to metal spacing	1

3.4 Design of Small-Scale and Medium-Scale Integrated Circuits

Gates are the basic building blocks in digital integrated circuits. Small-scale integrated circuits are essentially gate circuits, and medium-scale integrated circuits are circuits employing several gates. Gates, in turn, are based on inverters and can be realized from inverter circuits with some modifications, especially those modifications that allow for multiple inputs. This section will start with a discussion of inverters and gates.

NMOS Inverters

A resistive load NMOS inverter, its output characteristics, and its voltage transfer characteristic are shown in Fig. 3.4. The loadline is also shown on the output characteristics. Resistive load inverters are not widely used in ICs because resistors require long strips and, hence, large areas on the chip. A solution to this problem is to use active loads, since transistors are economical in chip area.

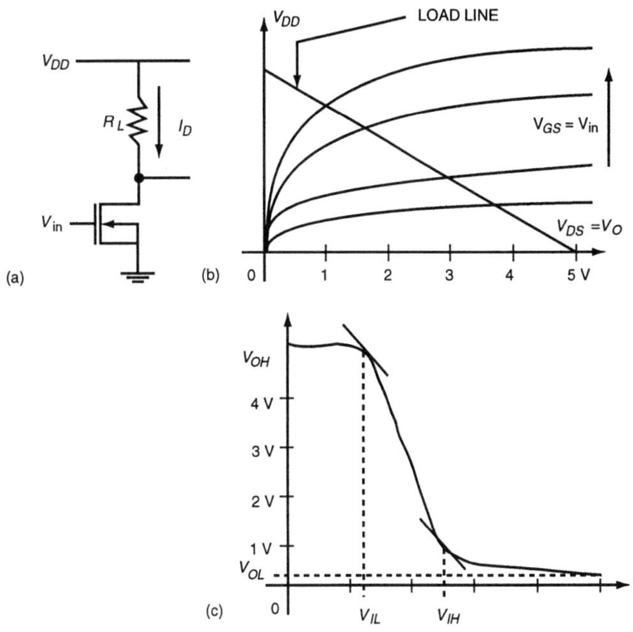

FIGURE 3.4 NMOS resistive load inverter: (a) resistive load NMOS inverter, (b) output characteristics, (c) transfer characteristic.

Figure 3.5 shows three NMOS inverters with three types of NMOS active loads: saturated enhancement load, linear enhancement load, and depletion load. One basis for comparison between these inverters is the geometric ratio K_R, which is defined as Z_{pu}/Z_{pd}. Z denotes the ratio of length to width of a transistor channel. The subscript pu refers to the pull-up or load device, whereas the subscript pd refers to the pull-down or driving transistor.

The saturated enhancement load inverter overcomes much of the area disadvantage of the resistive load inverter. When carrying the same current and having the same pull-down transistor as the resistive inverter, however, K_R is large for the saturated enhancement load inverter, indicating load transistor area minimization is still possible. This configuration yields a smaller logic swing relative to the resistive load inverter, however, because the load transistor stops conducting when its $V_{GS} = V_{DS}$ decreases to V_T. Thus, for this inverter, $V_{OH} = V_{DD} - V_T$.

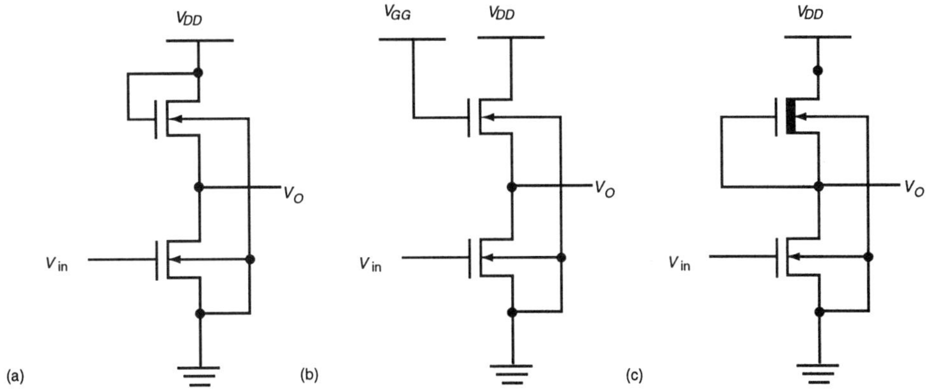

FIGURE 3.5 NMOS inverters with different types of active loads: (a) saturated enhancement load, (b) linear enhancement load, (c) depletion load.

In Fig. 3.5(b), because V_{GG} is greater than $V_{DD} + V_T$, V_{DS} is always smaller than $V_{GS} - V_T$; thus, the load always operates in the linear region. This results in a linear enhancement load NMOS inverter. The high value of V_{GG} also ensures that V_{GS} is always greater than V_T, so that the load remains on and V_{OH} pulls up to V_{DD}. The linear enhancement load configuration, however, requires a load transistor of larger area relative to the saturated enhancement load inverter, and requires additional chip area for the V_{GG} contact.

In the depletion NMOS load inverter of Fig. 3.5(c), $V_{GS} = 0$, thus the load device is always on and V_{OH} pulls all the way to V_{DD}. This configuration overcomes the area disadvantage without incurring a voltage swing penalty. It is, therefore, the preferred alternative. The performance of the NMOS inverters with the four different types of loads are graphically compared in Figs. 3.6(a) and 3.6(b). Both the loadlines and the voltage transfer characteristics were obtained from SPICE simulation. Figure 3.6(a) shows the loadlines superimposed on the output characteristics of the pull-down transistor, which is the same for the four inverters. R_L is 100 kΩ and each inverter has $V_{DD} = 5$ V, $V_{OL} = 0.2$ V and $I_{D\ \mathrm{max}} = 48\ \mu$A. Note that V_{OH} falls short of V_{DD} for the saturated enhancement load inverter but not for the others. Figure 3.6(b) shows the voltage transfer characteristics (VTC) for the four inverters. V_{OH} is again shown to be less than V_{DD} for the saturated enhancement load. Note, also, that the depletion load VTC more closely approaches the ideal inverter VTC than any of the others.

The loadlines of Fig. 3.5(a) are easy to generate. Consider, for example the depletion NMOS load. V_{GS} is fixed at 0 V, so that its output characteristic consists of only the curve for $V_{GS} = 0$. I_D is always the same for the load and driving transistor, but their V_{DS} add up to V_{DD}. Thus, when V_{DS} is high for one transistor, it is low for the other. The loadline is obtained by shifting the origin for V_{DS} for the load characteristic to V_{DD}, reflecting it about the vertical axis through V_{DD} and superimposing it on the V–I characteristics for the driving inverter.

The voltage transfer characteristics are best generated by computer simulation. Useful insights, however, can be gained from an analysis yielding the critical voltages V_{OH}, V_{OL}, V_{IH}, V_{IL}, and V_O for any specified V_{in}. The NMOS currents hold the key to such an analysis. Threshold voltages are different for enhancement and depletion NMOS transistors, but the drain current equations are the same. The drain current is given in the linear region and the saturated region, respectively, by

$$I_D = K_n\left[2(V_{GS} - V_T)V_{DS} - V_{DS}^2\right]; \quad V_{DS} \leq V_{GS} - V_T \tag{3.3}$$

$$I_D = K_n(V_{GS} - V_T)^2; \quad V_{DS} \geq V_{GS} - V_T \tag{3.4}$$

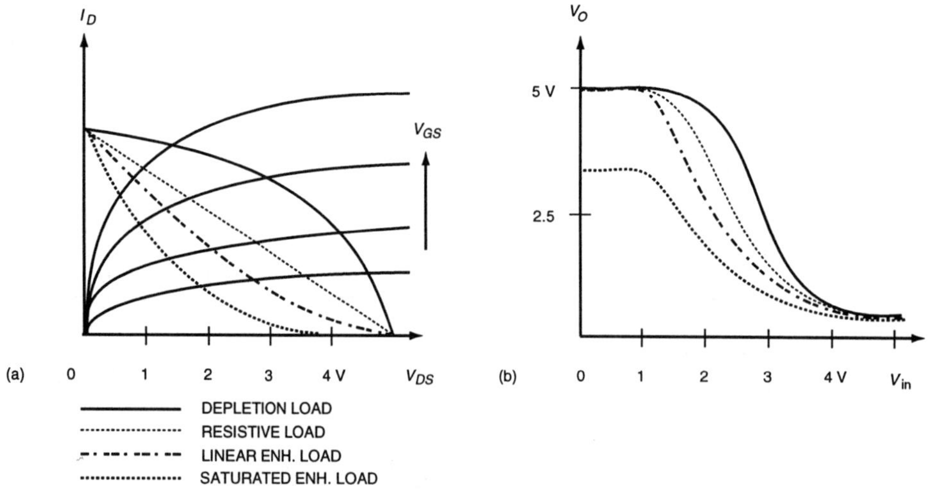

FIGURE 3.6 Performance of NMOS inverters with different types of loads: (a) output characteristics and load lines, (b) voltage transfer characteristics.

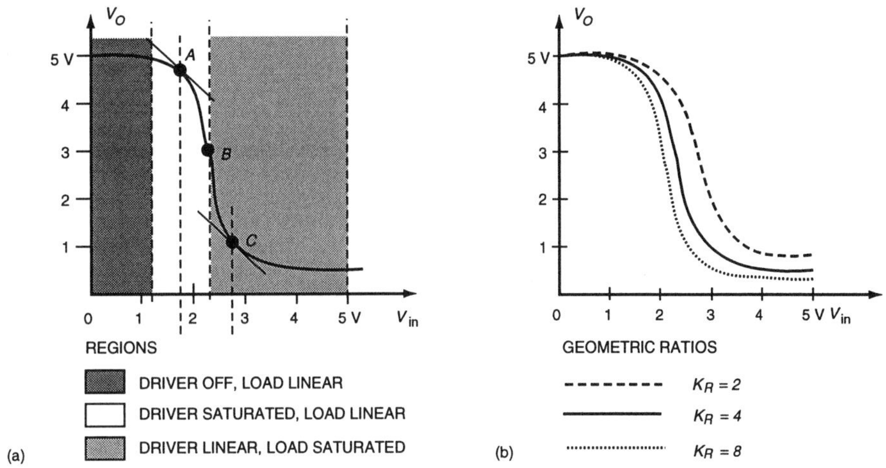

FIGURE 3.7 Regions of the VTC and VTCs for different geometric ratios for the depletion load NMOS inverters: (a) depletion load NMOS inverter VTC and its regions, (b) VTCs for different geometric ratios.

where

V_T = threshold voltage
$K_n = \frac{\mu_n C_{ox}}{2}\left(\frac{w}{L}\right)$ = transconductance
μ_n = electron channel mobility
C_{ox} = gate capacitance per unit area

Similar definitions apply to PMOS transistors.

Consider the VTC of Fig. 3.7(a) for a depletion load NMOS inverter. For the region $0 < V_{in} < V_T$ the driving transistor is off, so $V_{OH} = V_{DD}$. At A, V_{in} is small; thus, for the driving transistor, $V_{DS} = V_O > V_{in} - V_T = V_{GS} - V_T$. For the load $V_{DS} = V_{DD} - V_O$ is small. Hence, the driver is saturated and load is linear. Similar considerations lead to the conclusions as to the region in which each device operates, as noted in the figure. To find V_{IL} and V_{IH}, the drain currents for the appropriate region of operation for points A and C, respectively, are equated, for both transistors. Differentiating the resulting equations with respect to V_{in} and applying the condition that $dV_O/dV_{in} = -1$ yields the required critical voltages. Equating drain currents for saturated load and linear driver at $V_{in} = V_{DD}$ and solving yields V_{OL}. The output voltage V_O may be found at any value of V_{in} by equating the drain currents for the two transistors, appropriate for that region of V_{in}, and solving for V_O at the given V_{in}.

NMOS Gates

Only NOR and NAND need be considered because these are more economical in chip area than OR and AND, and any logic system can be implemented entirely with either NOR or NAND. By connecting driving transistors in parallel to provide the multiple inputs, the NMOS inverter is easily converted to a NOR gate as shown in Fig. 3.8(a). By connecting driving transistors in series as in Fig. 3.8(b), an NAND gate is obtained.

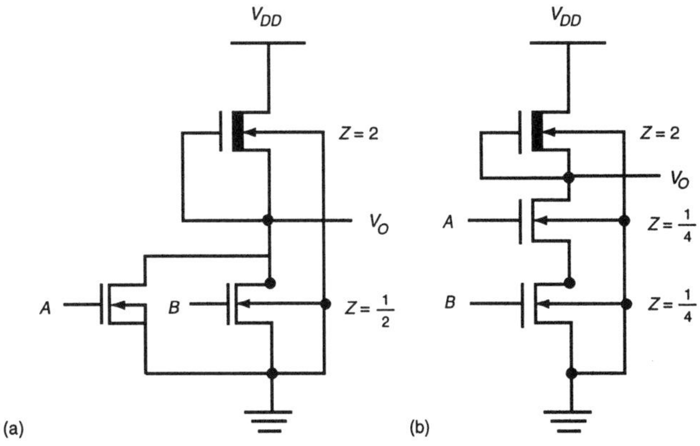

FIGURE 3.8 NMOS gates: (a) NMOS NOR gate, (b) NMOS NAND gate.

The NMOS, NOR and NAND gates are essentially modifications of the depletion load NMOS inverter of Fig. 3.5(c). They all have the same depletion load, and their performance should be similar. For the same value of V_{DD}, each has $V_{OH} = V_{DD}$, and they should have the same V_{OL} and the same drain current when $V_O = V_{OL}$. In Fig. 3.5(c), the depletion load inverter has $Z_{pu} = 2$, $Z_{pd} = \frac{1}{2}$ and $K_R = 4$. Thus, Z_{pu} is 2 for the NOR and the NAND gate. With only one driving transistor on in the NOR gate, the drain current should be sufficient to ensure that $V_O = V_{OL}$. Thus, for each of the driving transistors, $Z_I = \frac{1}{2}$, as for the depletion load inverter. For the NAND gate, the equivalent series combination of Z_{pd} (proportional to drain-source resistance) should also be $\frac{1}{2}$ to allow the same value of V_O, leading to $Z_I = \frac{1}{4}$ for each driving transistor in the NAND gate. Thus, K_R is 4 for the inverter, 4 for the NOR gate, and 8 for the NAND. As the number of inputs increases, K_R increases for the NAND but not for the NOR. It is clear that NMOS NAND gates are wasteful of chip area relative to NMOS NOR. Hence, NOR gates are preferred (and the NOR is the standard gate) in NMOS.

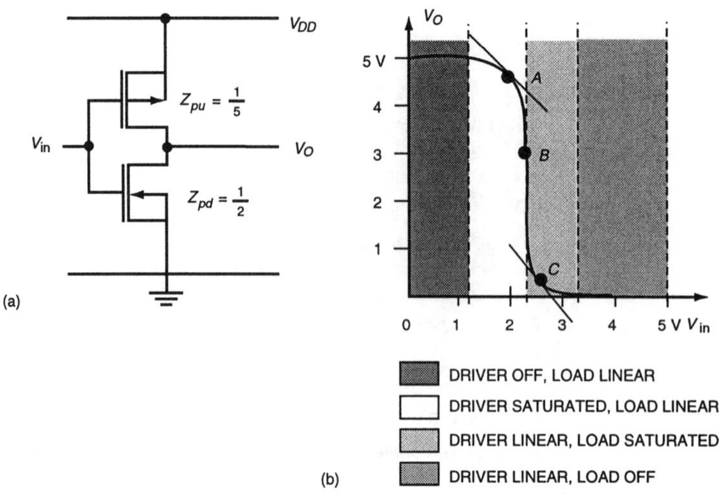

FIGURE 3.9 The CMOS inverter and its voltage transfer characteristics: (a) CMOS inverter, (b) CMOS transfer characteristic and its regions.

CMOS Inverters

As shown in Fig. 3.9(a), the CMOS inverter consists of an enhancement NMOS as the driving transistor, and a complementary enhancement PMOS load transistor. The driving transistor is off when V_{in} is low, and the load transistor is off when V_{in} is high. Thus, one of the two series transistors is always off (equivalently, drain current and power dissipation are zero) except during switching, when both transistors are momentarily on. The resulting low-power dissipation is an important CMOS advantage and makes it an attractive alternative in VLSI design.

NMOS circuits are ratioed in the sense that the pull up never turns off, and V_{OL} is determined by the inverter ratio. CMOS is ratioless in this sense, since V_{OL} is always the negative rail. If one desires equal sourcing and sinking currents, however, the pull-up device must be wider than the pull-down device by the ratio of the electron-to-hole mobilities, typically about 2.5 to 1. This also gives a symmetrical voltage transfer curve, with the voltage at which $V_{in} = V_O$ having a value of $V_{DD}/2$. This voltage is referred to as the inverter voltage V_{inv}.

The voltage transfer for the CMOS inverter is shown in Fig. 3.9(b). Note that the voltage transfer characteristic approaches that of the ideal logic inverter. These characteristics are best obtained with computer circuit simulation programs. As with the depletion load NMOS inverter, useful insights may be gained by performing an analytical solution. The analysis proceeds as previously described for the depletion load NMOS inverter. Note that the VTC of Fig. 3.9(b) has been divided into regions as in Fig. 3.7(a). In each region, the appropriate expressions for the load and driving transistor drain currents are equated so that V_O can be computed for any given V_{in}. To find V_{IL} and V_{IH}, the condition that $dV_O/dV_{in} = -1$ at such critical voltages is applied to the drain current equation. Note that the drain current equations for the PMOS are the same as for NMOS [Eqs. 3.3 and 3.4], except for reverse voltage polarities for the PMOS.

CMOS Gates

CMOS gates are based on simple modifications to the CMOS inverter. Figures 3.10(a) and 3.10(b) show that the CMOS NOR and NAND gates are essentially CMOS inverters in which the load and driving transistor are replaced by series or parallel combinations (as appropriate) of PMOS and NMOS transistors, respectively.

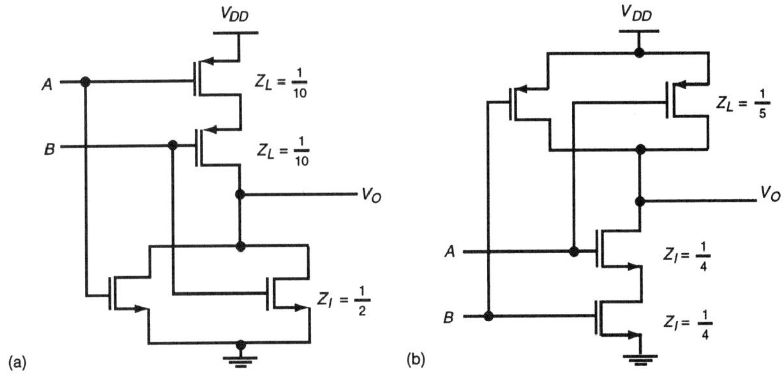

FIGURE 3.10 CMOS gates: (a) CMOS NOR gate, (b) CMOS NAND gate.

Suppose the NOR gate of Fig. 3.10(a) is to have the same V_{DD} and V_{inv} as the CMOS inverter of Fig. 3.9(a), then the equivalent Z_{pu} and Z_{pd} for the NOR gate should equal those for the inverter. Since only one of the parallel pull-down transistors needs be on in the NOR to ensure $V_O = 0$ V, $Z_I = Z_{\text{pd}} = \frac{1}{2}$, as for the inverter. For the series load, however, $Z_L = \frac{1}{10}$ to give equivalent $Z_{\text{pu}} = \frac{1}{5}$. If the NAND gate of Fig. 3.10(b) is to have the same V_{inv} as the said inverter, similar arguments lead to $Z_I = \frac{1}{4}$ and $Z_L = \frac{1}{5}$ for the NAND. Thus, $K_R = 0.4$ for the inverter, 0.2 for the NOR, and 0.8 (closer to unity) for NAND. Hence, NAND is the standard gate in CMOS. Another way of putting this is that for the given Z values, if the channel length L is constant, then the widths of the loads for the inverter, NOR, and NAND are in the ratio 1:2:1. Thus, the NOR requires more chip area, and this larger area requirement increases with the number of inputs.

Bipolar Gates

The major bipolar digital logic families are TTL, emitter-coupled logic (ECL), and integrated injection logic (I^2L). Within each logic family, there are sub-classifications, for example, the Schottky transistor logic (STL), and the integrated Schottky logic (ISL), which developed from the basic I^2L. Bipolar gates have faster switching speeds but greater power dissipation than CMOS gates. The most popular bipolar gate in SSI is the low-power Schottky TTL, which has moderate power dissipation and propagation delay. The fastest switching bipolar family is the ECL, but it has relatively high-power dissipation. The highest packing density is achieved with I^2L and its relatives with low-power dissipation and moderate switching speeds. A better comparison of logic families should be based on the power-delay product, which takes into account both power dissipation and propagation delay.

Medium-Scale Integrated Circuits

MSI circuits have between 10 and 100 transistors per chip. They are built from inverters and basic logic gates with hardly any modifications. They require minimal design effort beyond putting together and interconnecting logic gates. Examples of MSI circuits are flip-flops, counters, registers, adders, multiplexers, demultiplexers, etc.

3.5 LSI and VLSI Circuit Design

Semicustom design is a heavily utilized technique in LSI and VLSI design. In this technique, largely predesigned subcircuits or cells are interconnected to form the desired, larger circuit. Such subcircuits are usually highly regular in nature, so that the technique leads to highly regular circuits and layouts.

Multiphase Clocking

Multiphase clocking is an important technique that can be used to reduce device count in LSI and VLSI circuits. To illustrate the savings that can be realized with the technique, device count is compared for a conventional design

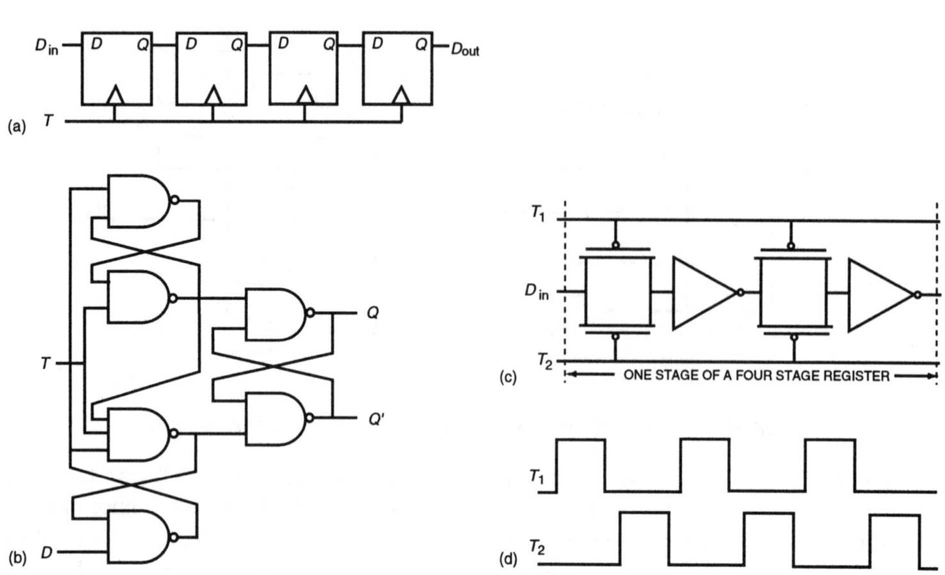

FIGURE 3.11 Conventional (static) and dynamic 4-b shift register: (a) conventional static shift register, (b) D flip-flop, (c) dynamic shift register, (d) two-phase clock pulses.

of a 4-b shift register employing *D* flip-flops based on CMOS NAND gates and a 4-b shift register employing two-phase clocks and CMOS technology.

Both designs are shown in Fig. 3.11. Figure 3.11(a) shows the conventional design for the shift register, which employs a single phase clock signal, whereas Fig. 3.11(b) shows the circuit realization of each D flip-flop with CMOS NAND gates [Taub and Schilling 1977]. The device count for each in this design is obtained as follows:

- 5 two-input CMOS NAND gates, 4 transistors each: 20 transistors.
- 1 three-input CMOS NAND gate: 6 transistors.
- Number of transistors per *D* flip-flop: 26.
- Total number of transistors for the 4-b register: 104.

The second design, which employs two-phase clocking, is shown in 3.11(c), whereas the nonoverlapping clock signals are shown in Fig. 3.11(d). Note that each flip-flop now consists of two CMOS transmission gates and two CMOS inverters. Thus, there are 8 transmission gates and 8 inverters in the 4-b shift register. Device count for this design is as follows:

- Number of transistors for 8 transmission gates: 16.
- Number of transistors for 8 CMOS inverters: 16.
- Total number of transistors for the 4-b register: 32.

In the preceding example, employing two-phase clocking helped to reduce device count to less than one-third of the requirement in the conventional static design. This gain, however, is partly offset by the need for more complex clocking and the fact that the shift register is now dynamic. To avoid loss of data due to leakage through off transistors, the clock must run above a minimum frequency. The times required to charge and discharge capacitive loads determine the upper clock frequency.

3.6 Increasing Packing Density and Reducing Power Dissipation in MOS Circuits

CMOS gates have much lower power dissipation than NMOS gates. This is a great advantage in LSI and VLSI design. Standard CMOS gates, however, require two transistors per input and, therefore, have higher device count than NMOS gates that require one driving transistor per input, plus one depletion load transistor, irrespective of the number of inputs [Mavor, Jack, and Denyer 1983]. This NMOS feature is put to advantage in applications such as semiconductor memories and programmable logic arrays, which will be discussed later. In addition to requiring a higher device count, it is necessary to isolate the PMOS and NMOS transistors in the CMOS and to employ metal interconnection between their drains, which are of opposite conductivity. Consequently, gate count per chip for NMOS is about half that of CMOS, using the same design rules.

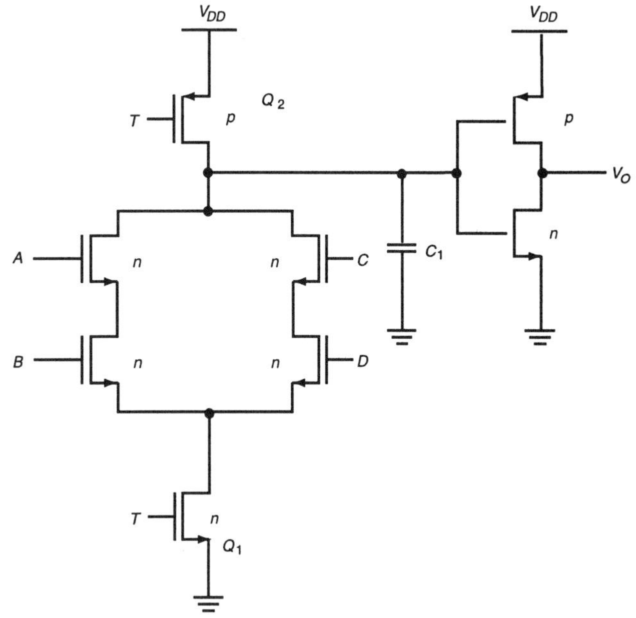

FIGURE 3.12 CMOS domino AND-OR logic.

Figure 3.12 shows a CMOS domino logic circuit in which clocking is employed in an unconventional CMOS circuit to provide both high density and low-power dissipation. When T is low, Q_1 is off, so there is no path to ground irrespective of the logic levels at the inputs A, B, C, D and E. Q_2 is on, so that the parasitic capacitance C_1 charges to V_{DD}. When T is high, Q_2 is off and Q_1 is on. Thus if both A and B, or both C and D, or all of A, B, C, and D are high, a path exists from C_1 to ground, and it discharges. Otherwise, C_1 remains high (but for slow leakage), and the valid logic $(A\ B) + (C + D)$ appears at the output F. Note that this circuit has only two load PMOS transistors, and only one driving transistor is required for each additional input. Thus, device count is minimized by using complex instead of simple logic functions. Each transistor, except those in the output inverter, may be minimum size, since they are required only to charge or discharge C_1. Power dissipation is low as for standard CMOS, because no steady-state current flows.

3.7 Gate Arrays

Gate arrays are a category of semicustom integrated circuits typically containing 100 to several thousand gate cells arranged in rows and columns on a chip. The gate cell may be a NAND, NOR, or other gate. Often, each gate cell is a set of components that could be interconnected to form the desired gate or gates. Identical gate cell pattern is employed, irrespective of chip function. Consequently, gate arrays can be largely processed in advance [Reinhard 1987]. Less effort is required for design with gate arrays since only the masks needed for interconnection are required to customize a chip for a particular application.

Figure 3.13 illustrates a gate array in various levels of detail and possible interconnections within a cell. The floor plan of Fig. 3.13(a) shows that there are 10 columns of cells with 10 cells per column, for a total of 100 cells in the chip. The cell layout of Fig. 3.13(b) shows that there are 4 NMOS and 4 PMOS transistors per cell. Thus there are a total of 800 transistors in the chip. The transistor channels are under the polysilicon and inside the diffusion areas. Figure 3.13(c) shows the cell layout with interconnection to form an NAND gate, whereas Fig. 3.13(d) shows the circuit equivalent of a cell.

Because of their simplicity, a significant amount of wiring is required for interconnections in gate arrays. Good computer software is essential for designing interconnections. In practice, wiring channels tend to fill up, so that it is difficult to utilize more than 70% of the cells on a chip [Alexander 1985]. The standard cell approach discussed next reduces this problem, to some extent, by permitting use of more complex logic functions or cells.

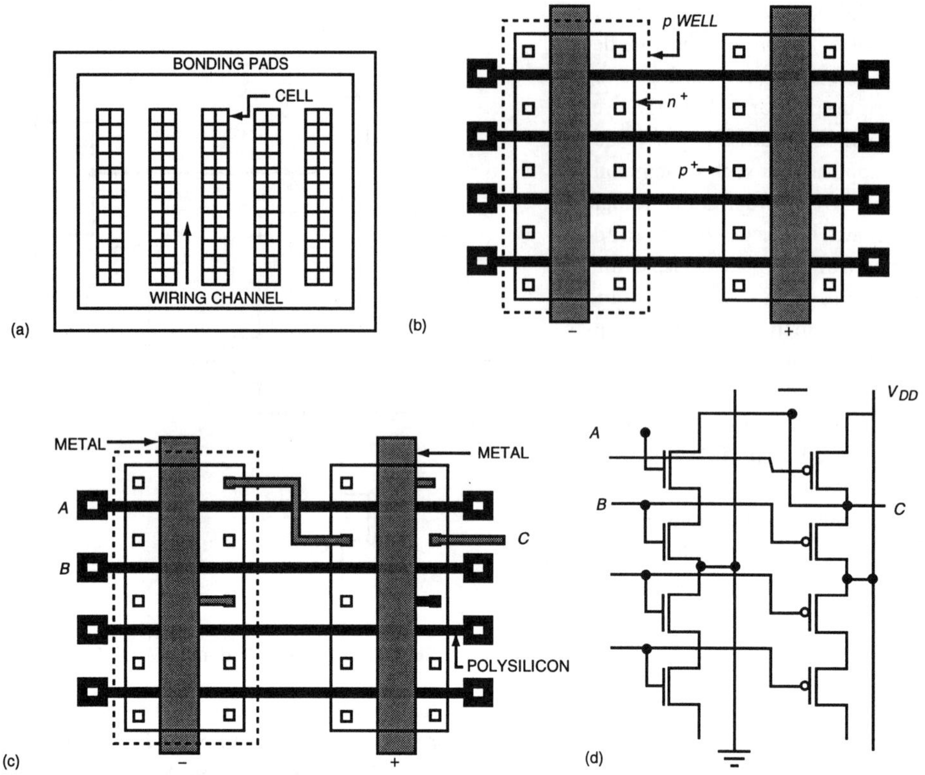

FIGURE 3.13 Gate array at various levels of detail: (a) cell structure, (b) transistor structure, (c) NAND gate interconnection, (d) equivalent circuit.

3.8 Standard Cells

In the **standard cell** approach, the IC designer selects from a library of predefined logic circuits or cells to build the desired circuit. In addition to the basic gates, the **cell library** usually includes more complex logic circuits such as exclusive-OR, AND-OR-INVERT, flip-flops, adders, read only memory (ROM), etc.

The standard cell approach to design is well suited to automated layout. The process consists of selecting cells from the library in accordance with the desired circuit functions, the relative placement of the cells, and their interconnections. The floor plan for a chip designed by this method will be similar to the floor plan for a gate array chip as shown in Fig. 3.13(a). Note, however, that the designer has control over the number and width of wiring channels in this case. Layout for a cell is always the same each time the cell is used, but the cells used and their relative placement is unique to a chip. Thus, every mask level is unique in this approach, and fabrication is more involved and more costly than in the gate array approach [Hodges and Jackson 1988].

3.9 Programmable Logic Devices

Programmable logic devices (PLDs) are a class of circuits widely used in LSI and VLSI design to implement two-level, sum-of-products, boolean functions. Multilevel logic can be realized with Weinberger arrays or gate matrices [Fabricius 1990, Weinberger 1967]. Included among PLDs are programmable logic arrays (PLAs), programmable array logic (PAL), and ROM. The AND-OR structure of the PLA, which can be used to implement any two-level function, is the core of all PLDs. The AND-OR function is often implemented with NOR–NOR or NAND–NAND logic.

PLDs have the advantage of leading to highly regular layout structure. The PLD consists of an AND plane followed by an OR plane. The logic function is determined by the presence or absence of contacts or connections at row and column intersections in a single conducting layer. Programming or establishment of appropriate contacts may be accomplished during fabrication. Alternatively, the PLDs may be user programmable by means of fuse links.

Figure 3.14 shows the three types of PLDs. Hollow diamonds at row/column intersections in an AND or OR plane indicates that the plane is programmable. Presence of solid diamonds in some row/column intersections indicate that the logic for that plane is already defined and fixed. The PLD is a PLA if both the AND and OR planes are programmable, a PAL if only the AND plane is programmable, and a ROM if only the OR plane (the decoder in this case) is programmable. Because PLAs are programmable in both planes, they permit more versatile logic realizations than PALs. Also, the PAL can be considered a special case of the PLA. Thus, only the PLA is discussed further.

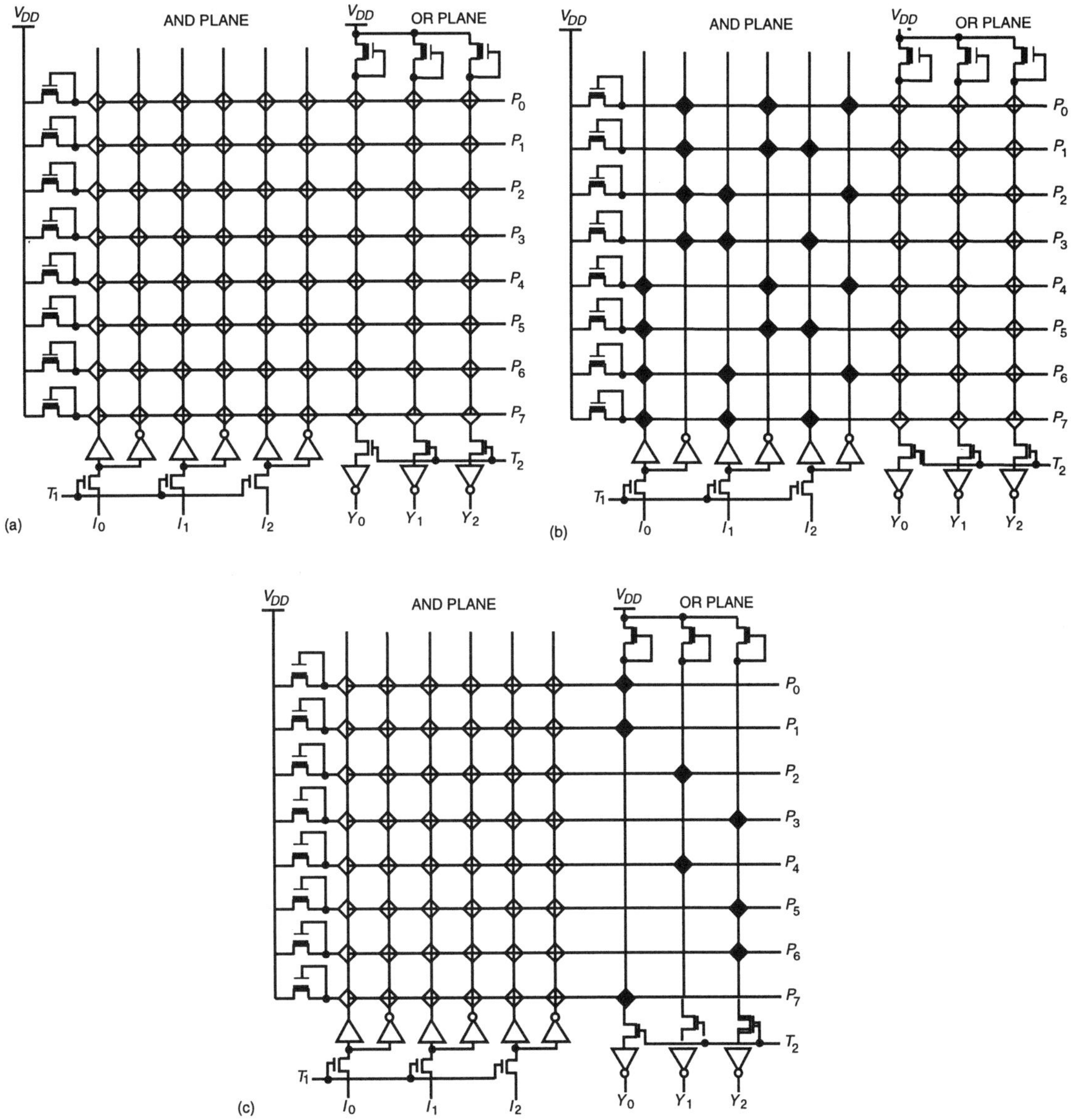

FIGURE 3.14 Types of programmable logic devices: (a) programmable logic array, (b) programmable read only memory, (c) programmable array logic.

Programmable Logic Array

PLAs provide an alternative to implementation of combinational logic that results in highly regular layout structures. Consider, for example, a PLA implementation of the following sum of product expressions:

$$Y_1 = \bar{I}_0\bar{I}_1 + I_0\bar{I}_2 \quad (3.5)$$

$$Y_2 = \bar{I}_0 I_1 I_2 + I_0\bar{I}_2 \quad (3.6)$$

$$Y_3 = \bar{I}_0\bar{I}_1\bar{I}_2 + \bar{I}_0 I_1 I_2 \quad (3.7)$$

The PLA has three inputs and three outputs. In terms of the AND and OR planes, the outputs of the AND plane are

$$P_1 = \bar{I}_0\bar{I}_2 \quad (3.8)$$

$$P_2 = \bar{I}_0\bar{I}_1\bar{I}_2 \quad (3.9)$$

$$P_3 = \bar{I}_0 I_1 I_2 \quad (3.10)$$

$$P_4 = I_0\bar{I}_2 \quad (3.11)$$

The overall output is the output of the OR plane and can be written in terms of the AND plane outputs as

$$Y_1 = P_1 + P_4 \quad (3.12)$$

$$Y_2 = P_3 + P_4 \quad (3.13)$$

$$Y_3 = P_2 + P_3 \quad (3.14)$$

Figure 3.15 shows the logic circuit consisting of the AND and the OR planes. Note that each product line in the AND plane is an NMOS NOR gate with one depletion load; the gate of each driving transistor is controlled by

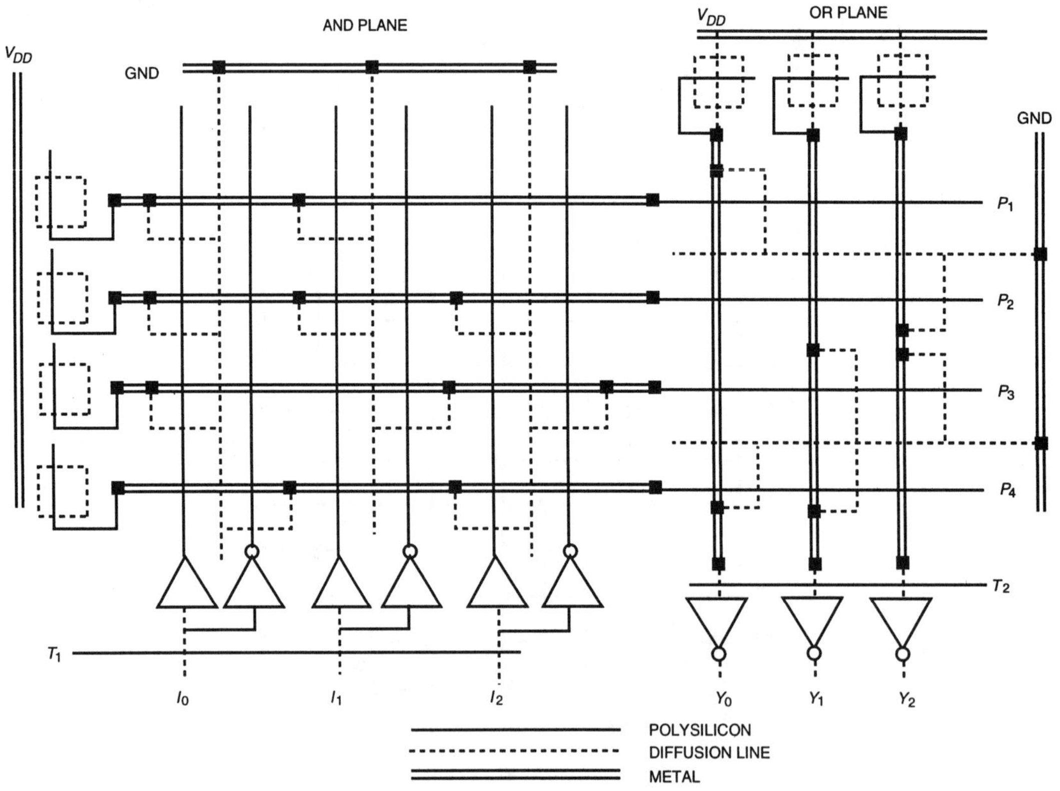

FIGURE 3.15 An NMOS PLA.

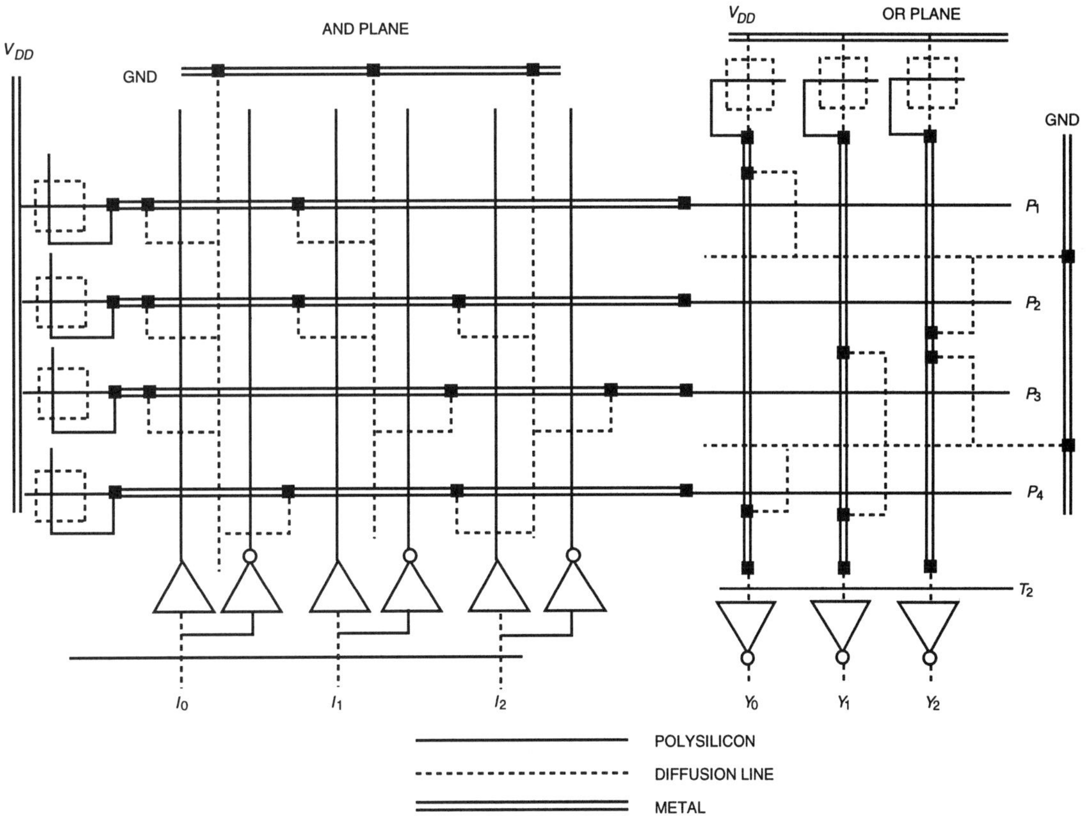

FIGURE 3.16 Stick diagram layout of the PLA shown in Fig. 3.15.

an input line. Likewise, each output line in the OR plane is an NMOS NOR gate with driving transistors whose gates are controlled by the product lines. Thus, the PLA employs a NOR–NOR implementation.

The personality matrix for a PLA [Lighthart, Aarts, and Beenker 1986] gives a good description of the PLA and how it is to be programmed. The personality matrix Q of the PLA of Fig. 3.15 is given by Eq. (3.15).

$$Q = \begin{bmatrix} 0 & 0 & x & 1 & 0 & 0 \\ 0 & 0 & 0 & 0 & 0 & 1 \\ 0 & 1 & 1 & 0 & 1 & 1 \\ 1 & x & 0 & 1 & 1 & 0 \end{bmatrix} \tag{3.15}$$

The first three columns comprise the AND plane of the matrix, whereas the last three columns comprise the OR plane of the three-input, three-output PLA. In the AND plane, element $q_{ij} = 0$ if a transistor is to link the product line P_i to the input line I_i; $q_{ij} = 1$ if a transistor is to link P_i to $\bar{I}_i$, and q_{ij} is a *don't care* if neither input is to be connected to P_i. In the OR plane, $q_{ij} = 1$ if product line P_i is connected to output Y_j and 0 otherwise.

Figure 3.16 shows the stick diagram layout of the PLA circuit of Fig. 3.15, and illustrates how the regular structure of the PLA facilitates its layout. The input lines to each plane are polysilicon, the output lines from each plane are metal, and the sources of driving transistors are connected to ground by diffused lines. The connecting transistors are formed by grounded, crossing diffusion lines.

3.10 Reducing Propagation Delays

Large capacitive loads are encountered in many ways in large integrated circuits. Bonding pads are required for interfacing the chip with other circuits, whereas probe pads are often required for testing. Both present large capacitive loads to their drivers. Interconnections within the chip are by means of metal or polysilicon lines. When long, such lines present long capacitive loads to their drivers. Although regular array structures, such as those of gate arrays, standard cells, and PLAs, are very convenient for semicustom design of LSI and VLSI, they have an inherent drawback with regard to propagation delay. Their row and column lines contact many devices and, hence, are very capacitive. The total delay of a long line may be reduced by inserting buffers along the line to restore the signal. Superbuffers are used for interfacing between small gates internal to the chip and large pad drivers and for driving highly capacitive lines.

Resistance–Capacitance (RC) Delay Lines

A long polysilicon line can be modeled as a lumped RC transmission line as shown in Fig. 3.17. Let Δx represent the length of a section of resistance R and capacitance C, and Δt be the time required for the signal to propagate along the section. Let $\Delta V = (V_{n-1} - V_n)\Delta x$, where V_n is the voltage at node n. The difference equation governing signal propagation along the line is [Fabricius 1990]

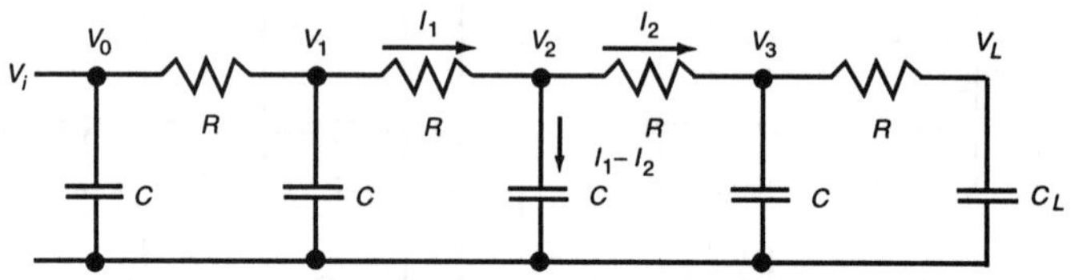

FIGURE 3.17 Lumped circuit model of a long polysilicon line.

$$RC\frac{\Delta V}{\Delta t} = \frac{\Delta^2 V}{\Delta x^2} \tag{3.16}$$

As the number of sections becomes very large, the difference equation can be approximated with the differential equation

$$RC\frac{dV}{dt} = \frac{d^2V}{dx^2} \tag{3.17}$$

For a delay line with N sections, matched load $C_L = C$, resistance R and capacitance C per unit length, and overall length L in micrometer, Horowitz [1983] showed that the propagation delay is

$$t_d = \frac{0.7N(n+1)L^2RC}{2N^2} \tag{3.18}$$

Note that the propagation delay is proportional to the square of the length of the line. As N tends to infinity, the propagation delay becomes

$$t_d = \frac{0.7RCL^2}{2} \tag{3.19}$$

To reduce the total delay restoring inverters can be inserted along a long line. Consider as an example, a 5-mm-long polysilicon line with $r = 20\ \Omega/\mu\text{m}$ and $C = 0.2\ \text{fF}/\mu\text{m}$. It is desired to find the respective propagation delays, if the number of inverters inserted in the line varies from zero to four. The delay of each inverter is proportional to the length of the segment it drives and is given by $t_I = 0.4$ ns when it is driving a 1-mm-long segment. In each case, the inverters used are spaced uniformly along the line. Let

K = number of inverters used
$K + 1$ = number of sections
ℓ = length per section, $\frac{5\text{mm}}{k+1}$
t_d = total delay

then

$$t_d = \left[(k+1)\left(\frac{0.7RC\ell^2}{2}\right) + 0.4\ell K\right] ns \tag{3.20}$$

From Eq. (3.20), the propagation delays can be calculated. The delay for each number of inverters as a percentage of the unbuffered line delay is also computed. The results are tabulated in Table 3.5.

TABLE 3.5 Improvement in Propagation Delay with Increase in Number of Line Buffered

No. of Inverters K	Total Delay t_d, ns	% of Unbuffered Line Delay
0 (unbuffered)	35.0	100
1	18.5	52.86
2	13.0	37.14
3	10.25	29.29
4	8.6	24.57

The results in the table show that the propagation delay decreases as the number of inverters is increased. The improvement in propagation delay, however, is less dramatic for each additional inverter than the one preceding it. The designer would stop increasing the number of inverters when the incremental gain no longer justifies an additional inverter. If the number of inverters is even, there is no inversion of the overall signal.

Superbuffers

Propagation delays can be reduced without excessive power consumption by using superbuffers. These are inverting or noninverting circuits that can source and sink larger currents and drive large capacitive loads faster than standard inverters. Unlike ratioed NMOS inverters in which the pull-up current drive capability is much less than the pull-down capability, superbuffers have symmetric drive capabilities. A superbuffer consists of a push-pull or totem pole output inverter driven by a conventional inverter. In an inverting superbuffer, the gates of both pull-down transistors in the driving and the totem pole inverters are driven by the input signal whereas the gate of the pull-up transistor in the output totem pole inverter is driven by the complement of the input signal.

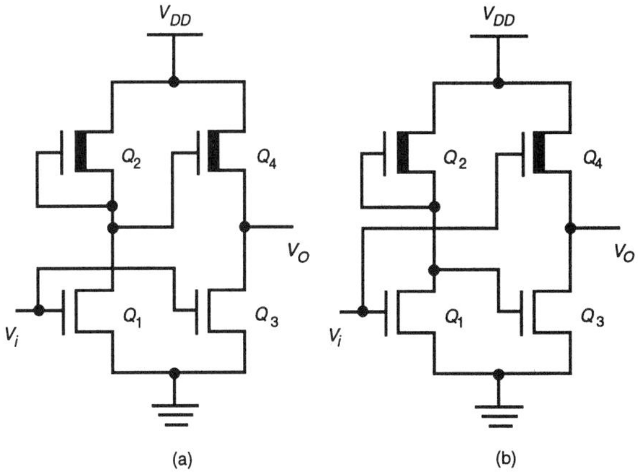

FIGURE 3.18 NMOS superbuffers: (a) inverting superbuffer, (b) noninverting superbuffer.

An inverting and a noninverting NMOS superbuffer is shown in Fig. 3.18. By designing for an inverter ratio (K_R) of 4, and driving the totem pole pull-up with twice the gate voltage of a standard depletion mode pull-up, the NMOS superbuffer can be rendered essentially ratioless. In standard NMOS inverters, the pull-up transistor has the slower switching speed. Consider the inverting superbuffer of Fig. 3.18(a). When the input voltage goes low, the output voltage of the standard inverter and the gate voltage of Q_4 goes high rapidly since the only load it sees is the small gate capacitance of Q_4. Thus, the totem pole output switches rapidly. Similar considerations will show that the noninverting super buffer also results in fast switching speeds.

The improvement in drive current capability of the NMOS superbuffer, relative to the standard (depletion load) NMOS inverter, can be estimated by comparing the average, output pull-up currents [Fabricius 1990]. The depletion load in the standard NMOS inverter is in saturation for $V_O < 2$ V and in linear region for $V_O > 2$ V. For the pull-up device, $V_{DS} = 5\text{ V} - V_O$. Thus, the pull-up transistor is in saturation when it has $3\text{ V} < V_{DS} < 5$ V and is in the linear region when $0\ V < V_{DS} < 3$ V. The average current will be estimated by evaluating I_D(sat) at $V_{DS} = 5$ V and $I_{D(\text{lin})}$ at $V_{DS} = 2.5$ V. Let $V_{TD} = -3$ V for the depletion mode transistor. Then for the standard NMOS inverter,

$$I_{D(\text{sat})} = K_{\text{pu}}(V_{GS} - V_{\text{TD}})^2 = K_{\text{pu}}(0+3)^2 = 9K_{\text{pu}} \tag{3.21}$$

$$I_{D(\text{lin})} = K_{\text{pu}}\left[2(V_{GS} - V_{\text{TD}})V_{DS} - V_{DS}^2\right] = K_{\text{pu}}[2(0+3)2.5 - 2.5^2] = 8.76K_{\text{pu}} \tag{3.22}$$

Thus the average pull-up current for the standard NMOS inverter is approximately $8.88K_{\text{pu}}$. For the totem pole output of the NMOS superbuffer, the average pull-up current is also estimated from drain currents at $V_{DS} = 5$ V and 2.5 V. Note that in this case, the pull-up transistor has $V_G = V_{DD} = 5$ V when it is on. Thus, $V_{GS} = V_{DS} = 5$ V so that it always operates in the linear region. The currents are

$$I_D(5\,\text{V}) = K_{\text{pu}}[2(5+3)5 - 5^2] = 55K_{\text{pu}} \tag{3.23}$$

$$I_D(2.5\,\text{V}) = K_{\text{pu}}[2(2.5+3)2.5 - 2.5^2] = 10.62K_{\text{pu}} \tag{3.24}$$

The average pull-up current for the totem pole output is $38.12K_{\text{pu}}$. The average totem pole pull-up current is approximately 4.3 times the average NMOS pull-up current. Consequently, the superbuffer will be roughly ratioless if designed for an inverter ratio of 4.

3.11 Output Buffers

Internal gates on a VLSI chip have load capacitances of about 50 fF or less and typical propagation delays of less than 1 ns. However, the chip output pins have to drive large capacitive loads of about 50 pF or more [Hodges and Jackson 1988]. For MOSFETs, the propagation delay is directly proportional to load capacitance. Thus, using a typical gate on the chip to drive an output pin would result in too long a propagation delay. Output buffers utilize a cascade of inverters of progressively larger drive capability to reduce the propagation delay.

An N-stage output buffer is illustrated in Fig. 3.19. Higher drive capability results from employing transistors of increasing channel width. As the transistor width increases from stage to stage by a factor of f, so does the current drive capability and the input capacitance. If C_G is the input or gate capacitance of the first inverter in the buffer, then the second inverter has an input capacitance of fC_G and the Nth inverter has an input capacitance of $f^{N-1}C_G$ and a load capacitance of $f^N C_G$, which is equal to C_L, the load capacitance at the output pin. The inverter on the left in the figure is a typical inverter on the chip with an input or gate capacitance of C_G and a propagation delay of τ. The first inverter in the buffer has an input capacitance of fC_G, but it has a current driving capability f times larger than the on chip inverter. Thus, it has a propagation delay of $f\tau$. The second inverter in the buffer has an input capacitance of f^2C_G and an accumulated delay of $2f\tau$ at its output. The Nth inverter has an input capacitance of $f^N C_G$, which is equal to the load capacitance at the output pin, and an accumulated propagation delay of $Nf\tau$, which is the overall delay of the buffer.

Let

$$\text{load capacitance} = YC_G = f^N C_G$$

$$\text{total delay} = t_B = Nf\tau$$

Then

$$N = \frac{\ln Y}{\ln f} \tag{3.25}$$

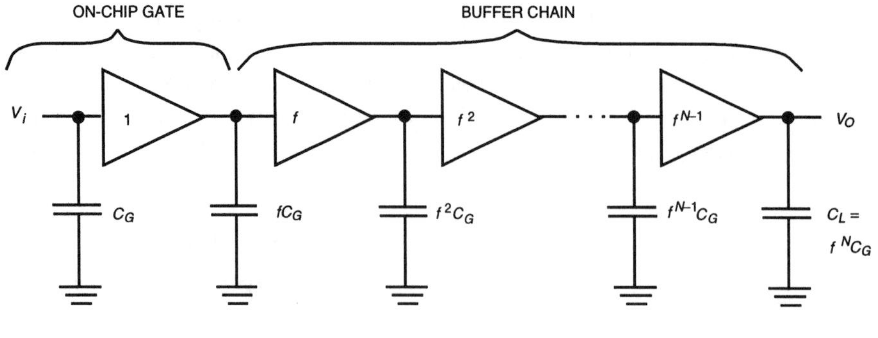

FIGURE 3.19 An N-stage output buffer chain.

and

$$t_B = \frac{\ln Y}{\ln f} f\tau \tag{3.26}$$

By equating to zero the first derivative of t_B with respect to f, it is found that t_B is minimum at $f = e = 2.72$, the base of the natural logarithms. This is not a sharp minimum [Moshen and Mead 1979], and values of f between 2 and 5 do not greatly increase the time delay.

Consider an example in which $C_G = 50$ fF and $\tau = 0.5$ ns for a typical gate driving an identical gate on the chip. Suppose this typical gate is to drive an output pin with load capacitance $C_L = 55$ pF, instead of an identical gate. If an output buffer is used,

$$Y = \frac{C_L}{C_G} = \frac{55 \text{ pF}}{50 \text{ fF}} = 1100$$

$$N = \ln Y = 7$$

$$t_B = 7e\tau = 9.5 \text{ ns}$$

If the typical chip gate is directly connected to the output pin, the propagation delay is $Y\tau = 550$ ns, which is extremely large compared with the 9.5 ns delay obtained when the buffer is used. This example illustrates the effectiveness of the buffer.

Defining Terms

Cell library: A collection of simple logic elements that have been designed in accordance with a specific set of design rules and fabrication processes. Interconnections of such logic elements are often used in semicustom design of more complex IC chips.

Custom design: A design method that aims at providing a unique implementation of the function needed for a specific application in a way that minimizes chip area and possibly other performance features.

Design rules: A prescription for preparing the photomasks used in IC fabrication so that optimum yield is obtained in as small a geometry as possible without compromising circuit reliability. They specify minimum device dimensions, minimum separation between features, and maximum misalignment of features on an IC chip.

Layout: An important step in IC chip design that specifies the position and dimension of features and components on the different layers of masks.

Masks: A set of photographic plates used to define regions for diffusion, metalization, etc., on layers of the IC wafer. Each mask consists of a unique pattern: the image of the corresponding layer.

Standard cell: A predefined logic circuit in a cell library designed in accordance with a specific set of design rules and fabrication processes. Standard cells are typically employed in semicustom design of more complex circuits.

Semicustom design: A design method in which largely predesigned subcircuits or cells are interconnected to form the desired more complex circuit or part of it.

References

Alexander, B. 1985. MOS and CMOS arrays. In *Gate Arrays: Design Techniques and Application.* ed. J.W. Read. McGraw–Hill, New York.

Fabricius, E.D. 1990. *Introduction to VLSI Design.* McGraw–Hill, New York.

Hodges, A.D. and Jackson, H.G. 1988. *Analysis and Design of Digital Integrated Circuits.* McGraw–Hill, New York.

Horowitz, M. 1983. Timing models for MOS pass networks. *Proceedings of the IEEE Symposium on Circuits and Systems*, pp. 198–201.

Keyes, R.W. 1975. Physical limits in digital electronics. *Proc. of IEEE* 63:740–767.

Lighthart, M.M., Aarts, E.H.L., and Beenker, F.P.M. 1986. Design for testability of PLAs using statistical cooling. *Proceedings of the 23rd ACM/IEEE Design Automation Conference*, pp. 339–345, June 29–July 2.

Mavor, J., Jack, M.A., and Denyer, P.B. 1983. *Introduction to MOS LSI Design*. Addison–Wesley, Reading, MA.

Mead, C.A. and Conway, L.A. 1980. *Introduction to VLSI Systems*. Addison–Wesley, Reading, MA.

Moshen, A.M. and Mead, C.A. 1979. Delay time optimization for driving and sensing signals on high capacitance paths of VLSI systems. *IEEE J. of Solid State Circ.* SC-14(2):462–470.

Pucknell, D.A. and Eshroghian, K. 1985. *Basic VLSI Design, Principles and Applications*. Prentice–Hall, Englewood Cliffs, NJ.

Reinhard, D.K. 1987. *Introduction to Integrated Circuit Engineering*. Houghton–Mifflin, Boston, MA.

Taub, H. and Schilling, D. 1977. *Digital Integrated Circuits*. McGraw–Hill, New York.

USC Information Sciences Inst. 1984. MOSIS scalable NMOS process, version 1.0. Univ. of Southern California, Nov., Los Angeles, CA.

USC Information Sciences Inst. 1988. MOSIS scalable and generic CMOS design rules, revision 6. Univ. of Southern California, Feb., Los Angeles, CA.

Weinberger, A. 1967. Large scale integration of MOS complex logic: A layout method. *IEEE J. of Solid-State Circ.* SC-2(4):182–190.

Further Information

IEEE Journal of Solid State Circuits.

IEEE Proceedings of the Custom Integrated Circuits Conference.

IEEE Transactions on Electron Devices.

Proceedings of the European Solid State Circuits Conference (ESSCIRC).

Proceedings of the IEEE Design Automation Conference.

4

Digital Logic Families

Robert J. Feugate, Jr.
Northern Arizona University

4.1 Introduction

Digital devices are constrained to two stable operating regions (usually voltage ranges) separated by a transition region through which the operating point may pass but not remain (see Fig. 4.1). Prolonged operation in the transition region does not cause harm to the devices, it simply means that the resulting outputs are unspecified. For the inverter shown in the figure, the output voltage is guaranteed to be greater than V_{OH} as long as the input voltage is below the specified V_{IH}. Note that the circuit is designed so that the input high voltage is lower than the output high voltage (vice versa for logic low voltages). This difference, called **noise margin**, permits interfering signals to corrupt the logic voltage within limits without producing erroneous operation. Logical conditions and binary numbers can be represented physically by associating one of the stable voltage ranges with one logic state or binary value and identifying the other stable voltage with the opposite state or value. By extension, then, it is possible to design electronic circuits that physically perform logical or arithmetic operations. As detailed examination of logic design is beyond the scope of this chapter, the reader is referred to any of the large number of digital logic textbooks.[1]

Digital logic components were the earliest commercially produced integrated circuits. *Resistor–transistor logic* (RTL) and a speedier variant *resistor–capacitor–transistor logic* (RCTL) were introduced in the early 1960s by Fairchild Semiconductor (now part of National Semiconductor). *Diode-transistor logic* (DTL) was introduced a few years later by Signetics Corporation. Although these families are often discussed in electronics texts as ancestors of later logic families, they have been obsolete for many years and are of historical interest only.

A primary performance characteristic of different logic families is their *speed-power product*, that is, the average propagation delay of a basic gate multiplied by the average power dissipated by the gate. Table 4.1 lists the speed-power product for several popular logic families. Note that propagation delays specified in manufacturer handbooks may be measured under different loading conditions, which must be taken into account in computing speed-power product.

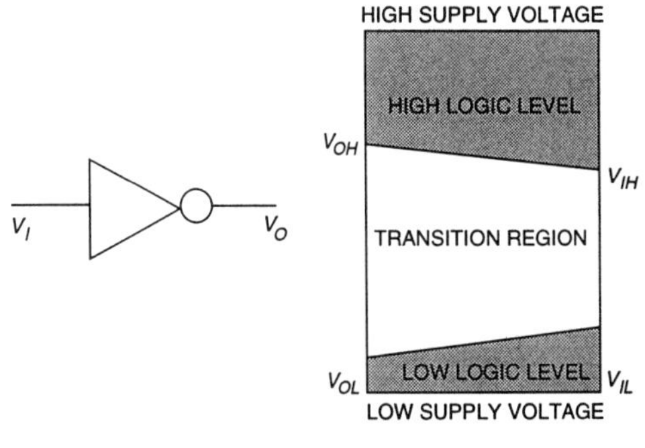

FIGURE 4.1 Logic voltage levels.

[1]Two excellent recent logic design texts are *Modern Digital Design*, by Richard Sandige, McGraw–Hill, New York, 1990 and *Contemporary Logic Design*, by Randy Katz, Benjamin Cummings, Redwood City, CA, 1994.

0-8493-0050-9/00/$0.00+$.50

TABLE 47.1 Speed-Product Comparison for Popular Discrete Logic Families[a]

	Power Dissipation/Gate			
Logic Family	Static at 100 kHz, mW	Typical Delay, mW	Product at 100 kHz, ns	pJ
Metal-gate CMOS	0.001	0.1	75	7.5
Silicon-gate CMOS	0.0000025	.17	10	1.7
Standard TTL	10	10	10	100
Low Power Schottky TTL	2	2	8	16
Schottky TTL	19	19	4.5	85.5
Adv. Schottky TTL	8.5	8.5	1	8.5
ALS TTL	1	1	2.5	2.5
10 K ECL	24	24	0.2	4.8

[a]Load capacitance is 50 pF, load resistance 500 Ω, except ECL is driving 50-Ω transmission.

4.2 Transistor–Transistor Logic

First introduced in the 1960s, transistor–transistor logic (TTL) was the technology of choice for discrete logic designs into the 1990s, when complementary metal oxide semi-conductor (CMOS) equivalents gained ascendancy. Because TTL has an enormous base of previous designs, is available in a rich variety of small-scale and medium-scale building blocks, is electrically rugged, offers relatively high-operating speed, and has well-known characteristics, transistor–transistor logic will continue to be an important technology for some time. Texas Instruments Corporation's 54/7400 TTL series became a de facto standard, with its device numbers and pinouts used by other TTL (and CMOS) manufacturers.

There are actually several families of TTL devices having circuit and semiconductor process variations that produce different speed-power characteristics. Individual parts are designated by a scheme combining numbers and letters to identify the exact device and the TTL family. Parts having 54 as the first digits have their performance specified over the military temperature range from −55 to +125°C, whereas 74-series parts are specified from 0 to +70°C. Letters identifying the family come next, followed by a number identifying the part's function. For example, 7486 identifies a package containing four, 2-input exclusive-OR gates from the standard TTL family, whereas 74ALS86 indicates the same function in the advanced low-power Schottky family. Additional codes for package type and, possibly, circuit revisions may also be appended. Generally speaking, devices from different TTL families can be intermixed, although attention must be paid to **fanout** (discussed subsequently) and noise performance.

The basic building block of standard TTL is the NAND gate shown in Fig. 4.2. If either input A or B is connected to a logic-low voltage, multiple-emitter transistor Q1 is saturated, holding the base of transistor Q2 low and keeping it cut off. Consequently, Q4 is starved of base current and is also off, while Q3 receives base drive current through resistor R2 and is turned on, pulling the output voltage up toward V_{cc}. Typical high output voltage for standard TTL circuits is 3.4 V. On the other hand, if both of the inputs are high, Q1 will move into the reverse-active region and current will flow out of its collector into the base of Q2. Q2 will turn on, raising Q4's base voltage until it is driven into saturation. At the same time, the Ohm's law voltage drop across R2 from Q2's collector current will lower the base voltage of Q3 below the value needed to keep D1 and Q3's base-emitter junction forward biased. Q3 cuts off, with the net result that the output will be drawn low, toward ground. In normal operation, then, transistors Q2, Q3, and Q4 must move between saturation and cutoff. Saturated transistors experience an accumulation in minority carriers in their base regions, and the time required for the transistor to leave saturation depends on how quickly excess carriers can be removed from the base. Standard TTL processes introduce gold as an impurity to create *trapping sites* in the base that speed the annihilation of minority electrons. Later TTL families incorporate circuit refinements and processing enhancements designed to reduce excess carrier concentrations and speed removal of accumulated base electrons.

Because the internal switching dynamics differ and because the pull-up and pull-down transistors have different current drive capabilities, TTL parts show a longer propagation time when driving the output from low to high than from high to low (see Fig. 4.3). Although the average propagation delay $[(t_{\rm phl} + t_{\rm plh})/2]$ is often used in

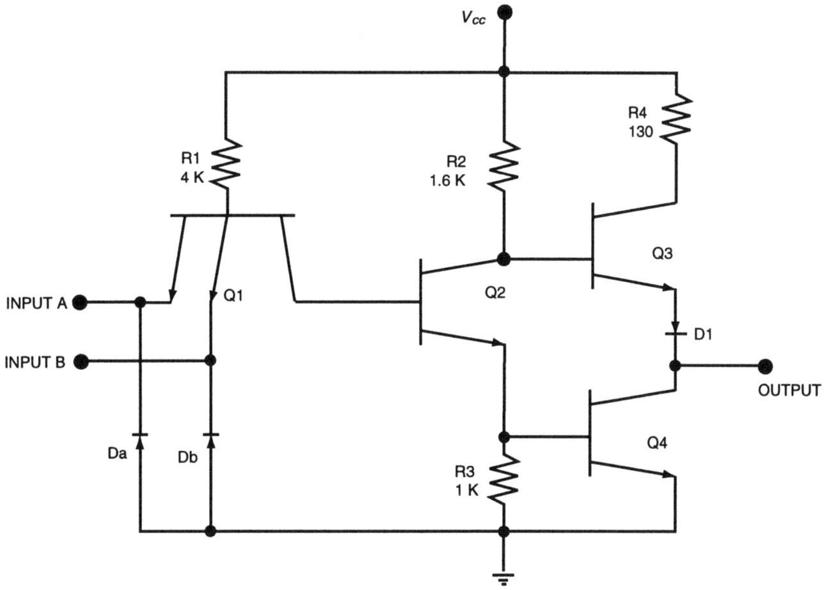

FIGURE 4.2 Two input standard TTL NAND gate.

speed comparisons, conservative designers use the slower propagation delay in computing circuit performance. NAND gate delays for standard TTL are typically 10 ns, with guaranteed maximum $t_{\rm phl}$ of 15 ns and $t_{\rm phl}$ of 22 ns.

When a TTL gate is producing a low output, the output voltage is the collector-to-emitter voltage of pull-down transistor Q4. If the collector current flowing in from the circuits being driven increases, the output voltage rises. The output current, therefore, must be limited to ensure that output voltage remains in the specified logic low range (less than V_{OL}). This places a maximum on the number of inputs that can be driven low by a given output stage; in other words, any TTL output has a maximum low-level fanout. Similarly, there is a maximum fanout for high logic levels. Since circuits must operate properly at either logic level, the smaller of the two establishes the maximum overall fanout. That is,

$$\text{fanout} = \text{minimum}[-I_{OH,\max}/I_{IH,\max}, -I_{OL,\max}/I_{IL,\max}]$$

where

$I_{OH,\max}$ = specified maximum output current for $V_{OH,\min}$
$I_{OL,\max}$ = specified maximum output current for $V_{OL,\max}$
$I_{IH,\max}$ = specified maximum input current, high level
$I_{IL,\max}$ = specified maximum input current, low level

The minus sign in the fanout computations arises because both input and output current reference directions are defined into the device. Most parts in the standard TTL family have a fanout of 10 when driving other standard TTL circuits. Buffer circuits with increased output current capability are available for applications such as clock signals that must drive an unusually large number of inputs. TTL families other than standard TTL have different input and output current specifications; fanout computations should be performed whenever parts from different families are mixed.

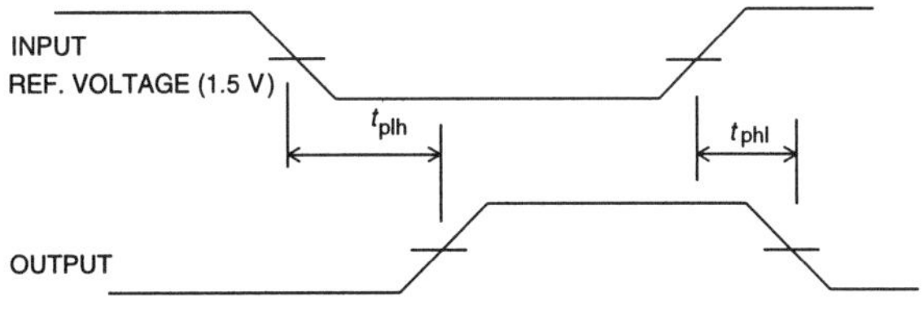

FIGURE 4.3 Propagation delay times.

One of the transistors in the conventional *totem pole* TTL output stage is always turned on, pulling the output pin toward either a logic high or low. Consequently, two or more outputs cannot be connected to the same node; if one output were attempting to drive a low logic voltage and the other a high, a low-resistance path will be created from between V_{cc} and ground. Excessive currents may damage the output transistors and, even if damage does not occur, the output voltage will settle in the forbidden transition region between V_{OH} and V_{OL}. Some TTL parts are available in versions with open-collector outputs, truncated output stages having only a pull down transistor (see Fig. 4.4). Two or more open-collector outputs can be connected to V_{cc} through a pull-up resistor, creating a wired-AND phantom logic gate. If any output transistor is turned on, the common point will be pulled low. It will only assume a high voltage if all output transistors are turned off. The pull-up resistor must be large enough to ensure that the I_{OL} specification is not exceeded when a single gate is driving the output low, but small enough to guarantee that input leakage currents will not pull the output below V_{OH} when all driving gates are off. Since increasing the pull-up resistor value increases the resistance–capacitance (RC) time constant of the wired-AND gate and slows the effective propagation delay, the pull-up resistor is usually selected near its minimum value. The applicable design equations are

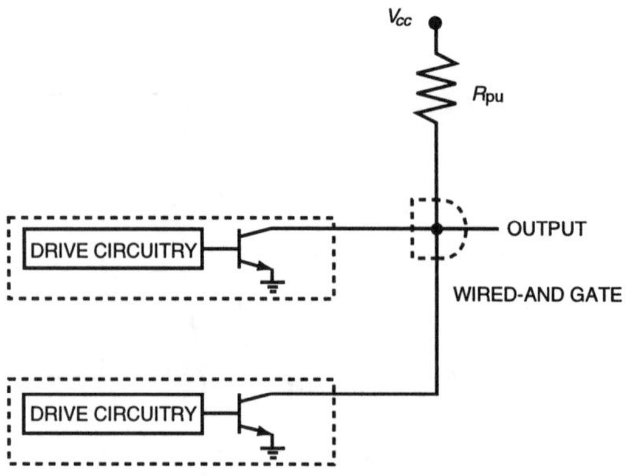

FIGURE 4.4 Wired-AND gate using open collector outputs and a pullup resistor.

$$(V_{cc} - V_{OL,\max})/(I_{OL,\max} + nI_{IL,\max}) < R_{\text{pu}} < (V_{cc} - V_{OH,\min})/(mI_{OH,\max} + nI_{IH,\max})$$

where m is the number of open collector gates connected to R_{pu} and n is the number of driven inputs. Reference direction for all currents is into the device pin.

Some TTL parts have three-state outputs: totem pole outputs that have the added feature that both output transistors can be turned off under the control of an output enable signal. When the output enable is in its active state, the part's output functions conventionally, but when the enable is inactive, both transistors are off, effectively disconnecting the output pin from the internal circuitry. Several three-state outputs can be connected to a common bus, and only one output is enabled at a time to selectively place its logic level on the bus.

Over the long history of TTL, a number of variants have been produced to achieve faster speeds, lower power consumption, or both. For example, the 54/74L family of parts incorporated redesigned internal circuitry using higher resistance values to lower the supply current, reducing power consumption to a little as 1/10 that of standard TTL. Speed was reduced as well, with delay times as much as three times longer. Conversely, the 54/74H family offered higher speeds than standard TTL, at the expense of increased power consumption. Both of these TTL families are obsolete, their places taken by various forms of Schottky TTL.

As noted earlier, TTL output transistors are driven well into saturation. Rather than using gold doping to hasten decay of minority carriers, Schottky transistors limit the forward bias of the collector-base junction and, hence, excess base minority carrier accumulation by paralleling the junction with a Schottky (gold-semiconductor) diode, as shown in Fig. 4.5. As the NPN transistor starts to saturate, the Schottky diode becomes forward biased. This clamps the collector-base junction to about 0.3 V, keeping the transistor out of hard saturation. Schottky TTL (S) is about three times faster than standard TTL, while consuming about 1.5 times as much power.

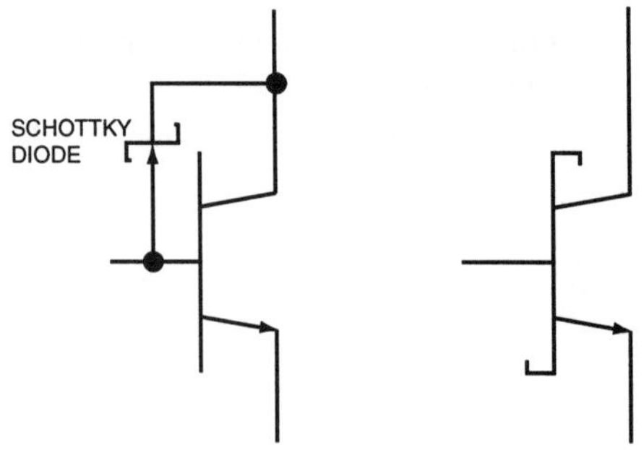

FIGURE 4.5 Schottky transistor construction and symbol.

Lower power schottky TTL (LS) combined an improved input circuit design with Schottky transistors to reduce power consumption to a fifth of standard TTL, while maintaining equal or faster speed. The redesigned input circuitry of LS logic changes the input logic thresholds, resulting in slightly reduced noise margin.

Still another TTL family is *Fairchild Advanced Schottky* TTL (FAST). The F parts combine circuit improvements with advanced processing techniques that reduce junction capacitances and increase transistor speeds. The result is a fivefold increase in speed compared to standard TTL, with lowered power consumption. FAST logic also claims better noise margin and better fanout than low-power Schottky.

As their names imply, advanced Schottky (AS) and advanced low-power Schottky (ALS) use improved fabrication processes to improve transistor switching speeds. Parts from these families are two to three times faster than their S and LS equivalents, while consuming half the power. Although the standard TTL, Schottky (S), and low-power Schottky (LS) families remain in volume production, the newer advanced Schottky families (AS, ALS, F) offer better speed-power products and are favored for new designs.

In making speed comparisons between different logic families, one should keep in mind that manufacturer's published specifications use different load circuits for different TTL families. For example, propagation delays of standard TTL are measured using a load circuit of a 15-pF capacitor paralleled by a 400-Ω resistor, whereas FAST uses 50 pF and 500 Ω. Since effective propagation delay increases approximately linearly with the capacitance of the load circuit, direct comparisons of raw manufacturer data is meaningless. Manufacturer's databooks include correction curves showing how propagation delay changes with changes in load capacitance.

As outputs change from one logic state to another, sharp transients occur on the power supply current. Decoupling capacitors are connected between the V_{cc} and ground leads of TTL logic packages to provide filtering, minimizing transient currents along the power supply lines and reducing noise generation. Guidelines for distribution of decoupling capacitors are found in manufacturer handbooks and texts on digital system design, for example, Barnes 1987.

Circuit boards using advanced TTL with very short **risetimes** (FAST and AS) require physical designs that reduce waveform distortion due to transmission line effects. A brief discussion of such considerations is given in the following section on emitter-coupled logic.

4.3 CMOS Logic

In 1962, Frank Wanlass of Fairchild Semiconductor noted that enhancement nMOS and pMOS transistors could be stacked totem pole fashion to form an extraordinarily simple inverter stage (see Fig. 4.6). If the input voltage is near V_{dd}, the nMOS pull-down transistor channel will be enhanced, whereas the pMOS transistor remains turned off. The voltage divider effect of a low nMOS channel resistance and very high pMOS channel resistance produces a low output voltage. On the other hand, if the input voltage is near ground, the nMOS transistor will be turned off, while the pMOS channel is enhanced to pull the output voltage high. Logic gates can be easily created by adding parallel pull-down transistors and series pull-up transistors. When a CMOS gate is in either of its

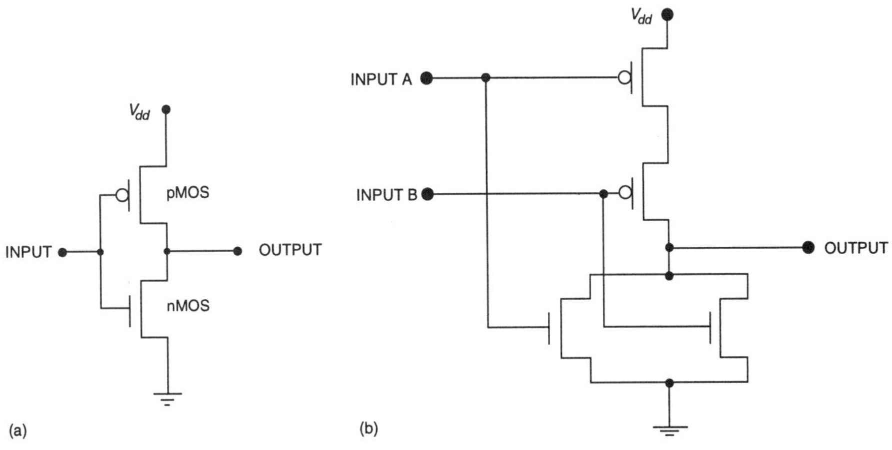

FIGURE 4.6 Elementary CMOS logic circuit: (a) complementary MOS inverter stage, (b) two-input CMOS NOR gate.

static states, the only current flowing is the very small leakage current through the off transistor plus any current needed by the driven inputs. Since the inputs to CMOS gates are essentially capacitances, input currents can be very small (on the order of 100 pA). Thus, the static power supply current and power dissipation of CMOS gates is far smaller than that of TTL gates. Since the early versions of discrete CMOS logic were much slower than TTL, CMOS was usually relegated to applications where power conservation was more important than performance.

In TTL logic, the high and low output logic levels and the input switching threshold (point at which a particular gate input interprets the voltage as a one rather than a zero) are established largely by forward bias voltages of PN junctions. Since these voltages do not scale with the power supply voltage, TTL operates properly over a rather restricted range (5.0 V $\pm$ 10%). In an all CMOS system, however, output voltage is determined by the voltage divider effect. Since the off-resistance of the inactive transistor's channel is far larger than the resistance of the active channel, CMOS logic high and low levels both are close to the power supply values (V_{dd} and ground) and scale with power supply voltage. The output transistors are designed to have matching channel resistances when turned on. Thus, as long as the supply voltage remains well above the transistor threshold voltage (about 0.7 V for modern CMOS), the input switching voltage also scales with supply voltage and is about one-half of V_{dd}. Unlike TTL, CMOS parts can operate over a very wide range of supply voltages (3–15 V) although lower voltages reduce noise immunity and speed. In the mid-1990s, integrated circuit manufacturers introduced new generations of CMOS parts optimized for operation in the 3-V range, intended for use in portable applications powered by two dry cell batteries.

Requiring only four transistors, the basic CMOS gate requires much less area than an LS TTL gate with its 11 diodes and bipolar transistors and 6 diffused resistors. This fact, coupled with its much higher static power consumption, disqualifies TTL as a viable technology for very large-scale integrated circuits. In the mid-1980s, CMOS quickly obsoleted the earlier nMOS integrated circuits and has become almost the exclusive technology of choice for complex digital integrated circuits.

Older discrete CMOS integrated circuits (RCA's 4000-series is the standard, with several manufacturers producing compatible devices) use aluminum as the gate material. The metal gate is deposited after creation of the drain and source regions. To guarantee that the gate overlies the entire channel even at maximum interlayer alignment tolerances, there is a substantial overlap of the gate and drain and source. Comparatively large parasitic capacitances result, slowing transistor (hence, gate) speed. In the mid-1980s several manufacturers introduced advanced discrete CMOS logic based on the self-aligning polysilicon gate processes developed for large-scale integrated circuits (ICs). Discrete silicon gate parts are often functional emulations of TTL devices, and have similar numbering schemes (e.g., 74HCxxx). Since silicon gate CMOS is much faster than metal gate CMOS (8 ns typical gate propagation delay vs 125 ns.) while offering similar static power consumption (1 mW per gate vs 0.6 mW), silicon gate CMOS is the technology of choice for new discrete logic designs. There are two distinct groups of modern CMOS parts: HC high-speed CMOS and AC or C advanced CMOS.

Even though parts from the 74HC and similar families may emulate TTL functions and have matching pin assignments, they cannot be intermingled with TTL parts. As mentioned, CMOS output voltages are much closer to the power supply voltages than TTL outputs. Gates are designed with this characteristic in mind and do not produce solid output voltages when driven by typical TTL high logic levels. Specialized advanced CMOS parts are produced with modified input stages specifically for interfacing with TTL circuits (for example, the HCT family). These parts exhibit slightly higher power consumption than normal silicon-gate CMOS.

The gate electrodes of CMOS transistors are separated from the channel by a layer of silicon dioxide that may be only a few hundred angstroms thick. Such a thin dielectric layer can be damaged by a potential difference of only 40–100 V. Without precautions, the normal electrostatic potentials that build up on the human body can destroy CMOS devices. Although damage from electrostatic discharge (ESD) can be minimized through proper handling procedures (for example, Matisof 1986), CMOS parts include ESD protection circuitry on all inputs (see Fig. 4.7). The input diodes serve to limit gate voltages to one diode drop above V_{dd} or below ground, whereas the resistor serves both to limit input current and to slow the rise time of very fast pulses.

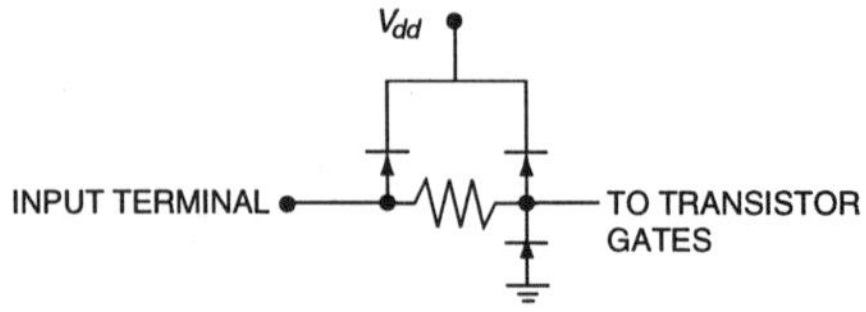

FIGURE 4.7 A typical CMOS input protection circuit.

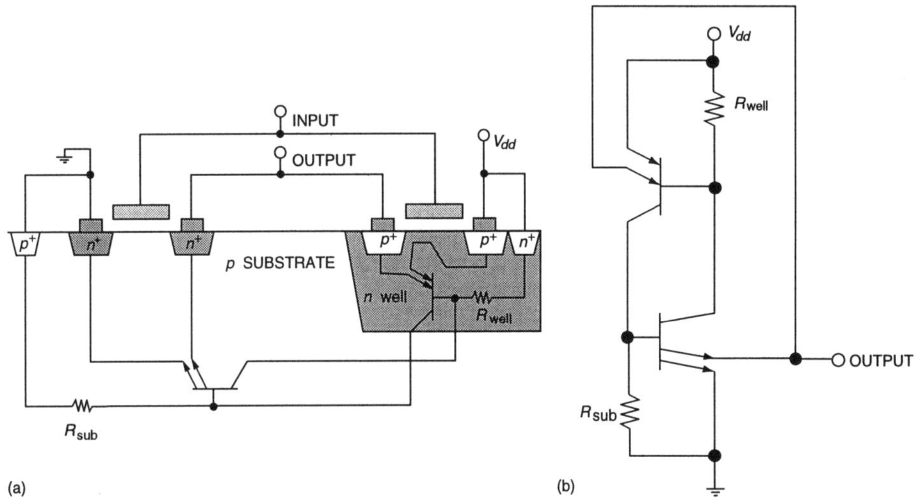

FIGURE 4.8 CMOS parasitic SCR: (a) cross-section showing parasitic bipolar transistors, (b) equivalent circuit.

As shown in Fig. 4.8, CMOS inverters and gates inherently have cross-connected parasitic bipolar transistors that form a silicon controlled rectifier (SCR). Suppose that external circuitry connected to the output pin pulls the pin low, below ground level. As this voltage approaches −0.5 V, the base-emitter junction of the parasitic NPN transistor will start to forward bias and the transistor will turn on. The resulting collector current is primarily determined by whatever external circuitry is pulling the output low. That current flows through R_{well} and lowers the base voltage of the parasitic PNP transistor. A large enough current will forward bias the PNP base emitter junction, causing PNP collector current to flow through R_{sub} and helping to maintain the forward bias of the NPN transistor. The SCR then enters a regenerative condition and will quickly assume a stable state in which both transistors remain on even after the initial driving stimulus is removed. In this **latchup** condition, substantial current flows from V_{dd} to ground. Normal operation of the CMOS gate is disrupted and permanent damage is likely. For latchup to occur, the product of the NPN and PNP transistor current gains must be greater than one, the output pin must be driven either below −0.5 V or above V_{dd} + 0.5 V by a source that supplies enough current to trigger regenerative operation, and V_{dd} must supply a sustaining current large enough to keep the SCR turned on. Although the potential for latchup cannot be avoided, CMOS manufacturers design input and output circuits that are latchup resistant, using configurations that reduce the gain of the parasitic transistors and that lower the substrate and p-well resistances. Nevertheless, CMOS users must ensure that input and output pins are not driven outside normal operating ranges. Limiting power supply current below the holding current level is another means of preventing latchup damage.

Fanout of CMOS outputs driving TTL inputs is limited by static current drive requirements and is calculated as for transistor–transistor logic. In all-CMOS systems, the static input current is just the leakage current flowing through the input ESD protection diodes, a current on the order of 100 pA. Consequently, static fanout of CMOS driving CMOS is so large as to be meaningless. Instead, fanout is limited by speed deterioration as the number of driven gates increases [Texas Instruments 1984]. If all driven inputs are assumed identical, an all-CMOS system can be represented by the approximate equivalent circuit of Fig. 4.9. The parallel input resistance R_{in} is on the order of 60 MΩ and can be neglected in later computations. The pull-up resistor R_{pu} can be approximately calculated from manufacturer data as $(V_{dd} - V_{OH})/I_{OH}$ (~50 Ω). The output will charge

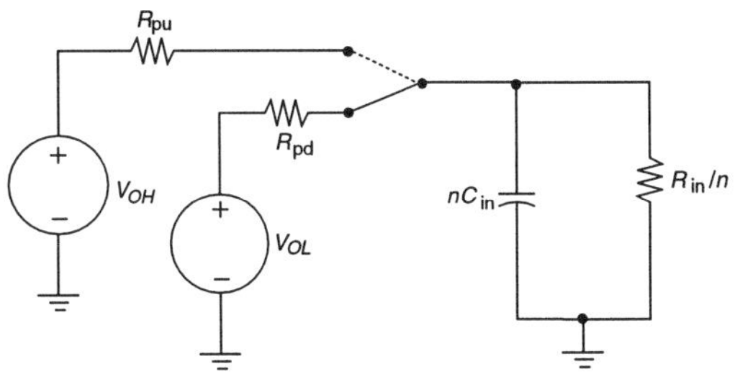

FIGURE 4.9 Equivalent circuit for a CMOS output driving nCMOS inputs.

exponentially from low to high with a time constant of $nR_{pu}C_{in}$, where n is the actual fanout. Since the low output voltage is nearly zero, time required to reach a logic high is given by

$$t = nR_{pu}C_{in}\ln(1 - V_{IH,\min}/V_{OH}).$$

Thus, once the maximum allowable risetime (that is, the maximum effective propagation delay) has been specified for an all-CMOS system, the dynamic fanout n can be computed. Although static power consumption of CMOS logic is quite small, outputs that are changing logic states will dissipate power each time they charge and discharge the capacitance loading the output. In addition, both the pull-up and pull-down transistors of the output will be simultaneously partially conducting for a short period on each switching cycle, creating a transient supply current beyond that needed for capacitance charging. Thus, the total power consumption of a CMOS gate is given by the relation

$$P_{\text{total}} = P_{\text{static}} + C_{\text{load}}V_{dd}^2 f$$

where

P_{total} = total power dissipated
P_{static} = static power dissipation
C_{load} = total equivalent load capacitance, $C_{pd} + C_{ext}$
C_{pd} = *Power dissipation capacitance*, a manufacturer specification representing the equivalent internal capacitance (the effect of transient switching current is included)
C_{ext} = total parallel equivalent load capacitance
V_{dd} = power supply voltage
f = output operating frequency

HCT and ACT parts include an additional static power dissipation term proportional to the number of TTL inputs begin driven ($n\Delta I_{dd}V_{dd}$, where ΔI_{dd} is specified by the manufacturer). CMOS average power consumption is proportional to the frequency of operation, whereas power consumption of TTL circuits is almost independent of operating frequency until about 1 MHz, when dynamic power dissipation becomes significant (see Fig. 4.10). Direct comparison of CMOS and TTL power consumption for a digital system is difficult, since not all outputs will be changing at the same frequency. Moreover, the relative advantages of CMOS become greater with increasing part complexity; CMOS and TTL power curves for a modestly complicated decoder circuit crossover at a frequency 10 times that of simple gates. Finally, TTL manufacturers do not publish C_{pd} specifications for their parts.

Like TTL parts, CMOS logic requires use of decoupling capacitors to reduce noise due to transient current spikes on power supply lines. Although such decoupling capacitors reduce electromagnetic coupling of noise into parallel signal leads, they do not eliminate an additional noise consideration that arises in both advanced CMOS and advanced Schottky due to voltages induced on the power supply leads from transient current spikes. Consider the multiple inverter IC package of Fig. 4.11, in which five outputs are switching from low to high simultaneously, while the remaining output stays low. In accord with Faraday's law, the switching current spike will induce transient voltage across L, which represents parasitic inductances from internal package leads and external ground supply traces. That same voltage will also lift the unchanging output above the ground reference. A sufficiently large voltage spike may make signal A exceed the input switching threshold and cause erroneous operation of following logic elements. Careful physical design of the power supply and ground leads can reduce external inductance and limit power supply sag and ground bounce. Other useful techniques include use of synchronous clocking and segregation of high drive signals, such as clock lines, into separate buffer packages, separated from reset and other asynchronous buffers.

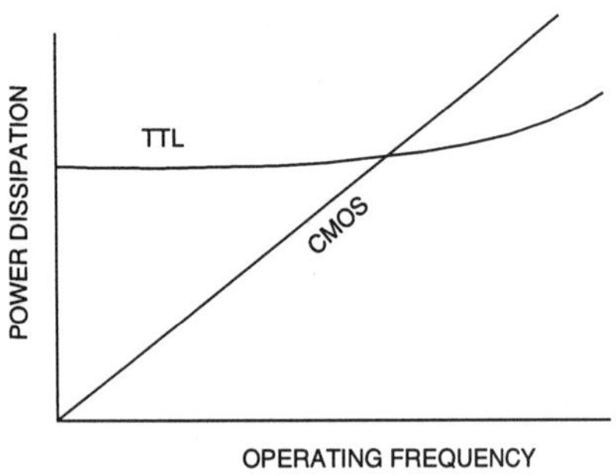

FIGURE 4.10 Power dissipation vs operating frequency for TTL and CMOS logic devices.

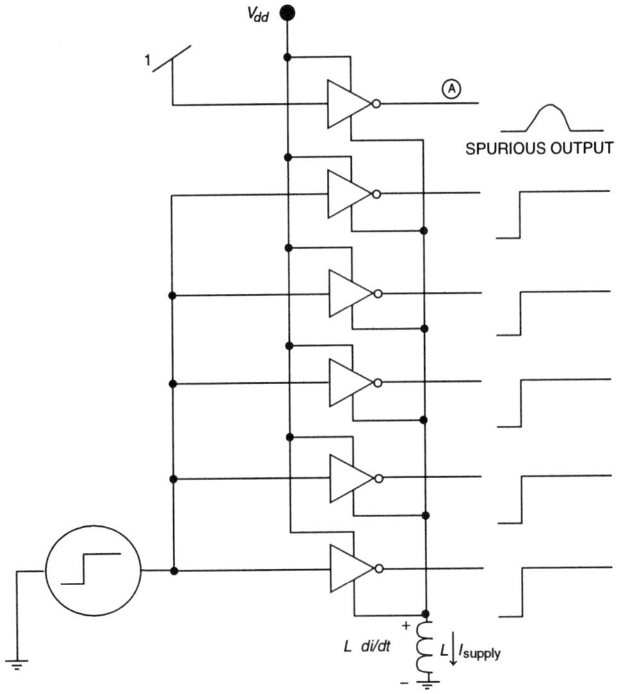

FIGURE 4.11 Output bounce due to ground lead inductance.

As has been suggested several times, effective propagation delay of a logic circuit is directly proportional to the capacitive loading of its output. Systems built from small- and medium-scale logic have many integrated circuit outputs with substantial load capacitance from interconnecting printed circuit traces. This is the primary impetus for using larger scale circuits: on-chip interconnections present far smaller capacitance and result in faster operation. Thus, a system built using a highly integrated circuit fabricated in a slower technology may be faster overall that the same system realized using faster small-scale logic.

Bipolar-CMOS (BiCMOS) logic is a hybrid technology for very large-scale integrated circuits that incorporates both bipolar and MOS transistors on the same chip. It is intended to offer the most of the circuit density and static power savings of CMOS while providing the electrical ruggedness and superior current drive capability of bipolar logic. BiCMOS also holds the opportunity for fabricated mixed systems using bipolar transistors for linear portions and CMOS for logic. It is a difficult task to produce high-quality bipolar transistors on the same chip as MOS devices, meaning that biCMOS circuits are more expensive and may have suboptimal performance.

4.4 Emitter-Coupled Logic

Emitter coupled logic (ECL) is the fastest variety of discrete logic. It is also one of the oldest, dating back to Motorola's MECL I circuits in 1962. Improved MECL III circuits were introduced in the late 1960s. Today, the 10 K and 100 K ECL families remain in use where very high speed is required. Unlike either TTL or CMOS, emitter-coupled logic uses a differential amplifier configuration as its basic topology (see Fig. 4.12). If all inputs to the gate are at a voltage lower than the reference voltage, the corresponding input transistors are cutoff and the entire bias current flows through Q1. The output is pulled up to V_{cc} through resistor R_a. If any input voltage rises above the reference, the corresponding transistor turns on, switching the current from Q1 to the input transistor and lowering the output voltage to $V_{cc} - I_{\text{bias}} R_a$. If the collector resistors and bias current are properly chosen, the transistors never enter saturation, and switching time is quite fast. Typical propagation delay for MECL 10 K gates is only 2.0 ns. To reduce the output impedance and improve current drive capability, actual ECL parts include emitter follower outputs [Fig. 4.12(b)]. They also include integral voltage references with temperature coefficients matching those of the differential amplifier.

Note that ECL circuits are normally operated with the collectors connected to ground and the emitter current source to a negative supply voltage. From the standpoint of supply voltage fluctuations, the circuit operates as a common base amplifier, attenuating supply variations at the output. The power supply current in both TTL and CMOS circuits exhibits sharp spikes during switching; in ECL gates, the bias current remains constant and simply shifts from one transistor to another. Lower logic voltage swings also reduce noise generation. Although the noise margin for ECL logic is less than TTL and CMOS, ECL circuits generate less noise.

Emitter-coupled logic offers both short propagation delay and fast risetimes. When interconnecting paths introduce a time delay longer than about one-half the risetime, the interconnection's transmission line behavior must be taken into account. Reflections from unterminated lines can distort the waveform with ringing.

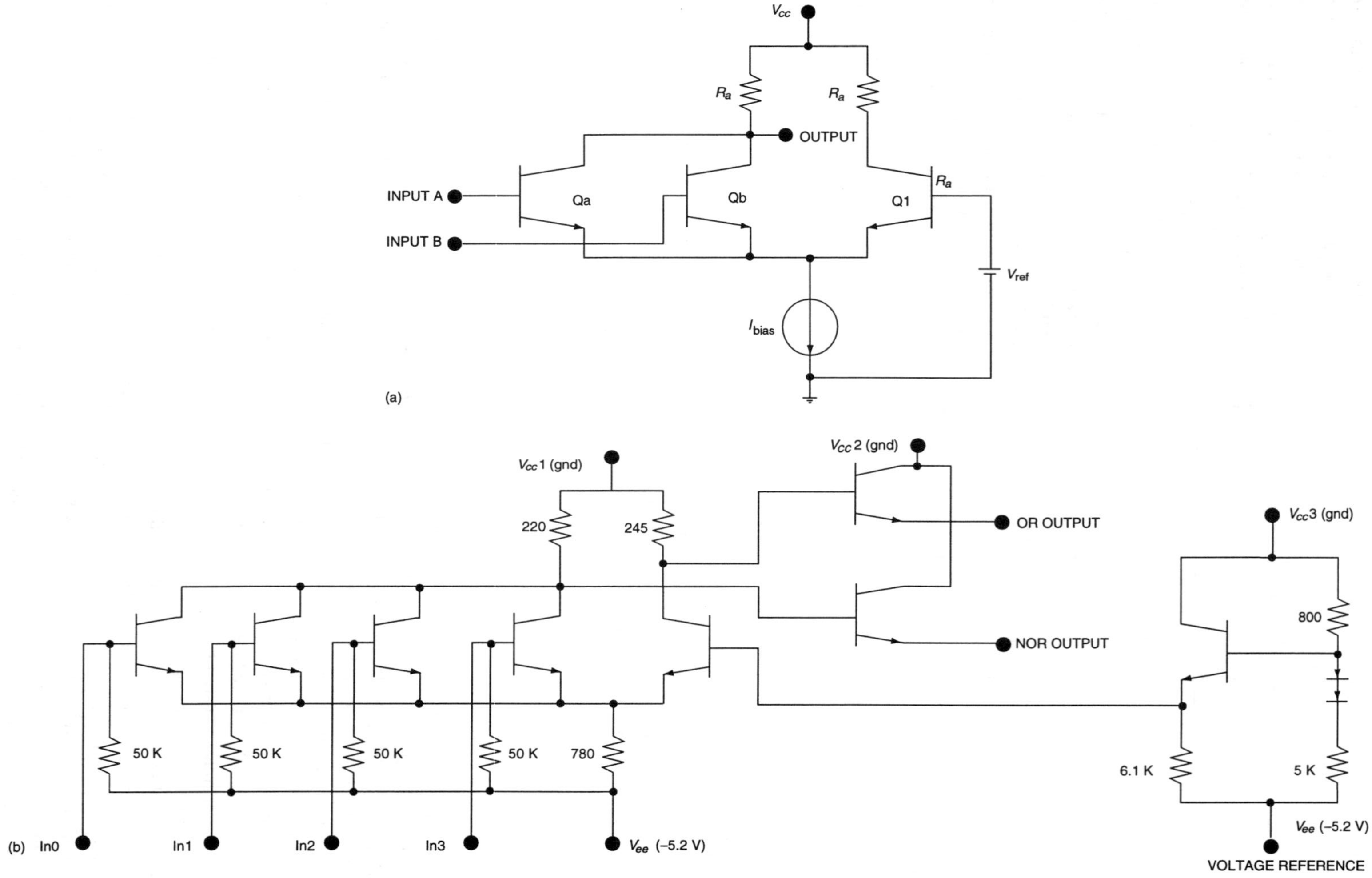

FIGURE 47.12 Emitter coupled logic (ECL) gates: (a) simplified schematic of ECL NOR gate, (b) commercial ECL gate.

The need to wait until reflections have damped to an acceptable value increases the effective propagation delay. The maximum open line length depends on the fanout of the driving output, the characteristic impedance of the transmission line, and the velocity of propagation of the line. Motorola databooks indicate that MECL III parts, with an actual fanout of 8 connected with fine-pitch printed circuit traces (100-Ω microstrip), can tolerate an unterminated line length of only 0.1 in. In high-speed ECL designs, physical interconnections must be made using properly terminated transmission lines (see Fig. 4.13). For an extended treatment of transmission lines in digital systems, see manufacturer's publications [Blood 1988] or Rosenstark [1994]. The necessity for using transmission lines with virtually all ECL III designs resulted the introduction in 1971 of the MECL 10 K family of parts, with 2-ns propagation delays and 3.5-ns risetimes. The slowed risetimes meant that much longer unterminated lines could be used (Motorola cites 2.6 in. for the configuration mentioned earlier). Thus, many designs could be implemented in 10 K ECL without transmission line interconnections. This relative ease of use and a wider variety of parts has made 10 K ECL and its variants the dominant ECL family. It should be noted that, although transmission line interconnections were traditionally associated with ECL, advanced CMOS and advanced Schottky TTL have rapid risetimes and are being used in very high-speed systems. Transmission line effects must be considered in the physical design of any high-performance system, regardless of the logic family being used.

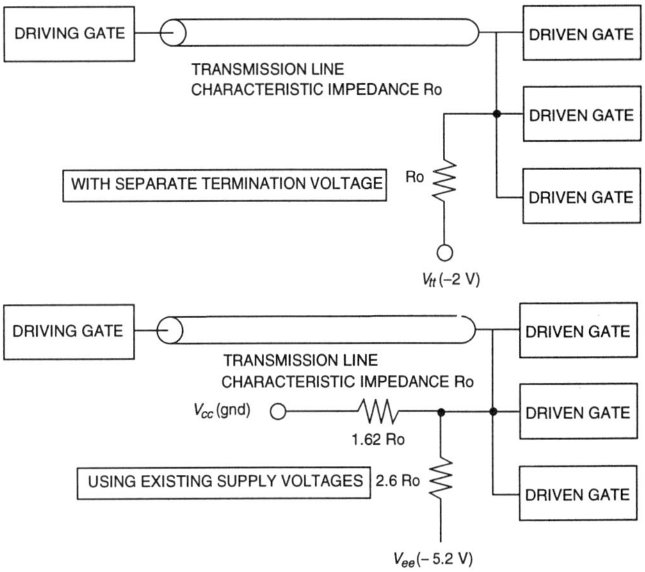

FIGURE 4.13 Parallel-terminated ECL interconnections.

Although ECL is often regarded as a power-hungry technology compared to TTL and CMOS, this is not necessarily true at very high-operating frequencies. At high frequencies, the power consumption for both of the latter families increases approximately linearly with frequency, whereas ECL consumption remains constant. Above about 250 MHz, CMOS power consumption exceeds that of ECL.

Gallium Arsenide

It was known even before the advent of the transistor that gallium arsenide constituted a semiconducting compound having significantly higher electron mobility than silicon. Holes, however, are substantially slower in GaAs than in silicon. As a consequence, unipolar GaAs transistors using electrons for charge transport are faster than their silicon equivalents and can be used to create logic circuits with very short propagation delays. Since gallium arsenide does not have a native oxide, it is difficult to fabricate high-quality MOS field effect transistors (MOSFETs). The preferred GaAs logic transistor is the *metal-semiconductor field-effect transistor* (MESFET) which is

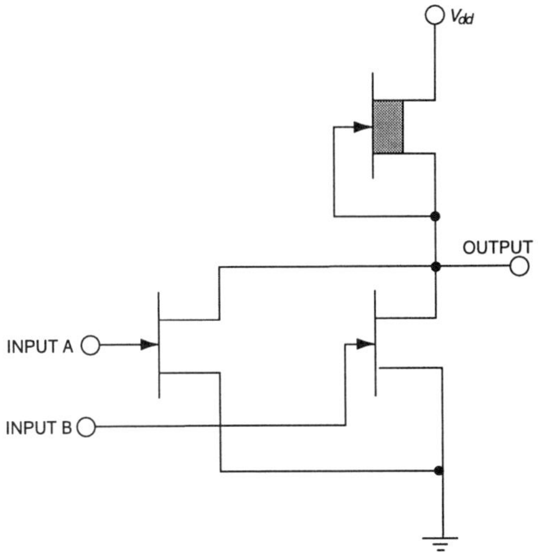

FIGURE 4.14 A GaAs enhancement/ depletion-mode MESFET NOR gate.

essentially a junction field effect transistor in which the gate is formed by a metal-semiconductor (Schottky) junction instead of a p–n diode. Figure 4.14 shows a NOR gate formed using two enhancement MESFETs as pull-down transistors with a depletion MESFET active load for the pull-up. At room temperatures, the underlying GaAs substrate has a very high resistivity (is semi-insulating), providing sufficient transistor-to-transistor isolation without the need for reverse-biased isolation junctions.

Gallium arsenide logic exhibits both short delay and fast risetimes. As noted in the ECL discussion, fast logic signals exhibit ringing if transmission line interconnections are not used. The waveform distortion due to ringing extends the effective propagation delay and can negate the speed advantage of the basic logic circuit. Furthermore, complex fabrication processes and limited volumes make gallium arsenide logic from 10 to 20 times more expensive than silicon. The dual requirements to control costs and to preserve GaAs speed advantage by limiting the number of off-chip interfaces make the technology better suited to large-scale rather small- and medium-scale logic. GaAs digital circuits are available from several vendors in the form of custom integrated circuits or gate arrays. Gate arrays are semifinished large-scale integrated circuits that have large numbers of unconnected gates. The customer specifies the final metallization patterns that will interconnect the gates to create the desired logic circuits. As of this writing, vendors were offering application-specific GaAs integrated circuits with more than 20,000 gates and worst-case internal gate delays on the order of 0.07 ns.

4.5 Programmable Logic

Although traditional discrete logic remains an important means of implementing digital systems, various forms of programmable logic have become increasingly popular. Broadly speaking, programmable logic denotes large- and very large-scale integrated circuits containing arrays of generalized logic blocks with user-configurable interconnections. In programming the device, the user modifies it to define the functions of each logic block and the interconnections between logic blocks. Depending on the complexity of the programmable logic being used, a single programmable package may replace dozens of conventional small- and medium-scale integrated circuits. Because most interconnections are on the programmable logic chip itself, programmable logic designs offer both speed and reliability gains over conventional logic. Effectively all programmable logic design is done using electronic design automation (EDA) tools that automatically generate device programming information from a description of the function to be performed. The functional description can be supplied in terms of a logic diagram, logic equations, or a hardware description language. EDA tools also perform such tasks as logic synthesis, simplification, device selection, functional simulation, and timing/speed analysis. The combination of electronic design automation and device programmability makes it possible to design and physically implement digital subsystems very quickly compared to conventional logic.

Although the definitions are somewhat imprecise, programmable logic is divided into two broad categories: programmable logic devices (PLDs) and field programmable gate arrays (FPGAs). The PLD category itself is often subdivided into complex PLDs (CPLDs) with numerous internal logic blocks and simpler PLDs. Although the general architecture of small PLDs is fairly standardized, that of FPGAs and CPLDs continues to evolve and differs considerably from one manufacturer to another. Programmable logic devices cover a very wide range, indeed, from simple PALs containing a few gates to field programmable gate arrays providing thousands of usable gate equivalents.

Programmable Array Logic

Programmable array logic (PAL) was first introduced by Monolithic Memories, Inc. (now part of Advanced Micro Devices), and although it is universally used, the acronym remains an AMD trademark. In their simplest form, PALs consist of a large array of logic AND gates fed by both the true (uninverted) and complemented (inverted) senses of the input signals. Any AND gate may be fed by either sense of any input. The outputs of the AND gates are hardwired to OR gates that drive the PAL outputs (see Fig. 4.15). By programming the AND inputs, the PAL user can implement any logic function provided that the number of product terms does not exceed the **fanin** to the OR gate. Modern PALs incorporate several logic macrocells that include a programmable inverter following the AND/OR combinatorial logic, along with a programmable flip-flop that can be used when registered

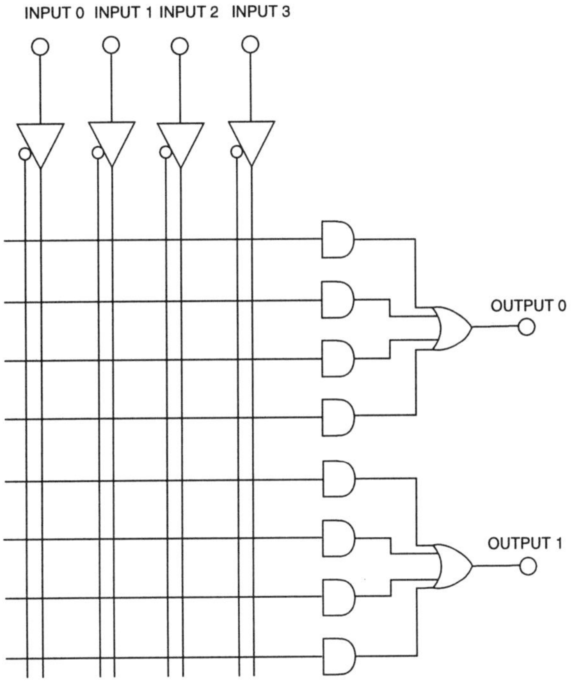

FIGURE 4.15 A simplified small PAL device.

outputs are needed. Additional AND gates are used to clock and reset the flip-flop. Macrocells usually also have a programmable three-state buffer driving the block output and programmable switches that can be used to feed macrocell outputs back into the AND–OR array to develop combinatorial logic with many product terms or to implement state machines.

PLDs are fabricated using either bipolar or CMOS technologies. Although CMOS parts have become quite fast, bipolar PALs offer higher speed (<5 ns input to output combinatorial delay) at the expense of higher power consumption. Different manufacturers accomplish interconnect and logic options programming in different ways. In some instances, the circuit is physically modified by blowing fuses or antifuses (an antifuse is an open circuit until fused, then it becomes a low-resistance circuit). CMOS programmable devices often use interconnections based on erasable programmable read only memory (EPROM) cells. EPROM-based parts can be programmed, then erased and reprogrammed. Although reprogrammability can be useful during the development phase, it is less important in production, and EPROM-based PLDs are also sold in cheaper, one-time programmable versions. Although not a major market factor, electrically reprogrammable PLDs are also available. When used in digital systems, PLDs require fanout computation, just as discrete TTL or CMOS logic. In addition, fast PLDs require the same consideration of transmission line effects.

Although it is possible to hand-develop programming tables for very simple PALs, virtually all design is now done using EDA tools supplied by PLD vendors and third parties. Although computer methods automate the design process to a considerable extent, they usually permit some user control over resource assignments. The user needs to have a thorough knowledge of a specific device's architecture and programming options to optimize the resulting design.

Programmable Logic Arrays

Field programmable logic arrays (FPLAs) resemble PALs with the additional feature that the connections from the AND intermediate outputs to the OR inputs are also programmable, rather than hardwired (Fig. 4.15). Like simpler PALs, FPLAs are produced in registered versions that have flip-flops between the OR gate and the output pin. Even though they are more complex devices, FPLAs were actually introduced before programmable array logic. Programmable AND/OR interconnections make them more flexible than PALs, since product terms (AND gate outputs) can drive more than one OR input and any number of product terms can be used to develop a particular output function. The additional layer of programmable interconnects consumes chip area, however, increasing cost. The extra level of interconnections also makes programming more complicated. The most important drawback to FPLAs, however, is lower speed. The programmable interconnections add significant series resistance, slowing signal propagation between the AND and OR gates and making FPLAs sluggish compared to PALs. Since programmable array logic is faster, simpler to program, and flexible enough for most uses, PALs are much more popular than field programmable logic arrays. Most FPLAs marketed today have registered outputs (that is, a flip-flop is placed between the OR gate and the macrocell output) that are fed back into to the AND/OR array. Termed *field programmable logic sequencers*, these parts are intended for implementing small state machines.

Field Programmable Gate Arrays and Complex PLDs

There is no universally accepted dividing line between complex PLDs and FPGAs. The Xilinx devices[2] that were first termed FPGAs consisted of a two-dimensional matrix of logic cells separated by horizontal and vertical channels containing routing paths and programmable path connections. Each logic cell included an eight-address lookup table for performing combinatorial logic and a D latch for sequential logic (Fig. 4.16). An alternative FPGA configuration, physically closer to ordinary gate arrays, is row-based architecture used by Actel. In earlier devices, Actel's logic blocks contain three, two-input multiplexers and an OR gate that can be configured to develop all logic functions of two variables and some functions of three or four variables or to make a D latch. More complex functions are formed by interconnecting two or more logic blocks. Channels between the rows carry segmented routing lines, while vertical interconnects run over the top of logic blocks. Later Actel devices have logic cells that are a mix of these combinatorial cells and sequential (flip-flop) cells.

Clearly, Actel and Xilinx logic cells differ considerably. These are but two examples of many FPGA architectures being produced. Although there is strong evidence that very simple, gate-level cells are not a good choice for FPGAs (see Brown et al. [1992]), the effectiveness of a cell architecture is a complex interplay of not only the logical structure of the cell but also the availability of routing paths, time-delay penalty incurred by routing switches, number of high drive/low skew lines for common signals (e.g., clocks), and, not the least, effectiveness of the electronic design automation support. This makes comparison of competing FPGAs a difficult and imprecise task and has motivated continuing interest in benchmarking techniques.

In either of the Actel or Xilinx architectures, the timing characteristics of a particular logic path depend on the placement of individual logic blocks along the path and the routing of block interconnections. This has a very important implication: later design modifications may change logic block placement and routing and cause significant, unpredictable timing changes. This timing variability is cited by some as the trait that distinguishes FPGAs from complex PLDs.

Not only do architectures differ from manufacturer to manufacturer, the method of interconnect programming also differs. Xilinx and some other manufacturers make interconnections using static random access memory (SRAM) cells with transmission gates; programming consists of loading the interconnect memory locations with bits indicating whether or not a connection exists. It is possible to reconfigure these FPGAs in-circuit by loading in new interconnection patterns. Since SRAM cells are volatile, however, the programming information is lost each time the part is turned off. An external read only memory is used to store programming data permanently. Logic built into the FPGA automatically reads programming data on power-up. Other vendors, like Actel, use antifuse interconnects to program their devices, in which case the programming information is permanently stored. FPGAs using flash memory technology are imminent; these would offer the permanence of antifuse devices along with the in-circuit reprogrammability of SRAM FPGAs.

The term *complex programmable logic device* (CPLD) generally denotes devices organized as large arrays of PAL macrocells, with fixed delay feedback of macrocells outputs into the AND–OR array (Fig. 4.17). CLPD time delays are more predictable than those of FPGAs, which can make them simpler to design with. A more important distinction between CPLDs and FPGAs is the ratio of combinatorial logic to sequential logic in their logic cells. The PAL-like structure of CPLDs means that the flip-flop in each logic macrocell is fed by the logical sum of several logical products. The products can include a large number of input and feedback signals. Thus, CPLDs are well suited to applications such as state machines that have logic functions including many variables. FPGAs, on the other hand, tend toward cells that include a flip-flop driven by a logic function of only three or so variables and are better suited to situations requiring many registers fed by relatively simple functions. Examples of large CPLDs are Altera Corporation's MAX7000 series parts, which may incorporate up to 128 macrocells and include 100 input/output pins.

Altera Corporation's FLEX8000 devices have a two-dimensional matrix of logic blocks, with horizontal and vertical routing channels. SRAM cells are used for logic and routing programming. Routing between blocks is done with fixed length lines, rather than the segmented paths of Xilinx parts. Each logic block includes eight logic

[2]At the time of writing, the FPGA/CPLD industry is highly volatile, with new business entries and different architectures appearing frequently. The manufacturers and architectures cited are intended to illustrate typical concepts; there are numerous other vendors and architectural variants available.

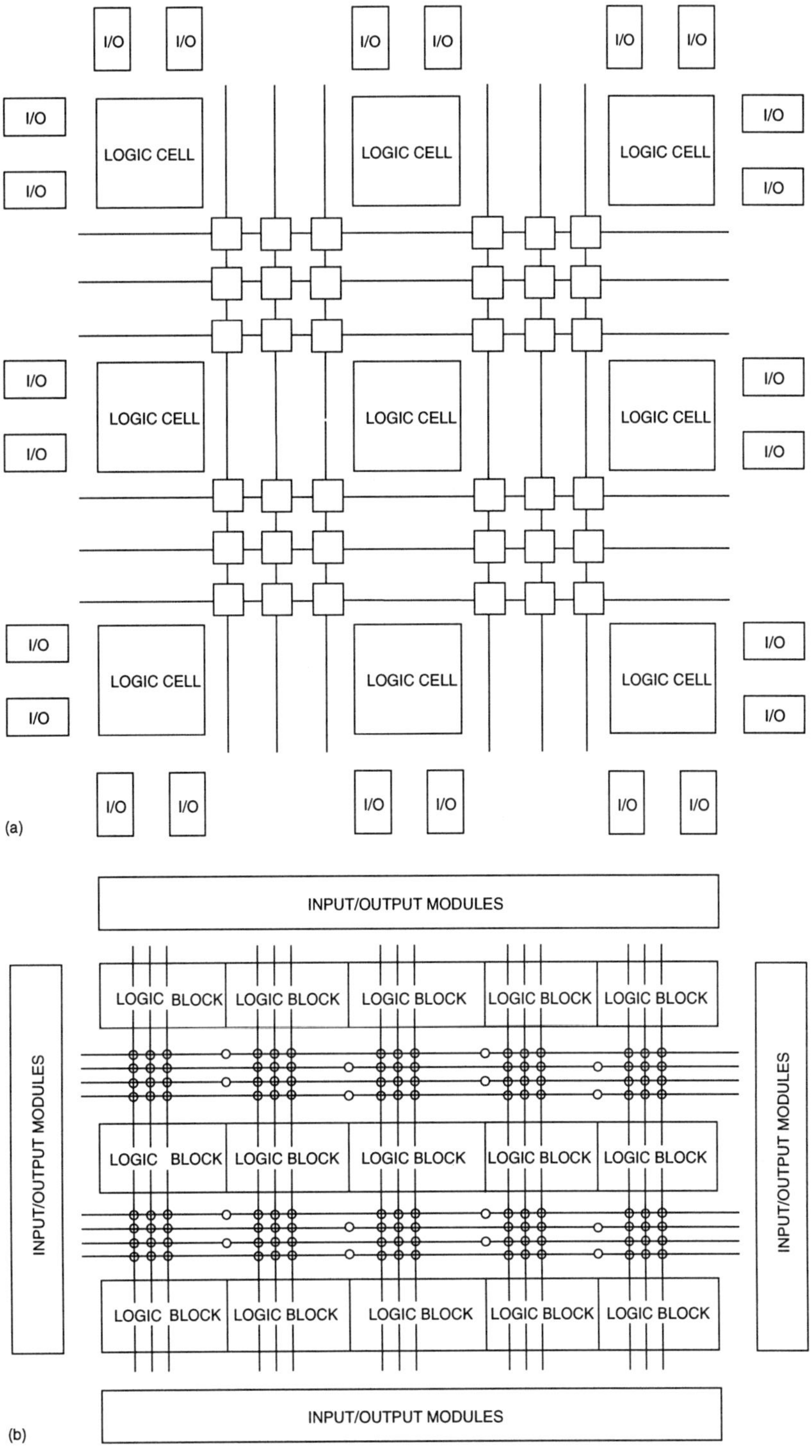

FIGURE 4.16 Two alternative field programmable gate array architectures: (a) Xilinx-like FPGA structure (small squares indicate programmable routing switches), (b) Actel-like FPGA structure (small circles indicate routing antifuses).

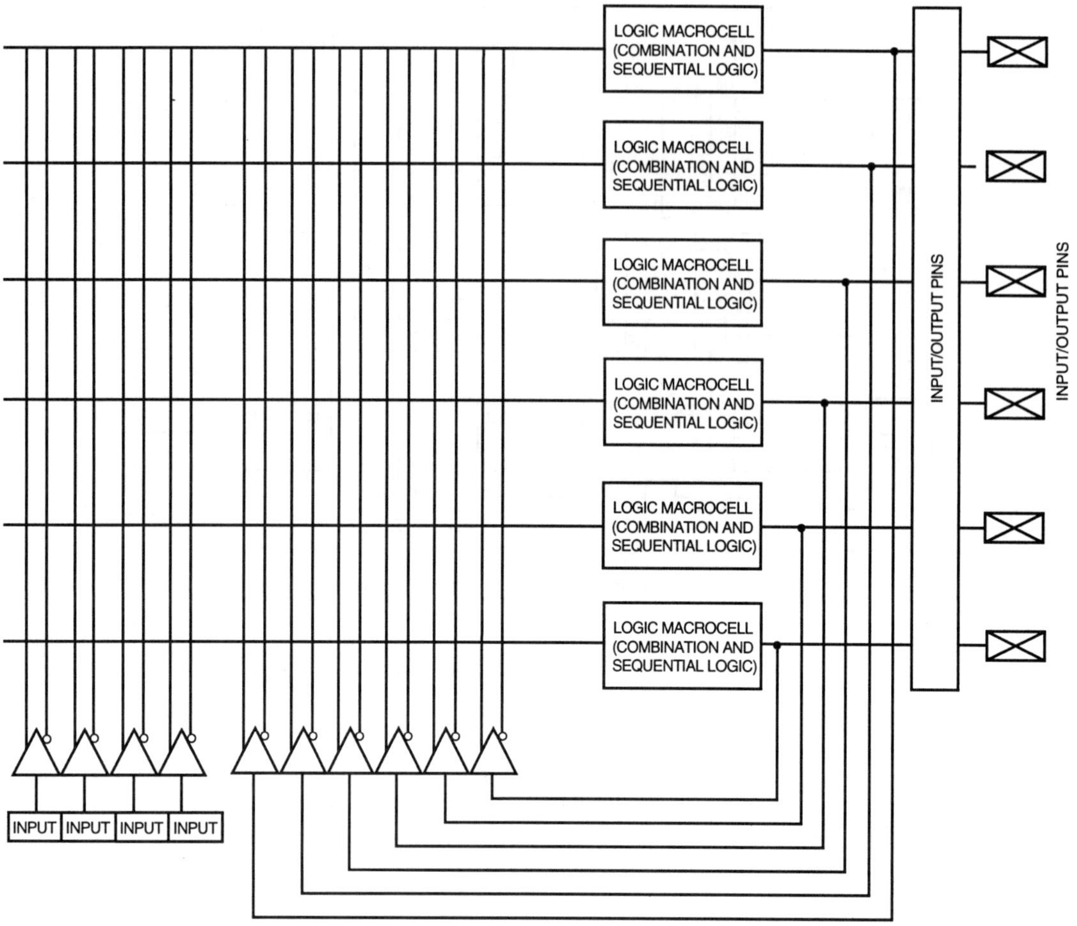

FIGURE 4.17 A simplified complex programmable logic device structure.

elements, each of which has a four input lookup table for combinatorial logic and a flip-flop, along with additional gating. Associated with each logic block is a local bus structure providing complete local interconnection along with separate data inputs for each lookup table. This architecture attempts to combine the flexible interconnection and predictable timing of a complex PLD with the register-rich structure of a FPGA.

Field programmable gate arrays and complex programmable logic devices concentrate far more logic per package than traditional discrete logic and smaller PLDs, making computer aided design essential. The overall procedure is as follows.

1. Describe system functionality to the EDA system, using schematics, boolean equations, truth tables, or high level programming languages. Subparts of the overall system may be described separately, using whatever description is most suitable to that part, and then combined. As systems become more complex, high level languages have become increasingly useful. Subsequent steps in the design process are largely automated functions of the EDA software.
2. Translate the functional description into equivalent logic cells and their interconnections.
3. Optimize the result of step 2 to produce a result well suited to the implementation architecture. The goal of the optimization and the strategies involved depend strongly on the target architecture. For instance, EDA tools for FPGAs with limited interconnections will seek to reduce the number of block interconnects.
4. Fit the optimized design to the target architecture. Fitting includes selection of an appropriate part, placing of logic (mapping of functional blocks to specific physical cells), and routing of cell interconnections.

5. Produce tables containing information required for device programming and back annotate interconnection netlists. Back annotation provides information about interconnection paths and fanout that is necessary to produce accurate timing estimates during simulation.
6. Perform functional and performance verification through simulation. Simulation is critical, because of the high ratio of internal logic to input–output ports in FPGA/CPLD designs. Logic errors due to timing problems or bugs in the compilation and fitting steps can be extremely difficult to locate and correct based purely on performance testing of a programmed device. Simulation, on the other hand, produces invaluable information about internal behavior.

Defining Terms

Fanin: The number of independent inputs to a given logic gate.

Fanout: (1) The maximum number of similar inputs that a logic device output can drive while still meeting output logic voltage specifications. (2) The actual number of inputs connected to a particular output.

Latchup: A faulty operating condition of CMOS circuits in which its parasitic SCRs produce a low resistance path between power supply rails.

Noise margin: The difference between an output logic high (low) voltage produced by a properly functioning output and the input logic high (low) voltage required by a driven input; noise margin provides immunity to a level of signal distortion less than the margin.

Risetime: The time required for a logic signal to transition from one static level to another; usually, risetime is measured from 10 to 90% of the final value.

References

Altera. 1993. *Data Book.* Altera Corp., San Jose, CA.

Alvarez, A. 1989. *BiCMOS Technology and Applications.* Kluwer Academic, Boston, MA.

Barnes, J.R. 1987. *Electronic System Design: Interference and Noise Control Techniques.* Prentice–Hall, Englewood Cliffs, NJ.

Blood, W.R. 1988. *MECL System Design Handbook.* Motorola Semiconductor Products, Phoenix, AZ.

Brown, S.D., Francis, R., Rose, J., and Vranesic, Z. 1992. *Field-Programmable Gate Arrays.* Kluwer Academic, Boston, MA.

Buchanan, J. 1990. *CMOS/TTL Digital System Design.* McGraw–Hill, New York.

Deyhimy, I. 1985. GaAs digital ICs promise speed and lower cost. *Computer Design* 35(11):88–92.

Jenkins, J.H. 1994. *Designing with FPGAs and CPLDs.* Prentice–Hall, Englewood Cliffs, NJ.

Kanopoulos, N. 1989. *Gallium Arsenide Digital Integrated Circuits: A Systems Perspective.* Prentice–Hall, Englewood Cliffs, NJ.

Leigh, B. 1993. Complex PLD & FPGA architectures, *ASIC & EDA.* (Feb.):44–50.

Matisof B. 1986. *Handbook of Electrostatic Controls (ESD).* Van Nostrand Reinhold, New York.

Matthew, P. 1984. *Choosing and Using ECL.* McGraw–Hill, New York.

Rosenstark, S. 1994. *Transmission Lines in Computer Engineering.* McGraw–Hill, New York.

Scarlett, J.A. 1972. *Transistor–Transistor Logic and Its Interconnections.* Van Nostrand Reinhold, London.

Schilling, D. and Belove, C. 1989. *Electronic Circuits: Discrete and Integrated,* 3rd ed. McGraw–Hill, New York.

Texas Instruments, 1984. *High-Speed CMOS Logic Data Book.* Texas Instruments, Dallas, TX.

Xilinx, 1993. *The Programmable Gate Array Data Book.* Xilinx, Inc., San Jose, CA.

Further Information

For the practitioner, the following technical journals are excellent sources dealing with current trends in both discrete logic and programmable devices.

Electronic Design, published 30 times a year by Penton Publishing, Inc. of Cleveland, OH, covers a very broad range of design-related topics. It is free to qualified subscribers.

EDN, published by Cahners Publishing Company, Newton, MA, appears 38 times annually. It also deals with the entire range of electronics design topics and is free to qualified subscribers.

Computer Design, Pennwell Publishing, Nashua, NH, appears monthly. It focuses more specifically on digital design topics than the foregoing. Like them, it is free to qualified subscribers.

Integrated System Design, a monthly periodical of the Verecom Group, Los Altos, CA, is a valuable source of current information about programmable logic and the associated design tools.

For the researcher, the *IEEE Journal of Solid State Circuits* is perhaps the best single source of information about advances in digital logic. The IEEE sponsors a number of conferences annually that deal in one way or another with digital logic families and programmable logic.

Manufacturer's databooks and application notes provide much current information about logic, especially with regard to applications information. They usually assume some background knowledge, however, and can be difficult to use as a first introduction. Recommended textbooks to provide that background include the following.

Brown, S.D., Francis, R., Rose, J., and Vranesic, Z. 1992. *Field-Programmable Gate Arrays.* Kluwer Academic, Boston, MA.

Buchanan, J. 1990. *CMOS/TTL Digital System Design.* McGraw–Hill, New York.

Jenkins, J.H. 1994. *Designing with FPGAs and CPLDs.* Prentice–Hall, Englewood Cliffs, NJ.

Matthew, P. 1984. *Choosing and Using ECL.* McGraw–Hill, New York.

Rosenstark, S. 1994. *Transmission Lines in Computer Engineering.* McGraw–Hill, New York.

5

Memory Devices

Shih-Lien L. Lu
Oregon State University

5.1 Introduction

Memory is an essential part of any computation system. It is used to store both the computation instructions and the data. Logically, memory can be viewed as a collection of sequential locations, each with an unique address as its label and capable of storing information. Accessing memory is accomplished by supplying the address of the desired data to the device.

Memory devices can be categorized according to their functionality and fall into two major categories, **read-only-memory (ROM)**, and write-and-read memory or random-access memory (RAM). There is also another subcategory of ROM: mostly read but sometimes write memory or flash ROM memory. Within the RAM category there are two types of memory devices differentiated by storage characteristics, **static RAM (SRAM)** and **dynamic RAM (DRAM)** respectively. DRAM devices need to be refreshed periodically to prevent the corruption of their contents due to charge leakage. SRAM devices, on the other hand, do not need to be refreshed.

Both SRAM and DRAM are volatile memory devices, which means that their contents are lost if the power supply is removed from these devices. Nonvolatile memory, the opposite of volatile memory, retains its contents even when the supply power is turned off. All current ROM devices, including mostly read sometimes write devices are nonvolatile memories. Except for a very few special memories, these devices are all interfaced in a similar way. When an address is presented to a memory device, and sometimes after a control signal is strobed, the information stored at the specified address is retrieved after a certain delay. This process is called a **memory read**. This delay, defined as the time taken from address valid to data ready, is called **memory read access time**. Similarly, data can be stored into the memory device by performing a **memory write**. When writing, data and an address are presented to the memory device with the activation of a write control signal. There are also other control signals used to interface. For example, most of the memory devices in packaged chip format has a chip select (or chip enable) pin. Only when this pin is asserted will the particular memory device be active. Once an address is supplied to the chip, internal address decoding logic is used to pinpoint the particular content for output. Because of the nature of the circuit structure used in implementing the decoding logic, a memory device usually needs to recover before a subsequent read or write can be performed. Therefore, the time between subsequent address issues is called **cycle time**. Cycle time is usually twice as long as the access time. There are other timing requirements for memory devices. These timing parameters play a very important role in interfacing the

0-8493-0050-9/00/$0.00+$.50

memory devices with computation processors. In many situations, a memory device's timing parameters affect the performance of the computation system greatly.

Some special memory structures do not follow the general accessing scheme of using an address. Two of the most frequently used are content addressable memory (CAM), and first-in-first-out (FIFO) memory. Another type of memory device, which accepts multiple addresses and produces several results at different ports, is called multiport memory. There is also a type of memory that can be written in parallel, but is read serially. It is referred to as video RAM or VDRAM since they are used primarily in graphic display applications. We will discuss these in more detail later.

5.2 Memory Organization

There are several aspects to memory organization. We will take a top down approach in discussing them.

Memory Hierarchy

The speed of memory devices has been lagging behind the speed of processors. As processors become faster and more capable, larger memory spaces are required to keep up with the every increasing software complexity written for these machines. Figure 5.1(a) illustrates the well-known Moore's law, depicting the exponential growth in central processing unit (CPU) and memory capacity. Although CPUs' speed continues to grow with the advancement of technology and design technique (in particular pipelining), due to the nature of increasing memory size, more time is needed to decode wider and wider addresses and to sense the information stored in the ever-shrinking storage element. The speed gap between CPU and memory devices will continue to grow wider. Figure 5.1(b) illustrates this phenomenon.

The strategy used to remedy this problem is called **memory hierarchy**. Memory hierarchy works because of the locality property of memory references due to the sequentially fetched program instructions and the conjugation of related data. In a hierarchical memory system there are many levels of memory hierarchies. A small amount of very fast memory is usually allocated and brought right next to the central processing unit to help match up the speed of the CPU and memory. As the distance becomes greater between the CPU and memory, the performance requirement for the memory is relaxed. At the same time, the size of the memory grows larger to accommodate the overall memory size requirement. Some of the memory hierarchies are *registers, cache, main memory,* and *disk*. Figure 5.2(a) illustrates the general memory hierarchy employed in a traditional system. When a memory reference is made, the processor accesses the memory at the top of the hierarchy. If the desired data is in the higher hierarchy, a "bit" is encountered and information is obtained quickly. Otherwise a *miss* is encountered. The requested information must be brought up from a lower level in the hierarchy. Usually memory space is divided into blocks so that it can be transferred between levels in groups. At the cache level a chunk is called a *cache block* or a *cache line*. At the main memory level a chunk is referred

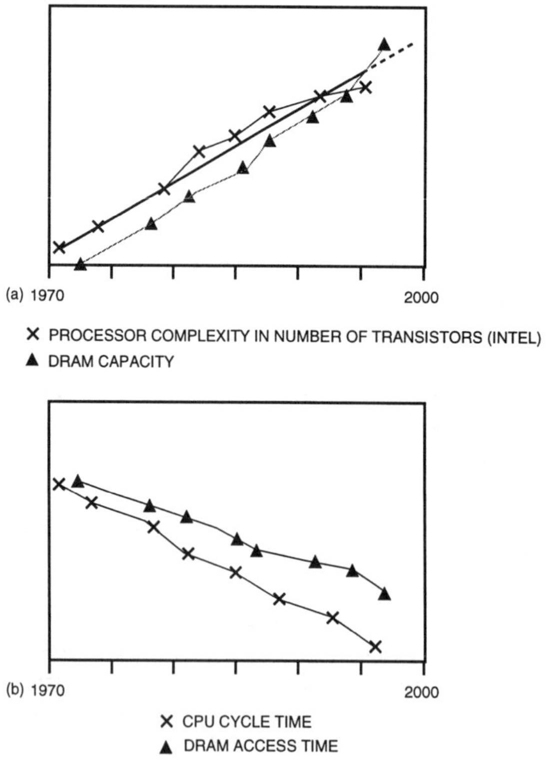

FIGURE 5.1 (a) Processor and memory development trend, (b) speed difference between RAM and CPU.

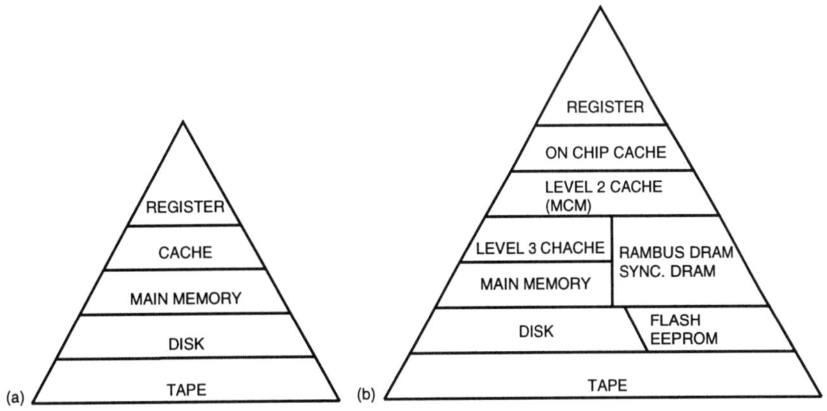

FIGURE 5.2 (a) Traditional memory hierarchy, (b) A current memory hierarchy.

to as a memory *page*. A miss in the cache is called a *cache miss* and a miss in the main memory is called a *page fault*. When a miss occurs, the whole block of memory containing the requested missing information is brought in from the lower hierarchy as mentioned before. If the current memory hierarchy level is full when a miss occurs, some existing blocks or pages must be removed and sometimes written back to a lower level to allow the new one(s) to be brought in. There are several different *replacement algorithms*. One of the most commonly used is the *least recently used* (LRU) replacement algorithm.

In modern computing systems, there may be several sublevels of cache within the hierarchy of cache. For example, the Intel Pentium™ PRO system has on chip cache (on the CPU chip), which is called level-1 (L1) cache. There is another level of cache, which resides in the same package (multichip module) with the CPU chip, which is called level-2 cache. There could also be level-3 cache on the mother board (system board) between the CPU chip(s) and main memory chips (DRAMs). Moreover, there are also newer memory devices such as synchronous RAM which provides enough speed to be interfaced with a processor directly through pipelining. Figure 5.2(b) shows a possible memory hierarchy that employs these newer devices. As mentioned before, a cache memory is larger than the registers but smaller than the main memory.

The general principle of memory hierarchy is that the farther away from the CPU it is, the larger its size, slower its speed, and the cheaper its price per memory unit becomes. Because the memory space addressable by the CPU is normally larger than necessary for a particular software program at a given time, disks are used to provide an economical supplement to main memory. This technique is called *virtual memory*. Besides disks there are tapes, optical drives, and other backup devices, which we normally call *backup storage*. They are used mainly to store information that is no longer in use, to protect against main memory and disk failures, or to transfer data between machines.

System Level Memory Organization

On the system level, we must organize the memory to accomplish different tasks and to satisfy the need of the program. In a computation system, addresses are supplied by the CPU to access the data or instruction. With a given address width a fixed number of memory locations may be accessed. This is referred to as the *memory address space*. Some processing systems have the ability to access another separate space called *input/output* (I/O) *address space*. Others use part of the memory space for I/O purposes. This style of performing I/O functions is called *memory–mapped I/O*. The memory address space defines the maximum size of the directly addressable memory that a computation system can access using memory type instructions. For example, a processor with address width of 16-b can access up to 64 K different locations (memory entries), whereas a 32-b address width can access up to 4 Gig different locations. However, sometimes we can use indirection to increase the address space. The method used by ×86 processors provides an excellent example of how this is done. The address used by the user or programmer in specifying a data item stored in the memory system is called a *logical address*. The address space accessed by the logical address is named *logical address space*. However, this logical address may not necessarily be used directly to index the physical memory. We called the memory space accessed by physical address the *physical address space*. When the logical space is larger than the physical space, then memory hierarchy is required to accommodate the difference of space sizes and store them in a lower hierarchy. In most of the current computing systems, a hard-disk is used as this lower hierarchy memory. This is termed *virtual memory system*. This mapping of logical address to physical address could be either linear or nonlinear. The actual address calculation to accomplish the mapping process is done by the CPU and the memory management unit (MMU). Thus far, we have not specified the exact size of a memory entry. A commonly used memory entry size is one byte. For historical reasons, memory is organized in bytes. A byte is usually the smallest unit of information transferred with each memory access. Wider memory entries are becoming more popular as the CPU continues to grow in speed and complexity. There are many modern systems that have a data width wider than a byte. A common size is a double word (32 b), for example, in current desktop computers. As a result, memory in bytes is organized in sections of multibytes. However, due to need for backward compatibility, these wide datapath systems are also organized to be byte addressable. The maximum width of the memory transfer is usually called *memory word length*, and the size of the memory in bytes is called *memory capacity*. Since there are different memory device sizes, the memory system can be populated with different sized memory devices. For example, a 4-megabyte main memory (physical memory) can be put together with 8, 4 megabit chips as shown in Fig. 5.3. It can also be designed with 16 512 K × 8 memory devices. Moreover, it can also be organized with a mixture of different sized devices. These memory chips are grouped together to form memory modules. Single inline memory module (SIMM) is a commonly used memory module which is widely used in current desktop computers. Similarly a memory space can also be populated by different type of memory devices. For example, out of the 4 MB space some may be SRAMs, some may be programmable read-only memory (PROMs) and some may be DRAMs. They are used in the system for different purposes. We will discuss the differences of these different type of memory devices later.

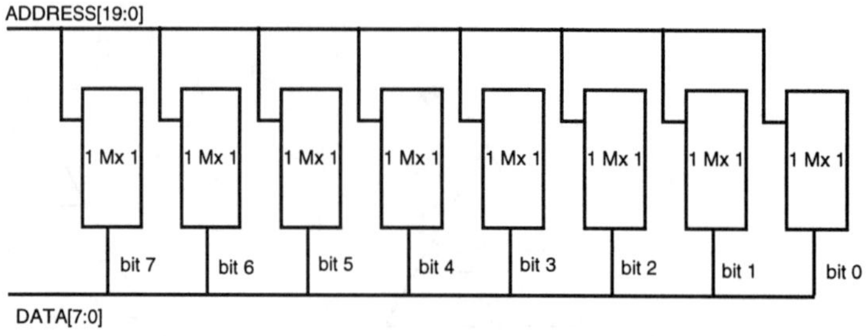

FIGURE 5.3 Eight 4-megabyte ×1 chips used to construct a 4 megabyte memory.

Memory Device Organization

Physically, within a memory device, cells are arranged in a two-dimensional array with each of the cell capable of storing 1 b of information. This matrix of cells is accessed by specifying the desired row and column addresses. The individual row enable line is generated using an address decoder while the column is selected through a multiplexer. There is usually a sense amplifier between the column bit line and the multiplexer input to detect the content of the memory cell it is accessing. Figure 5.4 illustrates this general memory cell array described with r bit of row address and c bit of column address. With the total number of $r + c$ address bits, this memory structure contains 2^{r+c} number of bits. As the size of memory array increases, the row enable lines, as well as the column bit lines, become longer. To reduce the capacitive load of a long row enable line, the row decoders, sense amplifiers, and column multiplexers are

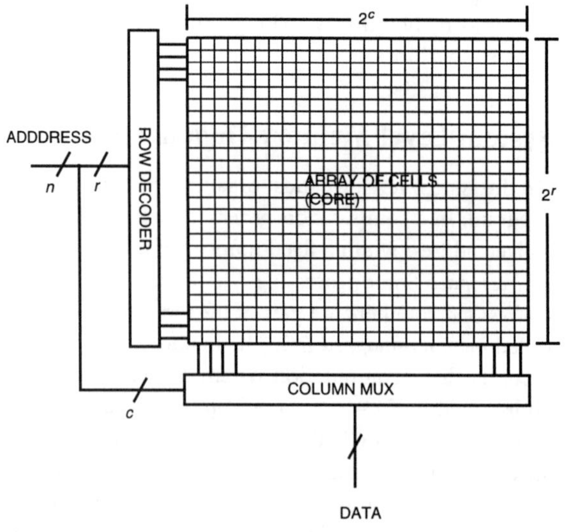

FIGURE 5.4 Generic memory structure.

often placed in the middle of divided matrices of cells, as illustrated in Fig. 5.5. By designing the multiplexer differently we are able to construct memory with different output width, for example, ×1, ×8, ×16, etc. In fact, memory designers go to great efforts to design the column multiplexers so that most of the fabrication masks may be shared for memory devices that have the same capacity but with different configurations. In large memory systems, with tens or hundreds of IC chips, it is more efficient to use 1-b wide (×1) memory IC chips. This tends to minimize the number of data pins for each chip, thereby reducing the total board area. Memory chips 1 b are disadvantageous in small systems since a minimum of eight chips are needed to implement the desired memory for a memory system with one byte width. Because of the limit of board size, often several memory chips are conjugated together to form a memory module on a specialized package. We call these memory modules; an example is SIMM.

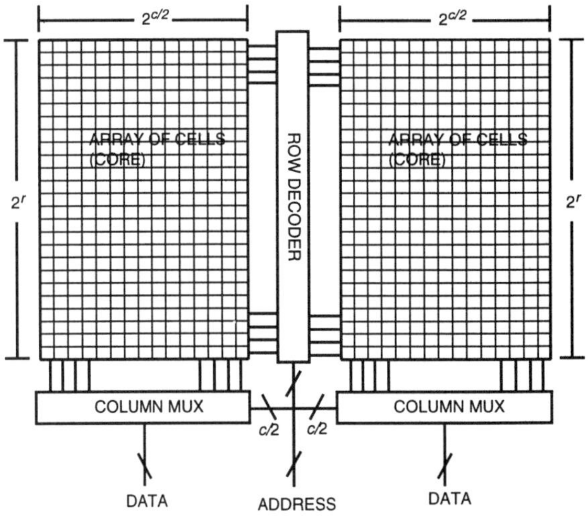

FIGURE 5.5 Divided memory structure.

5.3 Memory Device Types

As mentioned before, according to the functionality and characteristics of memory, we may divide memory devices into two major categories: ROM and RAM. We will describe these different type of devices in the following sections.

Read-Only Memory

In many systems, it is desirable to have the system level software (for example, basic input/output system [BIOS]) stored in a read-only format, because these types of programs are seldom changed. Many embedded systems also use read-only memory to store their software routines because these programs also are never changed during their lifetime, in general. Information stored in the read-only memory is permanent. It is retained even if the power supply is turned off. The memory can be read out reliably by a simple current sensing circuit without worrying about destroying the stored data. Figure 5.6 shows the general structure of a read-only memory (ROM). The effective switch position at the intersection of the word-line/bit-line determines the stored value. This switch could be implemented using different technologies resulting in different types of ROMs.

Masked Read-Only Memory (ROM)

The most basic type of this read-only-memory is called masked ROM, or simply ROM. It is programmed at manufacturing time using fabrication processing masks. ROM can be produced using different technologies, bipolar, complementary metal oxide semiconductor (CMOS), n-channel metal oxide semiconductor (nMOS), p-channel metal oxide semiconductor (pMOS), etc. Once they are programmed there are no means to change their contents. Moreover, the programming process is performed at the factory.

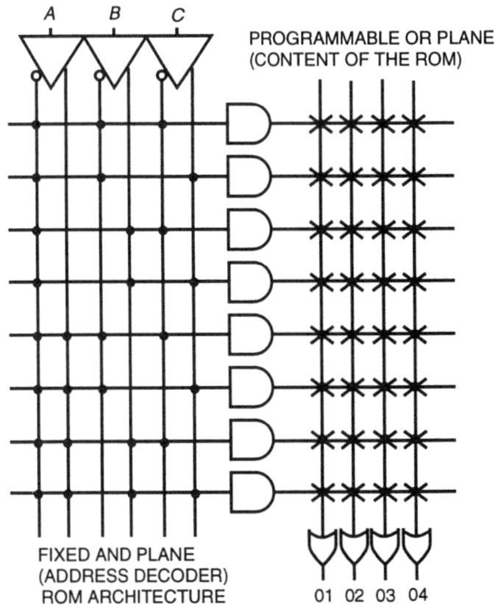

FIGURE 5.6 General structure of a ROM (an 8×4 ROM).

Programmable Read-Only Memory (PROM)

Some read-only memory is one-time programmable, but it is programmable by the user at the user's own site. This is called programmable read-only memory (PROM). It is also often referred to as write once memory (WOM). PROMs are based mostly on bipolar technology, since this technology supports it very well. Each of the single transistors in a cell has a fuse connected to its emitter. This transistor and fuse make up the memory cell. When a fuse is blown, no connection can be established when the cell is selected using the row line. Thereby a zero is stored. Otherwise, with the fuse intact, a logic one is represented. The programming is done through a programmer called a PROM programmer or PROM burner. Figure 5.7 illustrates the structure of a bipolar PROM cell and its cross section when fabricated.

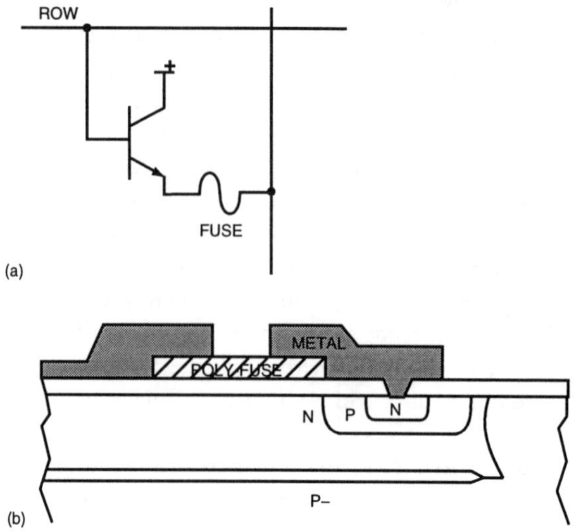

FIGURE 5.7 Bipolar PROM: (a) bipolar PROM cell, (b) cross section of a bipolar PROM cell.

Erasable Programmable Read-Only Memory (EPROM)

It is sometimes inconvenient to program the ROM only once. Thus, the erasable PROM is designed. This type of erasable PROM is called EPROM. The programming of a cell is achieved by avalanche injection of high-energy electrons from the substrate through the oxide. This is accomplished by applying a high drain voltage, causing the electrons to gain enough energy to jump over the 3.2-eV barrier between the substrate and silicon dioxide thus collecting charge at the floating gate. Once the applied voltage is removed, this charge is trapped on the floating gate. Erasing is done using an ultraviolet (UV) light eraser. Incoming UV light increases the energy of electrons trapped on the floating gate. Once the energy is increased above the 3.2-eV barrier, the electrons leave the floating gate and move toward the substrate and the selected gate. Therefore, these EPROM chips all have windows on their packages where erasing UV light can reach inside the packages to erase the content of cells. The erase time is usually in minutes. The presence of a charge on the floating gate will cause the MOS transistor to have a high threshold voltage. Thus, even with a positive select gate voltage applied at the second level of polysilicon the MOS remains to be turned off. The absence of a charge on the floating gate causes the MOS to have a lower threshold voltage. When the gate is selected the transistor will turn on and give the opposite data bit. Figure 5.8 illustrates the cross section of a EPROM cell with floating gate. EPROM technologies that migrate toward smaller geometries make floating-gate discharge (erase) via UV light exposure increasingly difficult. One problem is that the width of metal bit lines cannot reduce proportionally with advancing process technologies. EPROM metal width requirements limit bit-lines spacing, thus reducing the amount of high-energy photons that reach charged cells. Therefore, EPROM products built on submicron technologies will face longer and longer UV exposure time.

Electrical Erasable Read-Only Memory (EEPROM)

Reprogrammability is a very desirable property. However, it is very inconvenient to use a separate light-source eraser for altering the contents of the memory. Furthermore, even a few minutes of erase time is intolerable. For this reason, a new type of erasable PROM has been designed called electrical erasable PROM (EEPROM). EEPROM provides new applications where erasing is done without

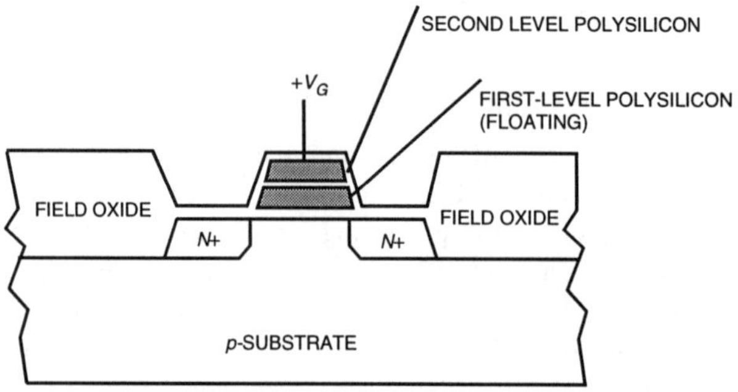

FIGURE 5.8 Cross section of a floating gate EPROM cell.

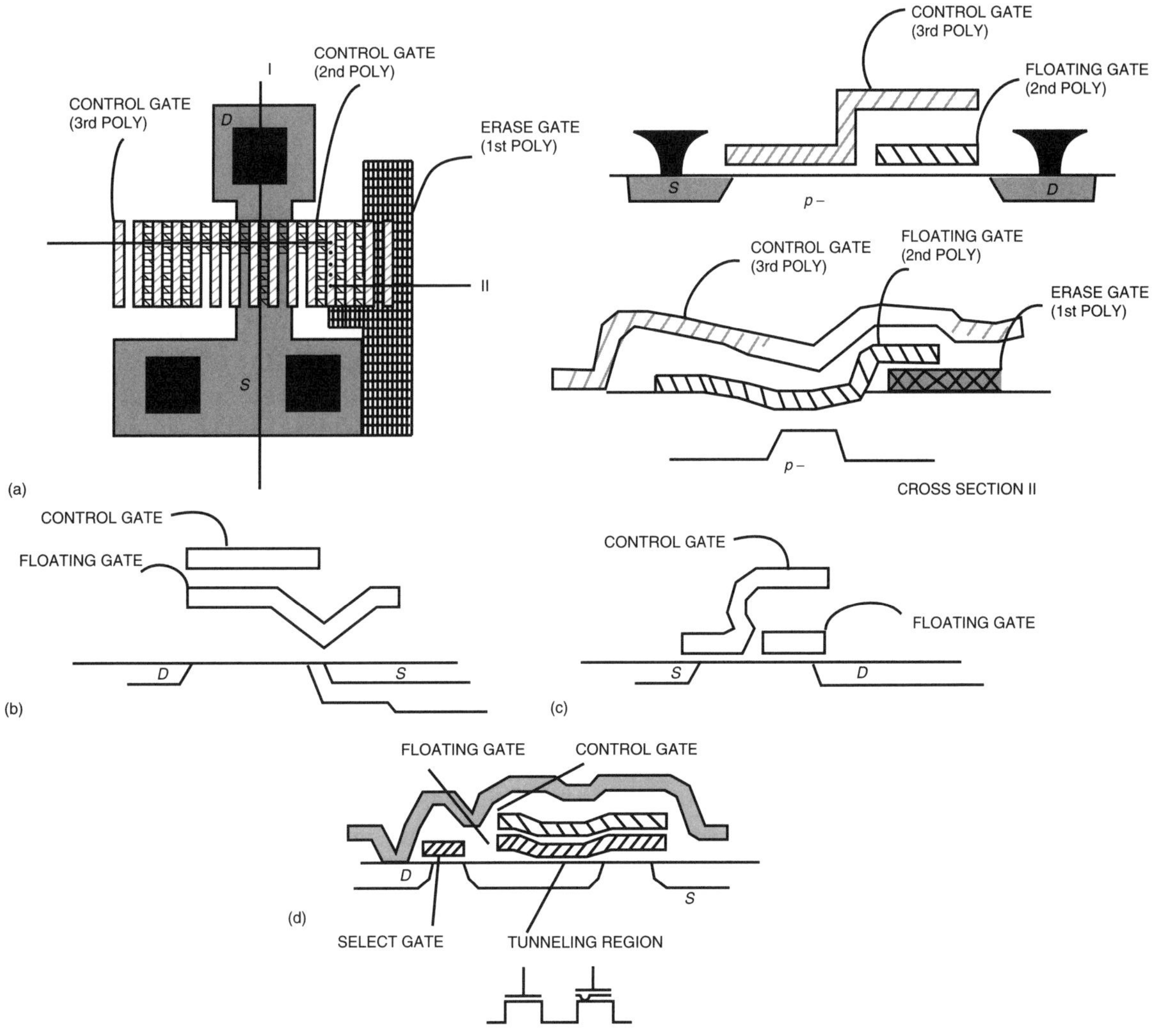

FIGURE 5.9 (a) Triple poly EEPROM cell layout and structure, (b) flotox EEPROM cell structure (source programming), (c) EEPROM with drain programming, (d) another source programming EEPROM.

removing the device from the system it resides in. There are a few basic technologies used in the processing of EEPROMs or electrical reprogrammable ROMs. All of them uses the Fowler–Nordheim tunneling effect to some extent. In this tunneling effect, cold electrons jump through the energy barrier at a silicon–silicon dioxide interface and into the oxide conduction band. This can only happen when the oxide thickness is of 100 Å or less, depending on the technology. This tunneling effect is reversible, allowing the reprogrammable ROMs to be used over and over again. One of the first electrical erasable PROMs is the electrical alterable ROM (EAROM) which is based on metal-nitrite-oxide silicon (MNOS) technology. The other is EEPROM, which is based on silicon floating gate technology used in fabricating EPROMs. The floating gate type of EEPROM is favored because of its reliability and density. The major difference between EPROM and EEPROM is in the way they discharge the charge stored in the floating gate. EEPROM must discharge floating gates electrically as opposed to using an UV light source in an EPROM device where electrons absorb photons from the UV radiation and gain enough energy to jump the silicon/silicon-dioxide energy barrier in the reverse direction as they return to the substrate. The solution for the EEPROM is to pass low-energy electrons through the thin oxide through high field (10^7 V/cm^2). This is known as the Fowler–Nordheim tunneling, where electrons can pass a short distance through the forbidden gap of the insulator and enter the conduction bank when the field applied is high enough. There are three common types of flash EEPROM cells. One uses the erase gate (three levels of polysilicon), the second and third use source and drain, respectively, to erase. Figures 5.9(a)–5.9(d) illustrate the cross sections of different EEPROMs. To realize a small EEPROM memory cell, the NAND structure was proposed in 1987. In this structure, cells are arranged in series. By using different patterns, an individual cell can be detected whether it is programmed or not. From the user's point of view, EEPROMs differs from RAM only in their write time and number of writes allowed before failure occurs. Early EEPROMs were hard to use because they have no latches for data and address to hold values during the long write operations. They also require a higher programming voltage, other than the operating voltage. Newer EEPROMs use charge pumps to generate the high programming voltage on the chip so the user does not need to provide a separate programming voltage.

Flash-EEPROM

A new alternative has been introduced recently, flash EEPROM. This type of erasable PROM lacks the circuitry to erase individual locations. When you erase them, they are erased completely. By doing so, many transistors may be saved, and larger memory capacities are possible. Note that sometimes you do not need to erase before writing. You can also write to an erased, but unwritten location, which results in an average write time comparable to an EEPROM. Another important thing to know is that writing zeros into a location charges each of the flash EEPROM's memory cells to the same electrical potential so that subsequent erasure will drain an equal amount of free charge (electrons) from each cell. Failure to equalize the charge in each cell prior to erasure can result in the overerasure of some cells by dislodging bound electrons in the floating gate and driving them out. When a floating gate is depleted in this way, the corresponding transistor can never be turned off again, thus destroying the flash EEPROM.

Random Access Memory (RAM)

RAM stands for random-access memory. It is really read-write memory because ROM is also random access in the sense that given a random address the corresponding entry is read. RAM can be categorized by content duration. A static RAM's contents will always be retained, as long as power is applied. A DRAM, on the other hand, needs to be refreshed every few milliseconds. Most RAMs by themselves are volatile, which means that without the power supply their content will be lost. All of the ROMs mentioned in the previous section are nonvolatile. RAM can be made nonvolatile by using a backup battery.

Static Random Access Memory (SRAM)

Figure 5.10 shows various SRAM memory cells (6T, 5T, 4T). The six transistor (6T) SRAM cell is the most commonly used SRAM. The crossed coupled inverters in a SRAM cell retain the information indefinitely, as long as the power supply is on since one of the pull-up transistors supplies current to compensate for the leakage current. During a read, bit and bitbar line are precharged while the word enable line is held low. Depending on the content of the cell, one of the lines is discharged slightly causing the precharged voltage to drop when the word enable line is strobed. This difference in voltage between the bit and bitbar lines is sensed by the sense amplifier, which produces the read result. During a write process, one of the bit/bitbar lines is discharged, and by strobing the word enable line the desired data is forced into the cell before the word line goes away.

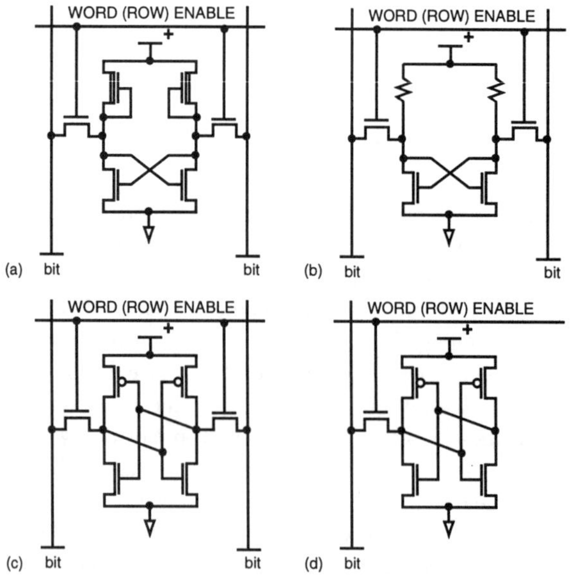

FIGURE 5.10 Different SRAM cells: (a) six-transistor SRAM cell with depletion transistor load, (b) four-transistor SRAM cell with polyresistor load, (c) CMOS six-transistor SRAM cell, (d) five-transistor SRAM cell.

Figure 5.11 gives the circuit of a complete SRAM circuit design with only one column and one row shown. One of the key design para-

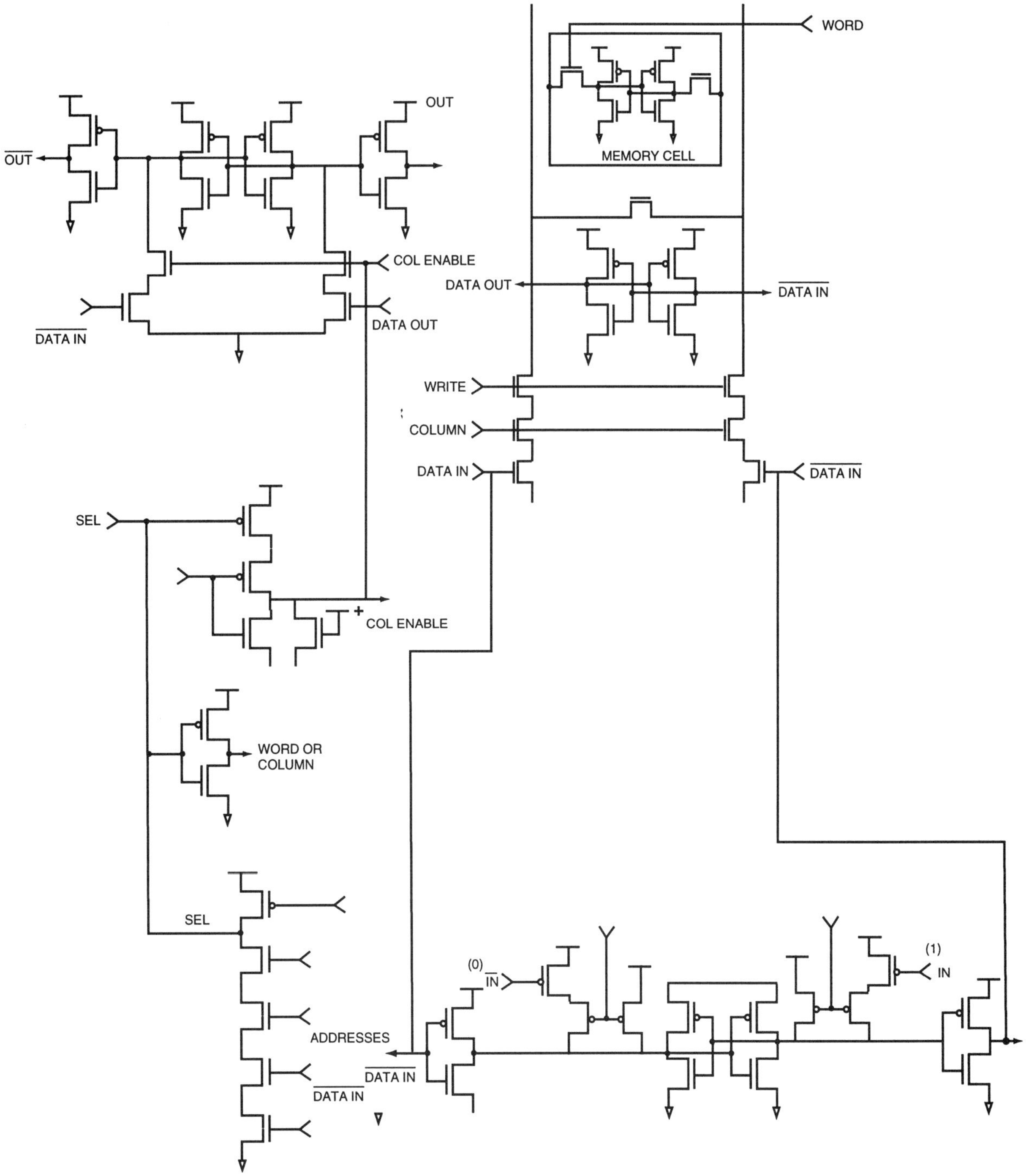

FIGURE 5.11 A complete circuit of an SRAM memory.

meters of a SRAM cell is to determine the size of
transistors used in the memory cell. We first need to determine the criteria used in sizing transistors in a CMOS 6-transistor/cell SRAM. There are three transistor sizes to choose in a 6-transistor CMOS SRAM cell due to symmetry. They are the pMOS pull-up size, the nMOS pull-down size, and the nMOS access transistor (also called the pass-transistor gate) size. There are two identical copies of each transistor, giving a total of six. Since

the sole purpose of the pMOS pull-up is to supply enough current in overcoming junction current leakage, we should decide this size first. This is also the reason why some SRAMs completely remove the two pMOS transistors and replace them with two 10-GΩ polysilicon resistors giving the 4T cell shown in Fig. 5.10. Since one of the goals is to make your cell layout as small as possible, pMOS pull-up is chosen to be minimum both in its length and width. Only when there is room available (i.e., if it does not increase the cell layout area), will the length of pMOS pull-up be increased. By increasing the length of pMOS pull-up transistors, the capacitance on the crossed-coupled inverters output nodes will be increased. This helps in protecting against soft errors a little. It also makes the cell slightly easier to write.

The next step is to choose the nMOS access transistor size. This is a rather complicated process. To begin we need to determine the length of this transistor. It is difficult to choose because, on one hand, we want it also to be minimum in order to reduce the cell layout size. However, on the other hand, a column of n SRAM bits (rows) has to have n of access transistors connected to the bit or bitbar line. If each of the cells leaks just a small amount of current, the leakage is multiplied by n. Thus, the bit or bitbar line, which one might think should be sitting at the bit-line pull-up (or precharge) voltage, is actually pulled down by this leakage. Thus, the bit or bitbar line high level is lower than the intended voltage. When this happens, the voltage difference between the bit and bitbar lines, which is seen by the sense amplifier during a read, is smaller than expected, perhaps catastrophically so. Thus, if the transistors used are not particularly leaky or n is small, a minimum sized nMOS is sufficient. Otherwise, a larger sized transistor should be used. Beside considering leakage, there are three other factors that may affect the transistor sizes of the two nMOSs. They are: (1) cell stability, (2) speed, and (3) layout area. The first factor, cell stability, is a DC phenomenon. It is a measure of the cell's ability to retain its data when reading and to change its data when writing. A read is done by creating a voltage difference between the bit and bitbar lines (which are normally precharged) for the sense amplifier to differentiate. A write is done by pulling one of the bit or bitbar lines down completely. Thus, one must design the size to satisfy the cell stability while achieving the maximum read and write speed and maintaining the minimum layout area.

Much work has been done in writing computer-aided design (CAD) tools that automatically size transistors for arrays of SRAM cells, and then do the polygon layout. Generally, these are known as SRAM macrogenerators or RAM compilers. These generated SRAM blocks are used as drop ins in many application specific intergrated circuits (ASICs). Standard SRAM chips are also available in many different organizations. Common ones are arranged in 4 b, bytes, and double words (32 b) in width.

There is also a special type of SRAM cell used in many modern computers to implement registers. These are called multiple port memories. In general, the contents can be read by many different requests at the same time. Figure 5.12 shows a dual-read port single-write port SRAM cell. When laying out SRAM cells, adjacent cells usually are mirrored to allow sharing of supply or ground lines. Figure 5.13 illustrates the layout of four adjacent SRAM cells using a generic digital process design rules. This block of four cells can be repeated in a two-dimensional array format to form the memory core.

Direct Random Access Memory (DRAM)

The main disadvantage of SRAM is in its size since it takes six transistors (or at least four transistors and two resistors) to construct a single memory cell. Thus, the DRAM is used to improve the capacity. There are different DRAM cell designs. There is the four-transistor DRAM cell, three-transistor DRAM cell, and the one-transistor DRAM cell. Figures 5.14 shows the corresponding circuits for these cells. Data writing is accomplished in a three-transistor cell by keeping the RD line low [see Fig. 5.14(b)] while strobing the WR

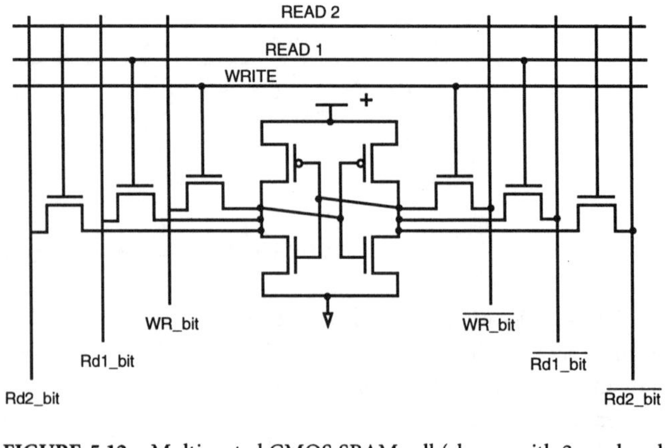

FIGURE 5.12 Multiported CMOS SRAM cell (shown with 2-read and 1-write).

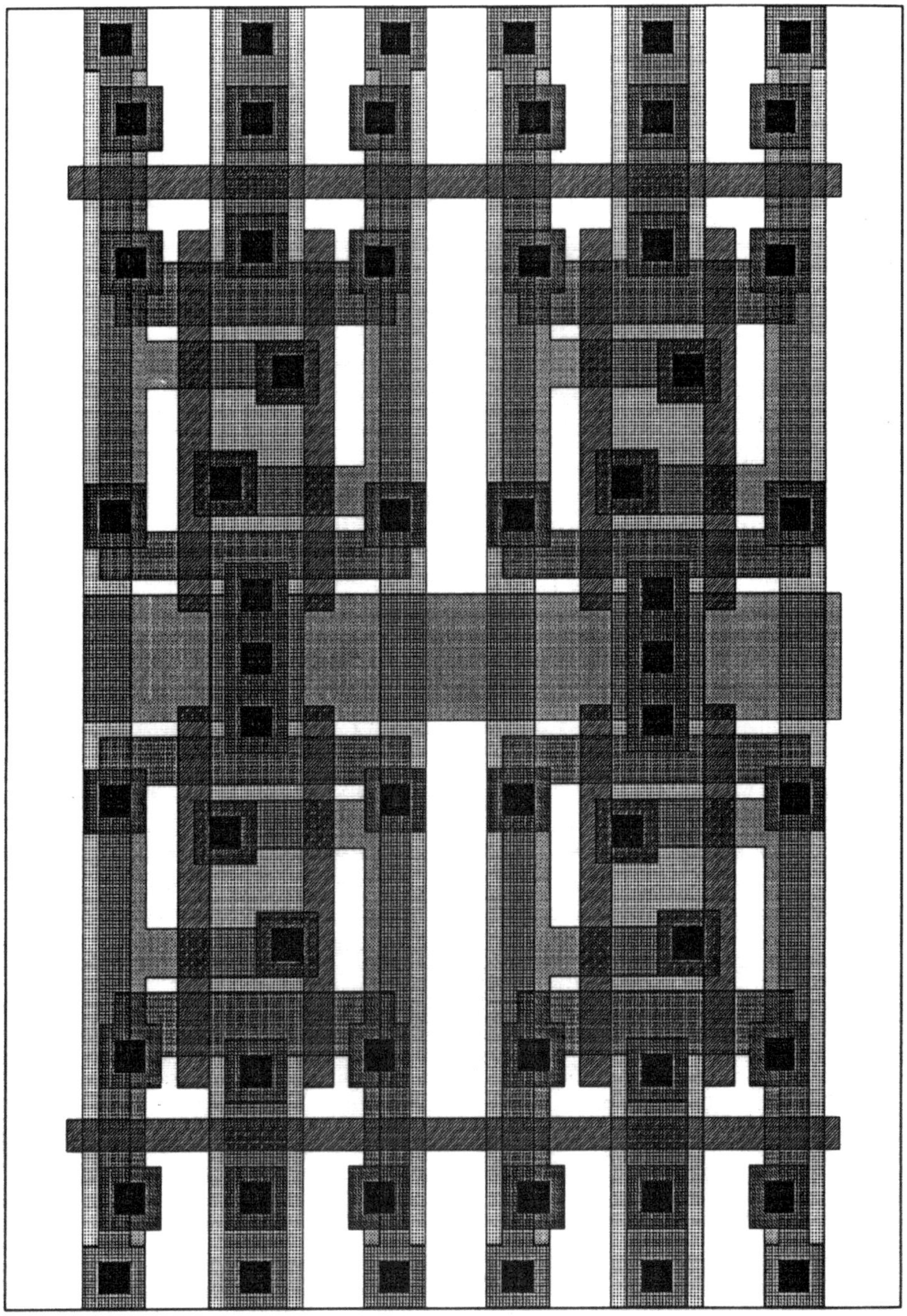

FIGURE 5.13 Layout example of four abutted 6–t SRAM cells.

line with the desired data to be written is kept on the bus. If a 1 is desired to be stored, the gate of T2 is charged turning on T2. This charge will remain on the gate of T2 for a while before the leakage current discharges it to a point where it cannot be used to turn on T2. When the charge is still there, a read can be performed by precharging the bus and strobing the RD line. If a 1 is stored, then both T2 and T3 are on during a read, causing the charge on the bus to be discharged. The lowering of voltage can be picked up by the sense amplifier. If a zero is stored, then there is no direct path from the bus to ground, thus the charge on the bus remains. To further reduce the area of a memory cell, a single transistor cell is often used and is most common in today's commercial DRAM cell. Figure 5.14(c) shows the single transistor cell with a capacitor. Usually, two columns of cells are mirror images of each other to reduce the layout area. The sense amplifier is shared, as shown in Fig. 5.15. In this one-transistor DRAM cell, there is a capacitor used to store the charge, which determines the content of the memory. The amount of the charge in the capacitor also determines the overall performance of the memory. A continuing goal is to downscale the physical area of this capacitor to achieve higher and higher density. Usually, as one reduces the area of the capacitor, the capacitance also decreases. One approach is to increase the surface area of the storage electrode without increasing the layout area by employing stacked capacitor structures, such as finned or cylindrical structures. Recently, a new technique used is to utilize a cylindrical capacitor structure with hemispherical grains. Figure 5.16 illustrates the cross section of a one-transistor DRAM cell with the cylindrical capacitor structure. Since the capacitor is charged by a source follower of the pass transistor, these capacitors can be charged maximally to a threshold voltage drop from the supply voltage. This reduces the total charge stored and affects performance, noise margin, and density. Frequently, to avoid this problem, the word lines are driven above the supply voltage when the data are written. Figure 5.17 shows typical layout of one-transistor DRAM cells.

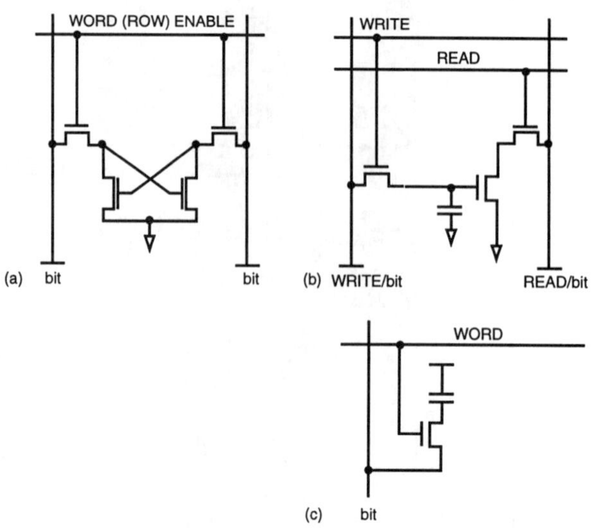

FIGURE 5.14 Different DRAM cells: (a) four-transistor DRAM cell, (b) three-transistor DRAM cell, (c) one-transistor DRAM cell.

The writing is done by putting either a 0 or 1 (the desired data to store) on the read/writing line. Then the row select line is strobed. A zero or one is stored in the capacitor as charge. A read is performed with precharging the read/write line then strobing the row select. If a zero is stored due to charge sharing, the voltage on the read/write line will decrease. Otherwise, the voltage will remain. A sense amplifier is placed at the end to pick up if there is a voltage change or not. DRAM differs from SRAM in another aspect. As the density of DRAM increases, the amount of charge stored in a cell reduces. It becomes more subject to noise. One type of noise is caused by radiation called alpha particles. These particles are helium nuclei that are present in the environment naturally or emitted from the package that houses the DRAM die. If an alpha particle hits a storage cell, it may change the state of the memory. Since alpha particles can be reduced, but not eliminated, some DRAMs institute error

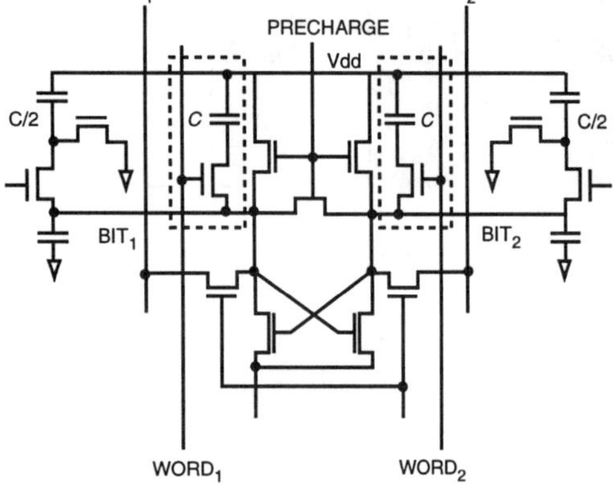

FIGURE 5.15 DRAM sense amplifier with 2 bit lines and 2 cells.

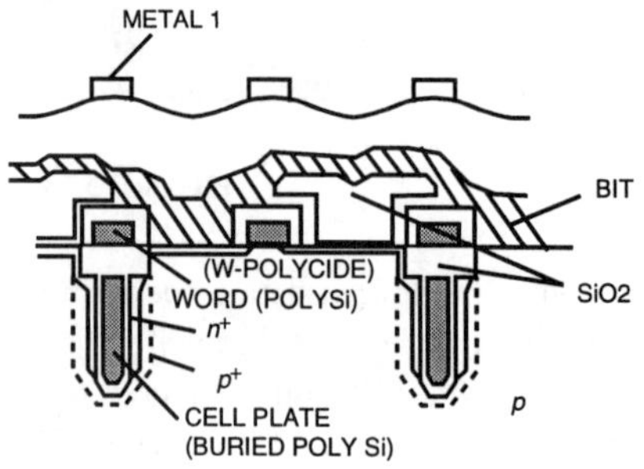

FIGURE 5.16 Cross section view of trenched DRAM cells.

detection and correction techniques to increase their reliability. Another difference between DRAMs and SRAMs is in the number of address pins needed for a given size RAM. SRAM chips require all address bits to be given at the same time. DRAMs, however, utilize time-multiplex address lines. Only half of the address bits are given at a given time. They are divided by rows and columns. An extra control signal is thus required. This is the reason why DRAM chips have two address strobe signals: row address strobe (RAS) and column address strobe (CAS).

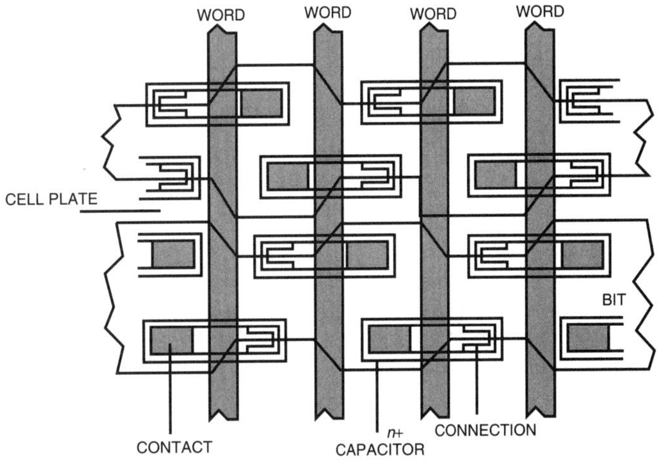

FIGURE 5.17 Physical layout of trenched DRAM cells.

Special Memory Structures

The current trend in memory devices is toward larger, faster, better-performance products. There is a complementary trend toward the development of special purpose memory devices. Several types of special purpose RAMs are offered for particular applications such as content addressable memory for cache memory, line buffers (FIFOs) for office automation machines, frame buffers for TV and broadcast equipment, and graphics buffers for computers.

Content Addressable Memory (CAM)

A special type of memory called content addressable memory (CAM) or associative memory is used in many applications such as cache memory and associative processor. A CAM stores a data item consisting of a tag and a value. Instead of giving an address, a data pattern is given to the tag section of the CAM. This data pattern is matched with the content of the tag section. If an item in the tag section of the CAM matches the supplied data pattern, the CAM outputs the value associated with the matched tag. Figure 5.18 illustrates the basic structure of a CAM. CAM cells must be both readable and writable just like the RAM cell. Figure 5.19 shows a circuit diagram for a basic CAM cell with a match output signal. This output signal may be used as an input for some logic to determine the matching process.

First-In–First-Out/Queue (FIFO/Queue)

A FIFO/queue is used to hold data while waiting. It serves as the buffering region for two systems, which may have different rates of consuming and producing data. A very popular application of FIFOs is in office automation

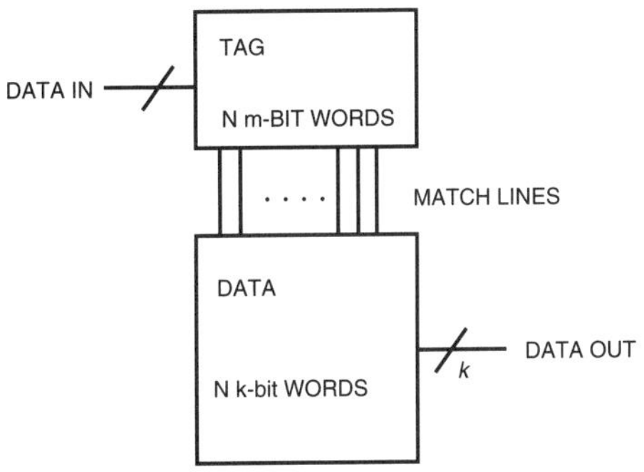

FIGURE 5.18 Functional view of a CAM.

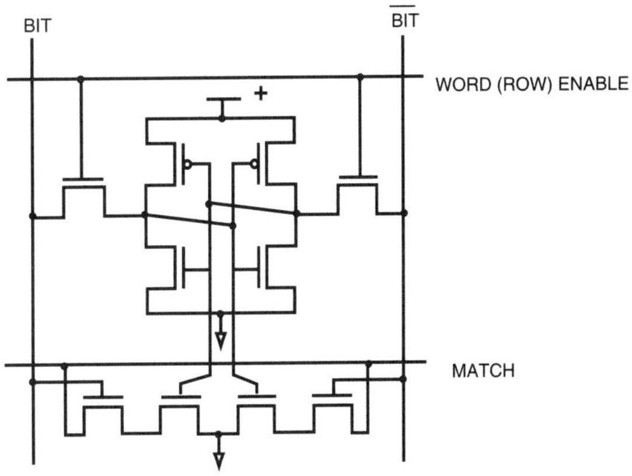

FIGURE 5.19 Static CMOS CAM cell.

equipment. These machines require high-performance serial access of large amount of data in each horizontal line such as digital facsimile machines, copiers, and image scanners. FIFO can be implemented using shift registers or RAMs with pointers.

Video RAM: Frame Buffer

There is a rapid growth in computer graphic applications. The technology that is most successful is termed raster scanning. In a raster scanning display system, an image is constructed with series of horizontal lines. Each of these lines are connected pixels of the picture image. Each pixel is represented with bits controlling the intensity. Usually there are three planes corresponding to each of the primary colors: red, green, and blue. These three planes of bit maps are called frame buffer or image memory. Frame buffer architecture affects the performance of a raster scanning graphic system greatly. Since these frame buffers needs to be read out serially to display the image line by line, a special type of DRAM memory called video memory is used. Usually these memory are dual ported with a parallel random access port for writing and a serial port for reading.

5.4 Interfacing Memory Devices

Besides capacity and type of devices, other characteristics of memory devices include its speed and the method of access. We mentioned in the Introduction memory access time. It is defined as the time between the address available to the divide and the data available at the pin for access. Sometimes the access time is measured from a particular control signal. For example, the time between the read control line ready to data ready is called the read command access time. The memory cycle time is the minimum time between two consecutive accesses. The memory write command time is measured from the write control ready to data stored in the memory. The memory latency time is the interval between the CPU issuing an address to data available for processing. The memory bandwidth is the maximum amount of memory capacity being transferred in a given time. Access is done with address, read/write control lines, and data lines. SRAM and ROMs are accessed similarly during read. Figure 5.20 shows the timing diagram of two SRAM read cycles. In both methods read cycle time is defined as the time period between consecutive read addresses. In the first method, SRAM acts as an asynchronous circuit. Given an address, the output of the SRAM changes and become valid after a certain delay, which is the read access time.

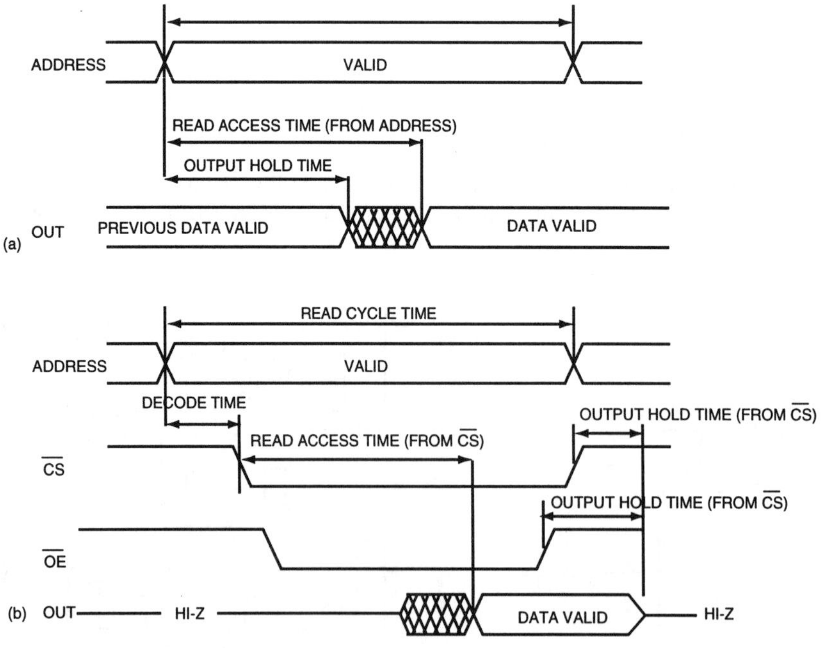

FIGURE 5.20 SRAM read cycle: (a) simple read cycle of SRAM (OE~, CE~ are all asserted and WE~ is low), (b) SRAM read cycle.

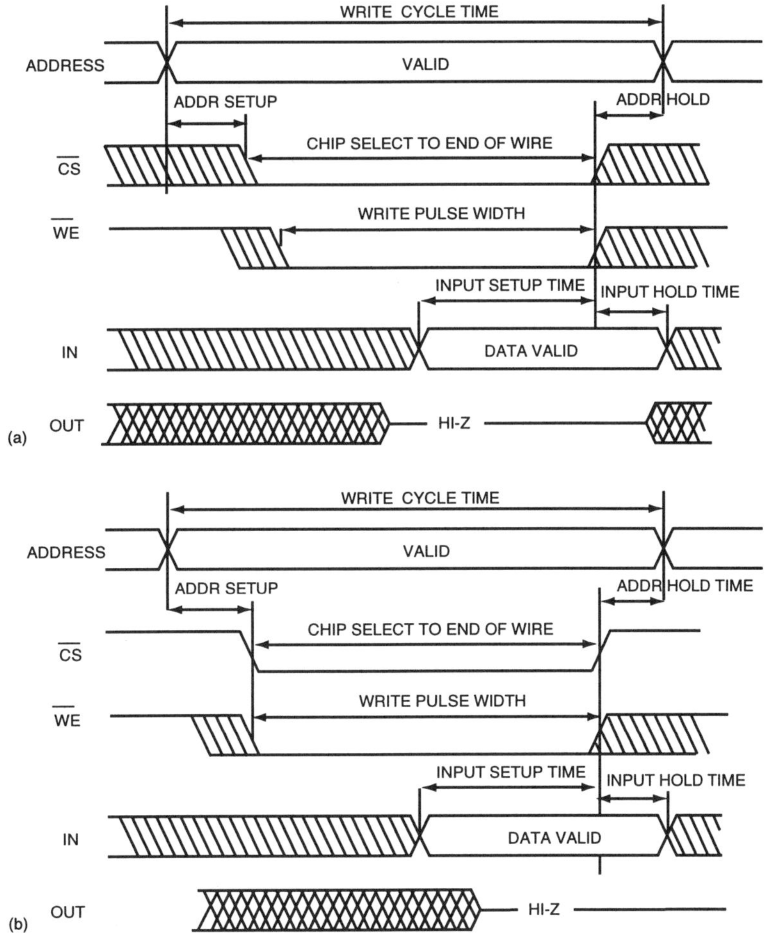

FIGURE 5.21 SRAM write cycles: (a) write enable controlled (b) chip enable controlled.

The second method uses two control signals, chip select and output enable, to initiate the read process. The main difference is in the data output valid time. With the second method data output is only valid after the output enable signal is asserted, which allows several devices to be connected to the data bus. Writing SRAM and electrically reprogrammable ROM is slightly different. Since there are many different programmable ROMs and their writing processes depend on the technology used, we will not discuss the writing of ROMs here. Figure 5.21 shows the timing diagram of writing typical SRAM chips. Figure 5.21(a) shows the write cycle using the write enable signal as the control signal, whereas Fig. 5.21(b) shows the write cycle using chip enable signals. Accessing DRAM is very different from SRAM and ROMs. We will discuss the different access modes of DRAMs in the following section.

Accessing DRAMs

DRAM is very different from SRAM in that its row and column address are time multiplexed. This is done to reduce the pins of the chip package. Because of time multiplexing there are two address strobe lines for DRAM address, RAS and CAS. There are many ways to access the DRAM. We list the five most common ways.

Normal Read/Write

When reading, a row address is given first, followed by the row address strobe signal RAS. RAS is used to latch the row address on chip. After RAS, a column address is given followed by the column address strobe CAS. After certain delay (read access time) valid data appear on the data lines. Memory write is done similarly to memory read with only the read/write control signal reversed. There are three cycles available to write a DRAM. They

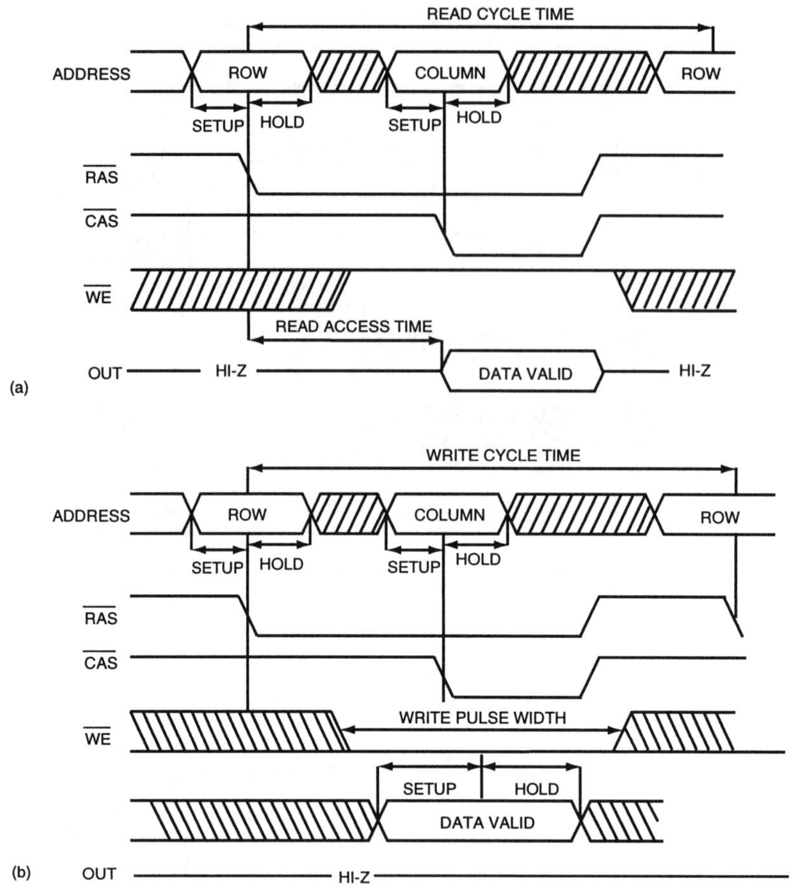

FIGURE 5.22 DRAM read and write cycles: (a) read, (b) write.

are early write, read-modify-write, and late write cycles. Figure 5.22 shows only the early write cycle of a DRAM chip. Other write cycles can be found in most of the DRAM databooks.

Fast Page Mode (FPM) or Page Mode

In page mode (or fast page mode), a read is done by lowering the RAS when the row address is ready. Then, repeatedly give the column address and CAS whenever a new one is ready without cycling the RAS line. In this way a whole row of the two-dimensional array (matrix) can be accessed with only one RAS and the same row address. Figure 5.23 illustrates the read timing cycle of a page mode DRAM chip.

Static Column

Static column is almost the same as page mode except the CAS signal is not cycled when a new column address is given—thus the static column name.

Extended Date Output (EDO) Mode

In page mode, CAS must stay low until valid data reach the output. Once the CAS assertion is removed, data are disabled and the output pins goes to open circuit. With EDO DRAM, an extra latch following the sense amplifier allows the CAS line to return to high much sooner, permitting the memory to start precharging earlier to prepare for the next access. Moreover, data are not disabled after CAS goes high. With burst EDO DRAM, not only does the CAS line return to high, it can also be toggled to step though the sequence in burst counter mode, providing even faster data transfer between memory and the host. Figure 5.24 shows a read cycle of an EDO page mode DRAM chip. EDO mode is also called hyper page mode (HPM) by some manufactures.

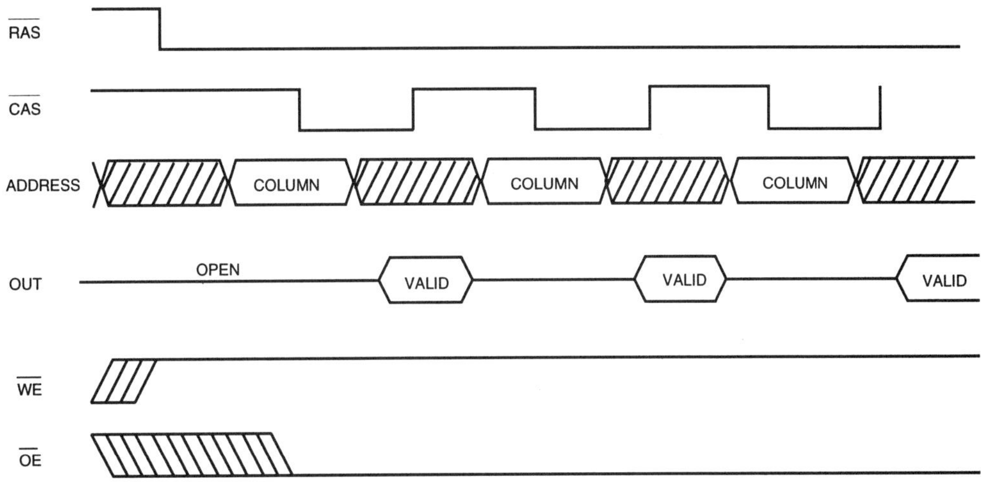

FIGURE 5.23 Page mode read cycle.

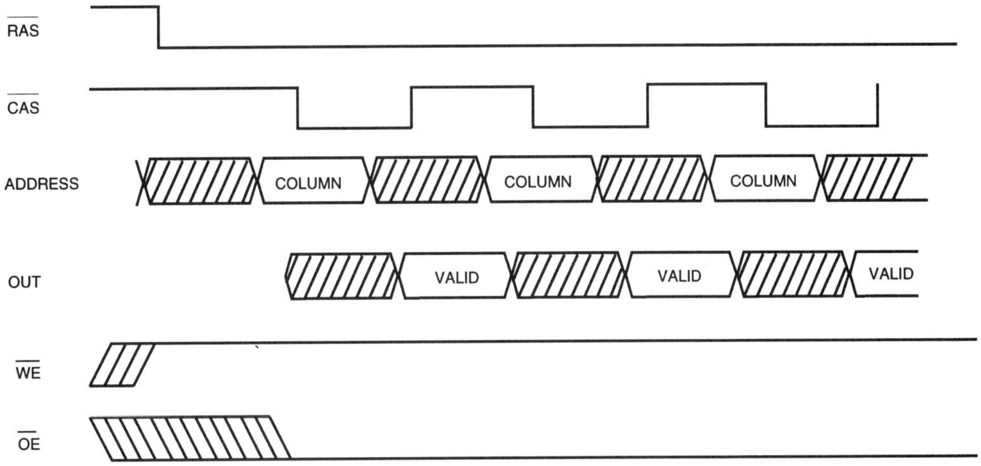

FIGURE 5.24 EDO-page mode read cycle.

Nibble Mode

In nibble mode after one CAS with a given column three more accesses are performed automatically without giving another column address (the address is assumed to be incremented from the given address).

Refreshing the DRAM

Row Address Strobe- (RAS-) Only Refresh

This type of refresh is done row by row. As a row is selected by providing the row address and strobing RAS, all memory cells in the row are refreshed in parallel. It will take as many cycles as the number of rows in the memory to refresh the entire device. For example, an 1M × 1 DRAM, which is built with 1024 rows and columns will take 1024 cycles to refresh the device. To reduce the number of refresh cycles, memory arrays are sometimes arranged to have less rows and more columns. The address, however, is nevertheless multiplexed as two evenly divided words (in the case of 1M × 1 DRAM the address word width is 10 b each for rows and columns). The higher order bits of address lines are used internally as column address lines and they are ignored during the refresh cycle. No CAS signal is necessary to perform the RAS-only refresh. Since the DRAM output buffer is enabled only when CAS is asserted, the data bus is not affected during the RAS-only refresh cycles.

Hidden Refresh

During a normal read cycle, RAS and CAS are strobed after the respective row and column addresses are supplied. Instead of restoring the CAS signal to high after the read, several RASs may be asserted with the corresponding refresh row address. This refresh style is called the hidden refresh cycles. Again, since the CAS is strobed and not restored, the output data are not affected by the refresh cycles. The number of refresh cycles performed is limited by the maximum time that CAS signal may be held asserted.

Column Address Strobe (CAS) Before RAS Refresh (Self-Refresh)

To simplify and speed up the refresh process, an on-chip refresh counter may be used to generate the refresh address to the array. In such a case, a separate control pin is needed to signal to the DRAM to initiate the refresh cycles. However, since in normal operation RAS is always asserted before CAS for read and write, the opposite condition can be used to signal the start of a refresh cycle. Thus, in modern self-refresh DRAMs, if the control signal CAS is asserted before the RAS, it signals the start of refresh cycles. We call this CAS-before-RAS refresh, and it is the most commonly used refresh mode in 1-Mb DRAMs. One discrepancy needs to be noted. In this refresh cycle the $\overline{\text{WE}}$ pin is a *don't care* for the 1-Mb chips. However, the 4 Mb specifies the CAS_Before_RAS refresh mode with $\overline{\text{WE}}$ pin held at high voltage. A CAS-before-RAS cycle with $\overline{\text{WE}}$ low will put the 4 Meg into the JEDEC-specified test mode (WCBR). In contrast, the 1 Meg test mode is entered by applying a high to the test pin.

All of the mentioned three refresh cycles can be implemented on the device in two ways. One method utilizes a distributed method, the second uses a wait and burst method. Devices using the first method refresh the row at a regular rate utilizing the CBR refresh counter to turn on rows one at a time. In this type of system, when it is not being refreshed, the DRAM can be accessed, and the access can begin as soon as the self-refresh is done. The first CBR pulse should occur within the time of the external refresh rate prior to active use of the DRAM to ensure maximum data integrity and must be executed within three external refresh rate periods. Since CBR refresh is commonly implemented as the standard refresh, this ability to access the DRAM right after exciting the self-refresh is a desirable advantage over the second method. The second method is to use an internal burst refresh scheme. Instead of turning on rows at regular interval, a sensing circuit is used to detect the voltage of the storage cells to see if they need to be refreshed. The refresh is done with a series of refresh cycles one after another until all rows are completed. During the refresh other access to the DRAM is not allowed.

5.5 Error Detection and Correction

Most DRAMs require a read parity bit for two reasons. First, alpha particle strikes disturb cells by ionizing radiation, resulting in lost data. Second, when reading DRAM, the cell's storage mechanism capacitively shares its charge with the bit-line through an enable (select) transistor. This creates a small voltage differential to be sensed during read access. This small voltage difference can be influenced by other close by bit-line voltages and other noises. To have even more reliable memory, error correction code may be used. One of the error correction methods is called the Hamming code, which is capable of correcting any 1-b error.

5.6 Conclusion

This chapter discusses several aspects of memory systems and devices. First, we looked at the organization of memory at different levels. Second, we discussed the different types of memory devices and their technologies. Last, we discussed the interface and method of access of these memory devices.

Defining Terms

Dynamic random access memory (DRAM): This memory is dynamic because it needs to be refreshed periodically. It is random access because it can be read and written.

Memory access time: The time between a valid address supplied to a memory device and data becoming ready at output of the device.

Memory cycle time: The time between subsequent address issues to a memory device.

Memory hierarchy: Organize memory in levels to make the speed of memory comparable to the processor.

Memory read: The process of retrieving information from memory.

Memory write: The process of storing information into memory.

ROM: Acronym for read-only memory.

Static random-access memory (SRAM): This memory is static because it needs not to be refreshed. It is random access because it can be read and written.

References

Alexandridis, N. 1993. *Design of Microprocessor-Based Systems.* Prentice–Hall, Englewood Cliffs, NJ.

Chang, S.S.L. 1980. Multiple-read single-write memory and its applications. *IEEE Trans. Comp.* C-29(8).

Chou, N.J. et. al. 1972. Effects of insulator thickness fluctuations on MNOS charge storage characteristics. *IEEE Trans. Elec. Dev.* ED-19:198.

Denning, P.J. 1968. The working set model for program behavior. *CACM* 11(5).

Flannigan, S. and Chappell, B. 1986. *J. Solid St. Cir.*

Fukuma, M. et al. 1993. Memory LSI reliability. *Proc. IEEE* 81(5). May.

Hamming, R.W. 1950. Error detecting and error correcting codes. *Bell Syst. J.* 29 (April).

Katz, R.H. et al. 1989. Disk system architectures for high performance computing. *Proc. IEEE* 77(12).

Lundstrom, K.I. and Svensson, C.M. 1972. Properties of MNOS structures. *IEEE Trans. Elec. Dev.* ED-19:826.

Masuoka, F. et al. 1984. A new flash EEPROM cell using triple poly-silicon technology. *IEEE Tech. Dig.* IEDM: 464–467.

Micro. *Micro DRAM Databook.*

Mukherjee, S. et al. 1985. A single transistor EEPROM cell and its implementation in a 512 K CMOS EEPROM. *IEEE Tech. Dig.* IEDM:616–619.

NEC. n.d. *NEC Memory Product Databook.*

Pohm, A.V. and Agrawal, O.P. 1983. *High-Speed Memory Systems.* Reston Pub., Reston, VA.

Prince, B. and Gunnar Due-Gundersen, G. 1983. *Semiconductor Memories.* Wiley New York.

Ross, E.C. and Wallmark, J.T. 1969. Theory of the switching behavior of MIS memory transistors. *RCA Rev.* 30:366.

Samachisa, G. et al. 1987. A 128 K flash EEPROM using double poly-silicon technology. *IEEE International Solid State Circuits Conference*, 76–77.

Sayers. et al. 1991. *Principles of Microprocessors.* CRC Press, Boca Raton, FL.

Scheibe, A. and Schulte, H. 1977. Technology of a new *n*-channel one-transistor EAROM cell called SIMOS. *IEEE Trans. Elec. Dev.* ED-24(5).

Seiichi Aritome, et al. 1993. Reliability issues of flash memory cells. *Proc. IEEE* 81(5).

Shoji, M. 1989. *CMOS Digital Circuit Technology.* Prentice–Hall, Englewood Cliffs, NJ.

Slater, M. 1989. *Design of Microprocessor-based Systems.*

Further Information

More information on basic issues concerning memory organization and memory hierarchy can be found in Pohm and Agrawal [1983]. Prince and Due-Gunderson [1983] provides a good background on the different memory devices. Newer memory technology can be found in memory device databooks such as Mukherjee et al. [1985] and the NEC databook. *IEEE Journal on Solid-State Circuits* publishes an annual special issue on the Internation Solid-State Circuits Conference. This conference reports the current state-of-the-art development on most memory devices such as DRAM, SRAM, EEPROM, and flash ROM. Issues related to memory technology can be found in the *IEEE Transactions on Electron Devices.* Both journals have an annual index, which is published at the end of each year (December issue).

6

Microprocessors

James G. Cottle
Hewlett-Packard

6.1 Introduction

In the simplest sense, a *microprocessor* may be thought of as a central processing unit (CPU) on a chip. Technical advances of microprocessors evolve quickly and are driven by progress in ultra large-scale integrated/very large-scale integrated (ULSI/VLSI) physics and technology; fabrication advances, including reduction in feature size; and improvements in architecture. Any book chapter devoted to the subject of microprocessor availability will rapidly become dated. Certain consistencies and predictability, however, will unlikely undergo a fundamental change in the next few years. I have chosen to concentrate on helpful information for the small systems developer. A general survey of the present state of the art of microprocessor architecture, available hardware, and family types is presented, along with suggestions for keeping up with rapid advances.

Developers of microprocessor-based systems are interested in cost, performance, power consumption, and ease of programmability. The latter of these is, perhaps, the most important element in bringing a product to market quickly. A key item in the ease of programmability is availability of program development tools because it is these that save a great deal of time in small system implementation and make life a lot easier for the developer. It is not unusual to find small electronic systems driven by microcontrollers and microprocessors that are far more complex than need be for the project at hand simply because the ease of development on these platforms offsets considerations of cost and power consumption. Development tools are, therefore, a tangible asset to the efficient implementation of systems, utilizing microprocessors.

Often, a more general purpose microprocessor has a companion *microcontroller* within the same family. The subject of microcontrollers is a subset of the broader area of *embedded systems*. A microcontroller contains the basic architecture of the parent microprocessor with additional on-chip special purpose hardware to facilitate easy small system implementation. Examples of special purpose hardware include analog to digital (A/D) and digital to analog (D/A) converters, timers, small amounts of on-chip memory, serial and parallel interfaces, and other input output specific hardware. These devices are most cost efficient for development of the small electronic control system, since all externally needed hardware is already designed onto the chip. Component count external to the microcontroller is therefore kept to a minimum. In addition, the advanced development tools of the parent microprocessor are often available for programming. Therefore, a potential product may be brought to market much faster than if the developer used a more general purpose microprocessor, which often requires a substantial amount of external support hardware. We will discuss some of the more common microcontrollers alongside of their parent microprocessors, where appropriate.

0-8493-0050-9/00/$0.00+$.50

6.2 Architecture Basics

In the early days of mainframe computers, only a few instructions (e.g., addition, load accumulator, store to memory) were available to the programmer and the CPU had to be patched, or its configuration changed, for various applications. Invention of microprogramming, by Maurice Wilkes in 1949, unbound the instruction set. Microprogramming made possible more complex tasks by manipulating the resources of the CPU (its registers, arithmetic logic unit, and internal buses) in a well-defined, but programmable, way. In this type of CPU the microprogram, or *control store*, contained the realization of the semantics of native machine instructions on a particular set of CPU internal resources. Microprogramming is still in strong use today. Within the microprogram resides the routines used to implement more complex instructions and addressing modes. It is this scheme that is still widely used and a key element of the complex instruction set computer(CISC) microprocessor.

Evolution from the original large mainframe computers with hard-wired logical paths and schemes for instruction manipulation toward more flexible and complex instruction sets and addressing modes was a natural one. It was driven by advances in semiconductor fabrication technology, memory design, and device physics. Advances in semiconductor fabrication technology have made it possible to place the CPU on a monolithic integrated circuit along with additional hardware. With the advent of programmable read-only memory, the microprogram was not even bound to the whims of the original designer. It could be reprogrammed at a later date if the manipulation of resources for a particular task needed to be modified or streamlined. In the simplest view, microprocessors became more and more complex with larger instruction sets and many addressing modes. All of these advances were welcomed by compiler writers, and the instruction set complexity was a natural evolution of advances in component reliability and density.

Complex Instruction Set Computer (CISC) and Reduced Instruction Set Computer (RISC) Processors

The strength of the microprocessor is in its ability to perform simple logical tasks at a high rate of speed. In *Complex instruction set computer* (CISC) microprocessors, small register sets, memory to memory operations, large instruction sets (with variable instruction lengths), and the use of microcode are typical. The basic simplified philosophy of the CISC microprocessor is that added hardware can result in an overall increase in speed. The penultimate CISC processor would have each high-level language statement mapped to a single native CPU instruction. Microcode simplifies the complexity somewhat but necessitates the use of multiple machine cycles to execute a single CISC instruction. After the instruction is decoded on a CISC machine, the actual implementation may require 10 or 12 machine cycles depending on the instruction and addressing mode used. The original trend in microprocessor development was toward increased complexity of the instruction set. Although there may be hundreds of native machine instructions, only a handful are actually used. Ironically, the CISC instruction set complexity evolves at a sacrifice in speed because its harder to increase the clock speed of a complex chip. Recently, recognition of this along with demands for increased clock speeds have yielded favor toward the reduced instruction set (RISC) microprocessor.

This trend reversal followed studies in the early 1970s that showed that although the CISC machines had plenty of instructions, only relatively few of these were actually being used by programmers. In fact, 85% of all programming consists of simple assignment instructions) (i.e., A=B). More recently RISC technology has begun to appear in the microprocessor market of personal computers. Most notably of the RISC machines in the PC market are those manufactured by Motorola (PowerPC 601, 604) and recently incorporated into the MacIntosh line machines manufactured by Apple Computer Corporation.

RISC machines have very few instructions and few machine cycles to implement them. What they do have is a lot of registers and a lot of parallelism. The ideal RISC machine attempts to accomplish a complete instruction in a single machine cycle. If this were the case, a 100-MHz microprocessor would execute 100 million instructions per second. There are typically many registers for moving data to accomplish the goal of reduced machine cycles. There is, therefore, a high degree of parallelism in a RISC processor. On the other hand, CISC machines have relatively few registers, depending on multiple machine cycle manipulation of data by the microprogram. In a CISC machine, the microprogram handles a lot of complexity in interpreting the native machine instruction. It is not uncommon for an advanced microprocessor of this type (e.g., the Motorola 68000 machine) to have over 100 native machine instructions and a slew of addressing modes.

The two basic philosophies, CISC and RISC, are ideal concepts. Both philosophies have their merits. In practice, the microprocessors of 1996 typically incorporate both philosophical schemes to enhance performance and speed. A general summary of RISC vs CISC is as follows. CISC machines depend on complexity of programming in the microprogram, thereby simplifying the compiler. Complexity in the RISC machine is realized by the compiler itself.

Logical Similarity

Microprocessors may often be highly specialized for a particular application. An example of this is the Texas Instruments TMS320C30, which is used specifically for digital signal processing. The external logical appearance of most all microprocessors, however, is the same. Although the package itself will differ, it contains connection pins for the various control signals coming from or going to the microprocessor. These are connected to the system power supply, external logic, and hardware to made up a complete computing system. External memory, math-processors, clock generators, and interrupt controllers are examples of such hardware and all must have an interface to the generic microprocessor. A diagram of a typical microprocessor pinout and the common grouping of these control signals is shown in Fig. 6.1. Externally, there are signals, which may be grouped as addressing, data, bus arbitration and control, coprocessor signals, interrupts, status, and miscellaneous connections. In some cases, multiple pins are used for connections such as the case for the address and data busses. For example, the INTEL 80286 microprocessor has 24 pins for address lines enabling an address space of 16 megabytes of memory. In addition, it has 16 data bus width pins. There are also examples of single pin connections on the 80286. Among the simplest are the power supply input (+5 V), ground, and reset. Single pins are also used for control signals.

An example of a control signal on the 80286 microprocessor is (Byte High Enable). This signal when asserted (the bar above the signal indicates assertion with a logic low) enables the 80286 to address single bytes of data in its 16 megabyte address space. The signal can be negated to prevent the upper byte from being transferred to memory.

Just as the logical connections of the microprocessor are the same from family to family, so are the basic instructions. For instance, all microprocessors have an add instruction, a subtract instruction, and a memory

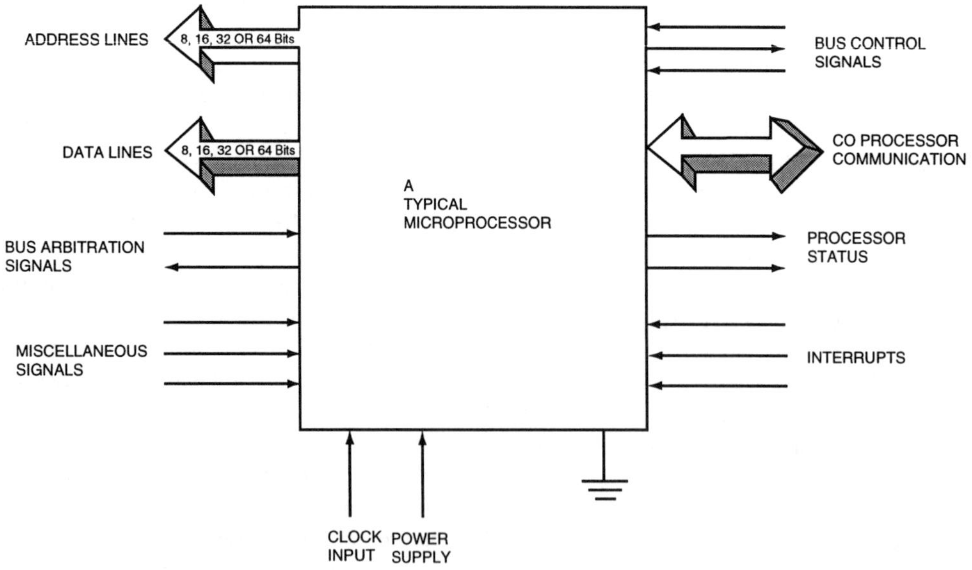

FIGURE 6.1 The pins connecting all microprocessors to external hardware are logically the same. They contain pins for addressing, data bus arbitration and control, coprocessor signals and interrupts, and those pins needed to connect to the power supply and clock generating chips.

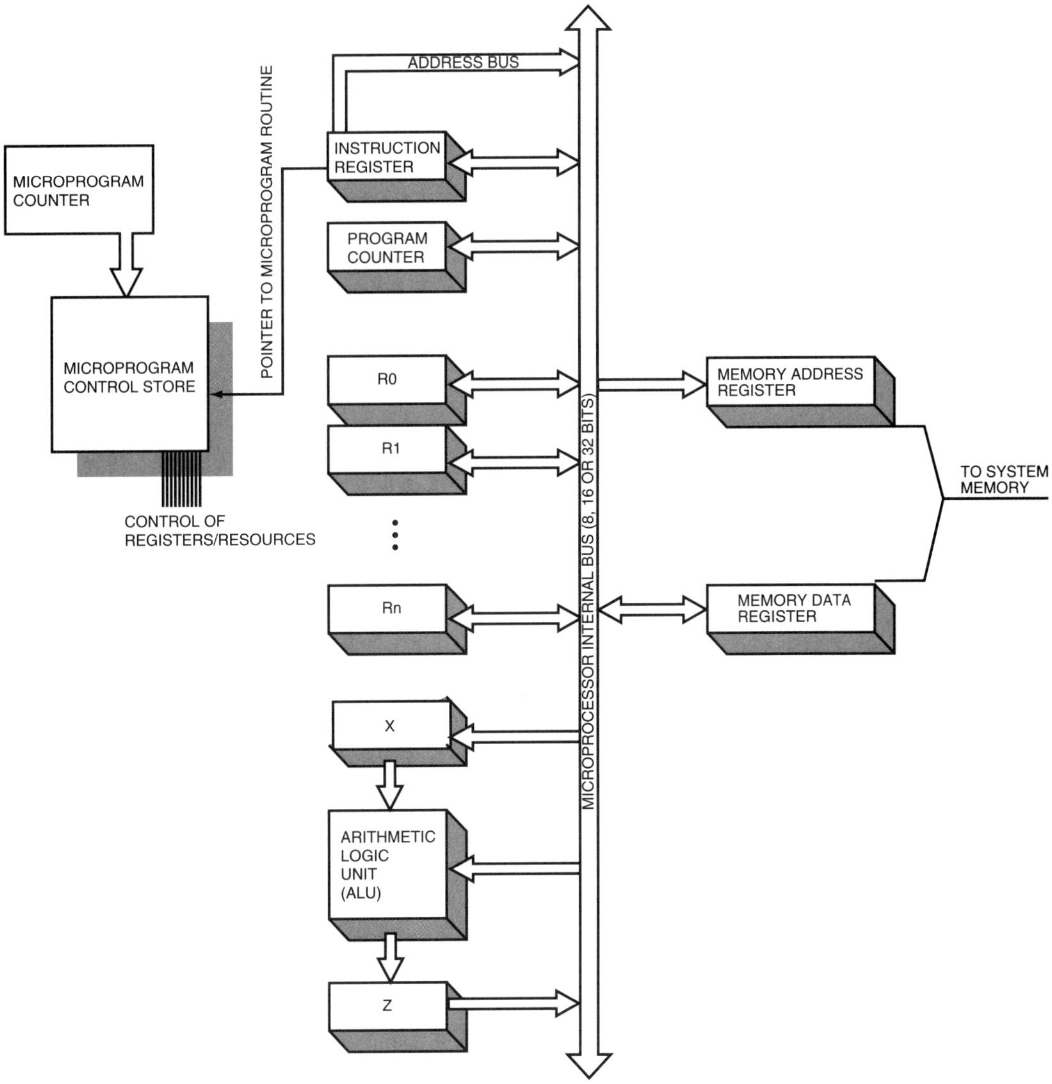

FIGURE 6.2 The basic single bus computer. This CISC machine is characterized by multiple registers linked by a bus, which carries both address and data information to and from the various registers (see text). Control lines gating the input and output of each register are governed by the information contained in the control store.

read instruction. Although logically, these instructions all are the same, the particular microprocessor realizes the semantics of the instruction (e.g., ADD) in a very unique way. This implementation of the ADD will depend on the internal resources of the CPU such as how many registers are available, how many internal buses there are, and whether the data and address information may be separated on travel along the same internal paths.

To understand how a microprogram realizes the semantics of a native machine language instruction in a CISC machine, refer to Fig. 6.2. The figure illustrates a very simple CISC machine with a single bus, which is used to transfer both memory and data information on the inside of the microprocessor. Only one instruction may be operated on at a time with this scheme, and it is far simpler than most microprocessors available today, but it illustrates the process of instruction implementation. The process begins by a procedure (a routine in control store called the instruction fetch) that fetches the next contiguous instruction in memory and places its contents in the instruction register for interpretation. The program counterregister contains the address of this instruction, and its contents are placed in the memory address register so that execution of a read command to the memory will cause, some time later, the contents of the next instruction to appear in the memory data register. These contents are then moved along the bus to the instruction register to be decoded. Meanwhile, the program counter is appropriately incremented and contains the address of the next instruction to be fetched. The contents of the instruction register for the current instruction contain the opcode and certain address information regarding the operands associated with the coded operation. For example, if the fetched instruction were an add, the instruction register would contain the coded bits indicating an add, and the information relevant to obtaining the operands for the addition. The number of bits for the machine is constant (such as 8-, 16-, or 32-b instructions) but the number of bits dedicated to the opcode and operand fields may vary to accommodate instructions with single or multiple operands, different addressing modes, and so forth. The opcode is almost always a coded address referring to a position in the microprogram of the microprocessor. For the case of the add instruction, the opcode indicates a place in microprogram memory that contains the logic code needed to implement the add. Several machine cycles will be necessary for the add instruction, including those steps needed to fetch the instruction itself. The number of these steps will depend on the particular microprocessor. That is, all machines have an add instruction. The microprogram contains the realization of the add on the specific microprocessor's architecture with its individual set of hardware resources.

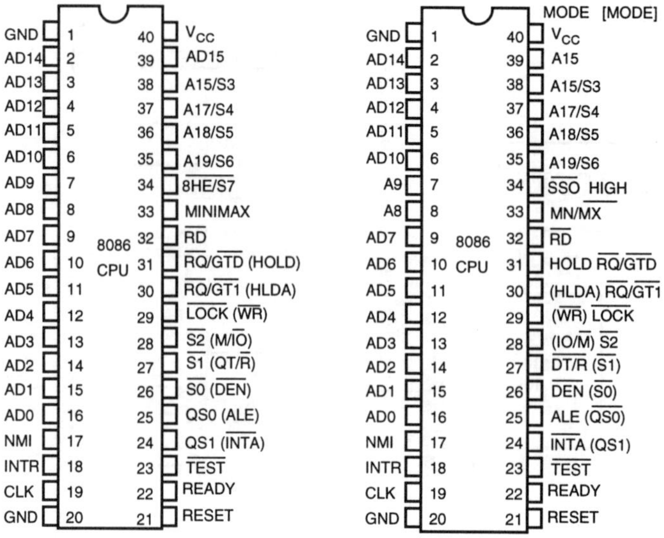

FIGURE 6.3 Microprocessor chip pinouts for the Intel 8086 (left) and the 8088 (right). Note the similarities with the generic diagram of Fig. 8.1.

Specific steps (micro-operations), which direct the microprocessor to fetch the next instruction from memory, are included with every native instruction. The instruction fetch procedure is an integral part of every native instruction and represents the minimum number of machine cycles that a microprocessor must go through to implement even the simplest instruction. Even a halt instruction, therefore, will required several machine cycles to implement.

The advantages of the CISC scheme are clear. With a large number of instructions available, the programmer of a compiler or high-level language interpreter has a relatively easy task and a lot of tools at hand. These include addressing mode schemes to facilitate relative addressing, indirect addressing, and modes such as autoincrementing or decrementing. In addition, there are typically instructions for memory to memory data transfers. Ease of programming, however, does not come without a price. The disadvantages of the CISC scheme are, therefore, relatively clear. Implementation of a particular instruction is bound to the microprogram, may take too many machine cycles, and may be relatively slow. The CISC architectural scheme also requires a lot of real estate on the semiconductor chip. Therefore, size and speed (which are really one and the same) are sacrificed for programming simplicity.

Until recently almost all CISC microprocessors incorporated the von Neumann architecture. The von Neumann machine was once the basis of most all CISC machines. This scheme is characterized by a common bus for data and address flow linking a small number of registers. The number of registers in the von Neumann machine varies depending on design, but typically consists of about 10 registers including those for general program data and addresses as well as special purposes such as instruction storage, memory addressing, and latching registers for the arithmetic logic unit (ALU). The basic architecture of the von-Neumann machine is illustrated in the figure below. Recent advances however have led to a departure of CISC machines from the basic single bus system for reasons which will be discussed shortly. A typical modern example is the Motorola 68030 architecture shown in Fig. 6.4.

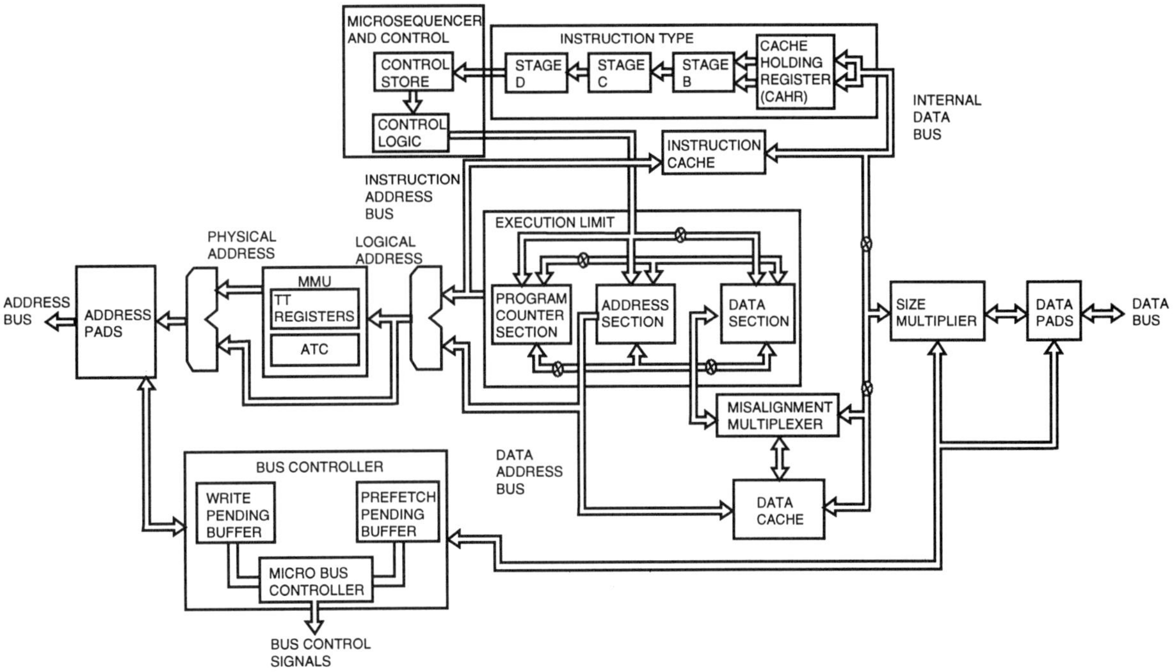

FIGURE 6.4 Internal architecture of the Motorola 68030 microprocessor. Note the multistage instruction type detection pointing to an address in control store. This is a typical characteristic of the CISC microprocessor. (*Source:* Motorola, Inc.)

Short Chronology of Microprocessor Development

The first single chip CPU was the Intel 4004 developed for calculators. It processed data in 4 bits, and its instructions were 8 bits long. Program and data were separate. The 4004 had 46 instructions, a 4 level stack, 12-b program counter, and 16, 4-b registers. Later in 1972, the 4004's successor, the 8008, was introduced, which was followed by the 8080 in 1974. The 8080 had a 16-b address bus and an 8-b data bus, seven, 8-b registers, a 16-b stack pointer to memory, and a 16-b program counter. It also had 256 input/output (I/O) ports so that I/O devices did not take up memory space and could be addressed more directly. The design was updated in 1976 (the 8085) to only require a +5 V supply.

Zilog in July 1976 introduced the Z-80, which was intended to be an improved 8080. It also used 8-b data and 16-b address, could execute all of the opcodes of the 8080 and added 80 more instructions. The register set was doubled and consisted of two banks, which could be switched. Two index registers (IX and IY) allowed for more complex memory instructions. Probably the most successful feature of the Z-80 was its memory interface. Until its introduction, dynamic random-access memory (RAM) had to be refreshed with rather complex external circuitry, which made small computing systems more complex and expensive. The Z-80 was the first chip to incorporate this refreshing capability on-chip, which increased its popularity among system developers. The Z-8 was an embedded processor similar to the Z-80 with on-chip RAM and read-only memory (ROM). It was available in clock speeds to 20 MHz and was used in a variety of small microprocessor-based control systems.

The next processor of note in the chronology was the 6800 in 1975. Although introduced by Motorola, MOS Technologies gained popularity through introducing its 650x series, principally, the 6502, which was used in early desktop computers (Commodores, Apples, and Ataris). The 6502 had very few registers and was principally an 8-b processor with a 16-b address bus. The Apple II, one of the first computers introduced to the mainstream consumer market, incorporated the 6502. Subsequent improvements in the Apple line of micros were downward compatible with the 6502 processor. The extension to the 6502 came in 1977 when Motorola introduced the 6809 with two 8-b accumulators, which could combine mathematics operations in a single 16-b combination. It had 59 instructions. Members of the 6800 family live on in embedded microcontrollers such as the 68HC05 and 68HC11. These microcontrollers are still popular for small control systems. The 68HC11 was extended to

16-b and named the 68HC16. Radiation hardened versions of the 68HC11 have been used in communications satellites.

Advanced micro devices (AMD) introduced a 4-b bit-sliced microprocessor, the Am2901. Bit-sliced processors were modular in that they could be assembled to form larger word sizes. The Am2901 had a 4-b ALU; 16, 4-b registers; and the hardware to connect carry/borrow signals between adjacent modules. In 1979, AMD developed the first floating point coprocessor for microprocessors. The AMD 9511 arithmetic circuit was used in some CP/M, Z-80-based systems and some systems based on the S-100 bus.

Around 1976, competition was heating up for the 16-b microprocessor market. The Texas Instruments (TI) TM9900 was one of the first truly 16-b microprocessors and was designed as a single chip version of the TI 990 minicomputer. The TM9900 had two 16-b registers, good interrupt handling capability, and a decent instruction set for compiler developers. An embedded version (the TMS 9940) was also produced by TI. In 1976, the stage was being set for IBM's choice of a microprocessor for its IBM-PC line of personal computers. Several 16-b microprocessors around at the time had much more powerful features and more straightforward open memory architectures (such as Motorola's 68000). It is rumoured that IBM's own engineers wanted to use the 68000 at the time IBM had already negotiated the rights to the Intel 8086. Apparently, the choice of the 8-b 8088 was a cost decision because the 8088 could have used the lower cost support chips associated with the 8085, whereas 68000 components were more expensive and not readily available.

Around 1976, Zilog introduced the Z-8000 (shortly after the 8086 by Intel). It was a 16-b microprocessor that had the capability of addressing up to 23-b of address data. The Z-8000 had 16, 16-b registers. The first 8 could be used as 16, 8-b registers, or all 16 could be used as 8, 32-b registers. This offered great flexibility in programming and for arithmetic calculations. Its instruction set included a 32-b multiply and divide instruction. It, also, like the Z-80 had memory refresh circuitry built into the chip. Probably most important, however, in the CPU chronology, is that the Z-8000 was the first microprocessor to incorporate two different modes of operation. One mode was strictly reserved for use by the operating system. The other mode was a general purpose user mode. The use of this scheme improved stability, in that the user could not crash the system as easily, and opened up the possibility of porting the chip towards multitasking, multiuser operating systems such as UNIX.

The Intel Family of Microprocessors

Intel was the first to develop a CPU on a chip in 1971 with the 4004 microprocessor. This chip, along with the 8008, was commissioned for calculators and terminal control. Intel did not expect the demand for these units to be high and several years later, developed a more general purpose microprocessor, the 8080, and a similar chip with more onboard hardware, the 8085. These were the industry's first truly general CPU's available to be integrated into microcomputing systems. The first 16-b chip, the 8086, was developed by Intel in 1978 and was the first industry entry into the realm of the 16-b processors. A companion mathematic coprocessor, the 8087, was developed for calculations requiring higher precision than the 16-b registers the 8086 would offer. Shortly after developing the 8086 and the 16-b address/8-b data version the 8088, IBM chose the 8088 for its IBM PC microcomputers. This decision was a tremendous boon to Intel and its microprocessor efforts. In some ways, it has also made the 80×86 family a victim of its early success, as all subsequent improvements and moves to larger data and address bus CPUs have had to contend with downward compatibility.

The 80186 and 80188 microprocessors were, in general, improvements to the 8086 and 8088 and incorporated more on-chip hardware for input and output support. They were never widely used, however, most likely due to the masking effect of the 8088's success in the IBM PC. The 80186 is architecturally identical to the 8086 but also contains a clock generator, a programmable controlled interrupt, three 16-b programmable timers, two programmable DMA controllers, a chip select unit, programmable control registers, a bus interface unit, and a 6-byte prefetch queue. The 80188 scheme is the same, with the exception that it only has an 8-b external data bus and a 4-byte prefetch queue.

None of the processors of the Intel family up to the beginning of the 1980s had the ability to address more than 1 megabyte of memory. The 80286, a 68-pin microprocessor, was developed to cater to the needs of systems and programs that were evolving, in a large part, due to the success of the 8088. The 80286 increased the available address space of the Intel microprocessor family to 16 megabytes of memory. Also, beginning with the 80286, the data and address lines external to the chip were not shared. In earlier chips, the address pins were multiplexed with the data lines. The internal architecture to accomplish this was a bit cumbersome, however, was kept so as to include downward compatibility with the earlier CPUs. Despite the scheme's unwieldiness, the 80286 was a huge success and is still found operating in IBM PC/AT computers and some of its PS/2 models.

In a decade, the evolution of the microprocessor had advanced from its earliest beginnings (with a 4-b CPU) to a true 16-b microprocessor. Many everday computing tasks were off loaded from mainframes to desktop machines. In 1985 Intel developed a true 32-b processor on a chip, the 80386. It was downward compatible with all object codes back to the 8008 and continued to lock Intel into the rather awkward memory model developed in the 80286. At Motorola, the 68000 in some ways had a far more simple and straightforward open address space and was a serious contender for heavy computing applications being ported to desktop machines from their larger, mainframe cousins. It is for this reason that even today the 68000 is found on science and engineering workstations and applications requiring compatibility with UNIX operating systems. The 80386 was, nevertheless, highly successful and is still in use on many desktop machines today. The 80386SX was a version of the 80386 developed with an identical package to the 80286 and meant as an upgrade for existing 80286 systems. It did so by upgrading the address but to 32 b but maintained the 16-b data bus of the 80286. A companion mathematic coprocessor [a floating point math unit (FPU)], the 80387, was developed for use with the 80386.

The 80386's success prompted other semiconductor companies (notably AMD and Cyrix) to piggyback on its success by offering clones of the processors and thus alternative sources for its end users and system developers. With Intel's addition of its 80486 in 1989, including full **pipelines**, on-chip **caching** and an integrated rather than separate floating point processor, competition for the chips, popularity was fierce. In late 1993, Intel could no longer protect the next subsequent name in the series (the 80586). It trademarked the Pentium name to its 80586 processor. Because of its popularity, the 80×86 line is the most widely cloned.

The Motorola Family of Microprocessors

Alongside of Intel's development of the 8080, Motorola developed the 6800, 8-b microprocessor. In the early 1970s, the 6800 was used in many embedded industrial control systems. It was not until 1979 however, that Motorola introduced its 16-b entry into the industry, the 68000. The 68000 was designed to be far more advanced than the 8086 microprocessor by Intel in several ways. All internal registers were 32-b wide, and it had the benefit of being able to address all 16 megabytes of external memory without the segmentation schemes utilized in the Intel series. This nonsegmented approach meant that the 68000 had no segment registers, and each of its instructions could address the complete complement of external memory.

The 68000 was the first 16-b microprocessor to incorporate 32-b internal registers. This asset allowed its selection by designers who set out to port sophisticated operating systems to desktop computer. In some ways, the 68000 was ahead of its time. If IBM had chosen the 68000 series as the core chip for its personal computers, the present state of the art of the desktop machine would be radically different. The 68000 was chosen by Apple for its MacIntosh computers. Other computer manufacturers, including Amiga and Atari chose it for its flexibility and its large internal registers.

In 1982, Motorola marketed another chip, the 68008, which was a stripped down version of the 68000 for low-end, low-cost products. The 68008 only had the capability to address 4 megabytes of memory, and its data bus was only 8-b wide. It was never very popular and certainly did not compete well with the 8088 by Intel (which was chosen for the IBM PC computers).

Advanced operating systems were ideal for the 68000 except that the chip had no capability for supporting virtual memory. For this reason, Motorola developed the 68010, which had the capability to continue an instruction after it had been suspended by a bus error. The 68012 was identical to the 68000 except that it had the capability to address 2 gigabytes of memory with its 30 address bus pins.

One of the most successful microprocessors introduced by Motorola was the 68020. It was introduced in 1984 and was the industry's first true 32-b microprocessor. Along with the 32-b registers standard to the 68000 series, it has the capability of addressing 4 gigabytes of memory and a true 32-b wide data bus. It is still widely used and is the heart of many scientific workstations such as those manufactured by Hewlett Packard and Sun

TABLE 6.1 Microprocessor Operating Parameters

Processor		Date (ship)	Bits (i/d)*	Clock MHz	SPEC-92 int	SPEC-92 fp	Pipeline stages	Cache	Vdd, v	Tech, μm	Metal layers	Power, W peak	Power, W typ	Size mm^2	Xsistor (10^6)
					CISC Processor Summary										
Intel	i386SX	1988	v/32	33	6.2	3.3	4	N/A	5.0	1.0	2	2		43	0.28
x86	i486	1989	v/32	33			5	8	5.0	1.0	2				1.2
	i486DX	1991	v/32	50	27.9	13.1	5	8	5.0	0.8	3	5.0	3.9	81	1.2
	i486DX2	1991	v/32	66	32.2	16.0	5	8	5.0	0.8	3	7.0	4.9	81	1.2
	i486DX4	1993	v/32	99	51	27	5	16	3.3	0.6	4		4		1.6
	P5	1993	v/32	66	78	63.6	5	8	5.0	0.8	3	16.0		296	3.1
	P54VRT	1994	v/32	75	89.1	68.5	5		2.9	0.6	4	5.2	2.4		
		1994	v/32	90	110	84.4	5	8	2.9	0.6	4	6.5	3.0		
	P54C	1994	v/32	100	122	93.2	5	8	3.3	0.6	4	5.0		163	3.1
	P54CQS	1995	v/32	120	172	108	5	8	3.3	0.35	4	10.0		163	3.1
	P54CS	1995	v/32	133	191	121	5	8	2.9	0.35	4				
	P55C	1995	v/32	155			5			0.35	4				
	P6	1995	v/32	150	276	220	14	8	3.1	0.6	4	29.2	23.2	306	5.5
		1995	v/32	200	366	283	14	8	3.1	0.35	4	35	28.1	195	5.5
		1995	v/32	166	327	261	14	8	3.1	0.35	4	29.4	23.4	195	5.5
Cyrix	5x86	1995	v/32	100			6	16	3.45	0.65	3	3.5	3	144	2.0
x86	(M1sc)	1995	v/32	120			6	16							2.0
	6X86	1995	v/32	100			7	16	3.3	0.65	3	10		394	3.0
	(M1)	1995	v/32	120			7	16	3.3	0.65	5			225	3.0
		1996	v/32	>133			7	16	3.3	0.5	5			169	3.0
NexGen	Nx586	1995	v/32	93			7–9	16		4.0	0.5	5.16	9.199	3.5	
	Nx686	1995	v/32	180				16			0.35	5			6
AMD	Am486	1995	v/32	120						0.5	3		3		
	Am586	1995	v/32	133				16		0.35	3				
	K5	1996	v/32	100	109	115	5	16	3.3	0.35	3			161	4.3
Motorola	68020	1985	v/32	25	NA	NA		NA	5.0	1.5	2				
	68030		v/32	50		NA	3	.25	5.0	1.2	2			55	0.27
68 K	68040	1989	v/32	25	21	15	6	4	5.0	0.8	3	6.0		164	1.2
	68060	1993	v/32	50	60	45	6	4	3.3	0.5	3	3.9		198	2.4
					RISC Processor Summary										
Power	601	93	32	50	40	60	4	32	3.6	0.6	4	9.1	6.5	121	2.8
PC	601+	94	32	100	105	125	4	32	3.3	0.5	5	5.6	4.0	74	2.8
	602	95	32	66	40		4	4	3.3	0.5	4		1.2	50	1.0
	603	94	32	80	75	85	4	8	3.3	0.5	4	3.0	2.2	85	1.6
	603e	95	32	100	120	105	4	16	3.3	0.5	4	3.5		98	2.6
	603eV	96	32	166	165	150	4	16	2.5	0.35	5		2.5	81	2.6
	604	94	32	100	128	120	4	16	3.3	0.5	4	13.0	9.0	196	3.6
		95	32	133	176	157	4	16	3.3	0.5	4		14	196	3.6
	604e	96	32	150	250					0.35					
	613	97		166	230										
	614	97	32	266	500										
	615	96	32	150	220 (i) PPC										
					138 (i) x86										
	620	95	64	133	225	300	4	32	3.3	0.5	4	30.0		311	6.9
	630	97?	64?	600											
Alpha	21064	92	64	200	133	200	7	8	3.3	0.75		30.0		234	1.68
	21064a	94	64	275	194	293	7	16	3.3	0.5	4	33.0		164	2.8
		95	64	300	220	300	7	16	3.3	0.5	4				2.8
	21066a	94	64	233	94	110	7	8	3.3	0.5	4	23.0		161	1.75
	21164	94	64	300	341	513	7	8	3.3	0.5	4	50.0		299	9.3
		95	64	333	400	570	7	8	3.3	0.5	4				9.3
	21164a	96	64	> 417	≈ 500		7	8	3.3	0.35					9.3
Sparc	CY7C601	90	32	40	21.8	22.8	5	NA	5	0.8	2	3			
	Micro	91	32	50	26.4	21.0		4	5	0.8	2	4		225	
	Weitek 2x	92	32	80	32.2	31.1		16		0.8	2	4			1.8
	Micro 2	94	32	85	64.0	54.6	5	16		0.5	3			233	2.3
		95	32	110	76	65	5	16		0.4					2.3
	Hyper	93	32	72	80	105	6	8		0.4					
	(Ross)	94	32	100	103	127	6	8		0.5	3			327	1.7

*Instruction/Data.

Microsystems and in many desktop computers still manufactured by Apple. The pinout of the 68020 is shown in Fig. N. Its actual package configuration, however, is a pin grid array (PGA) and is available in a 12.5-MHz version (MC68020RC12) and a 16-MHz version (MC68020RC16).

The 68020 contains an internal 256 byte cache memory. This is an instruction cache, holding up to 64 instructions of the long-word type. Direct access to this cache is not allowed. It serves only as an advance *prefetch queue* to enable the 68020 to execute tight loops of instructions with any further instruction fetches. Since an instruction fetch takes time to process, the presence of the 256-byte instruction cache in the 68020 is a significant speed enhancement.

The cache treatment was expanded in the 68030 to include a 256-byte data cache. In addition, the 68030 includes an onboard paged memory management unit (PMMU) to control access to virtual memory. This is the primary difference in the 68030 and the 68020. The PMMU is available as an extra chip (the 68851) for the 68020 but included on the same chip with the 68030. The 68030 also includes an improved bus interface scheme. Externally, the connections of the 68020 and the 68030 are very nearly the same. The 68030 is available in two speeds, the MC68030RC16 at 16 MHz and the MC68030RC20 with a 20-MHz clock.

The 68000 featured a supervisor and user mode. It was designed for expansion and could fetch the next instruction during an instruction's execution. (This represents 2-stage pipelining.) The 68040 had 6-stages of pipelining. The advances in the 680×0 series continued toward the 68060 in late 1994, which was a ***superscalar*** microprocessor similar to the Intel Pentium. It truly represents a merging of the two CISC and RISC philosophies of architecture. The 68060 10-stage pipeline translates 680×0 instructions into a decorded RISC-like form and uses a resource renaming scheme to reorder the execution of the instructions. The 68060 includes power saving features that can be shutdown and operates off of a 3.3-V power supply (again similar to the Intel Pentium processors).

RISC Processor Development

The major development efforts for RISC processors were led by the University of California at Berkeley and the Stanford University designs. Sun Microsystems developed the Berkeley version of the RISC processor [scalable processor architecture (SPARC)] for their high-speed workstations. This, however, was not the first RISC processor. It was preceeded by the MIPS R2000 (based on the Stanford University design), the Hewlett Packard PA-RISC CPU, and the AMD 29000.

The AMD 29000 is a RISC design, which follows the lead of the Berkeley scheme. It has a large set of registers spilt into local and global sets. The 64 global registers reduced instruction set processors were developed following the recognition that many of the CISC complex instructions were not being used.

Defining Terms

Cache: Small amount of *fast* memory, physically close to CPU, used as storage of a block of data needed immediately by the processor. Caches exist in a memory hierarchy. There is a small but very fast L1 (level one) cache; if that misses, then the access is passed on to the bigger but slower L2 (level two) cache, and if that misses, the access goes to the main memory (or L3 cache, if it exists).

Pipelining: A microarchitecture technique that divides the execution of an instruction into sequential steps. Pipelined CPUs have multiple instructions executing at the same time but at different stages in the machine. Or, the act of sending out an address before the data is actually needed.

Superscalar: Capable of executing multiple instructions in a given clock cycle. For example, the Pentium processor has two execution pipes (U and V) so it is superscalar level 2. The Pentium Pro processor can dispatch and retire three instructions per clock so it is superscalar level 3.

References

The Alpha 21164A: Continued performance leadership. 1995. *Microprocessor Forum*.

Internal architecture of the Alpha 21164 microprocessor. 1995. *CompCon 95*.

A 300 MHz quad-issue CMOS RISC microprocessor (21164). 1995. In *ISSC 95*, pp. 182–183.

A 200 MHz 64 b dual-issue CMOS microprocessor (21064). 1992. In *ISSC 92*, pp. 106–107.

Hobbit: A high performance, low-power microprocessor. 1993. *CompCon 93*, pp. 88–95.
MIPS R10000 superscalar microprocessor. 1995. *Hot Chips VII.*
The impact of dynamic execution techniques on the data path design of the P6 processor. 1995. *Hot Chips VII.*
A 0.6 μm BiCMOS processor with dynamic execution (P6). 1995. In *ISSC 95*, pp. 176–177.
A 3.3 v 0.6 μm BiCMOS superscalar processor (Pentium). 1994. In *ISSC 94*, pp. 202–203.
An overview of the Intel Pentium processor. 1993. In *CompCon 93*, pp. 60–62.
Superscalar architecture of the P5-×86 next generation processor. 1992. *Hot Chips IV.*
A 93 MHz x86 microprocessor with on-chip L2 cache controller (N586). 1995. In *ISSC 95*, pp. 172–173.
The AMD K5 processor. 1995. *Hot Chips VII.*
The PowerPC620 microprocessor: A high performance superscalar RISC microprocessor. *CompCon 95.*
A new powerPC microprocessor for the low power computing marker (602). 1995. *CompCon 95.*
133 MHz 64 b four-issue CMOS microprocessor (620). In *ISSC 95*. 1995. pp. 174–175.
The powerPC 604 RISC microprocessor. 1994. *IEEE Micro.* (Oct.).
The powerPC user instruction set architecture. 1994. *IEEE Micro.* (Oct.).
PowerPC 604. 1994. *Hot Chips VI.*
The powerPC 603 microprocessor: A low power design for portable applications. 1994. In *CompCon 94*, pp. 307–315.
A 3.0 W 75SPECint92 85SPECfp92 superscalar RISC microprocessor (603). 1994. In *ISSC 94*, pp. 212–214.
601 powerPC microprocessor. 1993. *Hot Chips V.*
The powerPC 601 microprocessor. 1993. In *Compcon 93*, pp. 109–116.
The great dark cloud falls: IBM's choice. In *Great Microprocessors of the Past and Present* (on-line) Sec. 3. http:// www.cpu info.berkeley. edu.

Further Information

In the fast changing world of the microprocessor, obsolecence is a fact of life. In fact , the marketing data shows that, with each introduction of a new generation, the time spent ramping up a new technology and ramping down the old gets shorter. There is more need for accuracy in design and development to avoid errors such as Intel experienced with their Pentium Processor with their floating point mathematic computations in 1994. For the small system developer or user of microprocessors, there is an important need to keep abreast of newer, more enhanced chips with better capabilities. Luckily, there is an excellent way to keep up to date.

The CPU Information Center at the University of California, Berkeley maintains an excellent, up to date compilation of microprocessors and microcontrollers, their architectures, and specifications on the World Wide Web (WWW). The site includes chip size and pinout information, tabular comparisons of microprocessor performance and architecture (such as that shown in the tables in this chapter), and references. It is updated regularly and serves as an excellent source for the small systems developer. Some of information contained in this chapter was obtained and used with permission from that Web site. If you have access to the World Wide Web over the Internet, point your Web Browser to http://www. infopad. berkeley.edu for an excellent, up to date summary. I would like to thank to Tom Burd for his permission to use information from his Web Site in preparing this chapter.

7

D/A and A/D Converters

Susan A.R. Garrod
Purdue University

Digital-to-analog (D/A) conversion is the process of converting digital codes into a continuous range of analog signals. *Analog-to-digital (A/D) conversion* is the complementary process of converting a continuous range of analog signals into digital codes. Such conversion processes are necessary to interface real-world systems, which typically monitor continuously varying analog signals, with digital systems that process, store, interpret, and manipulate the analog values.

D/A and A/D applications have evolved from predominately military-driven applications to consumer-oriented applications. Up to the mid-1980s, the military applications determined the design of many D/A and A/D devices. The military applications required very high performance coupled with hermetic packaging, radiation hardening, shock and vibration testing, and military specification and record keeping. Cost was of little concern, and "low power" applications required approximately 2.8 W. The major applications up the mid-1980s included military radar warning and guidance systems, digital oscilloscopes, medical imaging, infrared systems, and professional video.

The applications requiring D/A and A/D circuits in the 1990s have different performance criteria from those of earlier years. In particular, low power and high speed applications are driving the development of D/A and A/D circuits, as the devices are used extensively in battery-operated consumer products. The predominant applications include cellular telephones, hand-held camcorders, portable computers, and set-top cable TV boxes. These applications generally have low power and long battery life requirements, or they may have high speed and high resolution requirements, as is the case with the set-top cable TV boxes.

7.1 D/A and A/D Circuits

D/A and A/D conversion circuits are available as integrated circuits (ICs) from many manufacturers. A huge array of ICs exists, consisting of not only the D/A or A/D conversion circuits, but also closely related circuits such as sample-and-hold amplifiers, analog multiplexers, voltage-to-frequency and frequency-to-voltage converters, voltage references, calibrators, operation amplifiers, isolation amplifiers, instrumentation amplifiers, active filters, DC-to-DC converters, analog interfaces to digital signal processing systems, and data acquisition subsystems. Data books from the IC manufacturers contain an enormous amount of information about these devices and their applications to assist the design engineer.

The ICs discussed in this chapter will be strictly the D/A and A/D conversion circuits. Table 7.1 lists a small sample of the variety of the D/A and A/D converters currently available. The ICs usually perform either D/A or A/D conversion. There are serial interface ICs, however, typically for high-performance audio and digital signal processing applications, that perform both A/D and D/A processes.

0-8493-0050-9/00/$0.00+$.50

TABLE 7.1 D/A and A/D Integrated Circuits

D/A Converter ICs	Resolution, b	Multiplying vs Fixed Reference	Settling Time, μs	Input Data Format
Analog devices AD558	8	Fixed reference	3	Parallel
Analog devices AD7524	8	Multiplying	0.400	Parallel
Analog devices AD390	Quad, 12	Fixed reference	8	Parallel
Analog devices AD1856	16	Fixed reference	1.5	Serial
Burr–Brown DAC729	18	Fixed reference	8	Parallel
DATEL DACHF8	8	Multiplying	0.025	Parallel
National DAC0800	8	Multiplying	0.1	Parallel
A/D Converter ICs	**Resolution, b**	**Signal Inputs**	**Conversion Speed, μs**	**Output Data Format**
Analog devices AD572	12	1	25	Serial and Parallel
Burr–Brown ADC803	12	1	1.5	Parallel
Burr–Brown ADC701	16	1	1.5	Parallel
National ADC1005B	10	1	50	Parallel
TI, National ADC0808	8	8	100	Parallel
TI, National ADC0834	8	4	32	Serial
TI TLC0820	8	1	1	Parallel
TI TLC1540	10	11	21	Serial
A/D and D/A Interface ICs	**Resolution, b**	**Onboard Filters**	**Sampling Rate, kHz**	**Data Format**
TI TLC32040	14	Yes	19.2 (programmable)	Serial
TI 2914 PCM codec and filter	8	Yes	8	Serial

D/A and A/D Converter Performance Criteria

The major factors that determine the quality of performance of D/A and A/D converters are *resolution, sampling rate, speed,* and *linearity.*

The *resolution* of a D/A circuit is the smallest change in the output analog signal. In an A/D system, the resolution is the smallest change in voltage that can be detected by the system and that can produce a change in the digital code. The resolution determines the total number of digital codes, or *quantization levels,* that will be recognized or produced by the circuit.

The *resolution* of a D/A or A/D IC is usually specified in terms of the bits in the digital code or in terms of the least significant bit (LSB) of the system. An n-bit code allows for 2^n quantization levels, or $2^n - 1$ steps between quantization levels. As the number of bits increases, the step size between quantization levels decreases, therefore increasing the accuracy of the system when a conversion is made between an analog and digital signal. The system resolution can be specified also as the voltage step size between quantization levels. For A/D circuits, the resolution is the smallest input voltage that is detected by the system.

The *speed* of a D/A or A/D converter is determined by the time it takes to perform the conversion process. For D/A converters, the speed is specified as the *settling time.* For A/D converters, the speed is specified as the *conversion time.* The settling time for D/A converters will vary with supply voltage and transition in the digital code; thus, it is specified in the data sheet with the appropriate conditions stated.

A/D converters have a maximum *sampling rate* that limits the speed at which they can perform continuous conversions. The sampling rate is the number of times per second that the analog signal can be sampled and converted into a digital code. For proper A/D conversion, the minimum sampling rate must be at least two times the highest frequency of the analog signal being sampled to satisfy the Nyquist sampling criterion. The conversion speed and other timing factors must be taken into consideration to determine the maximum sampling rate of an A/D converter. **Nyquist A/D converters** use a sampling rate that is slightly more than twice the highest frequency in the analog signal. **Oversampling A/D converters** use sampling rates of N times rate, where N typically ranges from 2 to 64.

Both D/A and A/D converters require a voltage reference in order to achieve absolute conversion accuracy. Some conversion ICs have internal voltage references, whereas others accept external voltage references. For high-performance systems, an external precision reference is needed to ensure long-term stability, load regulation, and control over temperature fluctuations. External precision voltage reference ICs can be found in manufacturer's data books.

Measurement accuracy is specified by the converter's *linearity*. *Integral linearity* is a measure of linearity over the entire conversion range. It is often defined as the deviation from a straight line drawn between the endpoints and through zero (or the offset value) of the conversion range. Integral linearity is also referred to as *relative accuracy*. The *offset* value is the reference level required to establish the zero or midpoint of the conversion range. *Differential linearity* is the linearity between code transitions. Differential linearity is a measure of the *monotonicity* of the converter. A converter is said to be monotonic if increasing input values result in increasing output values.

The accuracy and linearity values of a converter are specified in the data sheet in units of the LSB of the code. The linearity can vary with temperature, and so the values are often specified at +25°C as well as over the entire temperature range of the device.

D/A Conversion Processes

Digital codes are typically converted to analog voltages by assigning a voltage weight to each bit in the digital code and then summing the voltage weights of the entire code. A general D/A converter consists of a network of precision resistors, input switches, and level shifters to activate the switches to convert a digital code to an analog current or voltage. D/A ICs that produce an analog current output usually have a faster settling time and better linearity than those that produce a voltage output. When the output current is available, the designer can convert this to a voltage through the selection of an appropriate output amplifier to achieve the necessary response speed for the given application.

D/A converters commonly have a fixed or variable reference level. The reference level determines the switching threshold of the precision switches that form a controlled impedance network, which in turn controls the value of the output signal. **Fixed reference D/A converters** produce an output signal that is proportional to the digital input. **Multiplying D/A** converters produce an output signal that is proportional to the product of a varying reference level times a digital code.

D/A converters can produce bipolar, positive, or negative polarity signals. A four-quadrant multiplying D/A converter allows both the reference signal and the value of the binary code to have a positive or negative polarity. The four-quadrant multiplying D/A converter produces bipolar output signals.

D/A Converter ICs

Most D/A converters are designed for general-purpose control applications. Some D/A converters, however, are designed for special applications, such as video or graphic outputs, high-definition video displays, ultra high-speed signal processing, digital video tape recording, digital attenuators, or high-speed function generators.

D/A converter ICs often include special features that enable them to be interfaced easily to microprocessors or other systems. Microprocessor control inputs, input latches, buffers, input registers, and compatibility to standard logic families are features that are readily available in D/A ICs. In addition, the ICs usually have laser-trimmed precision resistors to eliminate the need for user trimming to achieve full-scale performance.

A/D Conversion Processes

Analog signals can be converted to digital codes by many methods, including integration, **successive approximation**, parallel (flash) conversion, **delta modulation, pulse code modulation**, and **sigma–delta conversion**. Two of the most common A/D conversion processes are successive approximation A/D conversion and parallel or **flash A/D** conversion. Very high-resolution digital audio or video systems require specialized A/D techniques that often incorporate one of these general techniques as well as specialized A/D conversion processes. Examples of specialized A/D conversion techniques are pulse code modulation (PCM), and sigma–delta conversion. PCM is a common voice encoding scheme used not only by the audio industry in digital audio recordings but also by the telecommunications industry for voice encoding and multiplexing. Sigma–delta conversion is an oversampling A/D conversion where signals are sampled at very high frequencies. It has very high resolution and low distortion and is being used in the digital audio recording industry.

Successive approximation A/D conversion is a technique that is commonly used in medium- to high-speed data acquisition applications. It is one of the fastest A/D conversion techniques that requires a minimum amount of circuitry. The conversion times for successive approximation A/D conversion typically range from 10 to 300 μs for 8-b systems.

The successive approximation A/D converter can approximate the analog signal to form an n-bit digital code in n steps. The successive approximation register (SAR) individually compares an analog input voltage to the midpoint of one of n ranges to determine the value of 1 b. This process is repeated a total of n times, using n ranges, to determine the n bits in the code. The comparison is accomplished as follows. The SAR determines if the analog input is above or below the midpoint and sets the bit of the digital code accordingly. The SAR assigns the bits beginning with the most significant bit. The bit is set to a 1 if the analog input is greater than the midpoint voltage, or it is set to a 0 if it is less than the midpoint voltage. The SAR then moves to the next bit and sets it to a 1 or a 0 based on the results of comparing the analog input with the midpoint of the next allowed range. Because the SAR must perform one approximation for each bit in the digital code, an n-bit code requires n approximations.

A successive approximation A/D converter consists of four functional blocks, as shown in Fig. 7.1: the SAR, the analog comparator, a D/A converter, and a clock.

Parallel or flash A/D conversion is used in high-speed applications such as video signal processing, medical imaging, and radar detection systems. A flash A/D converter simultaneously compares the input analog voltage to $2^n - 1$ threshold voltages to produce an n-bit digital code representing the analog voltage. Typical flash A/D converters with 8-b resolution operate at 20–100 MHz.

The functional blocks of a flash A/D converter are shown in Fig. 7.2. The circuitry consists of a precision resistor ladder network, $2^n - 1$ analog comparators, and a digital priority encoder. The resistor network establishes threshold voltages for each allowed quantization level. The analog comparators indicate whether or not the input analog voltage is above or below the threshold at each level. The output of the analog comparators is input to the digital priority encoder. The priority encoder produces the final digital output code that is stored in an output latch.

An 8-b flash A/D converter requires 255 comparators. The cost of high-resolution A/D comparators escalates as the circuit complexity increases and as the number of analog converters rises by $2^n - 1$. As a low-cost alternative, some manufacturers produce modified flash A/D converters that perform the A/D conversion in two steps to reduce the amount of circuitry required. These modified flash A/D converters are also referred to as *half-flash* A/D converters, since they perform only half of the conversion simultaneously.

A/D Converter ICs

A/D converter ICs can be classified as general-purpose, high-speed, flash, and sampling A/D converters. The *general-purpose A/D converters* are typically low speed and low cost, with conversion times ranging from 2 μs to 33 ms. A/D conversion techniques used by these devices typically

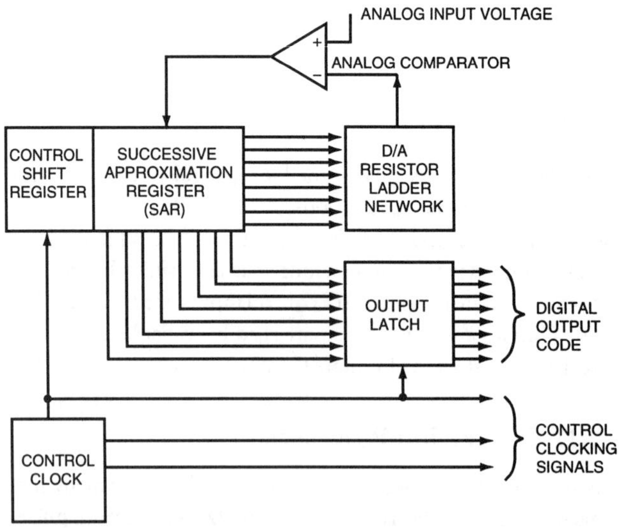

FIGURE 7.1 Successive approximation A/D converter block diagram. (*Source:* Garrod, S. and Borns, R. 1991. *Digital Logic: Analysis, Application, and Design*, p. 919. Copyright ©1991 by Saunders College Publishing, Philadelphia, PA. Reprinted by permission of the publisher.)

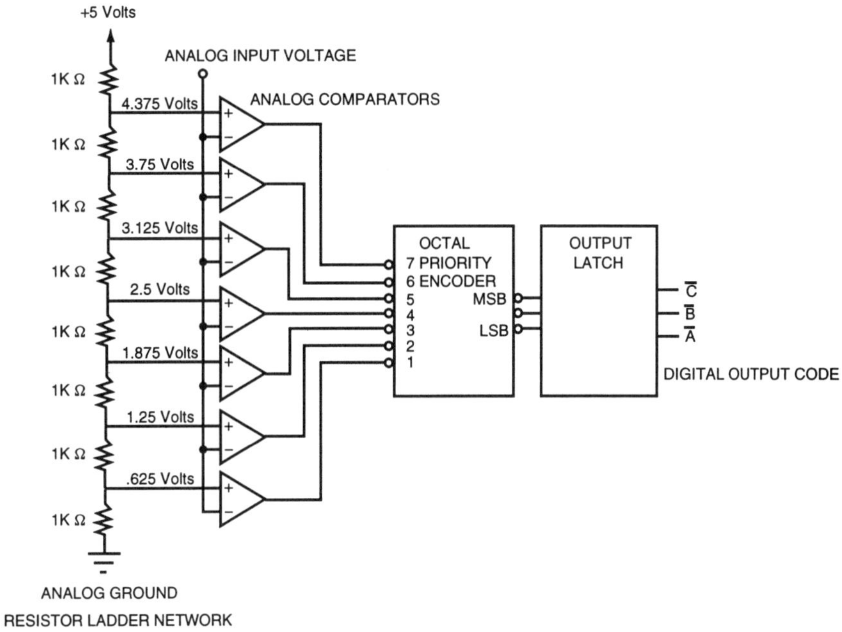

FIGURE 7.2 Flash A/D converter block diagram. (*Source:* Garrod, S. and Borns, R. *Digital Logic: Analysis, Application, and Design*, p. 928. Copyright ©1991 by Saunders College Publishing, Philadelphia, PA. Reprinted by permission of the publisher.)

include successive approximation, tracking, and integrating. The general-purpose A/D converters often have control signals for simplified microprocessor interfacing. These ICs are appropriate for many process control, industrial, and instrumentation applications, as well as for environmental monitoring such as seismology, oceanography, meteorology, and pollution monitoring.

High-speed A/D converters have conversion times typically ranging from 400 ns to 3 μs. The higher speed performance of these devices is achieved by using the successive approximation technique, modified flash techniques, and statistically derived A/D conversion techniques. Applications appropriate for these A/D ICs include fast Fourier transform (FFT) analysis, radar digitization, medical instrumentation, and multiplexed data acquisition. Some ICs have been manufactured with an extremely high degree of linearity, to be appropriate for specialized applications in digital spectrum analysis, vibration analysis, geological research, sonar digitizing, and medical imaging.

Flash A/D converters have conversion times ranging typically from 10 to 50 ns. Flash A/D conversion techniques enable these ICs to be used in many specialized high-speed data acquisition applications such as TV video digitizing (encoding), radar analysis, transient analysis, high-speed digital oscilloscopes, medical ultrasound imaging, high-energy physics, and robotic vision applications.

Sampling A/D converters have a sample-and-hold amplifier circuit built into the IC. This eliminates the need for an external sample-and-hold circuit. The throughput of these A/D converter ICs ranges typically from 35 kHz to 100 MHz. The speed of the system is dependent on the A/D technique used by the sampling A/D converter.

A/D converter ICs produce digital codes in a serial or parallel format, and some ICs offer the designer both formats. The digital outputs are compatible with standard logic families to facilitate interfacing to other digital systems. In addition, some A/D converter ICs have a built-in analog multiplexer and therefore can accept more than one analog input signal.

Pulse code modulation (PCM) ICs are high-precision A/D converters. The PCM IC is often refered to as a PCM *codec* with both encoder and decoder functions. The encoder portion of the codec performs the A/D conversion, and the decoder portion of the codec performs the D/A conversion. The digital code is usually formatted as a serial data stream for ease of interfacing to digital transmission and multiplexing systems.

PCM is a technique where an analog signal is sampled, quantized, and then encoded as a digital word. The PCM IC can include successive approximation techniques or other techniques to accomplish the PCM encoding. In addition, the PCM codec may employ nonlinear data compression techniques, such as companding, if it is necessary to minimize the number of bits in the output digital code. Companding is a logarithmic technique used to compress a code to fewer bits before transmission. The inverse logarithmic function is then used to expand the code to its original number of bits before converting it to the analog signal. Companding is typically used in telecommunications transmission systems to minimize data transmission rates without degrading the resolution of low-amplitude signals. Two standardized companding techniques are used extensively: A-law and μ-law. The A-law companding is used in Europe, whereas the μ-law is used predominantly in the United States and Japan. Linear PCM conversion is used in high-fidelity audio systems to preserve the integrity of the audio signal throughout the entire analog range.

Digital signal processing (DSP) techniques provide another type of A/D conversion ICs. Specialized A/D conversion such as *adaptive differential pulse code modulation* (ADPCM), sigma–delta modulation, *speech subband encoding, adaptive predictive speech encoding,* and *speech recognition* can be accomplished through the use of DSP systems. Some DSP systems require analog front ends that employ traditional PCM codec ICs or DSP interface ICs. These ICs can interface to a digital signal processor for advanced A/D applications. Some manufacturers have incorporated DSP techniques on board the single-chip A/D IC, as in the case of the DSP56ACD16 sigma–delta modulation IC by Motorola.

Integrating A/D converters are used for conversions that must take place over a long period of time, such as digital voltmeter applications or sensor applications such as thermocouples. The integrating A/D converter produces a digital code that represents the average of the signal over time. Noise is reduced by means of the signal averaging, or integration. Dual-slope integration is accomplished by a counter that advances while an input voltage charges a capacitor in a specified time interval, T. This is compared to another count sequence that advances while a reference voltage is discharging across the same capacitor in a time interval, *delta t*. The ratio of the charging count value to the discharging count value is proportional to the ratio of the input voltage to the reference voltage. Hence, the integrating converter provides a digital code that is a measure of the input voltage averaged over time. The conversion accuracy is independent of the capacitor and the clock frequency since they affect both the charging and discharging operations. The charging period, T, is selected to be the period of the fundamental frequency to be rejected. The maximum conversion rate is slightly less than $1/(2\,T)$ conversions per second. While this limits the conversion rate to be too slow for high-speed data acquisition applications, it is appropriate for long-duration applications of slowly varying input signals.

Grounding and Bypassing on D/A and A/D ICs

D/A and A/D converter ICs require correct grounding and capacitive bypassing in order to operate according to performance specifications. The digital signals can severely impair analog signals. To combat the electromagnetic interference induced by the digital signals, the analog and digital grounds should be kept separate and should have only one common point on the circuit board. If possible, this common point should be the connection to the power supply.

Bypass capacitors are required at the power connections to the IC, the reference signal inputs, and the analog inputs to minimize noise that is induced by the digital signals. Each manufacturer specifies the recommended bypass capacitor locations and values in the data sheet. The 1-μF tantalum capacitors are commonly recommended, with additional high-frequency power supply decoupling sometimes being recommended through the use of ceramic disc shunt capacitors. The manufacturers' recommendations should be followed to ensure proper performance.

Selection Criteria for D/A and A/D Converter ICs

Hundreds of D/A and A/D converter ICs are available, with prices ranging from a few dollars to several hundred dollars each. The selection of the appropriate type of converter is based on the application requirements of the system, the performance requirements, and cost. The following issues should be considered in order to select the appropriate converter.

1. What are the input and output requirements of the system? Specify all signal current and voltage ranges, logic levels, input and output impedances, digital codes, data rates, and data formats.
2. What level of accuracy is required? Determine the resolution needed throughout the analog voltage range, the dynamic response, the degree of linearity, and the number of bits encoding.
3. What speed is required? Determine the maximum analog input frequency for sampling in an A/D system, the number of bits for encoding each analog signal, and the rate of change of input digital codes in a D/A system.
4. What is the operating environment of the system? Obtain information on the temperature range and power supply to select a converter that is accurate over the operating range.

Final selection of D/A and A/D converter ICs should be made by consulting manufacturers to obtain their technical specifications of the devices. Major manufacturers of D/A and A/D converters include Analog Devices, Burr–Brown, DATEL, Maxim, National, Phillips Components, Precision Monolithics, Signetics, Sony, Texas Instruments, Ultra Analog, and Yamaha. Information on contacting these manufacturers and others can be found in an IC *Master Catalog.*

Defining Terms

Companding: A process designed to minimize the transmission bit rate of a signal by compressing it prior to transmission and expanding it upon reception. It is a rudimentary "data compression" technique that requires minimal processing.

Delta modulation: An A/D conversion process where the digital output code represents the change, or slope, of the analog input signal, rather than the absolute value of the analog input signal. A 1 indicates a rising slope of the input signal. A 0 indicates a falling slope of the input signal. The sampling rate is dependent on the derivative of the signal, since a rapidly changing signal would require a rapid sampling rate for acceptable performance.

Fixed reference D/A converter: The analog output is proportional to a fixed (nonvarying) reference signal.

Flash A/D: The fastest A/D conversion process available to date, also referred to as parallel A/D conversion. The analog signal is simultaneously evaluated by $2^n - 1$ comparators to produce an n-bit digital code in one step. Because of the large number of comparators required, the circuitry for flash A/D converters can be very expensive. This technique is commonly used in digital video systems.

Integrating A/D: The analog input signal is integrated over time to produce a digital signal that represents the area under the curve, or the integral.

Multiplying D/A: A D/A conversion process where the output signal is the product of a digital code multiplied times an analog input reference signal. This allows the analog reference signal to be scaled by a digital code.

Nyquist A/D converters: A/D converters that sample analog signals that have a maximum frequency that is less than the Nyquist frequency. The Nyquist frequency is defined as one-half of the sampling frequency. If a signal has frequencies above the Nyquist frequency, a distortion called *aliasing* occurs. To prevent aliasing, an *antialiasing filter* with a flat passband and very sharp rolloff is required.

Oversampling converters: A/D converters that sample frequencies at a rate much higher than the Nyquist frequency. Typical oversampling rates are 32 and 64 times the sampling rate that would be required with the Nyquist converters.

Pulse code modulation (PCM): An A/D conversion process requiring three steps: the analog signal is sampled, quantized, and encoded into a fixed length digital code. This technique is used in many digital voice and audio systems. The reverse process reconstructs an analog signal from the PCM code. The operation is very similar to other A/D techniques, but specific PCM circuits are optimized for the particular voice or audio application.

Sigma–delta A/D conversion: An *oversampling* A/D conversion process where the analog signal is sampled at rates much higher (typically 64 times) than the sampling rates that would be required with a Nyquist converter. Sigma–delta modulators integrate the analog signal before performing the delta modulation. The integral of the analog signal is encoded rather than the change in the analog signal, as is the case for traditional delta modulation. A digital sample rate reduction filter (also called a digital decimation filter) is used to provide an output sampling rate at twice the Nyquist frequency of the signal. The overall result of oversampling and digital sample rate reduction is greater resolution and less distortion compared to a Nyquist converter process.

Successive approximation: An A/D conversion process that systematically evaluates the analog signal in n steps to produce an n-bit digital code. The analog signal is successively compared to determine the digital code, beginning with the determination of the most significant bit of the code.

References

Analog Devices. 1989. *Analog Devices Data Conversion Products Data Book.* Analog Devices, Inc., Norwood, MA.

Burr–Brown. 1989. *Burr–Brown Integrated Circuits Data Book.* Burr–Brown, Tucson, AZ.

DATEL. 1988. *DATEL Data Conversion Catalog.* DATEL, Inc., Mansfield, MA.

Drachler, Will, and Bill, Murphy. 1995. New High-Speed, Low-Power Data-Acquisition ICs. *Analog Dialogue* 29(2):3–6. Analog Devices, Inc., Norwood, MA.

Garrod, S. and Borns, R. 1991. *Digital Logic: Analysis, Application and Design,* Chap. 16. Saunders College Publishing, Philadelphia, PA.

Jacob, J.M. 1989. *Industrial Control Electronics,* Chap. 6. Prentice–Hall, Englewood Cliffs, NJ.

Keiser, B. and Strange, E. 1995. *Digital Telephony and Network Integration,* 2nd ed. Van Nostrand Reinhold, New York.

Motorola. 1989. *Motorola Telecommunications Data Book.* Motorola, Inc., Phoenix, AZ.

National Semiconductor. 1989. *National Semiconductor Data Acquisition Linear Devices Data Book.* National Semiconductor Corp., Santa Clara, CA.

Park, S. 1990. *Principles of Sigma–Delta Modulation for Analog-to-Digital Converters.* Motorola, Inc., Phoenix, AZ.

Texas Instruments. 1986. *Texas Instruments Digital Signal Processing Applications with the TMS320 Family.* Texas Instruments, Dallas, TX.

Texas Instruments. 1989. *Texas Instruments Linear Circuits Data Acquisition and Conversion Data Book.* Texas Instruments, Dallas, TX.

Further Information

Analog Devices, Inc. has edited or published several technical handbooks to assist design engineers with their data acquisition system requirements. These references should be consulted for extensive technical information and depth. The publications include *Analog-Digital Conversion Handbook*, by the engineering staff of Analog Devices, published by Prentice–Hall, Englewood Cliffs, NJ, 1986; *Nonlinear Circuits Handbook, Transducer Interfacing Handbook,* and *Synchro and Resolver Conversion*, all published by Analog Devices Inc., Norwood, MA.

Engineering trade journals and design publications often have articles describing recent A/D and D/A circuits and their applications. These publications include *EDN Magazine, EE Times*, and *IEEE Spectrum*. Research-related topics are covered in *IEEE Transactions on Circuits and Systems* and also *IEEE Transactions on Instrumentation and Measurement.*

8

Application-Specific Integrated Circuits

Constantine N. Anagnostopoulos
Eastman Kodak Company

Paul P.K. Lee
Eastman Kodak Company

8.1 Introduction

Application specific integrated circuits (**ASICs**), also called custom ICs, are chips specially designed to (1) perform a function that cannot be done using standard components, (2) improve the performance of a circuit, or (3) reduce the volume, weight, and power requirement and increase the reliability of a given system by integrating a large number of functions on a single chip or a small number of chips.

ASICs can be classified into the following three categories: (1) full custom, (2) semicustom, and (3) programmable logic devices (PLDs). PLDs are treated in Section 62.1 of this volume and will, therefore, not be discussed in detail further here.

The first step of the process toward realizing a custom IC chip is to define the function the chip must perform. This is accomplished during system partitioning, at which time the system engineers and IC designers make some initial decisions as to which circuit functions will be implemented using standard, off the shelf, components and which will require custom ICs. After several iterations the functions that each custom chip has to perform are determined.

There usually are many different ways by which a given function may be implemented. For example, the function may be performed in either the analog or the digital domain. If a digital approach is selected, different execution strategies may be chosen. For example, a delay function may be implemented either by a shift register or by a random access memory. Thus, in the second step of the process, the system engineers and IC designers decide how the functions each of the chips has to perform will be executed. How a function is executed by a given chip constitutes the behavioral description of that chip.

Next, a design approach needs to be developed for each of the custom chips. Depending on the implementation method selected, either a full custom, a semicustom, or a user programmable device approach is chosen.

One important consideration in the design choice is cost and turn around time [Fey and Paraskevopoulos 1985]. Typically, a full custom approach takes the longest to design, and the cost per chip is very high unless the chip volume is also high, normally more than a few hundred thousands chips per year. The shortest design approach is using programmable devices but the cost per chip is the highest. This approach is best for prototype and limited production systems.

0-8493-0050-9/00/$0.00+$.50

8.2 Full Custom ASICs

In a typical full custom IC design, every device and circuit element on the chip is designed for that particular chip. Of course, common sense dictates that device and circuit elements proven to work well in previous designs are reused in subsequent ones whenever possible.

A full custom approach is selected to minimize the chip size or to implement a function that is not available or would not be optimum with semicustom or standard ICs. Minimizing chip size increases the fabrication yield and the number of chips per wafer. Both of these factors tend to reduce the cost per chip.

Fabrication yield is given by

$$Y = [(1 - e^{-AD})/AD]^2$$

where A is the chip or die area and D is the average number of defects per square centimeter per wafer. The number of die per wafer is given by

$$N = [\pi(R - A^{1/2})^2]/A$$

where R is the wafer radius and A is, again, the area of the die. Typically, wafer diameters range between 100 and 300 mm and the number of defects per square centimeter depends on the manufacturer and varies between 0.1 and 3. The fabrication cost per wafer varies between $500 for 100- mm wafers and a simple complementary metal oxide semiconductor (CMOS) process, to many thousands of dollars for a state-of-the-art BiCMOS process and large diameter wafers. Also, since wafers are processed in batches of at least 20 per batch, the cost of getting the first chip out is considerable.

Additional costs incurred involve testing of the die while still on the wafer and packaging a number of the good ones and testing them again. Then good die must be subjected to reliability testing to find out the expected lifetime of the chip at normal operating conditions [Hu 1992]. The cost of fabricating a few full custom chips can be substantially reduced by going through the MOS Information Sciences (MOSIS) Service of the Information Sciences Institute of the University of Southern California. This service makes it possible for a large number of users to share the cost of fabrication of a batch of wafers by having each wafer contain die from each of the users.

Fabrication of full custom ASICs is done at silicon foundries. Some of the foundries are captive, that is, they fabricate devices only for the system divisions of their own company. Others make available some of their lines to outside customers. There also are foundries that exclusively serve external customers. Foundries often provide design services as well. If users are interested in doing their own design, however, then the foundry provides a set of design rules for each of the processes they have available. The design rules describe in broad terms the fabrication technology, whether CMOS, bipolar, BiCMOS, or GaAs and then specify in detail the minimum dimensions that can be defined on the wafer for the various layers, the SPICE [Vladimirescu et al. 1981] parameters of each of the active devices, the range of values for the passive devices and other rules and limitations.

Designing a full custom chip is a complex task and can only be done by expert IC designers. Often a team of people is required both to reduce the design time [Fey and Paraskevopoulos 1986] and because one person may simply not have all the design expertise required. Also, sophisticated and powerful computer hardware and software are needed. Typically, the more sophisticated the computer aided design (CAD) tools are, the higher the probability that the chip will work the first time. Given the long design and fabrication cycle for a full custom chip and its high cost, it is important that as much as possible of the design be automated, that design rule checking should be utilized, and that the circuit simulation be as complete and accurate as possible.

During the design phase, a substantial effort should be made to incorporate on the chip additional circuitry to help verify that the chip is working properly after it is fabricated, or to help identify the section or small circuit responsible for the chip not working, or verify that an error occurred during the fabrication process. For digital full custom circuits, a large number of testability techniques have been developed [Williams and Mercer 1993], most of which are automated and easily incorporated in the design, given the proper software. For analog circuits there are no accepted universal testability methods. A unique test methodology for each circuit needs to be developed. Needless to say, full custom IC design is very risky.

Technology selection for a particular full custom ASIC depends on the functions the chip has to perform, the performance specifications, and its desired cost. The information in Table 8.1 can serve as a guide for selecting the most suitable of the mainstream technologies.

TABLE 8.1

Chip Technology	Analog	Digital	Mixed	Power Consumption	Circuit Density	Cost
Bipolar	best	poor	poor	high	low	fair
CMOS	poor	best	fair	very low	very high	low
BiCMOS	fair	good	best	low	high	high
GaAs	poor	good	poor	fair	fair	high

In the table, mixed indicates chips in which some of the circuits operate in the analog domain while others operate in the digital domain. The best choice for such circuits is usually a BiCMOS technology [Ting et al. 1994]. On the other hand, for a circuit that has to perform very demanding analog functions, bipolar technology is clearly the best choice. When a digital chip with a large number of gates is needed and final chip cost is an important consideration, it should be designed in a CMOS process.

8.3 Semicustom ASICs

The main distinguishing feature of semicustom ASICs, compared to full custom ones, is that the basic circuit building blocks, whether analog or digital, are already designed and proven to work. These basic circuits typically reside in libraries within a CAD system. The users simply select from the library the components needed, place them on their circuits, and interconnect them. Circuit simulation is also done at a much higher level than SPICE, and the designer is, therefore, not required to be familiar with either semiconductor or device physics.

Semicustom ICs are designed using either **gate arrays, standard cells**, analog arrays, **functional blocks** and PLDs such as field programmable gate arrays (FPGAs). Note that there does not exist a standard naming convention for these products within the industry. Often different manufacturers use different names to describe what are essentially very similar products. Gate arrays, standard cells, and FPGAs are used for digital designs. Analog arrays are used for analog designs and functional blocks are used for both.

Gate Arrays

A gate array consists of a regular array of transistors, usually arranged in two pairs of n- and p-channel, which is the minimum number required to form a NAND gate, and a fixed number of bonding pads, each incorporating an input/output (I/O) buffer. The major distinguishing feature of gate arrays is that they are partially prefabricated by the manufacturer. The designer customizes only the final contact and metal layers. Partially prefabricating the devices reduces delivery time and cost, particularly of prototype parts.

The layout of a simple 2048-gates CMOS gate array is shown in Fig. 8.1. The device consists of 16 columns of transistors and each column contains 128 pairs of n-channel and p-channel transistors. Between the columns are 18 vertical wiring channels, each containing 21 tracks. There are no active devices in the channels. There are 4 horizontal wiring tracks for each gate for a total of 512 horizontal tracks or *routes* for the whole array. The optimum number R of wiring tracks or routes per gate, of an array of a given number of gates, is given by the empirical formula [Fier and Heikkila 1982]

$$R = 3CG^{0.124}$$

Where C is the average number of connections per gate and G is the number of gates in the array. For a two-input NAND gate, shown in Fig. 8.2, the number of connections C is 3, the two inputs A and B and the output AB, and for a 2048-gates array, the preceding formula gives $R = 23$. In this device 25 routes per gate are provided.

In the perimeter of the device 68 bond pads are arranged. Of these, 8 pads are needed for power and ground connections. Another empirical formula [Fier and Heikkila 1982], based on Rent's rule, specifies that the number of I/O pads required to communicate effectively with the internal gates is given by

$$P = CG^{a}$$

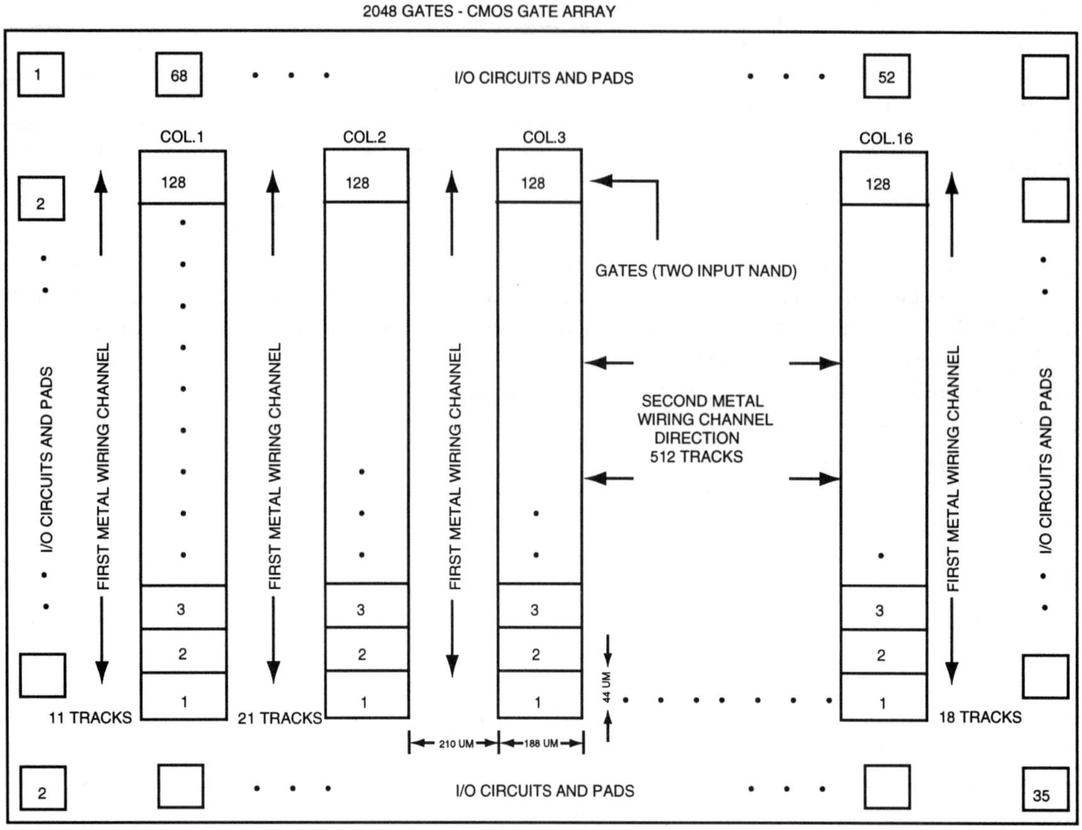

FIGURE 8.1 Schematic layout of a 2048-gates CMOS gate array.

where P is the sum of input and output pads, C is again the number of connections per gate, G is the number of gates in the array and a is Rent's exponent having a value between 0.5 and 0.7. For large-scale integrated (LSI) circuits, typically $a = 0.46$. Assuming this value for a, then P is equal to 134 for $G = 2048$ and $C = 3$. This gate array, therefore, with only 60 I/O pads should be pad limited for many designs.

As mentioned previously, the process of designing with gate arrays begins with the designer drawing, with the aid of a CAD system, the circuit schematic representing the function the chip must perform. This activity is called **schematic capture**. The schematic, containing circuit elements or *cells*, such as inverters, NAND and NOR gates, flip-flops, adders, etc., is formed by getting these components from a specified library in the computer. Each of the cells has a number of representations associated with it. It has a schematic used, of course, in drawing up the complete circuit schematic. It has a functional description that specifies what the element does, for example, an inverter takes an 1 input and produces a 0 in its output and vice versa. Another file describes its electrical characteristics, such as the time delay between when the inputs reach a certain state and the time the output responds to that state, often referred to as the propagation delay, and it has a physical representation.

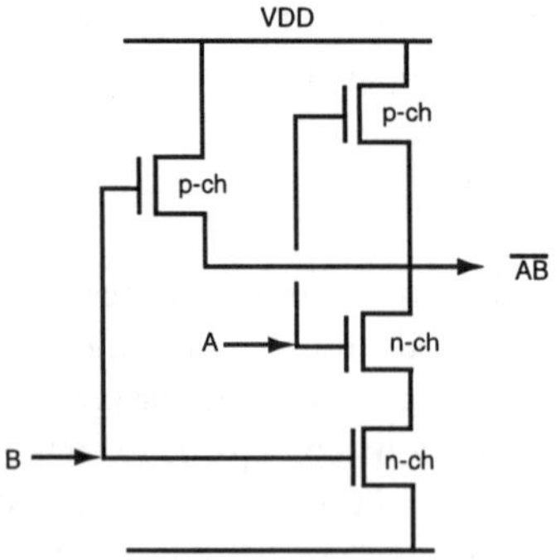

A	B	$\overline{AB}$
0	0	1
0	1	1
1	0	1
1	1	0

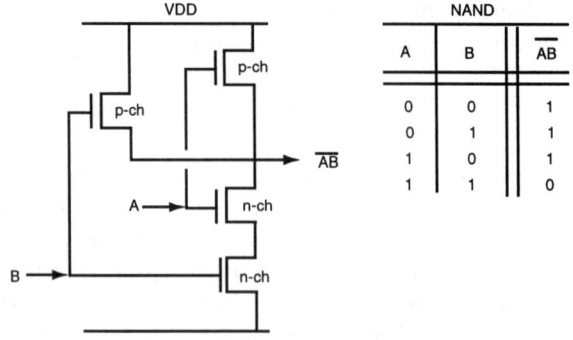

FIGURE 8.2 Electrical schematic of a CMOS NAND logic gate and its truth table.

In the array shown in Fig. 8.1, before the customized layers are placed on the device, each of

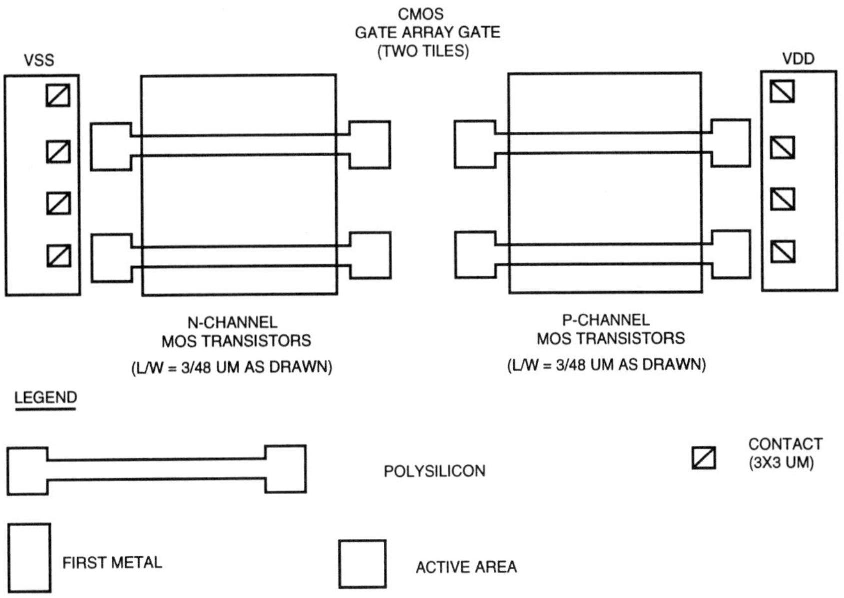

FIGURE 8.3 Schematic layout of one of the 2048 uncommitted gates in a device column.

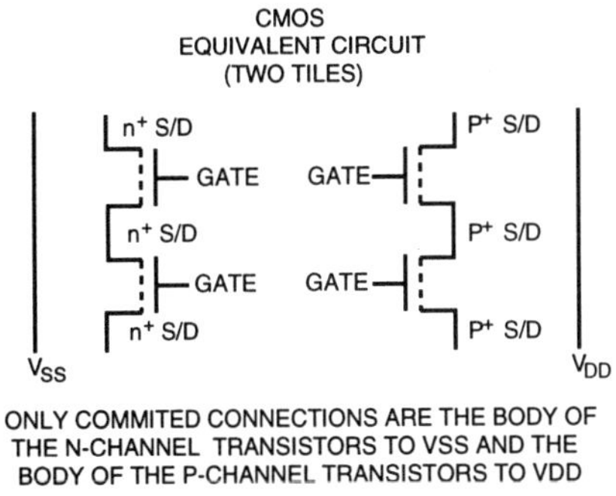

FIGURE 8.4 Equivalent electrical schematic of an uncommitted gate in the device column area of the gate array.

the 2048 gates appears schematically as shown in Fig. 8.3. There are two n-channel transistors on the left and two p-channel transistors on the right. There is also a ground or VSS buss line running vertically on the left and a VDD line on the right. Figure 8.4 shows the equivalent electrical schematic of the structure in Fig. 8.3. In Fig. 8.5 is shown only one pair of n- and p-channel transistors and locations within the device area where one of the customizing layers, the contacts, may be placed. The physical representation of a particular library cell contains information of how the uncommitted transistors in the device columns should be connected.

As an example, in Fig. 8.6 is shown a portion of one of the 16 device columns in the array. Figure 8.7 shows the location of the contact holes, shown as white squares, and the metal interconnects, for an inverter shown as the dark horizontal bars. An actual inverter circuit is realized by placing this inverter cell along

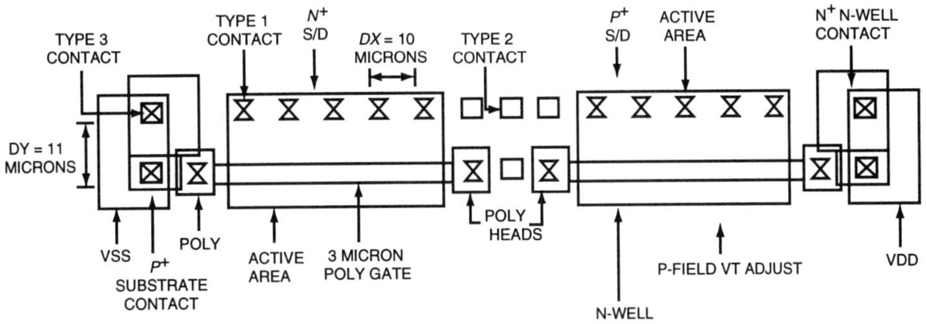

FIGURE 8.5 Layout of a single pair of n- and p-channel transistors in the device column showing how the contact locations are allocated.

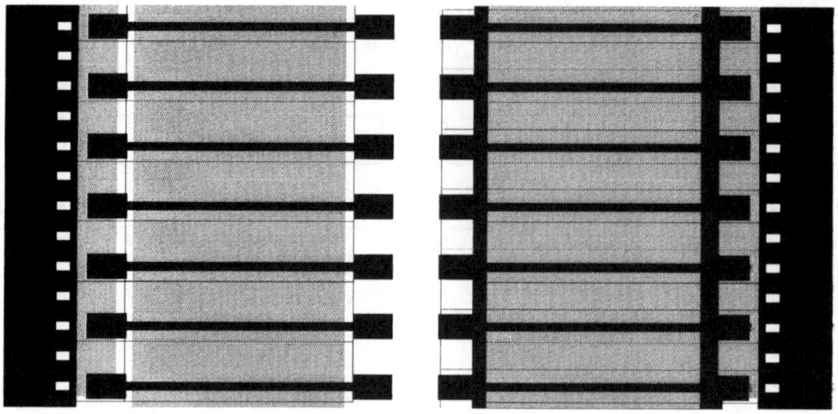

FIGURE 8.6 Layout of a portion of a device column.

FIGURE 8.7 Layout of a single inverter library cell.

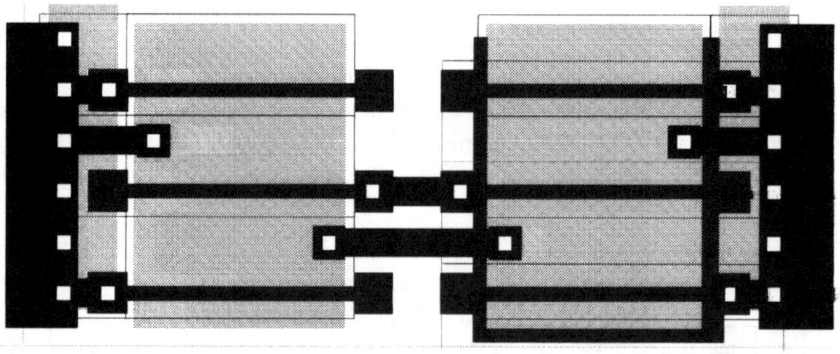

FIGURE 8.8 Completed layout of an inverter in device column.

anywhere in the column, as shown in Fig. 8.8. Figure 8.9 shows the electrical connections made in the gate array and in Fig. 8.9(b) is shown, for reference, the electric schematic for an inverter. Note that in Figs. 8.8 and 8.9, the polysilicon gates of the top and bottom n-channel transistors are connected to VSS and the corresponding gates of the the p-channel transistors are connected to VDD. This is done to isolate the inverter from interfering electrically with another logic gate that may be placed above or below it.

Returning to the design process, after schematic capture is completed, the designer simulates the entire circuit first to verify that it performs the logic functions the circuit is designed to perform and then using the electrical

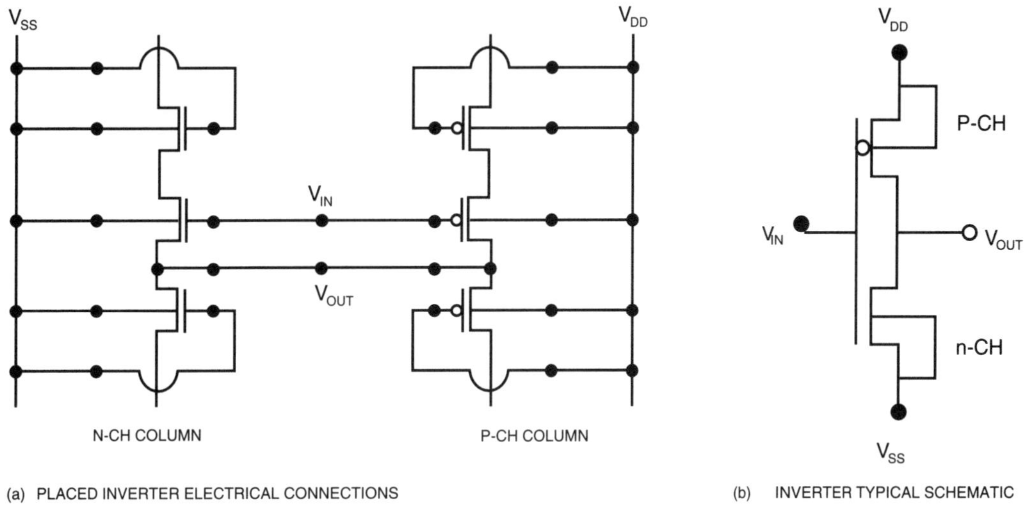

FIGURE 8.9 (a) Electrical connections of a placed inverter in a device column, (b) electrical schematic of an inverter.

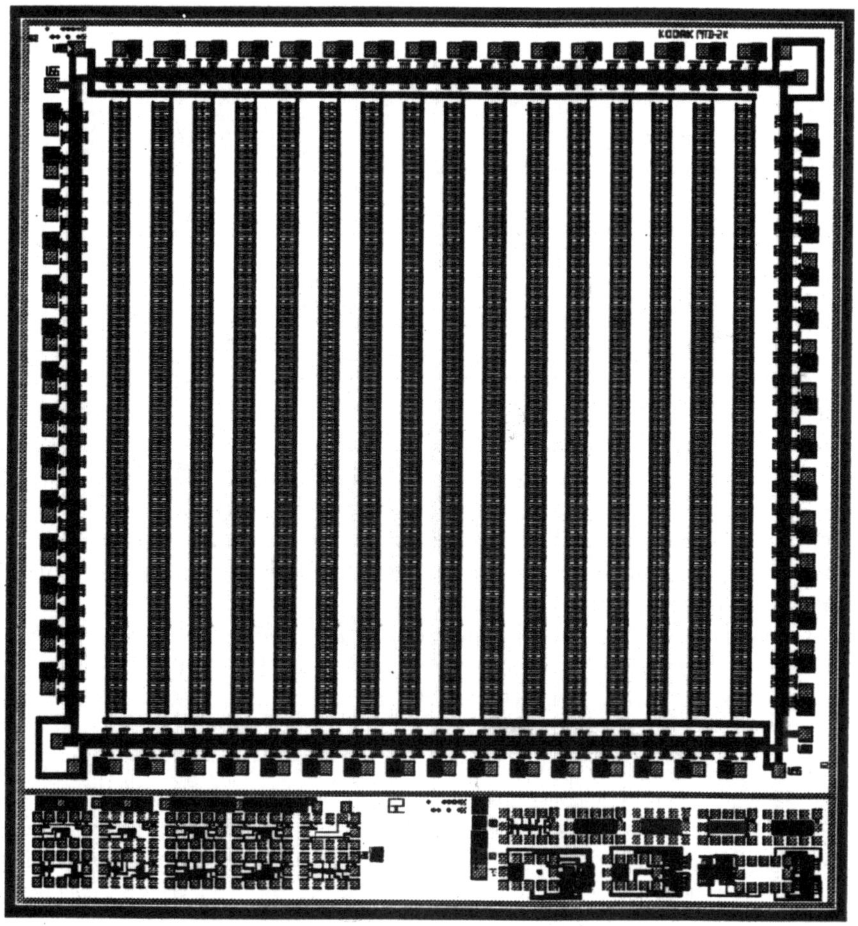

FIGURE 8.10 Layout of the actual 2048-gates gate array before customization.

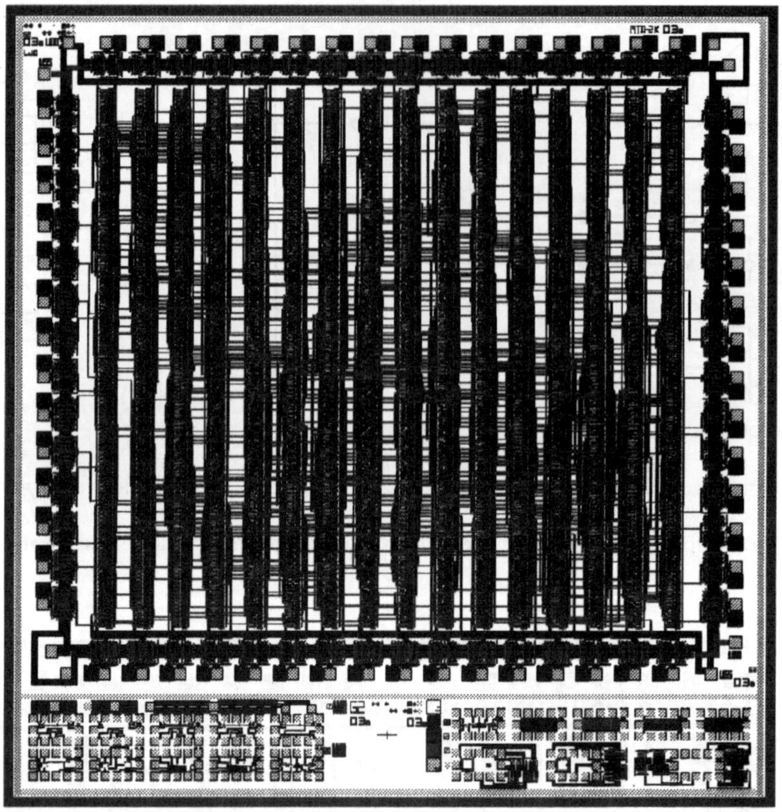

FIGURE 8.11 A custom circuit implementation on the 2048-gates gate array.

specifications files, a timing simulation is performed to make sure the circuit will operate at the clock frequency desired. However, the timing simulation is not yet complete. In the electrical characteristics file of each cell, a typical input and load capacitance is assumed. In an actual circuit the load capacitance may be quite different. Therefore, another timing simulation must be done after the actual capacitance values are found for each node.

To do this, first from the circuit schematic, a *netlist* is extracted. This list contains exact information about which node of a given element is connected to which nodes of other elements. This netlist is then submitted to the *place and route* program. This program places all of the elements or cells in the circuit on the gate array, in a fashion similar to what was done in Fig. 8.8, and attempts to connect them as specified in the netlist. The program may iterate the placement of cells and interconnection or routing until all cells listed in the netlist are placed on the gate array and interconnected as specified.

The routing portion of the place and route program is called a *channel router*. It derives its name by the method it uses to interconnect the cells in the array. Also, the gate array architecture, shown in Fig. 8.1, with its vertical wiring channels and a single horizontal wiring channel, is selected so that this type of router can be used. Figure 8.10 shows the actual 2048-gates gate array, before customization and Fig. 8.11 shows after it is completed with a custom circuit. In Fig. 8.12 a corner of the array is shown in higher magnification. The vertical darker lines are the first level metal interconnects and the horizontal lighter lines are the second level metal interconnects. This array requires four layers to be customized. These are the first and second level metals, contact, and the via. Vias are contacts exclusively between the two metal layers. Note that all internal wiring in the cells, like the inverter in Fig. 8.7, are made with first level metal and, therefore, the router is not allowed to route first level metal over the device areas.

Once the circuit is placed and routed, an extraction program is applied that does two things. First, it calculates the actual resistance and capacitance seen at each node of the circuit that is then fed back into the netlist for more accurate timing simulation; second, it extracts a new netlist of the circuit on the gate array that can be compared

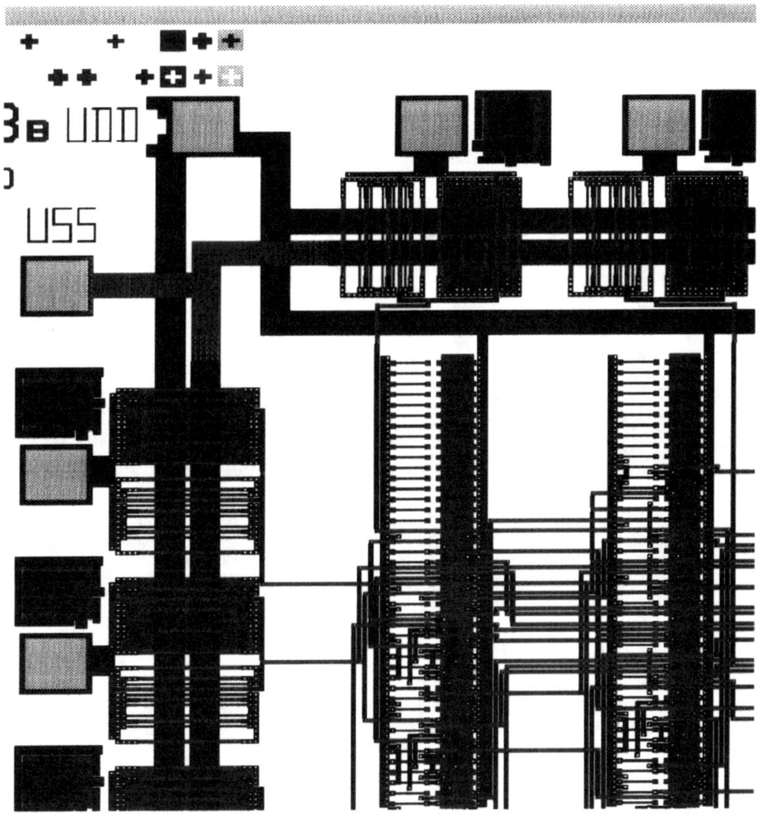

FIGURE 8.12 Closeup of a corner of the ASIC shown in Fig. 8.11.

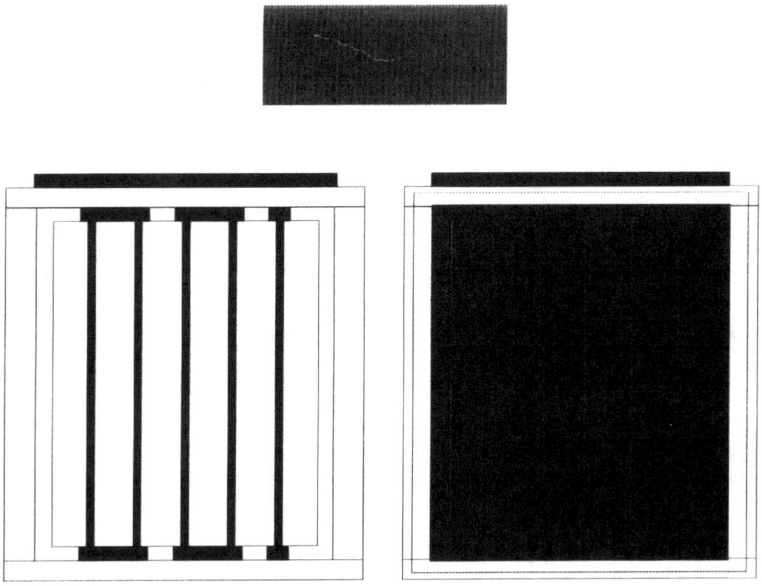

FIGURE 8.13 Layout of an uncommitted I/O buffer. In this figure, the bonding pad is the rectangular pattern at the top, the n-channel transistor is on the left and the p-channel is located on the right.

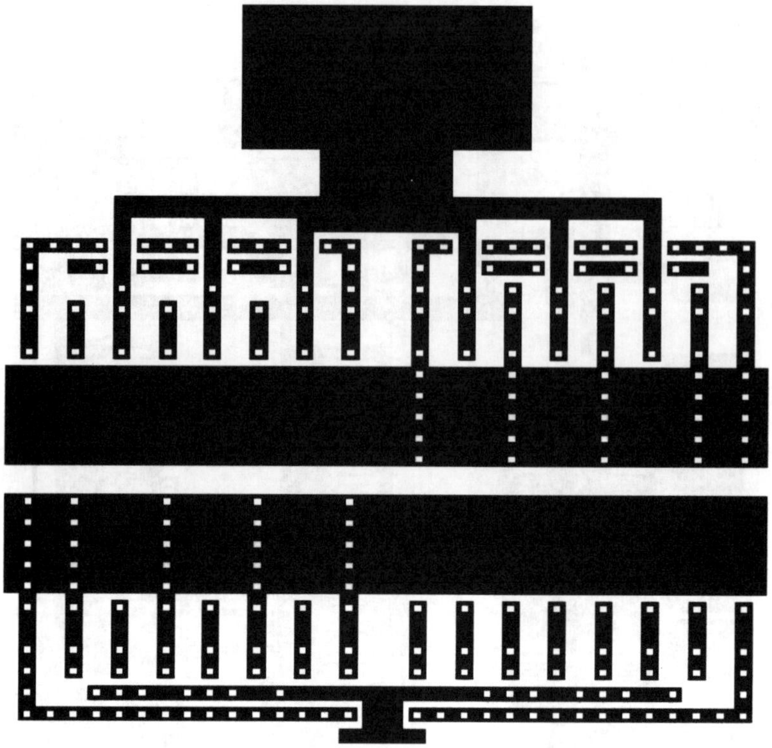

FIGURE 8.14 Layout of the contact and metal pattern of the output buffer library cell.

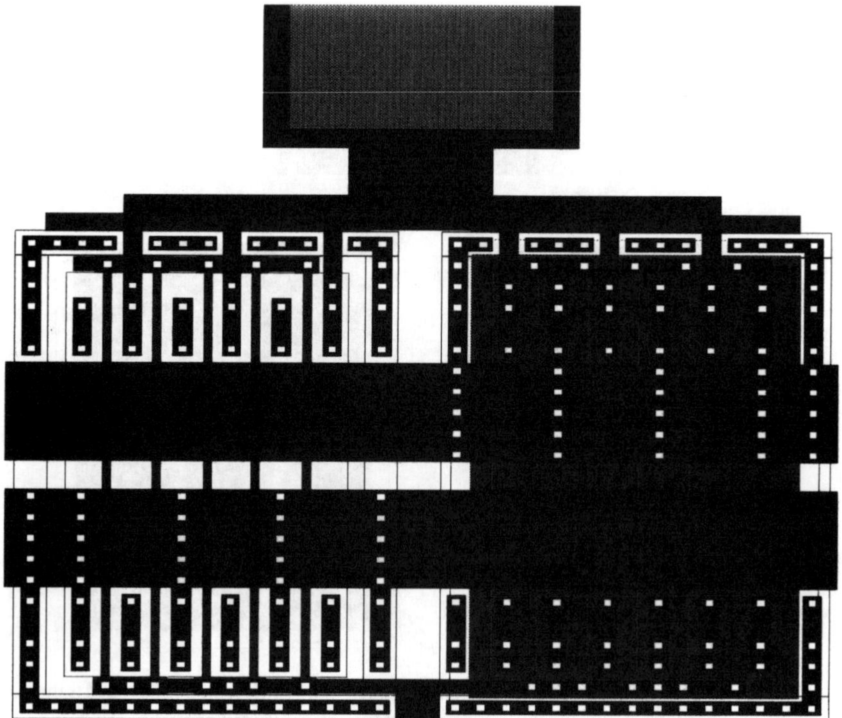

FIGURE 8.15 Completed layout of a full inverting output buffer.

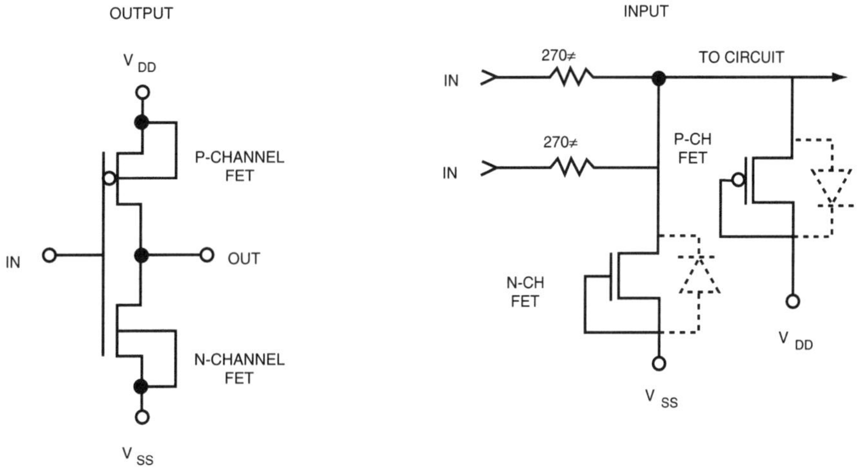

FIGURE 8.16 Electrical schematics of the input and output buffers. For TTL compatible output or input, or for other reasons, the buffers may use only some of the five segments of either the n- or p-channel transistors.

with the initial netlist. The two netlists should, of course, be identical. However, often manual intervention is required to finish either the placement or routing of a particularly difficult circuit and that could cause an inadvertent error to occur in the layout. The error is then detected by comparing this extracted netlist to the original one.

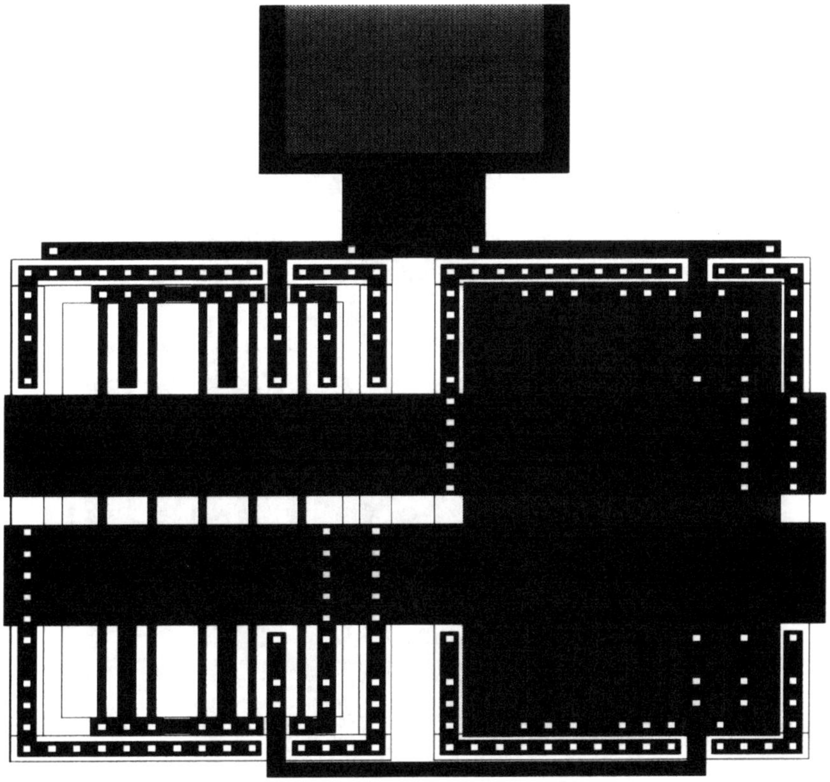

FIGURE 8.17 Completed layout of a simple noninverting input buffer.

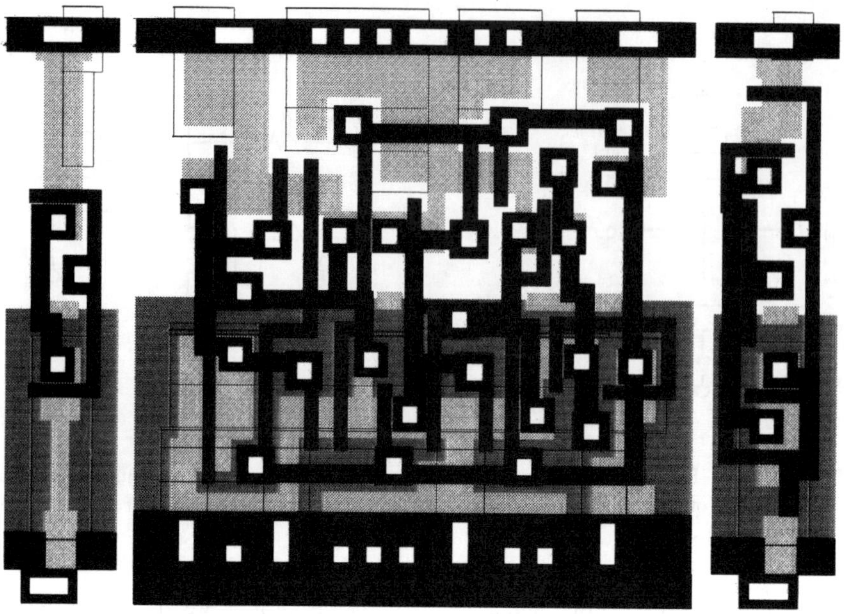

FIGURE 8.18 Three standard cells. A two-input NAND gate is shown on the left, a D type flip-flop is in the middle, and a simple inverter is shown on the right.

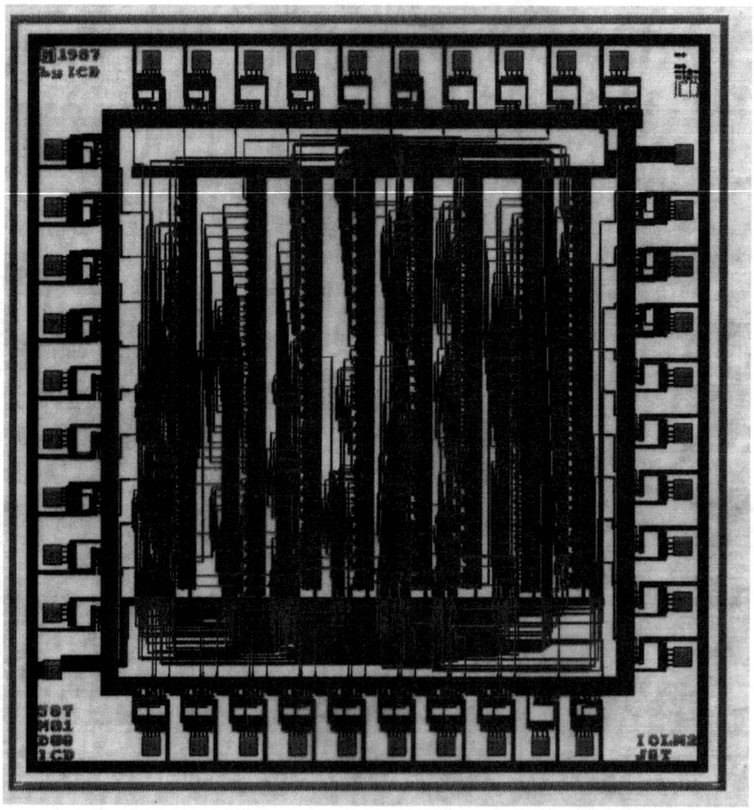

FIGURE 8.19 Layout of a completed standard cell ASIC.

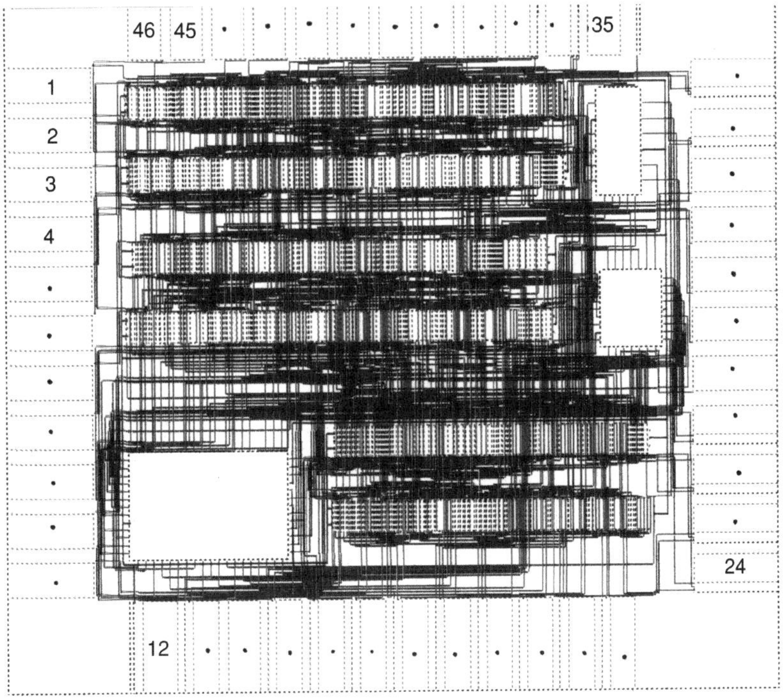

FIGURE 8.20 Layout of a digital functional block designed ASIC.

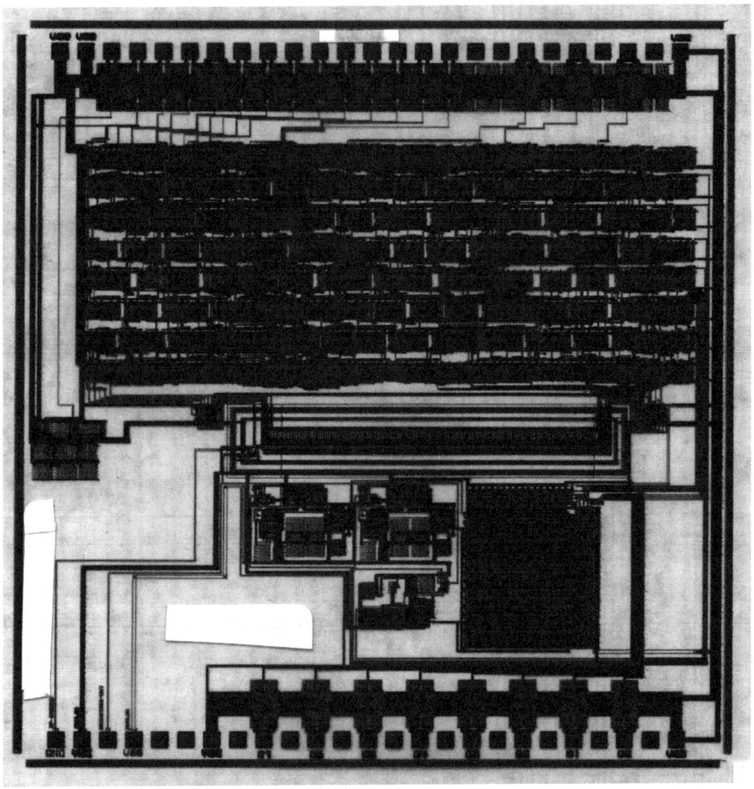

FIGURE 8.21 Layout of a mixed (combined analog and digital circuits on the same chip) functional block designed ASIC.

After the design is completed, **test vectors** must be generated. Two sets of test vectors are generated. The input set and the corresponding output set. Each test vector consists of a number of 1s and 0s. The input test vector has as many elements as the number of chip input pins and the output test vector, likewise, has as many elements as the number of output pins.

The input test vector is not, typically, similar to what the chip will see in actual operation. Instead, the goal is to select a set of input vectors that when applied to the chip will cause every internal node to change state at least once. The timing verification program is used for this task. The program keeps track of the nodes toggled, as each input vector is applied, and also saves the output vector. When the finished devices are received, the same input test vectors are applied in sequence, and the resulting output vectors are captured and compared to those obtained from the timing simulation runs. If the two match perfectly, the device is classified as good.

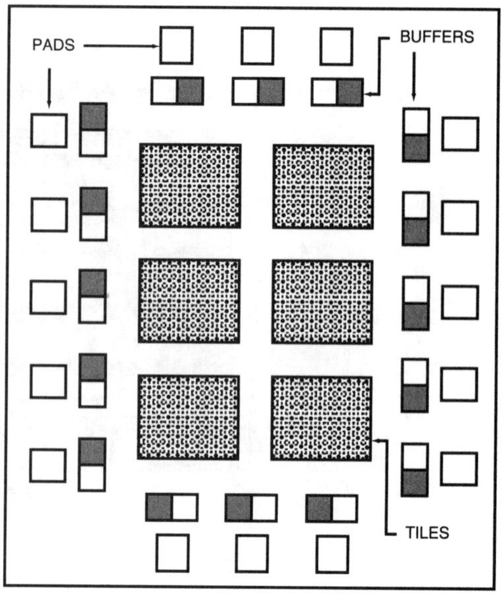

FIGURE 8.22 Architecture of an analog array.

The purpose of this test is to detect whether, because of a defect during fabrication, an internal gate is not operating properly. Usually if a test can cause about 90% of the internal nodes to switch state, it is considered to provide adequate coverage. It would be desirable to have 100% coverage, but often this is not practical because of the length of time it would take for the test to be completed. Finally, because the test is done at the frequency at which the chip will operate, only minimal functional testing is performed.

Just as in the device columns any logic gate can be placed anywhere in any of the columns, each of the pads can serve as either an input, an output, or a power pad, again by customizing the four layers. Figure 8.13 shows an I/O buffer at the end of the prefabrication process. Figure 8.14 shows the contact and metal pattern to make the uncommitted pad an inverting output buffer and Fig. 8.15 shows the two together. Figure 8.16 shows the circuit schematic for the output buffer as well as the schematic when the pad is customized as an input buffer, the layout of which is shown in Fig. 8.17.

The gate array architecture presented here is not the only one. A number of different designs have been developed and are currently used. For each of the architectures, a different place and route algorithm is then developed to take advantage of that particular design. Gate arrays have also taken advantage of developments in silicon technology, which primarily consists of shrinking of the minimum dimensions. As a result, arrays with several million transistors per chip are presently in use.

Standard Cells

Designing with gate arrays is fairly straightforward and is utilized extensively by system engineers. However, gate arrays are quite inefficient in silicon area use since the number of transistors, the number of routing tracks, and the number of I/O pads needed by a particular circuit does not always match the number of devices or pads on the array, which are, of course, fixed in number.

The standard cells design methodology was developed to address this problem. As was the case for gate arrays, a predesigned and characterized library of cells exists for the circuit designer to use. But, unlike gate arrays, the chip is layed out containing only the number of transistors and I/O pads specified by the particular circuit, resulting in a smaller, less expensive chip. Fabrication of a standard cells circuit, however, must start from the beginning of the process. This has two major drawbacks. First, the fabrication time for standard cells is several weeks longer than for gate arrays; second, the customer has to pay for the fabrication of the whole lot of wafers, whereas for gate arrays, only a small portion of the lot is processed for each customer. This substantially increases the cost of prototype parts.

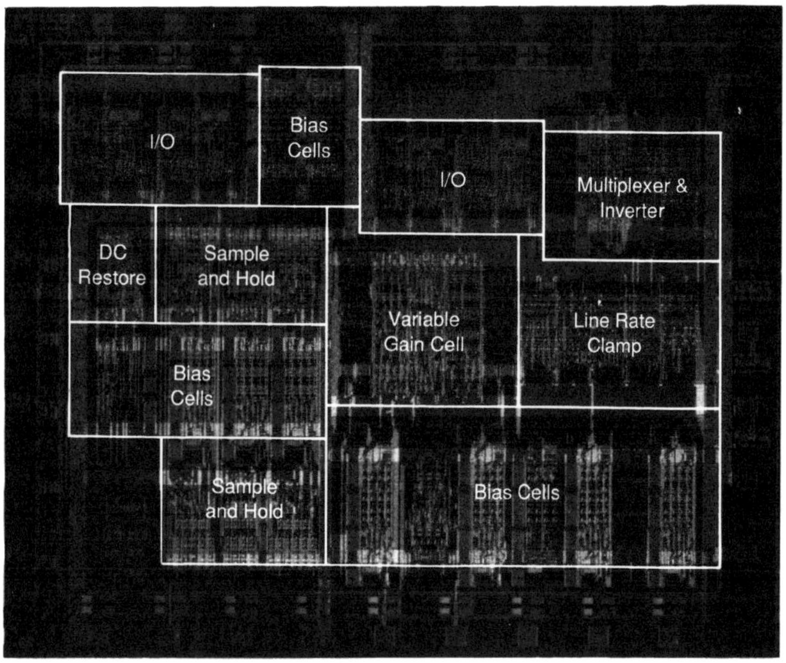

FIGURE 8.23 Example of an ASIC [Boisvert and Gaboury, 1992] implemented in a commercially available analog array.

Because standard cells are placed and routed automatically, the cells have to be designed with some restrictions, such as uniform height, preassigned locations for clocks and power lines, predefined input and output port locations, and others. Figure 8.18 shows the layout of three common cells from a standard cells library and Fig. 8.19 shows a completed standard cells chip. Its appearance is very similar to that of a gate array, but it contains no unused transistors or I/O pads and the minimum number of routing channels.

Functional Blocks

Advances in place and route software have made possible the functional blocks design methodology. Here the cells are no longer simple logic gates, but complete functions. The cells can be analog or digital and can have arbitrary sizes. Figure 8.20 shows an example of a digital functional block design. In the figure, the three white blocks are placed and routed automatically along with several rows of standard cells. In Fig. 8.21, a small linear charge-coupled device (CCD) image sensor, located in the center of the chip, and a few other analog circuits, located below it, are placed on the same chip with a block of standard cells that *C* produce all of the logic needed to run the chip. In this chip, care is taken as to both the placement of the various blocks and the routing because of the sensitivity of the analog circuits to noise, temperature, and other factors. Therefore, the designer had considerable input as to the placement of the various blocks and the routing, and the software is designed to allow this intervention, when needed.

Often the digital functional blocks are built by a class of software programs called **silicon compilers** or **synthesis tools.** With such tools circuit designers typically describe the functionality of their circuit at the behavioral level rather than the gate level, using specially developed computer languages called *hardware description languages* (HDL). These tools can quickly build memory blocks of arbitrary size, adders or multipliers of any length, PLAs, and many other circuits.

It should be obvious by now that the semicustom design approach makes heavy use of software tools. These tools eliminate the tedious and error prone task of hand layout, provide accurate circuit simulation, and help with issues such as testability. Their power and sophistication make transparent the complexity inherent in the design of integrated circuits.

Analog Arrays

Analog arrays are typically made in a bipolar process and are intended for the fabrication of high-performance analog circuits. Like gate arrays, these devices are prefabricated up to the contact level and, like all semicustom approaches, they are provided with a predesigned and characterized cell library. Because of the infinite variety of analog circuits, however, designers often design many of their own cells or modify the ones provided. Unlike digital circuits, the layout is done manually because analog circuits are more sensitive to temperature gradients, power supply voltage drops, crosstalk, and other factors and because the presently available software is not sufficiently sophisticated to take all of these factors into account.

The architecture of analog arrays is *tile* like, as shown in Fig. 8.22. Each tile is identical to all other tiles. Within each tile is contained a number of transistors, resistors, and capacitors of various sizes. The array also contains a fixed number of I/O pads, each of which can be customized with the last four layers to serve either as an input or output buffer. Figure 8.23 shows an actual circuit [Boisvert and Gaboury 1992] implemented in a commercially available analog array.

8.4 An ASIC Design Environment

The ASIC design environments consist of both software and hardware tools. Presently, UNIX workstations are the most common hardware platforms. Because of their increased capabilities and low cost, however, personal computers are also widely used for front end activities, such as schematic capture and documentation. More powerful servers, connected via ethernet or other local high-speed networks, do most of the compute intensive tasks, such as circuit simulation and place and route. An example of a typical ASIC design environment is shown in Fig. 8.24.

The major software tools needed for ASIC design are a layout editor for full custom designs, a place and route tool for semicustom circuits, a device and circuit simulator, like SPICE [Vladémirescu et al. 1981], for analog circuit simulation, logic and timing verification tools for digital circuit simulation, and synthesis tools for functional block design. Additionally, in silicon foundries there must exist software for design rule checking that verifies that the foundry's design rules have been adhered to in the design and layout of the chip and software that converts the layout data to the format required by the mask making tools. Finally, specialized hardware and software tools are needed to test the chips and verify that they contain no hidden faults. As mentioned earlier, for digital circuits, the test vectors that were produced at the time the design was finished are now used in these testers. For analog circuits the test procedures are not as well structured.

Rapid advances in the capabilities of both the software and hardware tools are currently taking place. The major effort is directed toward helping designers work

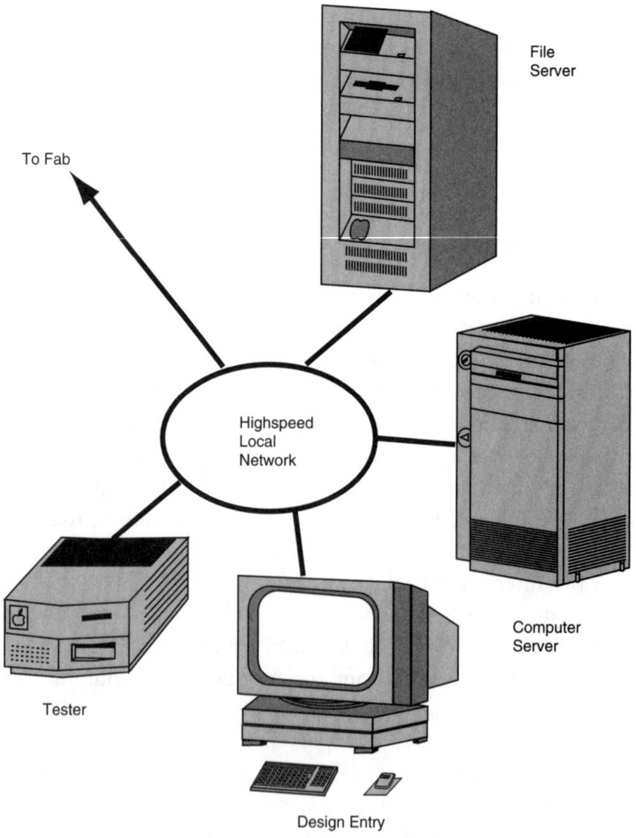

FIGURE 8.24 A typical ASIC design environment.

at yet higher levels of abstraction. Thus, instead of schematic capture, hardware description languages, such as VHDL, steer designers towards design capture. This allows digital ASICs to be designed by system engineer rather than expert ASIC designers. Furthermore, a chip captured with VHDL is independent of vendor-specific cell libraries and can be fabricated in different foundries, both of which are significant when chip cost and supply issues are important.

Defining Terms

ASICs: The acronym of *application specific integrated circuits*; another name for such chips is *custom integrated circuits*.

Functional blocks: ASICs designed using this methodology are more compact than either gate arrays or standard cells because the blocks can perform much more complex functions than do simple logic gates.

Gate arrays: Chips that contain uncommitted arrays of transistors and are prefabricated up to a certain step after which they are customized for the particular application.

Nonreccurring engineering (NRE): Costs the foundry charges the ASIC customer. These costs include engineering time, the cost of making the masks, the cost of fabricating one lot of wafers, and for packaging and testing the prototype parts.

Standard cells: A design methodology for realizing ASICs. Compared to gate arrays, they make more efficient use of silicon.

Schematic capture: The process by which the functionality of the chip is captured in an electrical schematic, usually with the aid of a computer, and using components from a cell library resident in that computer.

Silicon compilers, synthesis tools: Software programs that can construct an ASIC whose functionality is no longer described by a circuit schematic but in a special high level computer languages, generally called hardware description languages or HDL.

Test vectors: A test scheme that consists of pairs of input and output. Each input vector is a unique set of 1s and 0s applied to the chip inputs and the corresponding output vector is the set of 1s and 0s produced at each of the chip's output.

References

Boisvert, D.M. and Gaboury, M.J. 1992. An 8–10-bit, 1–40 MHz analog signal processor with configurable performance for electronic imaging applications. In *Proceedings IEEE International ASIC Conference and Exhibit*, pp. 396–400. Rochester, NY.

Fey, C.F. and Paraskevopoulos, D. 1985. Selection of cost effective LSI design methodologies. In *Proceedings of the IEEE Custom Integrated Circuits Conference*, pp. 148–153. Portland, OR.

Fey, C.F. and Paraskevopoulos, D. 1986. A model of design schedules for application specific ICs. In *Proceedings IEEE Custom Integrated Circuits Conference*, pp. 490–496. Rochester, NY.

Fier, D.F. and Heikkila, W.W. 1982. High performance CMOS design methodologies. In *Proceedings IEEE Custom Integrated Circuits Conference*, pp. 325–328. Rochester, NY.

Hu, C. 1992. IC reliability simulation. *IEEE J. of Solid State Circuits* 27(3):241–246; see also Proceedings of the Annual International Reliability Physics Symposium.

Ting, G., Guidash, R.M., Lee, P.P.K., and Anagnostopoulos, C. 1994. A low-cost, smart-power BiCMOS driver chip for medium power applications. In *Proceedings IEEE International ASIC Conference and Exhibit*, pp. 466–469. Rochester, NY.

Vladimirescu, A. et al. 1981. *SPICE Manual.* Dept. of Electrical Engineering and Computer Sciences, Univ. of California, Berkeley, CA, Oct.

Williams, T.W. and Mercer, M.R. 1993. Testing digital circuits and design for testability. In *Proceedings IEEE International ASIC Conference and Exhibit* (Tutorial Session), p. 10. Rochester, NY.

Further Information

The ASIC field has been expanding rapidly since about 1980. The best source of current information is the two major conferences where the more recent technology developments are reported. These are the IEEE Custom Integrated Circuits Conference (CICC) held annually in May and the IEEE International ASIC Conference and Exhibit, held annually in September. *IEEE Spectrum Magazine* lists these conferences in the calender of events. Apart from the regular technical sessions, educational sessions and exhibits by ASIC vendors are part of these two conferences.

An additional resource is the USC/Information Sciences Institute, 4676 Admiralty Way, Marina del Rey, CA 90292-6695, Telephone (213) 822-1511. See also advertisements in the IEEE Spectrum and Semiconductor industry magazines or newspapers.

Finally, a selected number of papers from the CICC are published every year, since 1984, in special issues of the *IEEE Journal of Solid State Circuits.*

Additional information can also be found in a number of other conferences and their proceedings, including the Design Automation Conference (DAC) usually held in June and the International Solid States Circuits Conference (ISSCC) usually held in February.

9

Digital Filters

Jonathon A. Chambers
Imperial College of Science, Technology, and Medicine

Sawasd Tantaratana
National Electronics and Computer Technology Center, Thailand

Bruce W. Bomar
The University of Tennessee Space Institute

9.1 Introduction

Digital filtering is concerned with the manipulation of discrete data sequences to remove noise, extract information, change the sample rate, and perform other functions. Although an infinite number of numerical manipulations can be applied to discrete data (e.g., finding the mean value, forming a histogram), the objective of digital filtering is to form a discrete output sequence $y(n)$ from a discrete input sequence $x(n)$. In some manner or another, each output sample is computed from the input sequence—not just from any one sample, but from many, in fact, possibly from all of the input samples. Those filters that compute their output from the present input and a finite number of past inputs are termed *finite impulse response* (FIR), whereas those that use all past inputs are *infinite impulse response* (IIR). This chapter will consider the design and realization of both FIR and IIR digital filters and will examine the effect of finite wordlength arithmetic on implementing these filters.

9.2 FIR Filters

A finite impulse response filter is a linear discrete time system that forms its output as the weighted sum of the most recent, and a finite number of past, inputs. Time-invariant FIR filters have finite memory, and their impulse response, namely, their response to a discrete input that is unity at the first sample and otherwise zero, matches the fixed weighting coefficients of the filter. Time-variant FIR filters, on the other hand, may operate at various sampling rates and/or have weighting coefficients that adapt in sympathy with some statistical property of the environment in which they are applied.

Fundamentals

Perhaps the simplest example of an FIR filter is the moving average operation described by the following linear constant-coefficient difference equation:

$$y(n) = \sum_{k=0}^{M} b_k x(n-k) \quad b_k = \frac{1}{M+1}$$

0-8493-0050-9/00/$0.00+$.50

where

$y(n)$ = output of the filter at integer sample index n
$x(n)$ = input to the filter at integer sample index n
b_k = filter weighting coefficients, $k = 0, 1, \ldots, M$
M = filter order

In a practical application, the input and output discrete-time signals will be sampled at some regular sampling time interval, T seconds, denoted $x(nT)$ and $y(nT)$, which is related to the sampling frequency by $f_s = 1/T$ samples per second. For generality, however, it is more convenient to assume that T is unity, so that the effective sampling frequency is also unity and the Nyquist frequency [Oppenheim and Schafer 1989], namely, the maximum analog frequency that when sampled at f_s will not yield an aliasing distortion, is one-half. It is then straightforward to scale, by multiplication, this normalized frequency range, that is, (0, 1/2), to any other sampling frequency.

The output of the simple moving average filter is the average of the $M + 1$ most recent values of $x(n)$. Intuitively, this corresponds to a smoothed version of the input, but its operation is more appropriately described by calculating the frequency response of the filter. First, however, the z-domain representation of the filter is introduced in analogy to the s- (or Laplace-) domain representation of analog filters. The z transform of a causal discrete-time signal $x(n)$ is defined by

$$X(z) = \sum_{n=0}^{\infty} x(n) z^{-n}$$

where

$X(z)$ = z transform of $x(n)$
z = complex variable

The z transform of a delayed version of $x(n)$, namely, $x(n - k)$ with k a positive integer, is found to be given by $z^{-k}X(z)$. This result can be used to relate the z transform of the output $y(n)$ of the simple moving average filter to its input

$$Y(z) = \sum_{k=0}^{M} b_k z^{-k} X(z) \qquad b_k = \frac{1}{M+1}$$

The z-domain transfer function, namely, the ratio of the output to input transform, becomes

$$H(z) = \frac{Y(z)}{X(z)} = \sum_{k=0}^{M} b_k z^{-k} \qquad b_k = \frac{1}{M+1}$$

Notice the transfer function $H(z)$ is entirely defined by the values of the weighting coefficients b_k, $k = 0, 1, \ldots, M$, which are identical to the discrete impulse response of the filter, and the complex variable z. The finite length of the discrete impulse response means that the transient response of the filter will only last for $M + 1$ samples, after which steady state will be reached. The frequency domain transfer function for the filter is found by setting $z = e^{j2\pi f}$, where $j = \sqrt{-1}$, and can be written as

$$H(e^{j2\pi f}) = \frac{1}{M+1} \sum_{k=0}^{M} e^{-j2\pi f k} = \frac{1}{M+1} e^{-j\pi f M} \frac{\sin[\pi f(M+1)]}{\sin(\pi f)}$$

The magnitude and phase response of the simple moving average filter, with $M = 7$, are calculated from $H(e^{j2\pi f})$ and shown in Fig. 9.1. The filter is seen clearly to act as a crude low-pass, smoothing filter with a **linear-phase** response. The sampling frequency periodicity in the magnitude and phase response is a property of discrete-time systems. The linear-phase response is due to the $e^{-j\pi f M}$ term in $H(e^{j2\pi f})$ and corresponds to a constant $M/2$ group delay through the filter. A phase discontinuity of +/− 180° is introduced each time the magnitude term

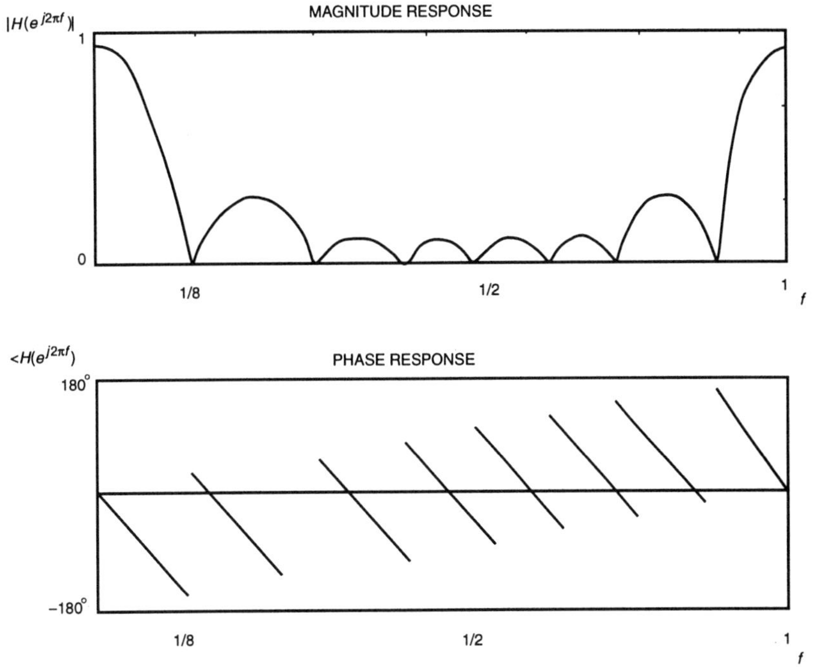

FIGURE 9.1 The magnitude and phase response of the simple moving average filter with $M = 7$.

changes sign. FIR filters that have center symmetry in their weighting coefficients have this constant, frequency independent group delay property that is very desirable in applications in which time dispersion is to be avoided, for example, in pulse transmission where it is important to preserve pulse shapes [Lee and Messerschmit 1994].

Another useful interpretation of the z-domain transfer function is obtained by re-writing $H(z)$ as follows:

$$H(z) = \frac{\sum_{k=0}^{M} b_k z^{M-k}}{z^M} = \frac{b_0 z^M + b_1 z^{M-1} + \cdots + b_{M-1} z + b_M}{z^M} = \frac{N(z)}{D(z)}$$

The z-domain transfer function is shown to be the ratio of two Mth-order polynomials in z, namely, $N(z)$ and $D(z)$. The values of z for which $N(z) = 0$ are termed the zeros of the filter, whereas those for which $D(z) = 0$ are the poles. The poles of such an FIR filter are at the origin, that is, $z = 0$, in the z plane. The positions of the zeros are determined by the weighting coefficients, that is, b_k, $k = 0, 1, \ldots, M$. The poles and zeros in the z plane for the simple moving average filter are shown in Fig. 9.2. The zeros, marked with a circle, are coincident with the unit circle, that is, the contour in the z plane for which $|z| = 1$, and match exactly the zeros in the magnitude response, hence their name; the discontinuities in the phase response are shown in Fig. 9.1. The zeros of an FIR filter may lie anywhere in the z plane because they do not impact on the stability of the filter; however, if the weighting coefficients are real and symmetric, or anti-symmetric, about their center value $M/2$, any complex zeros of the filter are constrained to lie as conjugate pairs coincident with the unit circle or as quartets of roots off the unit circle with the form $[\rho e^{j\theta}, \rho e^{-j\theta}, (1/\rho)e^{j\theta}, (1/\rho)e^{-j\theta}]$ where ρ and θ are, respectively, the radius and angle of the first zero. Zeros that lie within the unit circle are termed *minimum phase*, whereas those which lie outside the unit circle are called *maximum phase*. This distinction describes the contribution made by a particular zero to the overall phase response of a filter. A minimum-phase FIR filter that has all its zeros within the unit circle can have identical magnitude response to a maximum-phase FIR that has all its zeros outside the unit circle for the special case when they have an equal number of zeros with identical angles but reciprocal radii. An example of this would be second-order FIR filters with z-domain transfer functions $H_{\min}(z) = 1 + 0.5z^{-1} + 0.25z^{-2}$ and $H_{\max}(z) = 0.25 + 0.5z^{-1} + z^{-2}$. Notice that the center symmetry in the coefficients is lost, but the minimum- and maximum-phase weighting coefficients are simply reversed. Physically, a minimum-phase FIR filter corresponds

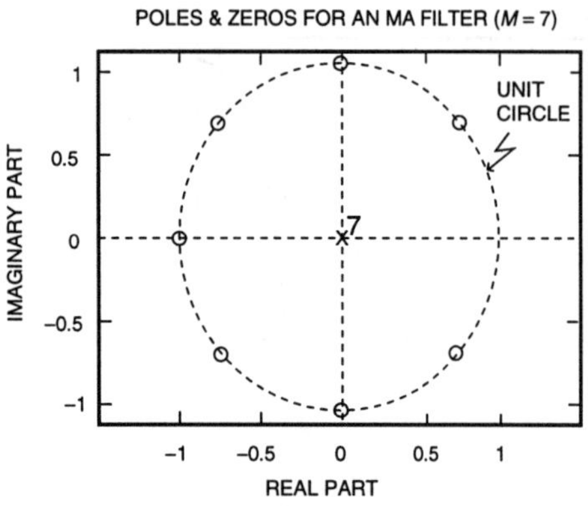

FIGURE 9.2 The pole-zero plot for the simple moving average filter with $M = 7$.

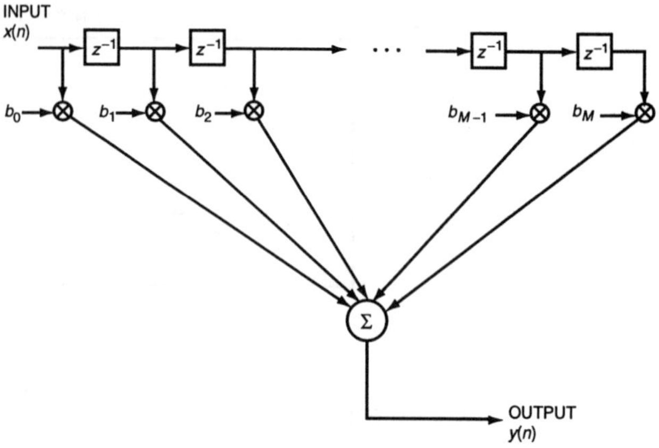

FIGURE 9.3 Direct form structure of an FIR filter.

to a system for which the energy is rapidly transferred from its input to its output, hence, the large initial weighting coefficients; whereas a maximum-phase FIR filter is slower to transfer the energy from its input to output; as such, its larger coefficients are delayed. Such FIR filters are commonly used for modeling multipath mobile communication transmission paths.

The characteristics of the frequency response of an FIR filter are entirely defined by the values of the weighting coefficients b_k, $k = 0, 1, \ldots, M$, which match the impulse response of the filter, and the order M. Various techniques are available for designing these coefficients to meet the specifications of some application. However, next consider the structures available for realizing an FIR filter.

Structures

The structure of an FIR filter must realize the z-domain transfer function given by

$$H(z) = \sum_{k=0}^{M} b_k z^{-k}$$

where z^{-1} is a unit delay operator. The building blocks for such filters are, therefore, adders, multipliers, and unit delay elements. Such elements do not have the disadvantages of analog components such as capacitors, inductors, operational amplifiers, and resistors, which vary with temperature and age. The direct or tapped-delay line form, as shown in Fig. 9.3, is the most straightforward realization of an FIR filter. The input $x(n)$ is delayed, scaled by the weighting coefficients b_k, $k = 0, 1, \ldots, M$, and accumulated to yield the output.

An equivalent, but transposed, structure of an FIR filter is shown in Fig. 9.4, which is much more modular and well suited to integrated circuit realization. Each module within the structure calculates a partial sum so that only a single addition is calculated at the output stage.

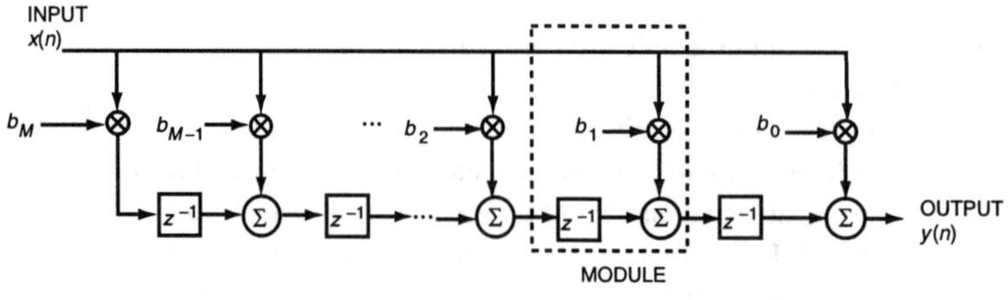

FIGURE 9.4 Modular structure of an FIR filter.

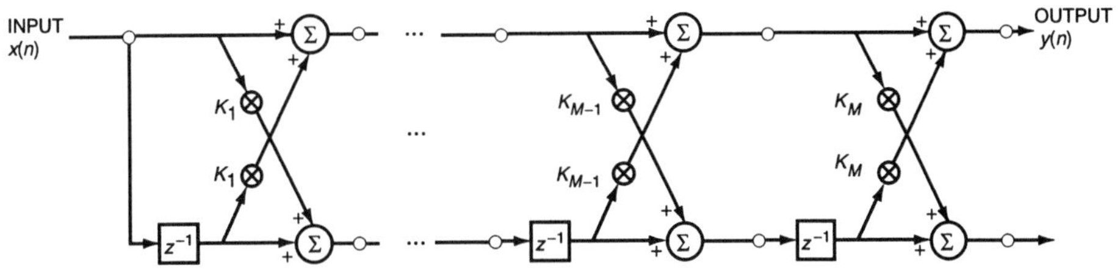

FIGURE 9.5 Lattice structure of an FIR filter.

Other structures are possible that exploit symmetries or redundancies in the filter weighting coefficients. For example, FIR filters with linear phase have symmetry about their center coefficient $M/2$; approximately one-half of the coefficients of an odd length, that is, M even, half-band FIR filter, namely, a filter that ideally passes frequencies for $0 \leq f \leq 1/4$ and stops frequencies $1/4 < f \leq 1/2$, are zero. Finally, a lattice structure as shown in Fig. 9.5 may be used to realize an FIR filter. The multiplier coefficients within the lattice k_j, $j = 1, \ldots, M$, are not identical to the weighting coefficients of the other FIR filter structures but can be found by an iterative procedure. The attraction of the lattice structure is that it is staightforward to test whether all its zeros lie within the unit circle, namely, the minimum-phase property, and it has low sensitivity to quantization errors. These properties have motivated the use of lattice structures in speech coding applications [Rabiner and Schafer 1978].

Design Techniques

Linear-phase FIR filters can be designed to meet various filter specifications, such as low-pass, high-pass, bandpass and bandstop filtering. For a low-pass filter, two frequencies are required, namely, the maximum frequency of the pass band below which the magnitude response of the filter is approximately unity, denoted the pass-band corner frequency f_p, and the minimum frequency of the stop band above which the magnitude response of the filter must be less than some prescribed level, named the stop-band corner frequency f_s. The difference between the pass- and stop-band corner frequencies is the *transition-bandwidth*. Generally, the order of FIR filter M required to meet some design specification will increase with a reduction in the width of the transition band. There are three established techniques for coefficient design:

- Windowing
- Frequency sampling
- Optimal approximations

The windowing design method calculates the weighting coefficients by sampling the ideal impulse response of an analog filter and multiplying these values by a smoothing window to improve the overall frequency domain response of the filter. The frequency sampling technique samples the ideal frequency domain specification of the filter and calculates the weighting coefficients by inverse transforming these values. However, better results can generally be obtained with the optimal approximations method and with the increasing availability of desktop and portable computers with fast microprocessors, large quantities of memory and sophisticated software packages, such as the matrix laboratory (MATLAB®) with its own dedicated signal processing toolbox; it is the preferred method for weighting coefficient design. The impulse response and magnitude response for a 40th-order optimal half-band FIR low-pass filter designed with the Parks–McClellan algorithm are shown in Fig. 9.6 together with the ideal frequency domain design specification. Notice the zeros in the impulse response.

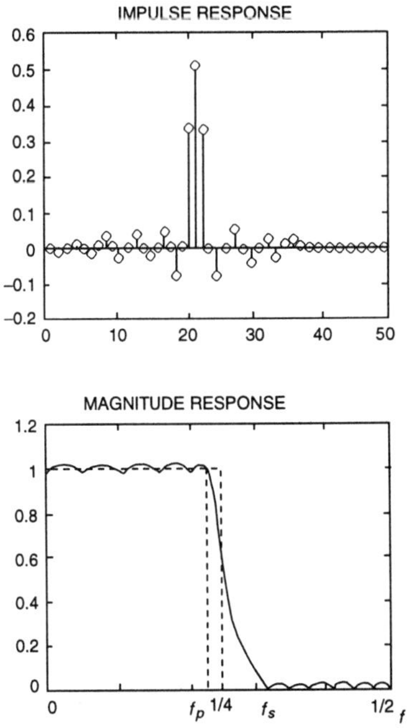

FIGURE 9.6 Impulse and magnitude response of an optimal 40th-order half-band FIR filter.

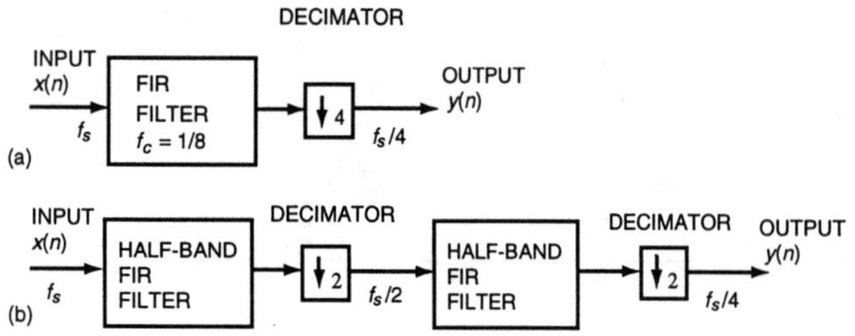

FIGURE 9.7 **Multirate FIR** low-pass **filter** structures: (a) method 1, (b) method 2.

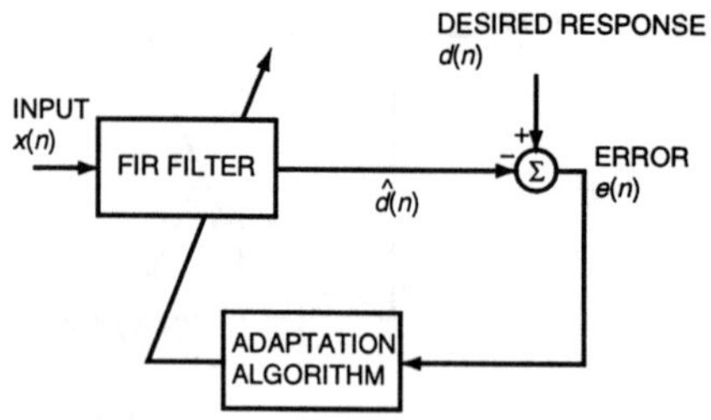

FIGURE 9.8 Adaptive FIR filter structure.

This algorithm minimizes the maximum deviation of the **magnitude response** of the design filter from the ideal magnitude response. The magnitude response of the design filter alternates about the desired specification within the pass band and above the specification in the stop band. The maximum deviation from the desired specification is equalized across the pass and stop bands; this is characteristic of an optimal solution.

The optimal approximation approach can also be used to design *discrete-time differentiators* and *hilbert transformers*, namely, phase shifters. Such filters find extensive application in digital modulation schemes.

Multirate and Adaptive FIR Filters

FIR filter structures can be combined very usefully with multirate processing elements. Such a filtering scheme is, however, time variant and as such may introduce some distortion during processing, but with careful design this can be minimized. Consider the design of a low-pass filter with $f_c = 1/8$; frequencies above this value are essentially to be eliminated. The frequency range between f_c and the Nyquist frequency, that is, 1/2, therefore contains no useful information. The sampling frequency could, provided no aliasing distortion is introduced, therefore be usefully reduced by a factor of 4. Further processing, possibly adaptive, of the signal could then proceed at this reduced rate, hence, reducing the demands on the system design. This operation can be achieved in two ways: A time-invariant low-pass FIR filter, with $f_c = 1/8$, can be designed to operate at the original sampling rate, and its output can be down sampled, termed decimated, by a factor of four; or, more efficiently, the filtering can be performed in two stages, using the same simple half-band filter design twice, each operating at a lower sampling rate. These methods are shown diagrammatically in Fig. 9.7. The second scheme has considerable computational advantages and, because of the nature of half-band filters, it is also possible to move the decimators in front of the filters. With the introduction of modulation operations it is possible to use the same approach to achieve high-pass and band-pass filtering.

The basic structure of an **adaptive FIR filter** is shown in Fig. 9.8. The input to the adaptive filter $x(n)$ is used to make an estimate $\cap d(n)$ of the desired response $d(n)$. The difference between these quantities $e(n)$, is used by the adaptation algorithm to control the weighting coefficients of the FIR filter. The derivation of the desired response signal depends on the application; in channel equalization, as necessary for reliable communication with data modems, it is typically a training sequence known to the receiver. The weighting coefficients are adjusted to minimize some function of the error, such as the mean square value. The most commonly used adaptive algorithm is the least mean square (lms) algorithm that approximately minimizes the mean square error. Such filters have the ability to adapt to time-varying environments where fixed filters are inappropriate.

Applications

The absence of drift in the characteristics of digitally implemented FIR filters, their reproducability, multirate realizations, and their ability to adapt to time-varying environments has meant that they have found many applications, particularly in telecommunications, for example, in receiver and transmitter design, speech compression and coding, and channel multiplexing. The primary advantages of fixed coefficient FIR filters are their unconditional stability due to the lack of feedback within their structure and their exact linear-phase characteristics.

Nonetheless, for applications that require sharp, selective, filtering in standard form, they do require relatively large orders. For some applications, this may be prohibitive and, therefore, recursive; IIR filters are a valuable alternative.

9.3 Infinite Impulse Response (IIR) Filters

A digital filter with impulse response having infinite length is called an *infinite impulse response* (IIR) filter. An important class of IIR filters can be described by the difference equation

$$y(n) = b_0x(n) + b_1x(n-1) + \cdots + b_Mx(n-M) - a_1y(n-1) - a_2y(n-2) - \cdots - a_Ny(n-N) \quad (9.1)$$

where $x(n)$ is the input, $y(n)$ is the output of the filter, and $\{a_1, a_2, \ldots, a_N\}$ and $\{b_0, b_1, \ldots, b_M\}$ are real-value coefficients.

We denote the impulse response by $h(n)$, which is the output of the system when it is driven by a unit impulse at $n = 0$, with the system being initially at rest. The system function $H(z)$ is the z transform of $h(n)$. For the system in Eq. (9.1), it is given by

$$H(z) = \frac{Y(z)}{X(z)} = \frac{b_0 + b_1z^{-1} + \cdots + b_Mz^{-M}}{1 + a_1z^{-1} + a_2z^{-2} + \cdots + a_Nz^{-N}} \quad (9.2)$$

where N is called the filter order. Equation (9.2) can be put in the form of poles and zeros as

$$H(z) = b_0z^{N-M}\frac{(z-q_1)(z-q_2)\cdots(z-q_M)}{(z-p_1)(z-p_2)\cdots(z-p_N)} \quad (9.3)$$

The poles are at $p_1, p_2, \ldots, p_N$. The zeros are at $q_1, q_2, \ldots, q_M$, as well as $N-M$ zeros at the origin.

The frequency response of the IIR filter is the value of the system function evaluated on the unit circle on the complex plane, that is, with $z = e^{j2\pi f}$, where f varies from 0 to 1, or from $-1/2$ to $1/2$. The variable f represents the digital frequency. For simplicity, we write $H(f)$ for $H(z)|_{z=\exp(j2\pi f)}$. Therefore,

$$H(f) = b_0e^{j2\pi(N-M)f}\frac{(e^{j2\pi f}-q_1)(e^{j2\pi f}-q_2)\cdots(e^{j2\pi f}-q_M)}{(e^{j2\pi f}-p_1)(e^{j2\pi f}-p_2)\cdots(e^{j2\pi f}-p_N)} \quad (9.4)$$

$$= |H(f)|e^{j\theta(f)} \quad (9.5)$$

where $|H(f)|$ is the magnitude response and $\theta(f)$ is the phase response.

Compared to an FIR filter, an IIR filter requires a much lower order than an FIR filter to achieve the same requirement of the magnitude response. However, whereas an FIR filter is always stable, an IIR filter can be unstable if the coefficients are not properly chosen. Assuming that the system (9.1) is causal, then it is stable if all of the poles lie inside the unit circle on the z plane. Since the phase of a stable causal IIR filter cannot be made linear, FIR filters are chosen over IIR filters in applications where linear phase is essential.

Realizations

Equation (9.1) suggests a realization of an IIR filter as shown in Fig. 9.9(a), which is called direct form I. By rearranging the structure, we can obtain direct form II, as shown in Fig. 9.9(b). Through transposition, we can obtain transposed direct form I and transposed direct form II, as shown in Fig 9.9(c) and 9.9(d).

The system function can be put in the form

$$H(z) = \prod_{i=0}^{K}\frac{b_{i0} + b_{i1}z^{-1} + b_{i2}z^{-2}}{1 + a_{i1}z^{-1} + a_{i2}z^{-2}} \quad (9.6)$$

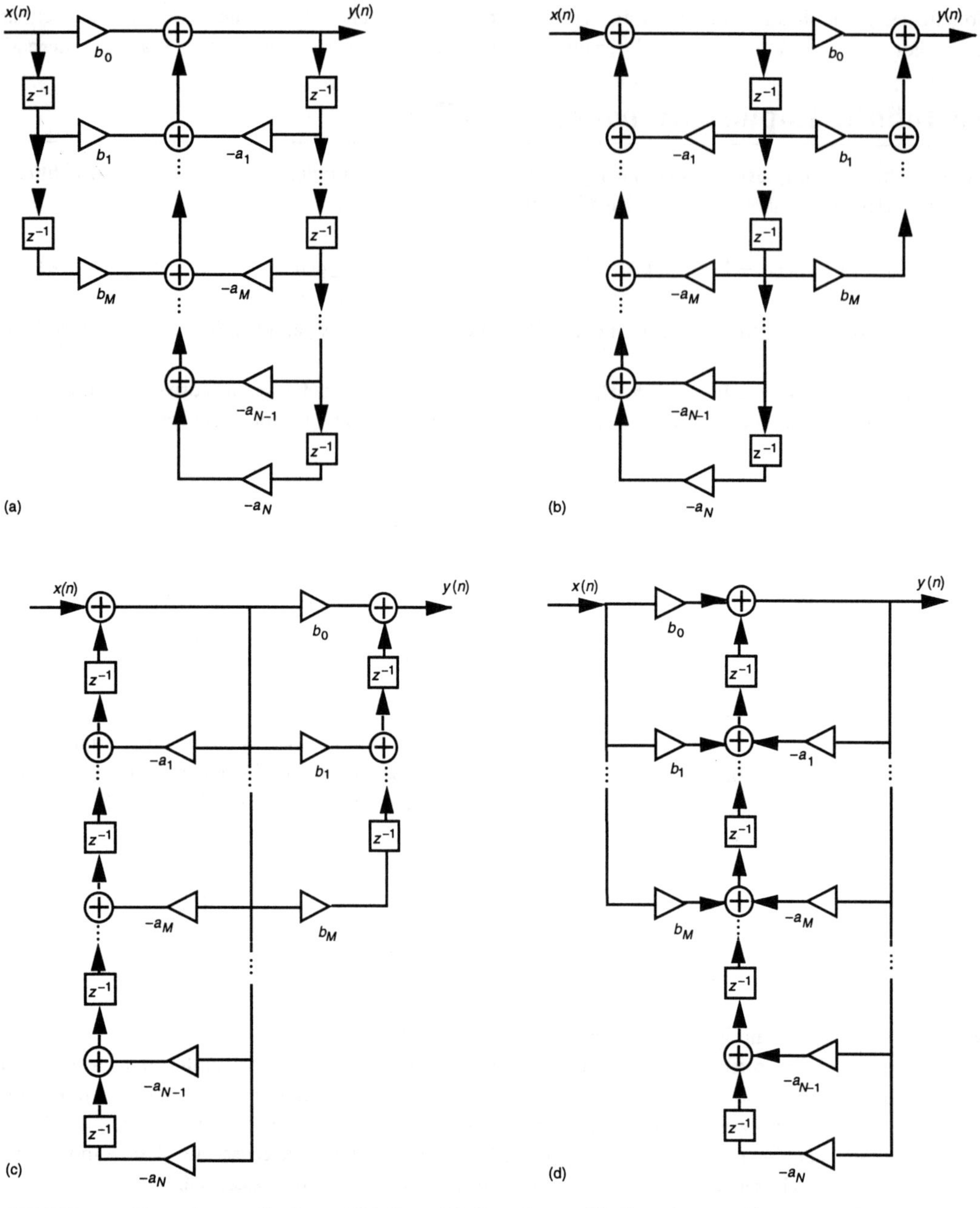

FIGURE 9.9 Direct form realizations of IIR filters: (a) direct form I, (b) direct form II, (c) transposed direct form I, (d) transposed direct form II.

by factoring the numerators and denominators into second-order factors, or in the form

$$H(z) = c_0 + \sum_{i=1}^{K} \frac{b_{i0} + b_{i1}z^{-1}}{1 + a_{i1}z^{-1} + a_{i2}z^{-2}} \tag{9.7}$$

by partial fraction expansion. The value of K is $N/2$ when N is even, and it is $(N + 1)/2$ when N is odd. When N is odd, one of a_{i2} must be zero, as well as one of b_{i2} in Eq. (9.6) and one of b_{i1} in Eq. (9.7). According to

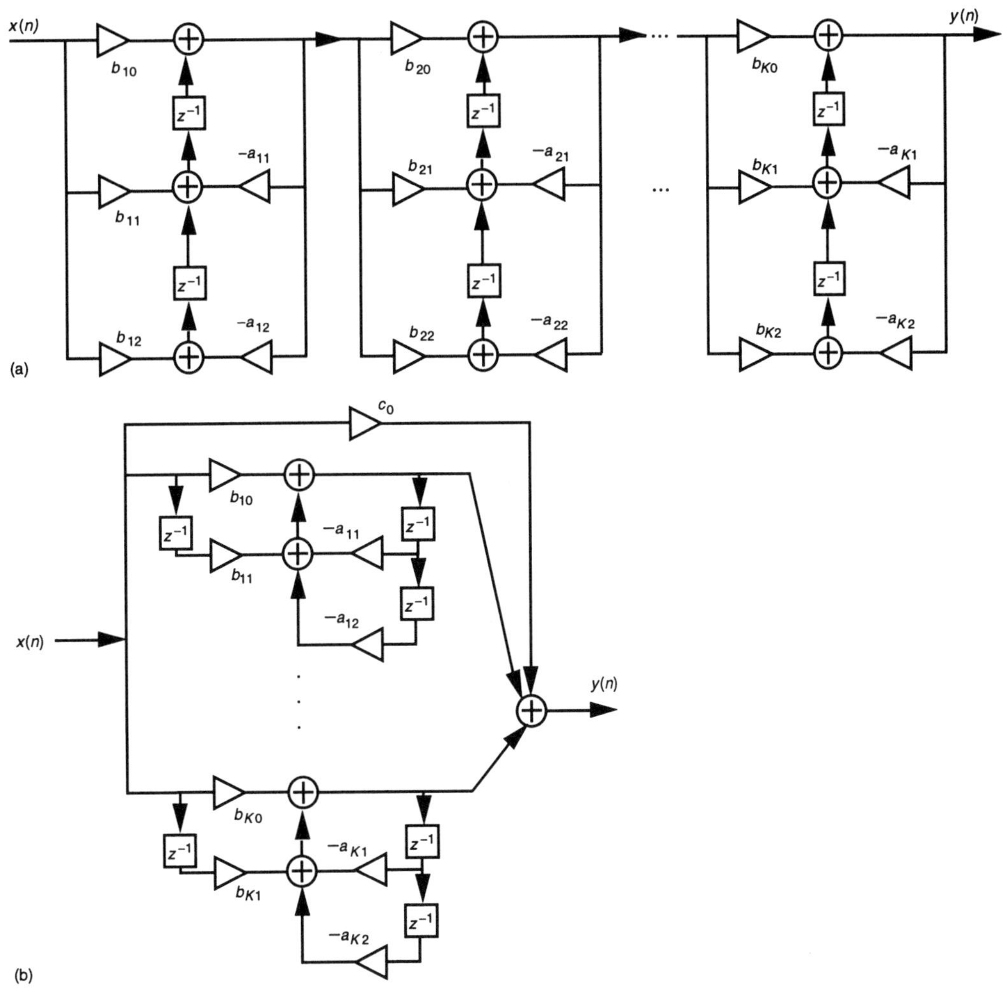

FIGURE 9.10 Realizations of IIR filters: (a) cascade form, (b) parallel form.

Eq. (9.6), the IIR filter can be realized by K second-order IIR filters in cascade, as shown in Fig. 9.10(a). According to Eq. (9.7), the IIR filter can be realized by K second-order IIR filters and one scaler (i.e., c_0) in parallel, as depicted in Fig. 9.10(b). Each second-order subsystem can use any of the structures given in Fig. 9.9.

There are other realizations for IIR filters, such as state-space structure, wave structure, and lattice structure. See the references for details. In some situations, it is more convenient or suitable to use software realizations that are implemented by programming a general purpose microprocessor or a digital signal processor.

IIR Filter Design

Designing an IIR filter involves choosing the coefficients to satisfy a given specification, usually a magnitude response specification. We assume that the specification is in the form depicted by Fig. 9.11, where the magnitude square must be in the range $[1/(1 + \varepsilon^2), 1]$ in the pass band and it must be no larger than δ^2 in the stop band. The pass-band and the stop-band edges are denoted by f_p and f_s, respectively. No constraint is imposed on the response in the transition band, which lies between a pass band and a stop band.

There are various IIR filter design methods: design using an analog prototype filter, design using digital frequency transformation, and computer-aided design. In the first method, an analog filter is designed to meet the (analog) specification and the analog filter transfer function is transformed to digital system function. The

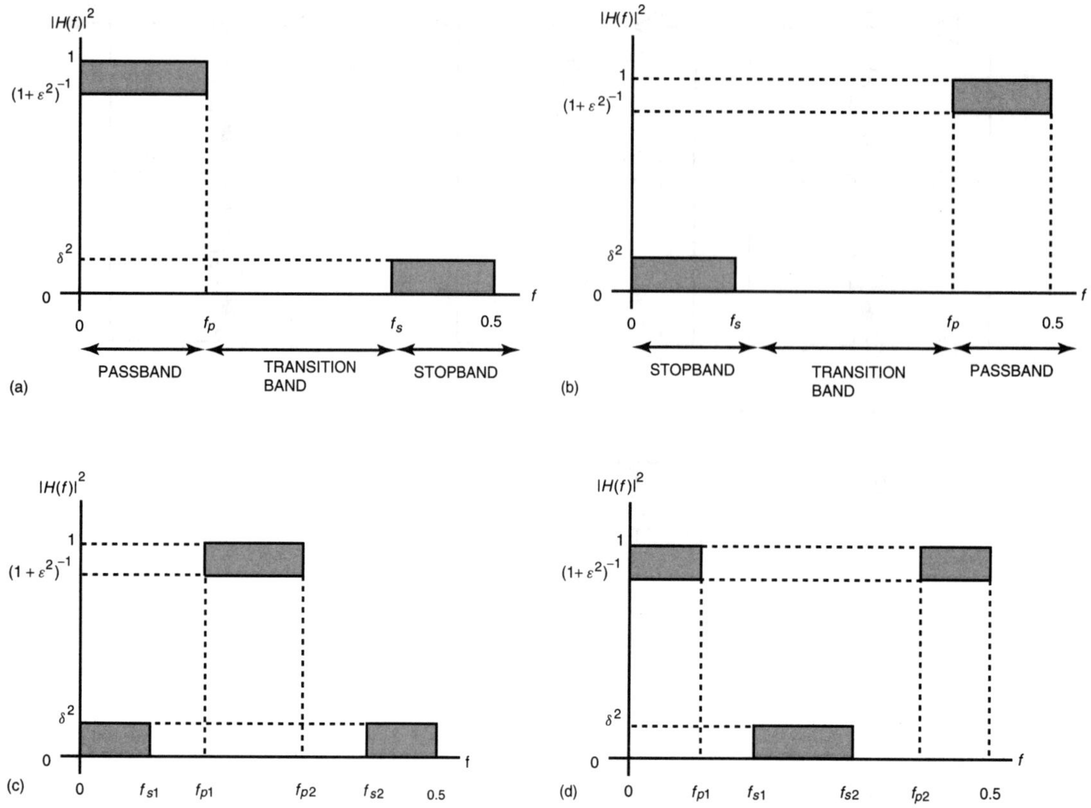

FIGURE 9.11 Specifications for digital IIR filters: (a) low-pass filter, (b) high-pass filter, (c) bandpass filter, (d) bandstop filter.

second method assumes that some digital low-pass filter is available and the desired digital filter is obtained from the digital low-pass filter by a digital frequency transformation. The last method involves algorithms that choose the coefficients so that the response is as close (in some sense) as possible to the desired filter. The first two methods are simple to do, and they are suitable for designing standard filters (low-pass, high-pass, bandpass, and bandstop filters). A computer-aided design requires computer programming, but it can be used to design standard and nonstandard filters. We will focus only on the first method and present some design examples, as well as a summary of some analog filters, in the following sections.

Analog Filters

Here, we summarize three basic types of analog low-pass filters that can be used as prototype for designing IIR filters. For each type, we give the transfer function, its magnitude response, and the order N needed to satisfy the (analog) specification. We will use $H_a(s)$ to denote the transfer function of an analog filter, where s is the complex variable in the Laplace transform. Each of these filters have all its poles on the left-half s plane, so that it is stable. We will use the variable λ to represent the analog frequency in radians/second. The frequency response $H_a(\lambda)$ is the transfer function evaluated at $s = j\lambda$.

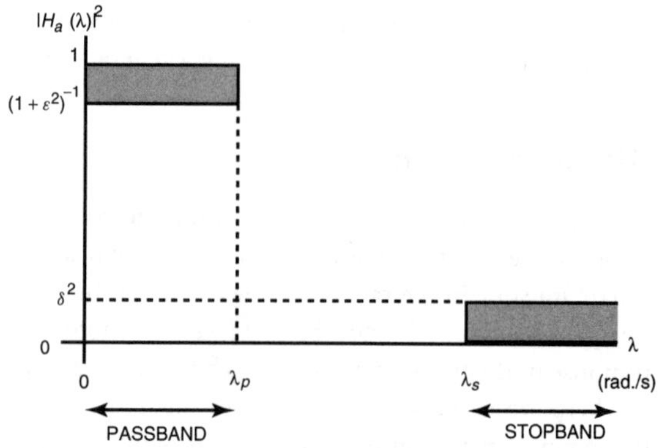

FIGURE 9.12 Specification for an analog low-pass filter.

The analog low-pass filter specification, as shown in Fig. 9.12, is given by

$$\begin{aligned} (1+\varepsilon^2)^{-1} \le |H_a(\lambda)|^2 \le 1 \quad & \text{for } 0 \le (\lambda/2\pi) \le (\lambda_p/2\pi)\,\text{Hz} \\ 0 \le |H_a(\lambda)|^2 \le \delta^2 \quad & \text{for } (\lambda_s/2\pi) \le (\lambda/2\pi) < \infty\,\text{Hz} \end{aligned} \tag{9.8}$$

where λ_p and λ_s are the pass-band edge and stop-band edge, respectively.

Butterworth Filters

The transfer function of an Nth-order Butterworth filter is given by

$$H_a(s) = \begin{cases} \displaystyle\prod_{i=1}^{N/2} \frac{1}{(s/\lambda_c)^2 - 2Re(s_i)(s/\lambda_c) + 1}, & N = \text{even} \\ \displaystyle\frac{1}{(s/\lambda_c)+1} \prod_{i=1}^{(N-1)/2} \frac{1}{(s/\lambda_c)^2 - 2Re(s_i)(s/\lambda_c) + 1}, & N = \text{odd} \end{cases} \tag{9.9}$$

where $s_i = \exp\{j[1+(2i-1)/N]\pi/2\}$ and λ_c is the frequency where the magnitude drops by 3 dB. The magnitude response square is

$$|H_a(\lambda)|^2 = [1 + (\lambda/\lambda_c)^{2N}]^{-1} \tag{9.10}$$

Figure 9.13(a) shows the magnitude response $|H_a(\lambda)|$. To satisfy the specification in Eq. (9.8), the filter order is

$$N = \text{integer} \ge \frac{\log\left[\varepsilon/(\delta^{-2}-1)^{\frac{1}{2}}\right]}{\log[\lambda_p/\lambda_s]} \tag{9.11}$$

and the value of λ_c is chosen from the following range:

$$\lambda_p \varepsilon^{-1/N} \le \lambda_c \le \lambda_s(\delta^{-2}-1)^{-1/(2N)} \tag{9.12}$$

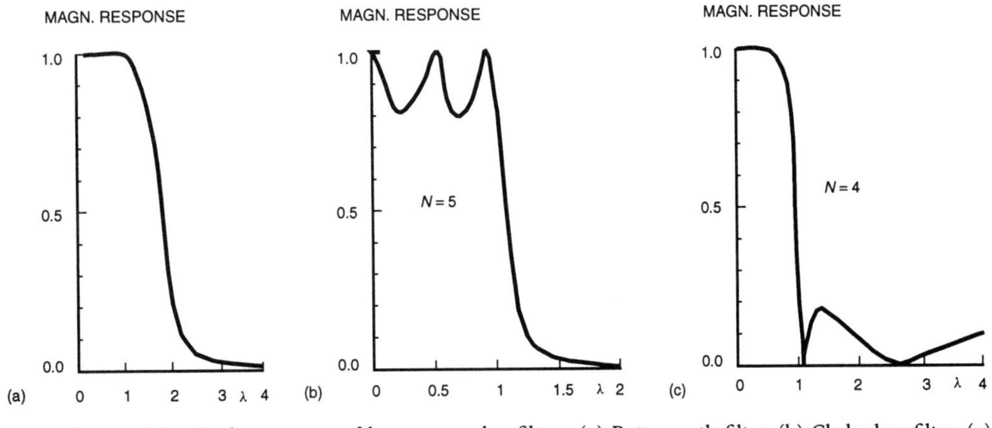

FIGURE 9.13 Magnitude responses of low-pass analog filters: (a) Butterworth filter, (b) Chebyshev filter, (c) inverse Chebyshev filter.

Chebyshev Filters (Type I Chebyshev Filters)

The Nth-order Chebyshev filter has a transfer function given by

$$H_a(s) = C\prod_{i=1}^{N}\frac{1}{(s-p_i)} \tag{9.13}$$

where

$$p_i = -\lambda_p \sinh(\phi)\sin\left(\frac{2i-1}{2N}\pi\right) + j\lambda_p \cosh(\phi)\cos\left(\frac{2i-1}{2N}\pi\right) \tag{9.14}$$

$$\phi = \frac{1}{N}\ln\left[\frac{1+(1+\varepsilon^2)^{\frac{1}{2}}}{\varepsilon}\right] \tag{9.15}$$

and where $C = -\prod_{i=1}^{N} p_i$ when N is odd and $(1+\varepsilon^2)^{-\frac{1}{2}}\prod_{i=1}^{N} p_i$ when N is even. Note that C normalizes the magnitude so that the maximum magnitude is 1. The magnitude square can be written as

$$|H_a(\lambda)|^2 = \left[1+\varepsilon^2 T_N^2(\lambda/\lambda_p)\right]^{-1} \tag{9.16}$$

where $T_N(x)$ is the Nth-degree Chebyshev polynomial of the first kind that is given recursively by $T_0(x) = 1$, $T_1(x) = x$, and $T_{n+1}(x) = 2xT_n(x) - T_{n-1}(x)$ for $n \geq 1$. Figure 9.13(b) shows an example of the magnitude response square. Notice that there are equiripples in the pass band. The filter order required to satisfy Eq. (9.8) is

$$N \geq \frac{\log\left\{\left[(\delta^{-2}-1)^{\frac{1}{2}}/\varepsilon\right] + \left[(\delta^{-2}-1)/\varepsilon^2 - 1\right]^{\frac{1}{2}}\right\}}{\log\{(\lambda_s/\lambda_p) + [(\lambda_s/\lambda_p)^2 - 1]^{\frac{1}{2}}} \tag{9.17}$$

which can be computed knowing ε, δ, λ_p, and λ_s.

Inverse Chebyshev Filters (Type II Chebyshev Filters)

For inverse Chebyshev filters, the equiripples are inside the stop band, as opposed to the pass band in the case of Chebyshev filters. The magnitude response square of the inverse Chebyshev filter is

$$|H_a(\lambda)|^2 = \left[1+(\delta^{-2}-1)/T_N^2(\lambda_s/\lambda)\right]^{-1} \tag{9.18}$$

Figure 9.13(c) depicts an example of Eq. (9.18). The value of $|H_a(\infty)|$ equals 0 if N is odd, and it equals δ if N is even. The transfer function giving rise to Eq. (9.18) is

$$H_a(s) = \begin{cases} C\displaystyle\prod_{i=1}^{N}\frac{(s-q_i)}{(s-p_i)}, & N \text{ is even} \\ \dfrac{C}{(s-p_{(N+1)/2})}\displaystyle\prod_{\substack{i=1 \\ i\neq(N+1)/2}}^{N}\frac{(s-q_i)}{(s-p_i)}, & N \text{ is odd} \end{cases} \tag{9.19}$$

where

$$p_i = \frac{\lambda_s}{\alpha_i^2+\beta_i^2}(\alpha_i - j\beta_i); \qquad q_i = j\frac{\lambda_s}{\cos\left(\frac{2i-1}{2N}\pi\right)} \tag{9.20}$$

$$\alpha_i = -\sinh(\phi)\sin\left(\frac{2i-1}{2N}\pi\right); \qquad \beta_i = \cosh(\phi)\cos\left(\frac{2i-1}{2N}\pi\right) \tag{9.21}$$

$$\phi = \frac{1}{N}\cosh^{-1}(\delta^{-1}) = \frac{1}{N}\ln\left[\delta^{-1} + (\delta^{-2} - 1)^{\frac{1}{2}}\right] \tag{9.22}$$

and where

$$C = \prod_{i=1}^{N}(p_i/q_i)$$

when N is even and

$$C = -p_{(N+1)/2} \prod_{i=1, i\neq(N+1)/2}^{N} (p_i/q_i)$$

when N is odd. The filter order N required to satisfy Eq. (9.8) is the same as that for the Chebyshev filter, which is given by Eq. (9.17).

Comparison

In comparison, the Butterworth filter requires a higher order than both types of Chebyshev filters to satisfy the same specification. There is another type of filters called elliptic filters (Cauer filters) that have equiripples in the pass band as well as in the stop band. Because of the lengthy expressions, this type of filters is not given here (see the references). The Butterworth filter and the inverse Chebyshev filter have better (closer to linear) phase characteristics in the pass band than Chebyshev and elliptic filters. Elliptic filters require a smaller order than Chebyshev filters to satisfy the same specification.

Design Using Bilinear Transformations

One of the most simple way of designing a digital filter is by way of transforming an analog low-pass filter to the desired digital filter. Starting from the desired digital filter specification, the low-pass analog specification is obtained. Then, an analog low-pass filter $H_a(s)$ is designed to meet the specification. Finally, the desired digital filter is obtained by transforming $H_a(s)$ to $H(z)$. There are several types of transformation. The all-around best one is the bilinear transformation, which is the subject of this subsection.

In a bilinear transformation, the variable s in $H_a(s)$ is replaced with a bilinear function of z to obtain $H(z)$. Bilinear transformations for the four standard types of filters, namely, low-pass filter (LPF), high-pass filter (HPF), bandpass filter (BPF), and bandstop filter (BSF), are shown in Table 9.1. The second column in the table gives the relations between the variables s and z. The value of T can be chosen arbitrarily without affecting the resulting design. The third column shows the relations between the analog frequency λ and the digital frequency f, obtained from the relations between s and z by replacing s with $j\lambda$ and z with $\exp(j2\pi f)$. The fourth and fifth columns show the required pass-band and stop-band edges for the analog LPF. Note that the allowable variations in the pass band and stop band, or equivalently the values of ε and δ, for the analog low-pass filter remain the same as those for the desired digital filter. Notice that for the BPF and BSF, the transformation is performed in two steps: one for transforming an analog LPF to/from an analog BPF (or BSF), and the other for transforming an analog BPF (or BSF) to/from a digital BPF (or BSF). The values of W and $\tilde{\lambda}_0$ are chosen by the designer. Some convenient choices are: (1) $W = \tilde{\lambda}_{p2} - \tilde{\lambda}_{p1}$ and $\tilde{\lambda}_0^2 = \tilde{\lambda}_{p1}\tilde{\lambda}_{p2}$, which yield $\lambda_p = 1$; (2) $W = \tilde{\lambda}_{s2} - \tilde{\lambda}_{s1}$ and $\tilde{\lambda}_0^2 = \tilde{\lambda}_{s1}\tilde{\lambda}_{s2}$, which yield $\lambda_s = 1$. We demonstrate the design process by the following two examples.

Design Example 1

Consider designing a digital LPF with a pass-band edge at $f_p = 0.15$ and stop-band edge at $f_s = 0.25$. The magnitude response in the pass band is to stay within 0 and -2 dB, although it must be no larger than -40 dB

TABLE 52.1 Bilinear Transformations for Standard Digital Filters

	Transformation (between s and z)	Frequency Relations (between f and λ)	Analog LPF Passband Edge	Analog LPF Stopband Edge
Digital LPF to/from analog LPF	$s = \frac{2}{T}\frac{1-z^{-1}}{1+z^{-1}}$ $z = \frac{(2/T)+s}{(2/T)-s}$	$f = \frac{1}{\pi}\tan^{-1}\left(\frac{\lambda T}{2}\right)$ $\lambda = \frac{2}{T}\tan(\pi f)$	$\lambda_p = \frac{2}{T}\tan(\pi f_p)$	$\lambda_s = \frac{2}{T}\tan(\pi f_s)$
Digital HPF to/from analog LPF	$s = \frac{T}{2}\frac{1+z^{-1}}{1-z^{-1}}$ $z = \frac{s+(T/2)}{s-(T/2)}$	$f = \begin{cases} -\frac{1}{2}+\frac{1}{\pi}\tan^{-1}\left(\frac{2\lambda}{T}\right), & \lambda \geq 0 \\ \frac{1}{2}+\frac{1}{\pi}\tan^{-1}\left(\frac{2\lambda}{T}\right), & \lambda \leq 0 \end{cases}$ $\lambda = \frac{T}{2}\tan[\pi(f+0.5)],\ \lambda \geq 0$	$\lambda_p = \frac{T}{2}\tan\left[\pi\left(f_p+\frac{1}{2}\right)\right]$	$\lambda_s = \frac{T}{2}\tan\left[\pi\left(f_s+\frac{1}{2}\right)\right]$
Digital BPF to/from analog LPF	$s = \frac{\bar{s}^2+\bar{\lambda}_0^2}{W\bar{s}}$ where $\bar{s} = \frac{2}{T}\frac{1-z^{-1}}{1+z^{-1}}$	$\lambda = \frac{\bar{\lambda}^2-\bar{\lambda}_0^2}{W\bar{\lambda}}$ where $\bar{\lambda} = \frac{2}{T}\tan(\pi f)$	$\lambda_p = \max\left\{\left\lvert\frac{\bar{\lambda}_0^2-\bar{\lambda}_{p1}^2}{W\bar{\lambda}_{p1}}\right\rvert, \left\lvert\frac{\bar{\lambda}_0^2-\bar{\lambda}_{p2}^2}{W\bar{\lambda}_{p2}}\right\rvert\right\}$ where $\bar{\lambda}_{p1} = \frac{2}{T}\tan(\pi f_{p1})$ $\bar{\lambda}_{p2} = \frac{2}{T}\tan(\pi f_{p2})$	$\lambda_s = \min\left\{\left\lvert\frac{\bar{\lambda}_0^2-\bar{\lambda}_{s1}^2}{W\bar{\lambda}_{s1}}\right\rvert, \left\lvert\frac{\bar{\lambda}_0^2-\bar{\lambda}_{s2}^2}{W\bar{\lambda}_{s2}}\right\rvert\right\}$ where $\bar{\lambda}_{s1} = \frac{2}{T}\tan(\pi f_{s1})$ $\bar{\lambda}_{s2} = \frac{2}{T}\tan(\pi f_{s2})$
Digital BSF to/from analog LPF	$s = \frac{W\bar{s}}{\bar{s}^2+\bar{\lambda}_0^2}$ where $\bar{s} = \frac{2}{T}\frac{1-z^{-1}}{1+z^{-1}}$	$\lambda = \frac{W\bar{\lambda}}{\bar{\lambda}_0^2-\bar{\lambda}^2}$ where $\bar{\lambda} = \frac{2}{T}\tan(\pi f)$	$\lambda_p = \max\left\{\left\lvert\frac{W\bar{\lambda}_{p1}}{\bar{\lambda}_0^2-\bar{\lambda}_{p1}^2}\right\rvert, \left\lvert\frac{W\bar{\lambda}_{p2}}{\bar{\lambda}_0^2-\bar{\lambda}_{p2}^2}\right\rvert\right\}$ where $\bar{\lambda}_{p1} = \frac{2}{T}\tan(\pi f_{p1})$ $\bar{\lambda}_{p2} = \frac{2}{T}\tan(\pi f_{p2})$	$\lambda_s = \min\left\{\left\lvert\frac{W\bar{\lambda}_{s1}}{\bar{\lambda}_0^2-\bar{\lambda}_{s1}^2}\right\rvert, \left\lvert\frac{W\bar{\lambda}_{s2}}{\bar{\lambda}_0^2-\bar{\lambda}_{s2}^2}\right\rvert\right\}$ where $\bar{\lambda}_{s1} = \frac{2}{T}\tan(\pi f_{s1})$ $\bar{\lambda}_{s2} = \frac{2}{T}\tan(\pi f_{s2})$

in the stop band. Assuming that the analog Butterworth filter is to be used as the prototype filter, we proceed the design as follows.

1. Compute ε, δ, λ_p, λ_s, and the analog filter order. An attenuation of -2 dB means $10\log_{10}(1+\varepsilon^2)^{-1} = -2$, which yields $\varepsilon = 0.7648$, and attenuation of -40 dB means $10\log_{10}(\delta^2) = -40$, which gives $\delta = 0.01$. From Table 9.1, the analog pass-band and stop-band edges are $\lambda_p = (2/T)\tan(\pi f_p)$ and $\lambda_s = (2/T)\tan(\pi f_s)$. For convenience, we let $T = 2$, which makes $\lambda_p = 0.5095$ and $\lambda_s = 1.0$. Now, we can calculate the required order of the analog Butterworth filter from Eq. (9.11), which yields $N \geq 7.23$; thus, we choose $N = 8$.
2. Obtain the analog LPF transfer function. From Eq. (9.12), we can pick a value of λ_c between 0.5269 and 0.5623. Let us choose $\lambda_c = 0.54$. With $N = 8$, the transfer function is calculated from Eq. (9.9) as

$$H_a(s) = \frac{7.2302 \times 10^{-3}}{(s^2+0.2107s+0.2916)(s^2+0.6s+0.2916)} \bullet \frac{1}{(s^2+0.8980s+0.2916)(s^2+1.0592s+0.2916)} \tag{9.23}$$

3. Obtain the digital filter. Using the transformation in Table 9.1, we transform Eq. (9.23) to the desired digital filter by replacing s with $(2/T)(z-1)/(z+1) = (z-1)/(z+1)$ (since we let $T = 2$). After substitution

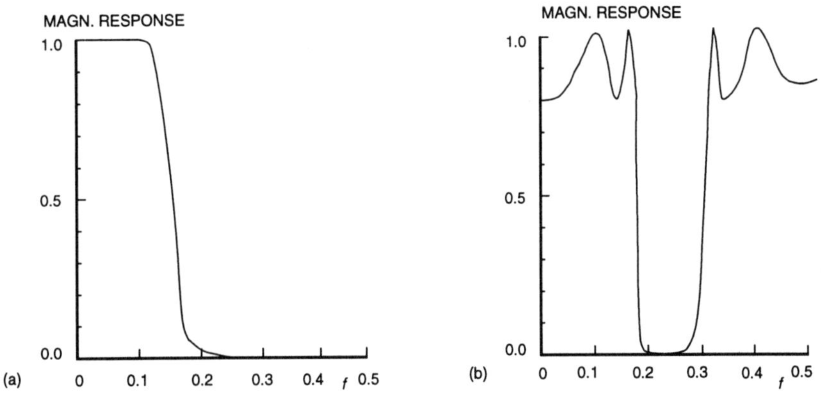

FIGURE 9.14 Magnitude responses of the designed digital filters in the examples: (a) LPF (example 1), (b) BSF (example 2).

and simplification, the resulting digital LPF has a system function given by

$$H(z) = H_a(s)|_{s=(z-1)(z+1)} = \frac{4.9428 \times 10^{-4}(z^2 + 2z + 1)^4}{(z^2 - 0.9431z + 0.7195)(z^2 - 0.7490z + 0.3656)} \bullet \frac{1}{(z^2 - 0.6471z + 0.1798)(z^2 - 0.6027z + 0.0988)} \tag{9.24}$$

The magnitude response $|H(f)|$ is plotted in Fig. 9.14(a). The magnitude is 0.8467 at the pass-band edge $f_p = 0.15$, and it is 0.0072 at the stop-band edge $f_s = 0.25$, both of which are within the specification.

Design Example 2

Now, suppose that we wish to design a digital BSF with pass-band edges at $f_{p1} = 0.15$ and $f_{p2} = 0.30$ and stop-band edges at $f_{s1} = 0.20$ and $f_{s2} = 0.25$. The magnitude response in the pass bands is to stay within 0 and −2 dB, although it must be no larger than −40 dB in the stop band. Let us use the analog type I Chebyshev filter as the prototype filter. Following the same design process as the first example, we have the following.

1. Compute $\varepsilon, \delta, \lambda_p, \lambda_s$, and the analog filter order. From the first example, we have $\varepsilon = 0.7648$ and $\delta = 0.01$. From Table 9.1 and letting $T = 2$, we compute the analog pass-band and stop-band edges: $\tilde{\lambda}_{p1} = \tan(\pi f_{p1}) = 0.5095$, $\tilde{\lambda}_{p2} = \tan(\pi f_{p2}) = 1.3764$, $\tilde{\lambda}_{s1} = \tan(\pi f_{s1}) = 0.7265$, $\tilde{\lambda}_{s2} = \tan(\pi f_{s2}) = 1.0$. We choose $W = \tilde{\lambda}_{p2} - \tilde{\lambda}_{p1} = 0.8669$ and $\tilde{\lambda}_0^2 = \tilde{\lambda}_{p1}\tilde{\lambda}_{p2} = 0.7013$, so that $\lambda_p = 1.0$ and $\lambda_s = \min\{3.6311, 2.9021\} = 2.9021$. From Eq. (9.17), the required analog filter order is $N \geq 3.22$; thus, we choose $N = 4$.
2. Obtain the analog LPF transfer function. From Eq. (55.13) and with $N = 4$, we find the transfer function of the analog Chebyshev filter as

$$H_a(s) = \frac{1.6344 \times 10^{-1}}{(s^2 + 0.2098s + 0.9287)(s^2 + 0.5064s + 0.2216)} \tag{9.25}$$

3. Obtain the digital filter. Using Table 9.1, we transform Eq. (9.25) to $H(z)$ by replacing s with $W\bar{s}/(\bar{s}^2 + \tilde{\lambda}_0^2) = W(z^2-1)/[(z-1)^2 + \tilde{\lambda}_0^2(z+1)^2]$ since $T = 2$ and $\bar{s} = (z-1)/(z+1)$. After substitution and simplification, the resulting digital BSF is

$$H(z) = \frac{1.7071 \times 10^{-1}(z^4 - 0.7023z^3 + 2.1233z^2 - 0.7023z + 1)^2}{(z^4 - 0.5325z^3 + 1.1216z^2 - 0.4746z + 0.8349)} \bullet \frac{1}{(z^4 - 0.3331z^3 + 0.0660z^2 - 0.0879z + 0.3019)} \tag{9.26}$$

The magnitude response is plotted in Fig. 9.14(b), which satisfies the specification.

9.4 Finite Wordlength Effects

Practical digital filters must be implemented with finite precision numbers and arithmetic. As a result, both the filter coefficients and the filter input and output signals are in discrete form. This leads to four types of finite wordlength effects.

Discretization (quantization) of the filter coefficients has the effect of perturbing the location of the filter poles and zeroes. As a result, the actual filter response differs slightly from the ideal response. This *deterministic* frequency response error is referred to as *coefficient quantization error.*

The use of finite precision arithmetic makes it necessary to quantize filter calculations by rounding or truncation. *Roundoff noise* is that error in the filter output that results from rounding or truncating calculations within the filter. As the name implies, this error looks like low-level noise at the filter output.

Quantization of the filter calculations also renders the filter slightly nonlinear. For large signals this nonlinearity is negligible and roundoff noise is the major concern. For recursive filters with a zero or constant input, however, this nonlinearity can cause spurious oscillations called *limit cycles.*

With fixed-point arithmetic it is possible for filter calculations to overflow. The term *overflow oscillation,* sometimes also called *adder overflow limit cycle,* refers to a high-level oscillation that can exist in an otherwise stable filter due to the nonlinearity associated with the overflow of internal filter calculations.

In this section we examine each of these finite wordlength effects for both fixed-point and floating-point number represention.

Number Representation

In digital signal processing, $(B + 1)$-bit fixed-point numbers are usually represented as two's-complement signed fractions in the format

$$b_0.b_{-1}b_{-2}\cdots b_{-B}$$

The number represented is then

$$X = -b_0 + b_{-1}2^{-1} + b_{-2}2^{-2} + \cdots + b_{-B}2^{-B} \tag{9.27}$$

where b_0 is the sign bit and the number range is $-1 \leq X < 1$. The advantage of this representation is that the product of two numbers in the range from -1 to 1 is another number in the same range.

Floating-point numbers are represented as

$$X = (-1)^s m\, 2^c \tag{9.28}$$

where s is the sign bit, m is the *mantissa,* and c is the *characteristic* or *exponent.* To make the representation of a number unique, the mantissa is *normalized* so that $0.5 \leq m < 1$.

Fixed-Point Quantization Errors

In fixed-point arithmetic, a multiply doubles the number of significant bits. For example, the product of the two 5-bit numbers 0.0011 and 0.1001 is the 10-bit number 00.00011011. The extra bit to the left of the decimal point can be discarded without introducing any error. However, the least significant four of the remaining bits must ultimately be discarded by some form of quantization so that the result can be stored to five bits for use in other calculations. In the preceding example this results in 0.0010 (quantization by rounding) or 0.0001 (quantization by truncating). When a sum of products calculation is performed, the quantization can be performed either after each multiply or after all products have been summed with double-length precision.

We will examine the case of fixed-point quantization by rounding. If X is an exact value then the rounded value will be denoted $Q_r(X)$. If the quantized value has B bits to the right of the decimal point, the quantization step size is

$$\Delta = 2^{-B} \tag{9.29}$$

Since rounding selects the quantized value nearest the unquantized value, it gives a value which is never more than $\pm\Delta/2$ away from the exact value. If we denote the rounding error by

$$\varepsilon_r = Q_r(X) - X \tag{9.30}$$

then

$$-\frac{\Delta}{2} \leq \varepsilon_r \leq \frac{\Delta}{2} \tag{9.31}$$

The error resulting from quantization can be modeled as a random variable uniformly distributed over the appropriate error range. Therefore, calculations with roundoff error can be considered error-free calculations that have been corrupted by additive white noise. The mean of this noise for rounding is

$$m_{\varepsilon_r} = E\{\varepsilon_r\} = \frac{1}{\Delta}\int_{-\frac{\Delta}{2}}^{\frac{\Delta}{2}} \varepsilon_r \, d\varepsilon_r = 0 \tag{9.32}$$

where $E\{\,\}$ represents the operation of taking the expected value of a random variable. Similarly, the variance of the noise for rounding is

$$\sigma_{\varepsilon_r}^2 = E\left\{(\varepsilon_r - m_{\varepsilon_r})^2\right\} = \frac{1}{\Delta}\int_{-\frac{\Delta}{2}}^{\frac{\Delta}{2}} (\varepsilon_r - m_{\varepsilon_r})^2 \, d\varepsilon_r = \frac{\Delta^2}{12} \tag{9.33}$$

Floating-Point Quantization Errors

With floating-point arithmetic it is necessary to quantize after both multiplications and additions. The addition quantization arises because, prior to addition, the mantissa of the smaller number in the sum is shifted right until the exponent of both numbers is the same. In general, this gives a sum mantissa that is too long and so must be quantized.

We will assume that quantization in floating-point arithmetic is performed by rounding. Because of the exponent in floating-point arithmetic, it is the relative error that is important. The relative error is defined as

$$\varepsilon_r = \frac{Q_r(X) - X}{X} = \frac{\varepsilon_r}{X} \tag{9.34}$$

Since $X = (-1)^s \, m \, 2^c$, $Q_r(X) = (-1)^s \, Q_r(m) \, 2^c$ and

$$\varepsilon_r = \frac{Q_r(m) - m}{m} = \frac{\varepsilon}{m} \tag{9.35}$$

If the quantized mantissa has B bits to the right of the decimal point, $|\varepsilon| < \Delta/2$ where, as before, $\Delta = 2^{-B}$. Therefore, since $0.5 \leq m < 1$,

$$|\varepsilon_r| < \Delta \tag{9.36}$$

If we assume that ε is uniformly distributed over the range from $-\Delta/2$ to $\Delta/2$ and m is uniformly distributed over 0.5–1,

$$
\begin{aligned}
m_{\varepsilon_r} &= E\left\{\frac{\varepsilon}{m}\right\} = 0 \\
\sigma^2_{\varepsilon_r} &= E\left\{\left(\frac{\varepsilon}{m}\right)^2\right\} = \frac{2}{\Delta}\int_{\frac{1}{2}}^{1}\int_{-\frac{\Delta}{2}}^{\frac{\Delta}{2}} \frac{\varepsilon^2}{m^2} d\varepsilon\, dm = \frac{\Delta^2}{6} = (0.167)2^{-2B}
\end{aligned} \tag{9.37}
$$

From Eq. (9.34) we can represent a quantized floating-point value in terms of the unquantized value and the random variable ε_r using

$$
Q_r(X) = X(1 + \varepsilon_r) \tag{9.38}
$$

Therefore, the finite-precision product X_1X_2 and the sum $X_1 + X_2$ can be written

$$
fl(X_1X_2) = X_1X_2(1 + \varepsilon_r) \tag{9.39}
$$

and

$$
fl(X_1 + X_2) = (X_1 + X_2)(1 + \varepsilon_r) \tag{9.40}
$$

where ε_r is zero mean with the variance of Eq. (9.37).

Roundoff Noise

To determine the roundoff noise at the output of a digital filter we will assume that the noise due to a quantization is stationary, white, and uncorrelated with the filter input, output, and internal variables. This assumption is good if the filter input changes from sample to sample in a sufficiently complex manner. It is not valid for zero or constant inputs for which the effects of rounding are analyzed from a limit cycle perspective.

To satisfy the assumption of a sufficiently complex input, roundoff noise in digital filters is often calculated for the case of a zero-mean white noise filter input signal $x(n)$ of variance σ_x^2. This simplifies calculation of the output roundoff noise because expected values of the form $E\{x(n)x(n-k)\}$ are zero for $k \neq 0$ and give σ_x^2 when $k = 0$. If there is more than one source of roundoff error in a filter, it is assumed that the errors are uncorrelated so that the output noise variance is simply the sum of the contributions from each source. This approach to analysis has been found to give estimates of the output roundoff noise that are close to the noise actually observed in practice.

Another assumption that will be made in calculating roundoff noise is that the product of two quantization errors is zero. To justify this assumption, consider the case of a 16-bit fixed-point processor. In this case a quantization error is of the order 2^{-15}, whereas the product of two quantization errors is of the order 2^{-30}, which is negligible by comparison.

The simplest case to analyze is a finite impulse response filter realized via the convolution summation

$$
y(n) = \sum_{k=0}^{N-1} h(k)x(n-k) \tag{9.41}
$$

When fixed-point arithmetic is used and quantization is performed after each multiply, the result of the N multiplies is N times the quantization noise of a single multiply. For example, rounding after each multiply gives, from Eqs. (9.29) and (9.33), an output noise variance of

$$
\sigma_o^2 = N\frac{2^{-2B}}{12} \tag{9.42}
$$

Virtually all digital signal processor integrated circuits contain one or more double-length accumulator registers that permit the sum-of-products in Eq. (9.41) to be accumulated without quantization. In this case only a single quantization is necessary following the summation and

$$\sigma_o^2 = \frac{2^{-2B}}{12} \tag{9.43}$$

For the floating-point roundoff noise case we will consider Eq. (9.41) for $N = 4$ and then generalize the result to other values of N. The finite-precision output can be written as the exact output plus an error term $e(n)$. Thus,

$$\begin{aligned} y(n) + e(n) = (\{[h(0)x(n)(1+\varepsilon_1(n)) + h(1)x(n-1)(1+\varepsilon_2(n))][1+\varepsilon_3(n)] \\ + h(2)x(n-2)(1+\varepsilon_4(n))\}\{1+\varepsilon_5(n)\} + h(3)x(n-3)(1+\varepsilon_6(n)))(1+\varepsilon_7(n)) \end{aligned} \tag{9.44}$$

In Eq. (9.44) $\varepsilon_1(n)$ represents the error in the first product, $\varepsilon_2(n)$ the error in the second product, $\varepsilon_3(n)$ the error in the first addition, etc. Notice that it has been assumed that the products are summed in the order implied by the summation of Eq. (9.41).

Expanding Eq. (9.44), ignoring products of error terms, and recognizing $y(n)$ gives

$$\begin{aligned} e(n) = h(0)x(n)[\varepsilon_1(n) + \varepsilon_3(n) + \varepsilon_5(n) + \varepsilon_7(n)] + h(1)x(n-1)[\varepsilon_2(n) + \varepsilon_3(n) + \varepsilon_5(n) + \varepsilon_7(n)] \\ + h(2)x(n-2)[\varepsilon_4(n) + \varepsilon_5(n) + \varepsilon_7(n)] + h(3)x(n-3)[\varepsilon_6(n) + \varepsilon_7(n)] \end{aligned} \tag{9.45}$$

Assuming that the input is white noise of variance σ_x^2 so that $E\{x(n)x(n-k)\}$ is zero for $k \neq 0$, and assuming that the errors are uncorrelated,

$$E\{e^2(n)\} = [4h^2(0) + 4h^2(1) + 3h^2(2) + 2h^2(3)]\sigma_x^2\sigma_{\varepsilon_r}^2 \tag{9.46}$$

In general, for any N,

$$\sigma_o^2 = E\{e^2(n)\} = \left[Nh^2(0) + \sum_{k=1}^{N-1}(N+1-k)h^2(k)\right]\sigma_x^2\sigma_{\varepsilon_r}^2 \tag{9.47}$$

Notice that if the order of summation of the product terms in the convolution summation is changed, then the order in which the $h(k)$ appear in Eq. (9.47) changes. If the order is changed so that the $h(k)$ with smallest magnitude is first, followed by the next smallest, etc., then the roundoff noise variance is minimized. Performing the convolution summation in nonsequential order, however, greatly complicates data indexing and so may not be worth the reduction obtained in roundoff noise.

Analysis of roundoff noise for IIR filters proceeds in the same way as for FIR filters. The analysis is more complicated, however, because roundoff noise arising in the computation of internal filter variables (state variables) must be propogated through a transfer function from the point of the quantization to the filter output. This is not necessary for FIR filters realized via the convolution summation since all quantizations are in the output calculation. Another complication in the case of IIR filters realized with fixed-point arithmetic is the need to scale the computation of internal filter variables to avoid their overflow. Examples of roundoff noise analysis for IIR filters can be found in Weinstein and Oppenheim [1969] and Oppenheim and Weinstein [1972] where it is shown that differences in the filter realization structure can make a large difference in the output roundoff noise. In particular, it is shown that IIR filters realized via the parallel or cascade connection of first- and second-order subfilters are almost always superior in terms of roundoff noise to a high-order direct form (single difference equation) realization. It is also possible to choose realizations that are optimal or near optimal in a roundoff noise sense [Mullins and Roberts 1976; Jackson, Lindgren, and Kim 1979]. These realizations generally require more computation to obtain an output sample from an input sample, however, and so suboptimal realizations with slightly higher roundoff noise are often preferable [Bomar 1985].

Limit Cycles

A limit cycle, sometimes referred to as a *multiplier roundoff limit cycle*, is a low-level oscillation that can exist in an otherwise stable filter as a result of the nonlinearity associated with rounding (or truncating) internal filter calculations [Parker and Hess 1971]. Limit cycles require recursion to exist and do not occur in nonrecursive FIR filters.

As an example of a limit cycle, consider the second-order filter realized by

$$y(n) = Q_r\left\{\frac{7}{8}y(n-1) - \frac{5}{8}y(n-2) + x(n)\right\} \tag{9.48}$$

where $Q_r\{\ \}$ represents quantization by rounding. This is a stable filter with poles at $0.4375 \pm j0.6585$. Consider the implementation of this filter with four-bit (three bits and a sign bit) two's complement fixed-point arithmetic, zero initial conditions [$y(-1) = y(-2) = 0$], and an input sequence $x(n) = \frac{3}{8}\delta(n)$ where $\delta(n)$ is the unit impulse or unit sample. The following sequence is obtained:

$$y(0) = Q_r\left\{\frac{3}{8}\right\} = \frac{3}{8} \qquad y(1) = Q_r\left\{\frac{21}{64}\right\} = \frac{3}{8} \qquad y(2) = Q_r\left\{\frac{3}{32}\right\} = \frac{1}{8}$$

$$y(3) = Q_r\left\{-\frac{1}{8}\right\} = -\frac{1}{8} \qquad y(4) = Q_r\left\{-\frac{3}{16}\right\} = -\frac{1}{8} \qquad y(5) = Q_r\left\{-\frac{1}{32}\right\} = 0$$

$$y(6) = Q_r\left\{\frac{5}{64}\right\} = \frac{1}{8} \qquad y(7) = Q_r\left\{\frac{7}{64}\right\} = \frac{1}{8} \qquad y(8) = Q_r\left\{\frac{1}{32}\right\} = 0$$

$$y(9) = Q_r\left\{-\frac{5}{64}\right\} = -\frac{1}{8} \qquad y(10) = Q_r\left\{-\frac{7}{64}\right\} = -\frac{1}{8} \qquad y(11) = Q_r\left\{-\frac{1}{32}\right\} = 0$$

$$y(12) = Q_r\left\{\frac{5}{64}\right\} = \frac{1}{8} \qquad \cdots$$

Notice that although the input is zero except for the first sample, the output oscillates with amplitude $\frac{1}{8}$ and period 6.

Limit cycles are primarily of concern in fixed-point recursive filters. As long as floating-point filters are realized as the parallel or cascade connection of first- and second-order subfilters, limit cycles will generally not be a problem since limit cycles are practically not observable in first- and second-order systems implemented with 32-bit floating-point arithmetic [Bauer 1993]. It has been shown that such systems must have an extremely small margin of stability for limit cycles to exist at anything other than underflow levels, which are at an amplitude of less than 10^{-38} [Bauer 1993].

There are at least three ways of dealing with limit cycles when fixed-point arithmetic is used. One is to determine a bound on the maximum limit cycle amplitude, expressed as an integral number of quantization steps. It is then possible to choose a wordlength that makes the limit cycle amplitude acceptably low. Alternately, limit cycles can be prevented by randomly rounding calculations up or down [Buttner 1976]. This approach, however, is complicated to implement. The third approach is to properly choose the filter realization structure and then quantize the filter calculations using magnitude truncation [Bomar 1994]. This approach has the disadvantage of slightly increasing roundoff noise.

Overflow Oscillations

With fixed-point arithmetic it is possible for filter calculations to overflow. This happens when two numbers of the same sign add to give a value having magnitude greater than one. Since numbers with magnitude greater than one are not representable, the result overflows. For example, the two's complement numbers 0.101 ($\frac{5}{8}$) and 0.100 ($\frac{4}{8}$) add to give 1.001, which is the two's complement representation of $-\frac{7}{8}$.

The overflow characteristic of two's complement arithmetic can be represented as $R\{\ \}$ where

$$R\{X\} = \begin{cases} X - 2, & X \geq 1 \\ X, & -1 \leq X < 1 \\ X + 2, & X < -1 \end{cases} \tag{9.49}$$

For the example just considered, $R\{\frac{9}{8}\} = -\frac{7}{8}$.

An overflow oscillation, sometimes also referred to as an adder overflow limit cycle, is a high-level oscillation that can exist in an otherwise stable fixed-point filter due to the gross nonlinearity associated with the overflow of internal filter calculations [Ebert, Mazo, and Taylor 1969]. Like limit cycles, overflow oscillations require recursion to exist and do not occur in nonrecursive FIR filters. Overflow oscillations also do not occur with floating-point arithmetic due to the virtual impossibility of overflow.

As an example of an overflow oscillation, once again consider the filter of Eq. (9.48) with four-bit fixed-point two's complement arithmetic and with the two's complement overflow characteristic of Eq. (9.49):

$$y(n) = Q_r\left\{R\left\{\frac{7}{8}y(n-1) - \frac{5}{8}y(n-2) + x(n)\right\}\right\} \tag{9.50}$$

In this case we apply the input

$$x(n) = \left\{-\frac{3}{4}, -\frac{5}{8}, 0, 0, 0, \ldots\right\} \tag{9.51}$$

giving the output sequence

$$y(0) = Q_r\left\{R\left\{-\frac{3}{4}\right\}\right\} = Q_r\left\{-\frac{3}{4}\right\} = -\frac{3}{4} \qquad y(1) = Q_r\left\{R\left\{-\frac{41}{32}\right\}\right\} = Q_r\left\{\frac{23}{32}\right\} = \frac{3}{4}$$

$$y(2) = Q_r\left\{R\left\{\frac{9}{8}\right\}\right\} = Q_r\left\{-\frac{7}{8}\right\} = -\frac{7}{8} \qquad y(3) = Q_r\left\{R\left\{-\frac{79}{64}\right\}\right\} = Q_r\left\{\frac{49}{64}\right\} = \frac{3}{4}$$

$$y(4) = Q_r\left\{R\left\{\frac{77}{64}\right\}\right\} = Q_r\left\{-\frac{51}{64}\right\} = -\frac{3}{4} \qquad y(5) = Q_r\left\{R\left\{-\frac{9}{8}\right\}\right\} = Q_r\left\{\frac{7}{8}\right\} = \frac{7}{8}$$

$$y(6) = Q_r\left\{R\left\{\frac{79}{64}\right\}\right\} = Q_r\left\{-\frac{49}{64}\right\} = -\frac{3}{4} \qquad y(7) = Q_r\left\{R\left\{-\frac{77}{64}\right\}\right\} = Q_r\left\{\frac{51}{64}\right\} = \frac{3}{4}$$

$$y(8) = Q_r\left\{R\left\{\frac{9}{8}\right\}\right\} = Q_r\left\{-\frac{7}{8}\right\} = -\frac{7}{8} \qquad \cdots$$

This is a large-scale oscillation with nearly full-scale amplitude.

There are several ways to prevent overflow oscillations in fixed-point filter realizations. The most obvious is to scale the filter calculations so as to render overflow impossible. However, this may unacceptably restrict the filter dynamic range. Another method is to force completed sums-of-products to saturate at ± 1, rather than overflowing [Ritzerfeld 1989]. It is important to saturate only the completed sum, since intermediate overflows in two's complement arithmetic do not affect the accuracy of the final result. Most fixed-point digital signal processors provide for automatic saturation of completed sums if their *saturation arithmetic* feature is enabled. Yet another way to avoid overflow oscillations is to use a filter structure for which any internal filter transient is guaranteed to decay to zero [Mills, Mullis, and Roberts 1978]. Such structures are desirable anyway, since they tend to have low roundoff noise and be insensitive to coefficient quantization [Barnes 1979].

Coefficient Quantization Error

Each filter structure has its own finite, generally nonuniform grids of realizable pole and zero locations when the filter coefficients are quantized to a finite wordlength. In general the pole and zero locations desired in a filter do not correspond exactly to the realizable locations. The error in filter performance (usually measured in terms of a frequency response error) resulting from the placement of the poles and zeroes at the nonideal but realizable locations is referred to as coefficient quantization error.

Consider the second-order filter with complex-conjugate poles

$$\lambda = r\, e^{\pm j\theta} = \lambda_r \pm j\lambda_i = r\cos(\theta) \pm jr\sin(\theta) \tag{9.52}$$

and transfer function

$$H(z) = \frac{1}{1 - 2r\cos(\theta)z^{-1} + r^2 z^{-2}} \tag{9.53}$$

realized by the difference equation

$$y(n) = 2\lambda_r y(n-1) - r^2\, y(n-2) + x(n) \tag{9.54}$$

Quantizing the difference equation coefficients results in a nonuniform grid of realizable pole locations in the z plane. The nonuniform grid is defined by the intersection of vertical lines corresponding to quantization of $2\lambda_r$ and concentric circles corresponding to quantization of $-r^2$. The sparseness of realizable pole locations near $z = \pm 1$ results in a large coefficient quantization error for poles in this region. By contrast, quantizing coefficients of the normal realization [Barnes 1979] corrresponds to quantizing λ_r and λ_i resulting in a uniform grid of realizable pole locations. In this case large coefficient quantization errors are avoided for all pole locations.

It is well established that filter structures with low roundoff noise tend to be robust to coefficient quantization, and vice versa [Jackson 1976, Rao 1986]. For this reason, the normal (uniform grid) structure is also popular because of its low roundoff noise.

It is well known that in a high-order polynomial with clustered roots the root location is a very sensitive function of the polynomial coefficients. Therefore, filter poles and zeroes can be much more accurately controlled if higher order filters are realized by breaking them up into the parallel or cascade connection of first- and second-order subfilters. One exception to this rule is the case of linear-phase FIR filters in which the symmetry of the polynomial coefficients and the spacing of the filter zeroes around the unit circle usually permits an acceptable direct realization using the convolution summation.

Realization Considerations

Linear-phase FIR digital filters can generally be implemented with acceptable coefficient quantization sensitivity using the direct convolution sum method. When implemented in this way on a digital signal processor, fixed-point arithmetic is not only acceptable but may actually be preferable to floating-point arithmetic. Virtually all fixed-point digital signal processors accumulate a sum of products in a double-length accumulator. This means that only a single quantization is necessary to compute an output. Floating-point arithmetic, on the other hand, requires a quantization after every multiply and after every add in the convolution summation. With 32-bit floating-point arithmetic these quantizations introduce a small enough error to be insignificant for most applications.

When realizing IIR filters, either a parallel or cascade connection of first- and second-order subfilters is almost always preferable to a high-order direct form realization. With the availability of low-cost floating-point digital signal processors, like the Texas Instruments TMS320C32, it is highly recommended that floating-point arithmetic be used for IIR filters. Floating-point arithmetic simultaneously eliminates most concerns regarding scaling, limit cycles, and overflow oscillations. Regardless of the arithmetic employed, a low roundoff noise structure should be used for the second-order sections. The use of a low roundoff noise structure for the second-order sections also tends to give a realization with low coefficient quantization sensitivity. First-order sections are not as critical in determining the roundoff noise and coefficient sensitivity of a realization, and so can generally be implemented with a simple direct form structure.

Defining Terms

Adaptive FIR filter: A finite impulse response structure filter with adjustable coefficients. The adjustment is controlled by an adaptation algorithm such as the least mean square (lms) algorithm. They are used extensively in adaptive echo cancellers and equalizers in communication sytems.

Causality: The property of a system that implies that its output can not appear before its input is applied. This corresponds to an FIR filter with a zero discrete-time impulse response for negative time indices.

Discrete time impulse response: The output of an FIR filter when its input is unity at the first sample and otherwise zero.

Group delay: The group delay of an FIR filter is the negative derivative of the phase response of the filter and is, therefore, a function of the input frequency. At a particular frequency it equals the physical delay that a narrow-band signal will experience passing through the filter.

Linear phase: The phase response of an FIR filter is linearly related to frequency and, therefore, corresponds to constant group delay.

Linear, time invariant (LTI): A system is said to be LTI if superposition holds, that is, its output for an input that consists of the sum of two inputs is identical to the sum of the two outputs that result from the individual application of the inputs; the output is not dependent on the time at that the input is applied. This is the case for an FIR filter with fixed coefficients.

Magnitude response: The change of amplitude, in steady state, of a sinusoid passing through the FIR filter as a function of frequency.

Multirate FIR filter: An FIR filter in which the sampling rate is not constant.

Phase response: The phase change, in steady state, of a sinusoid passing through the FIR filter as a function of frequency.

References

Antoniou, A. 1993. *Digital Filters Analysis, Design, and Applications*, 2nd ed. McGraw–Hill, New York.

Barnes, C.W. 1979. Roundoff noise and overflow in normal digital filters. *IEEE Trans. Circuits Syst.* CAS-26(3):154–159.

Bauer, P.H. 1993. Limit cycle bounds for floating-point implementations of second-order recursive digital filters. *IEEE Trans. Circuits Syst.–II* 40(8):493–501.

Bomar, B.W. 1985. New second-order state-space structures for realizing low roundoff noise digital filters. *IEEE Trans. Acoust., Speech, Signal Processing* ASSP-33(1):106–110.

Bomar, B.W. 1994. Low-roundoff-noise limit-cycle-free implementation of recursive transfer functions on a fixed-point digital signal processor. *IEEE Trans. Industrial Electronics* 41(1):70–78.

Buttner, M. 1976. A novel approach to eliminate limit cycles in digital filters with a minimum increase in the quantization noise. In *Proceedings of the 1976 IEEE International Symposium on Circuits and Systems*, pp. 291–294. IEEE, NY.

Cappellini, V., Constantinides, A.G., and Emiliani, P. 1978. *Digital Filters and their Applications*. Academic Press, New York.

Ebert, P.M., Mazo, J.E., and Taylor, M.G. 1969. Overflow oscillations in digital filters. *Bell Syst. Tech. J.* 48(9): 2999–3020.

Gray, A.H. and Markel, J.D., 1973. Digital lattice and ladder filter synthesis. *IEEE Trans. Acoustics, Speech and Signal Processing* ASSP-21:491–500.

Herrmann, O. and Schuessler, W. 1970. Design of nonrecursive digital filters with minimum phase. *Electronics Letters* 6(11):329–330.

IEEE DSP Committee. 1979. *Programs for Digital Signal Processing*. IEEE Press, New York.

Jackson, L.B. 1976. Roundoff noise bounds derived from coefficient sensitivities for digital filters. *IEEE Trans. Circuits Syst.* CAS-23(8):481–485.

Jackson, L.B., Lindgren, A.G., and Kim, Y. 1979. Optimal synthesis of second-order state-space structures for digital filters. *IEEE Trans. Circuits Syst.* CAS-26(3):149–153.

Lee, E.A. and Messerschmitt, D.G. 1994. *Digital Communications*, 2nd ed. Kluwer, Norwell, MA.

Macchi, O. 1995. *Adaptive Processing: The Least Mean Squares Approach with Applications in Telecommunications*, Wiley, New York.

Mills, W.T., Mullis, C.T., and Roberts, R.A. 1978. Digital filter realizations without overflow oscillations. *IEEE Trans. Acoust., Speech, Signal Processing*. ASSP-26(4):334–338.

Mullis, C.T. and Roberts, R.A. 1976. Synthesis of minimum roundoff noise fixed-point digital filters. *IEEE Trans. Circuits Syst.* CAS-23(9):551–562.

Oppenheim, A.V. and Schafer, R.W. 1989. *Discrete-Time Signal Processing*. Prentice–Hall, Englewood Cliffs, NJ.

Oppenheim, A.V. and Weinstein, C.J. 1972. Effects of finite register length in digital filtering and the fast fourier transform. *Proc. IEEE* 60(8):957–976.

Parker, S.R. and Hess, S.F. 1971. Limit-cycle oscillations in digital filters. *IEEE Trans. Circuit Theory* CT-18(11):687–697.

Parks, T.W. and Burrus, C.S. 1987. *Digital Filter Design*. Wiley, New York.

Parks, T.W. and McClellan, J.H. 1972a. Chebyshev approximations for non recursive digital filters with linear phase. *IEEE Trans. Circuit Theory* CT-19:189–194.

Parks, T.W. and McClellan, J.H. 1972b. A program for the design of linear phase finite impulse response filters. *IEEE Trans. Audio Electroacoustics* AU-20(3):195–199.

Proakis, J.G. and Manolakis, D.G. 1992. *Digital Signal Processing Principles, Algorithms, and Applications*, 2nd ed. MacMillan, New York.

Rabiner, L.R. and Gold, B. 1975. *Theory and Application of Digital Signal Processing*. Prentice–Hall, Englewood Cliffs, NJ.

Rabiner, L.R. and Schafer, R.W. 1978. *Digital Processing of Speech Signals*. Prentice–Hall, Englewood Cliffs, NJ.

Rabiner, L.R. and Schafer, R.W. 1974. On the behavior of minimax FIR digital Hilbert transformers. *Bell Sys. Tech. J.* 53(2):361–388.

Rao, D.B.V. 1986. Analysis of coefficient quantization errors in state-space digital filters. *IEEE Trans. Acoust., Speech, Signal Processing* ASSP-34(1):131–139.

Ritzerfeld, J.H.F. 1989. A condition for the overflow stability of second-order digital filters that is satisfied by all scaled state-space structures using saturation. *IEEE Trans. Circuits Syst.* CAS-36(8):1049–1057.

Roberts, R.A. and Mullis, C.T. 1987. *Digital Signal Processing*. Addison–Wesley, Reading, MA.

Vaidyanathan, P.P. 1993. *Multirate Systems and Filter Banks*. Prentice–Hall, Englewood Cliffs, NJ.

Weinstein, C. and Oppenheim, A.V. 1969. A comparison of roundoff noise in floating-point and fixed-point digital filter realizations. *Proc. IEEE* 57(6):1181–1183.

Further Information

Additional information on the topic of digital filters is available from the following sources.

IEEE Transactions on Signal Processing, a monthly publication of the Institute of Electrical and Electronics Engineers, Inc, Corporate Office, 345 East 47 Street, NY.

IEEE Transactions on Circuits and Systems—Part II: Analog and Digital Signal Processing, a monthly publication of the Institute of Electrical and Electronics Engineers, Inc, Corporate Office, 345 East 47 Street, NY.

The Institute of Electrical and Electronics Engineers holds an annual conference at worldwide locations called the International Conference on Acoustics, Speech and Signal Processing, ICASSP, Corporate Office, 345 East 47 Street, NY.

IEE Transactions on Vision, Image and Signal Processing, a monthly publication of the Institute of Electrical Engineers, Head Office, Michael Faraday House, Six Hills Way, Stevenage, UK.

Signal Processing, a publication of the European Signal Processing Society, Switzerland, Elsevier Science B.V., Journals Dept., P.O. Box 211, 1000 AE Amsterdam, The Netherlands.

In addition, the following books are recommended.

Bellanger, M. 1984. *Digital Processing of Signals: Theory and Practice*. Wiley, New York.

Burrus, C.S. et al. 1994. *Computer-Based Exercises for Signal Processing Using MATLAB*. Prentice–Hall, Englewood Cliffs, NJ.

Jackson, L.B. 1986. *Digital Filters and Signal Processing.* Kluwer Academic, Norwell, MA, 1986.

Oppenheim, A.V. and Schafer, R.W. 1989. *Discrete-Time Signal Processing.* Prentice–Hall, Englewood Cliffs, NJ.

Widrow, B. and Stearns, S. 1985. *Adaptive Signal Processing.* Prentice–Hall, Englewood Cliffs, NJ.

10

Multichip Module Technology

Paul D. Franzon
North Carolina State University

10.1 Introduction

Multichip module (MCM) technology allows bare integrated circuits and passive devices to be mounted together on a common interconnecting substrate. An example is shown in Fig. 10.1. In this photograph, eight chips are **wire-bonded** onto an MCM. MCM technology, however, is not just about packaging chips together; it can lead to new capabilities and unique performance and cost advantages. The purposes of this chapter are as follows: to explain the different multichip module technologies, to show how MCMs can be employed to improve the price/performance of systems, and to indicate some of the more important issues in MCM-system design.

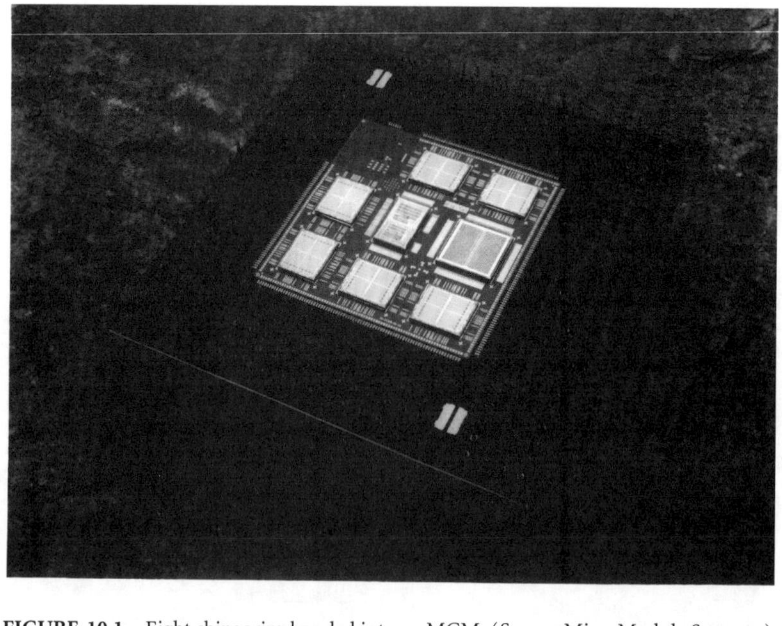

FIGURE 10.1 Eight chips wire-bonded into an MCM. (*Source:* MicroModule Systems.)

0-8493-0050-9/00/$0.00+$.50

10.2 Multichip Module Technology Definitions

In its broadest sense, a multichip module is an assembly in which more than one integrated circuit (IC) are bare mounted on a common substrate. This definition is often narrowed in such a way as to imply that the common substrate provides a higher wiring density than a conventional **printed circuit board (PCB)**. The main physical components are shown in Fig. 10.2 and can be described as follows:

1. The substrate technology provides the interconnect between the chips (ICs or die) and any discrete circuit elements, such as resistors, capacitors, and inductors.
2. The chip connection technology provides the means whereby signals and power are transferred between the substrate and the chips.
3. The MCM package technology is the housing that surrounds the MCM and allows signals, and power and heat to be conducted to the outside world.

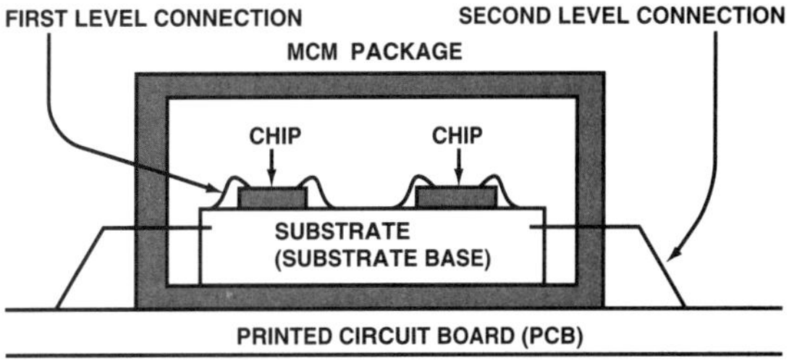

FIGURE 10.2 Physical technology components to an MCM.

What cannot be shown in Fig. 10.2 are several other components that are important to the success of an MCM:

1. The test technology used to ensure correct function of the bare die, the MCM substrate, and the assembled MCM.
2. The repair technology used to replace failed die, if any are detected after assembly.
3. The design technology.

There are many different versions of these technology components. The substrate technology is broadly divided into three categories:

1. **Laminate MCMs (MCM-L)** is shown in (Fig. 10.3). Essentially fine-line PCBs, MCM-Ls are usually constructed by first patterning copper conductors on fiberglass/resin-impregnated sheets as shown in Fig. 10.4. These sheets are laminated together under heat and pressure. Connections between conductors on different sheets are made through via holes drilled in the sheets and plated. Recent developments in MCM-L technology have emphasized the use of flexible laminates. Flexible laminates have the potential for permitting finer lines and vias than fiberglass-based laminates.

2. **Ceramic MCMs (MCM-C)** are mostly based on pin grid array (PGA) technology. Typical cross

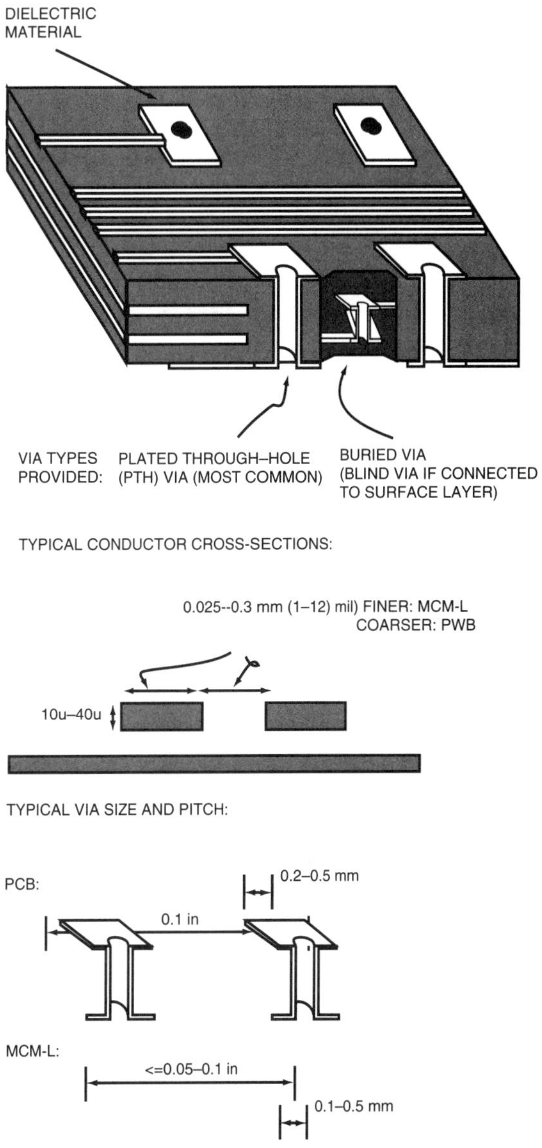

FIGURE 10.3 Typical cross sections and feature sizes in printed circuit board and MCM-L technologies.

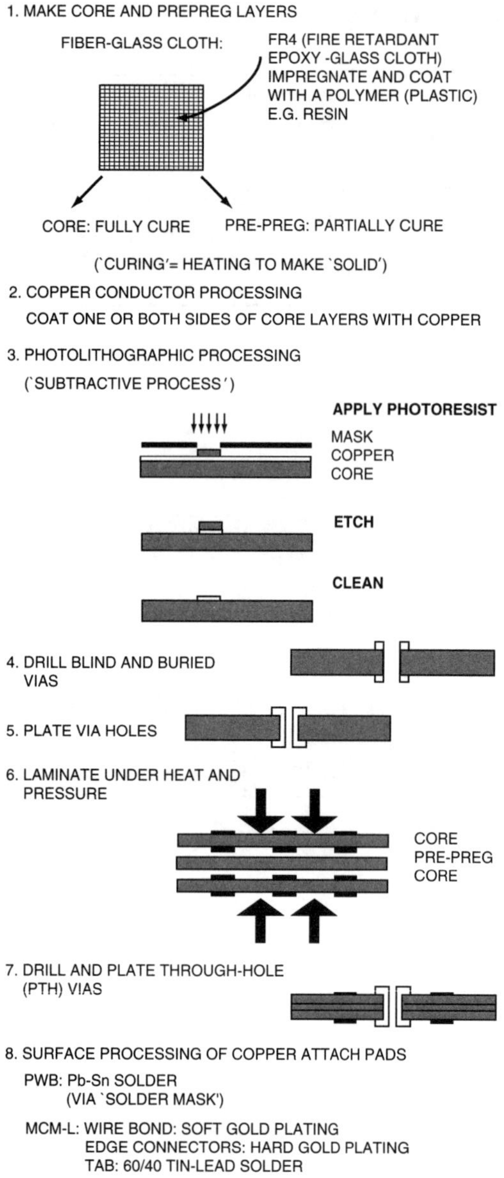

FIGURE 10.4 Basic manufacturing steps for MCM-Ls, using fiberglass/resin prepreg sheets as a basis.

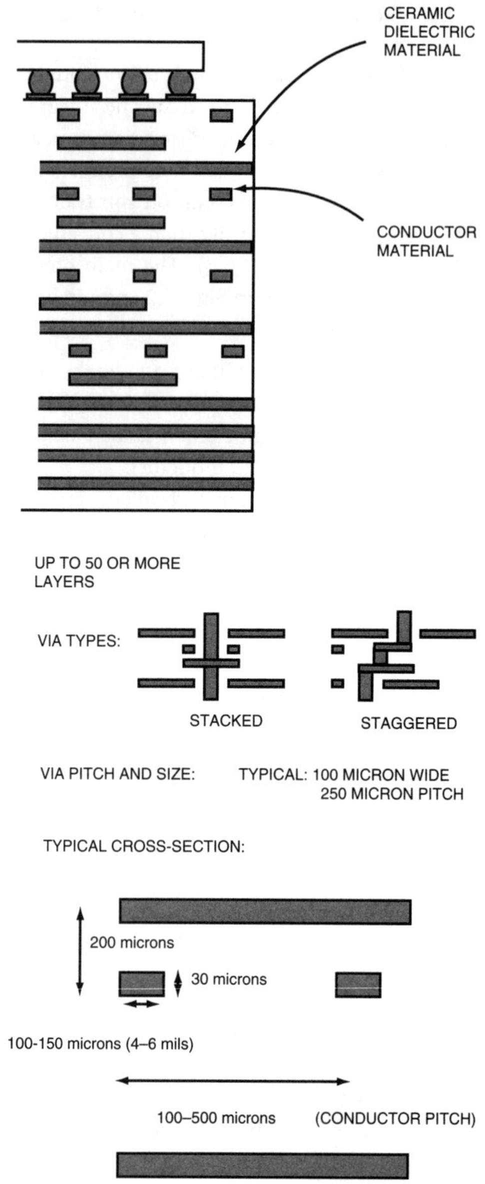

FIGURE 10.5 Typical cross sections and feature sizes in MCM-C technology.

sections and geometries are illustrated in Fig. 10.5 and the basic manufacturing steps are outline in Fig. 10.6. MCM-Cs are made by first casting a uniform sheet of prefired ceramic material, called green tape, then printing a metal ink onto the green tape, then punching and metal filling holes for the vias, and finally cofiring the stacked sheets together under pressure in an oven. In addition to metal, other inks can be used, including ones to print discrete resistors and capacitors in the MCM-C.

3. **Thin-film (deposited) MCMs (MCM-D)** are based on chip metalization processes. MCM-Ds have very fine feature sizes, giving high wiring densities (Fig. 10.7). MCM-Ds are made one layer at a time, using successive photolithographic definitions of metal conductor patterns and vias and the deposition of insulating layers, usually polyimide (Fig. 10.8). Often MCM-Ds are built on a silicon substrate, allowing capacitors, resistors, and transistors to be built, cheaply, as part of the substrate.

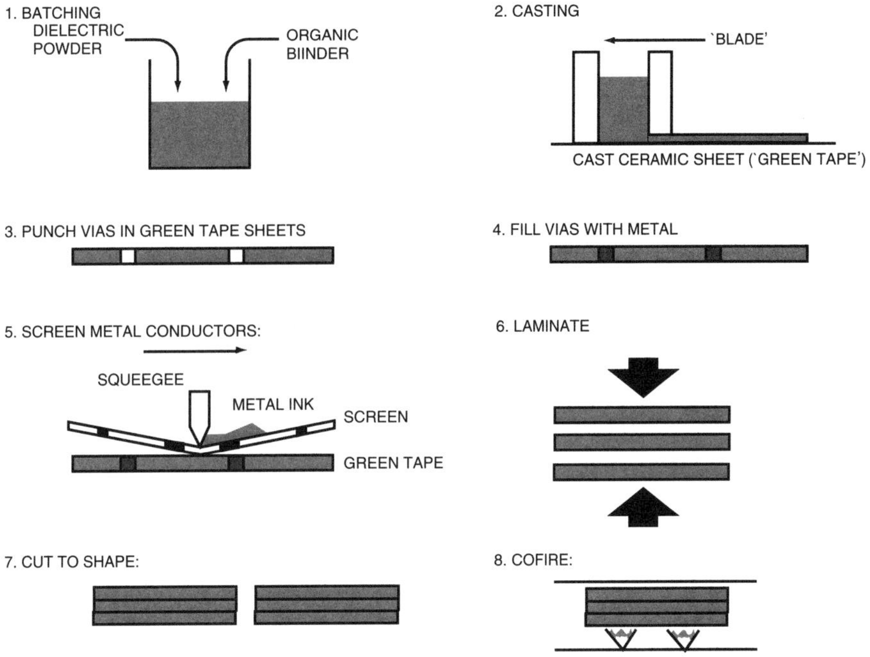

FIGURE 10.6 Basic manufacturing steps for MCM-Cs.

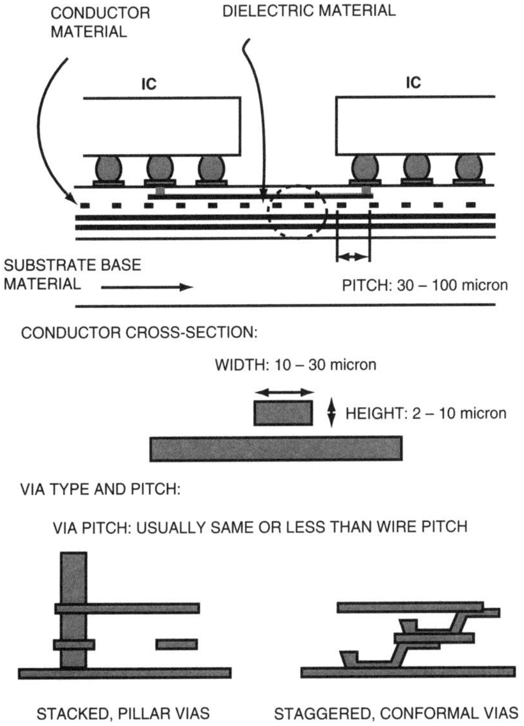

FIGURE 10.7 Typical cross sections and feature sizes in MCM-D technology.

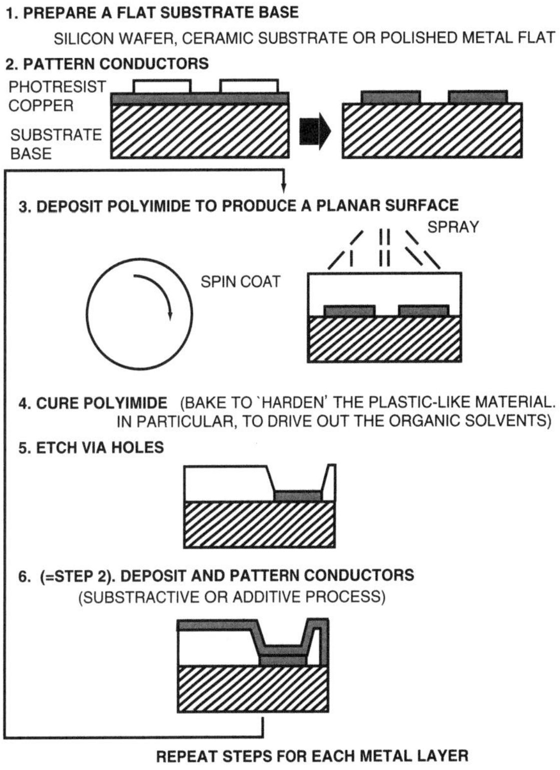

FIGURE 10.8 Basic manufacturing steps for MCM-Cs.

Table 10.1 gives one comparison of the alternative substrate technologies. MCM-Ls provide the lowest wiring and via density but are still very useful for making a small assembly when the total wire count in the design is not too high. MCM-Ds provide the highest wiring density and are useful in designs containing high pin-count chips; however, it is generally the most expensive technology on a per-unit-area basis. MCM-Cs fall somewhere in between, providing intermediate wiring densities at intermediate costs.

TABLE 10.1 Rough Comparison of the Three Different Substrate Technologies: 1 mil = 1/1000 in ≈ 25 μm

	Min. Wire Pitch, μm	Via Size, μm	Approx. Cost per Part $/in^2	Nonrecurring Costs, $
PCB	300	500	0.3	100
MCM-L	150	200	4	100–10,000
MCM-C	200	100	12	25,000
MCM-D	30	25	20	15,000

Current research promises to reduce the cost of MCM-Ls and MCM-D substrates. MCM-Ls based on flexible board technologies should be both cheaper and provide denser wiring than current fibreglass MCM-L technologies. Though the nonrecurring costs of MCM-Ds will always be high, due to the requirement to make fine feature photolithographic masks, the cost per part is projected to decrease as more efficient manufacturing techniques are brought to fruition.

The chip to substrate connection alternatives are presented in Fig. 10.9. Currently, over 95% of the die mounted in MCMs or single chip packages are wire bonded. Most bare die come suitable for wire-bonding, and wire bonding techniques are well known. **Tape automated bonding (TAB)** is an alternative to wire bonding but has only enjoyed mixed success. With TAB assembly, the chip is first attached to the TAB frame inner leads. These leads are then shaped (called *forming*); then the outer leads can be attached to the MCM. TAB has a significant advantage over wire bonding in that the TAB-mounted chips can be more 2easily tested than bare die. The high tooling costs, however, for making the TAB frames, makes it less desirable in all but high volume production chips. The wire pitch in wire bonding or TAB is generally 75 μm or more.

With **flip-chip solder-bump** attachment, solder bumps are deposited over the area of the silicon wafer. The wafer is then diced into individual die and flipped over the MCM substrate. The module is then placed in a reflow oven, where the solder makes a strong connection between the chip and the substrate. Flip-chip attachment is gaining in popularity due to the following unique advantages:

- Solder bumps can be placed over the entire area of the chip, allowing chips to have thousands of connections. For example, a 1-cm^2 chip could support 1600 solder bumps (at a conservative 250-μm pitch) but only 533 wire bonds.
- Transistors can be placed under solder bumps but cannot be placed under wire-bond or TAB pads. The reason arises from the relative attachment process. Making a good wire-bond or TAB lead attachment requires compression of the bond into the chip pad, damaging any transistors beneath a pad. On the other hand, a good soldered connection is made through the application of heat only. As a result, the chip can be smaller by an area equivalent to the total pad ring area. For example, consider a 100-mm^2 chip with a 250-μm pad ring. The area consumed by the pad ring would total 10-mm^2, or 10% of the chip area. This area could be used for other functions if the chip were flipped and solder bumped. Alternatively, the chip could be made smaller and cheaper.
- The electrical parasitics of a solder bump are far better than for a wire-bond or TAB lead. The latter generally introduce about 1 nH of inductance and 1 pF of capacitance into a circuit. In contrast, a solder bump introduces about 10 pH and 10 nF. The lower parasitic inductance

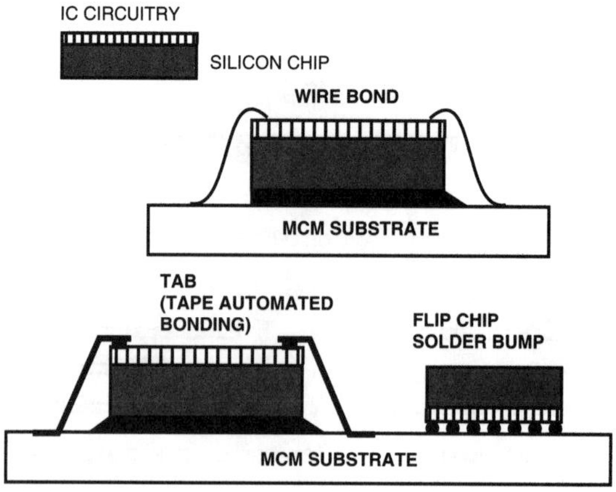

FIGURE 10.9 Chip attach options.

and capacitance make solder bumps attractive for high-frequency radio applications, for example, in the 5.6-GHz communications band and in high clock rate digital applications; MCMs clocking at over 350 MHz are on the drawing board as of this writing.

- The costs of flip-chip are comparable to wire bonding. Solder-bumping a wafer today costs about $400 and assembly only requires alignment and an oven. In contrast, wire-bonding costs about 1 cent per bond. A wafer containing one hundred 400-pin chips would cost the same with either approach.

Flip-chip solder-bump, however, does have some disadvantages. The most significant follows from the requirement that the solder bump pads are larger than wire-bond pads (60–80 μm vs approximately 50 μm). The solder-bump pads need to be larger in order to make the bumps taller. Taller bumps can more easily absorb the stresses that arise when the module is heated and cooled during assembly. As the thermal coefficient of expansion (TCE) of silicon is usually different from that of the MCM substrate, these two elements expand and contract at different rates, creating stresses in the connecting bumps. For the same reason, it is better to place the bumps in the center of the chip, rather than around the edge. By reducing the distance over which differential expansion and contraction occur, the stresses are reduced.

The larger bump pad sizes requires that the chips be designed specifically for solder-bumping, or that the wafers be postprocessed to distribute solder bumps pads over their surface, and wire (*redistribute*) these pads to the conventional wire-bond pads.

Another potential disadvantage arises from the lead content of the solder bump. Containing radioactive isotopes, most lead is a source of alpha particles that can potentially change the state of nearby transistors not shielded by aluminum. The effects of alpha particles are mainly of concern to dynamic random access memories (DRAMs) and dynamic logic.

10.3 Design, Repair, and Test

An important adjunct to the physical technology is the technology required to test and repair the die and modules. An important question that has to be answered for every MCM is how much to test the die before assembly vs how much to rely on postassembly test to locate failed die. This is purely a cost question. The more the bare die are tested, the more likely that the assembled MCM will work. If the assembled MCM does not work, then it must either be repaired (by replacing a die) or discarded. The question reduces to the one that asks what level of bare die test provides a sufficiently high confidence that the assembled MCM will work, or is the assembled module cheap enough to throw away is a die is faulty.

TABLE 10.2 Levels of Known Good Die and Their Impact

Known Good Die Level	Test Cost Impact	Test Escape Impact
Wafer level functional and parametric	Low	<1–2% fallout for mature ICs possibly >5% fallout for new ICs
At-pin speed sorted	Medium	Min. fallout for new digital ICs
Burned-in	High	Burn-in important for memories
Dynamically burned-in with full testing	Highest	Min. memory fallout

In general, there are four levels of bare die test and burn-in, referred to as four levels of **known good die (KGD)**. In Table 10.2, these test levels are summarized, along with their impact. The lowest KGD level is to just use the same tests normally done at wafer level, referred to as the *wafer sort tests.* Here the chips are normally subject to a low-speed functional test combined with some parametric measurements (e.g., measurement of transistor curves). With this KGD level, test costs for bare die are limited to wafer test costs only. There is some risk, however, that the chip will not work when tested as part of the MCM. This risk is measured as the *test escape rate.* With conventional packaging, the chips are tested again, perhaps at full speed, once they are packaged, making the test escape rate zero.

If the MCM contains chips that must meet tight timing specifications (e.g., a workstation) or the MCM must meet high reliability standards (e.g., for aerospace), however, then higher KGD test levels are required. For example, a workstation company will want the chips to be speed-sorted. Speed-sorting requires producing a test fixture that can carry high-speed signals to and from the tester. The test fixture type usually used at wafer sort, generally referred to as a *probe card*, is usually not capable of carrying high-speed signals. Instead, it is necessary to build a more expensive, high-speed test fixture, or temporarily mount the bare die in some temporary package for testing. Naturally, these additional expenses increase the cost of providing at-pin speed sorted die.

Some applications require even higher levels of known good die. Aerospace applications usually demand burn-in of the die, particularly for memories, so as to reduce the chance of infant mortality. There are two levels of burned-in KGD. In the lowest level, the die are stressed at high temperatures for a period of time and then tested. In the higher level, the die are continuously tested while in the oven.

How do you decide what test level is appropriate for your MCM? The answer is driven purely by cost; the cost of test vs the cost of repair. For example, consider a 4-chip MCM using mature ICs. Mature ICs tend to have very high yields, and the process engineers have learned how to prevent most failure modes. As a result, the chances are very small that a mature IC would pass the waferlevel functional test and fail in the MCM. With a test escape rate of 2%, there is only a 8% ($1 - 0.98^4$) chance that each MCM would need to have a chip replaced after assembly. If the repair costs \$30 per replaced chip, then the average excess repair cost per MCM is \$2.40. It is unlikely that a higher level of KGD would add only \$0.60 to the cost of a chip. Thus, the lowest test level is justified.

On the other hand, consider an MCM containing four immature ICs. The functional and parametric wafer-level tests are poor at capturing speed faults. In addition, the process engineers have not had a chance to learn how to maximize the chip yield and how to best detect potential problems. If the test escape rate was 5%, there would be a 40% chance that a 4-chip MCM would require a chip to be replaced. The average repair cost would be \$12 per MCM in this scenario, and the added test cost of obtaining speed-sorted die would be justified.

For high-speed systems, speed sorting also greatly influences the speed rating of the module. Today, micro-processors, for example, are graded according to their clock rate. Faster parts command a higher price. In an MCM the entire module will be limited by the slowest chip on the module, and when a system is partitioned into several smaller chips, it is highly likely that a module will contain a slow chip if they are not tested.

For example, consider a set of chips that are manufactured in equal numbers of slow, medium, and fast parts. If the chips are speed sorted before assembly into the MCM, there should be 33% slow modules, 33% medium modules and 33% fast modules. If not speed sorted, these ratios change drastically as shown in Fig. 10.10. For a 4-chip module assembled from unsorted die, there will be 80% slow systems, 19% medium systems, and only 1% fast systems. This dramatic reduction in fast modules also justifies the need for speed-sorted die.

Design technology is also an important component to the MCM equation. Most MCM computer-aided design (CAD) tools have their basis in the PCB design tools. The main differences have been the inclusion of features to permit the use of the wide range of physical geometries possible in an MCM (particularly via geometries), as well as the ability to use bare die. A new format, the *die interchange format*, has been specifically developed to handle physical information concerning bare die (e.g., pad locations).

There is more to MCM design technology, however, than just making small changes to the physical design tools (those tools that actually produce the MCM wiring patterns). Design correctness is more important in an MCM than in a PCB. For example, a jumper wire is difficult to place on an MCM in order to correct for an error. Thus, recent tool developments have concentrated on improving the designers ability to ensure that the multichip system is designed correctly before it is built. These developments include new simulation libraries that allow the designer to simulate the entire chip set before building the MCM, as well as tools that automate the electrical and thermal design of the MCM.

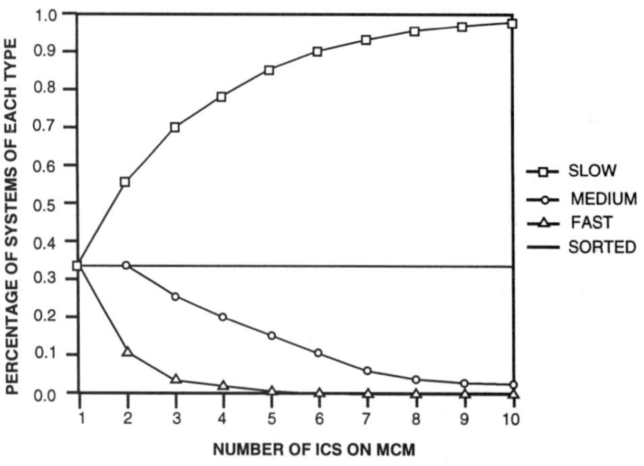

FIGURE 10.10 The effect of speed sorting on MCM performance.

10.4 When to Use Multichip Modules

There are a number of scenarios given as follows that typically lead to consideration of an MCM alternative.

1. You must achieve a smaller form factor than is possible with single chip packaging. Often integrating the digital components onto an MCM-L provides the most cost effective way in which to achieve the required size. MCM-D provides the greatest form factor reduction when the utmost smallest size is needed. For example, an MCM-D might be used to integrate a number of ICs so that they can fit into one existing socket. MicroModule Systems integrated four ICs in an MCM-D to make a Intel 286 to 486 PC upgrade component that was the same size as the original 286 package. If the design has a large number of discrete capacitors and resistors, then using an MCM-D technology with integrated capacitors and resistors might be the cheapest alternative. Motorola often uses MCM-Cs for this purpose, whereas AT&T uses MCM-Ds with integrated capacitors and resistors. In both cases, the capacitors and resistors are printed onto the MCM substrate using the same techniques previously used to make so-called *hybrid components*. Size reduction is currently the most common driver of commercial use of MCMs.

2. The alternative is a single die that would either be too large to manufacture or would be so large as to have insufficient yield. The design might be custom or semicustom. In this case, partitioning the die into a number of smaller die and using an MCM to achieve the interconnect performance of the large die is an attractive alternative. For example, Silicon Graphics once implemented a 4-chip graphics coprocessor, using 4 application specific integrated circuits (ASICs) on an MCM-C (essentially a multicavity pin grid array [PGA]). The alternative was a single full custom chip, which would have taken a lot longer to design. But they needed a single chip form factor in order to fit the coprocessor into the board space allocated.

Recently, teams of researchers have been investigating the possibility of building a *megachip* in a multichip module. The prospect is that a multichip module containing several chips be built to provide the performance and size of a single, very large, highly integrated chip, but at a lower cost. Cost reduction arises due to the far better yields (percentage of die that work) of a set of small die over a larger die. This use of MCM technology is of particular interest to the designers of high-performance computing equipment. This concept is explored further in Dehkordi et al. [1995] and Franzon et al. [1995].

3. You have a technology mix that makes a single IC expensive or impossible and electrical performance is important. For example, you need to interface a complementary-metal-oxide-semiconductor (CMOS) digital IC with a GaAs microwave monolithic integrate circuit (MMIC) or high bandwidth analog IC. Or, you need a large amount of static random access memory (SRAM) and a small amount of logic. In these cases, an MCM might be very useful. An MCM-L might be the best choice if the number of interchip wires are small. A high layer count MCM-C or MCM-D might be the best choice if the required wiring density is large. For example, AT&T wanted to integrate several existing die into one new die. All of the existing die had been designed for different processes. They determined that it was cheaper to integrate the exisiting die into a small MCM, rather than redesign all of the die into a common IC process.

4. You are pad limited or speed limited between two ICs in the single chip version. For example, many computer designs benefit by having a very wide bus (256 bits or more) between the two levels of cache. In this case, an MCM allows a large number of very short connections between the ICs. For example, in the Intel Pentium Pro (P6) processor, there is a 256-bit bus between the central processing unit and the second level cache chip. Intel decided that an MCM-C was required to route this bus between the two chips and achieve the required bus delay. The combination of MCM-D with flip-chip solder-bump technology provides the highest performance means of connecting several die. The greatest benefit is obtained if the die are specially designed for the technology.

5. You are not sure that you can achieve the required electrical performance with single chip packaging. An example might be a bus operating above 100 MHz. If your simulations show that it will be difficult to guarantee the bus speed, then an MCM might be justified. Another example might be a mixed signal design in which noise control might be difficult. In general MCMs offer superior noise levels and interconnect speeds to single-chip package designs.

6. The conventional design has a lot of solder joints and reliability is important to you. An MCM design has far fewer solder joints, resulting in less field failures and product returns. (Flip-chip solder bumps have shown themselves to be far more reliable than the solder joints used at board level.)

Though there are many other cases where the use of MCM technology makses sense, these are the main ones that have been encountered so far. If an MCM is justified, the next question might be to decide what ICs need to be placed on the MCM. A number of factors follow that need to be considered in this decision:

1. It is highly desirable that the final MCM package be compatible with single-chip packaging assembly techniques to facilitate manufacturability.
2. Although wires between chips on the MCM are inexpensive, off-MCM pins are expensive. The MCM should contain as much of the wiring as is feasible. On the other hand, an overly complex MCM with a high component count will have a poor final yield.
3. In a mixed signal design, separating the noisy digital components from the sensitive analog components is often desirable. This can be done by placing the digital components on an MCM. If an MCM-D with an integrated decoupling capacitor is used, then the on-MCM noise might be low enough to allow both analog and digital components to be easily mixed. Be aware in this case, however, that the on-MCM ground will have different noise characteristics than the PCB ground.

In short, most MCM system design issues are decided on by careful modeling of the system-wide cost and performance. Despite the higher cost of the MCM package itself, often cost savings achieved elsewhere in the system can be used to justify the use of an MCM.

10.5 Issues in the Design of Multichip Modules

The most important issue in the design of multichip modules is obtaining bare die. The first questions to be addressed in any MCM project are the questions as to whether the required bare die are available in the quantities required, with the appropriate test level, with second sourcing, at the right price, etc. Obtaining answers to these questions is time consuming as many chip manufacturers still see their bare die sales as satisfying a niche market. If the manufactured die are not properly available, then it is important to address alternative chip sources early.

The next most important issue is the test and verification plan. There are a number of contrasts with using a printed circuit board. First, as the nonrecurring engineering cost is higher for a multichip module, the desirability for first pass success is higher. Complete prefabrication design verification is more critical when MCMs are being used, so more effort must be spent on logic and electrical simulation prior to fabrication.

It is also important to determine, during design, how faults are going to be diagnosed in the assembled MCM. In a prototype, you wish to be able locate design errors before redoing the design. In a production module, you need to locate faulty die or wire bonds/solder bumps if a faulty module is to be repaired (typically by replacing a die). It is more difficult to physically probe lines on an MCM, however, than on a PCB. A fault isolation test plan must be developed and implemented. The test plan must be able to isolate a fault to a single chip or chip-to-chip interconnection. It is best to base such a plan on the use of chips with **boundary scan** (Fig. 10.11). With boundary scan chips, test vectors can be scanned in serially into registers around each chip. The MCM can then be run for one clock cycle, and the results scanned out. The results are used to determine which chip or interconnection has failed. If boundary scan is not available, and repair is viewed as necessary, then an alternative means for sensitizing between-chip faults is needed.

The decision as to whether a test is considered necessary is based purely on cost and test-escape considerations. Sandborn and Moreno [1994] and Ng in Donane and Franzon [1992, Chap. 4] provide more information. In general, if an MCM consists only of inexpensive, mature die, repair is unlikely to be worthwhile. The cost of repair (generally \$20–\$30 per repaired MCM + the chip cost) is likely to be more than the cost of just throwing away the failed module. For MCMs with only one expensive, low-yielding die, the same is true, particularly if it is confirmed that the cause of most failures is that die. On the other hand, fault diagnosis and repair is usually desirable for modules containing multiple high-value die. You do not wish to throw all of these die away because only one failed.

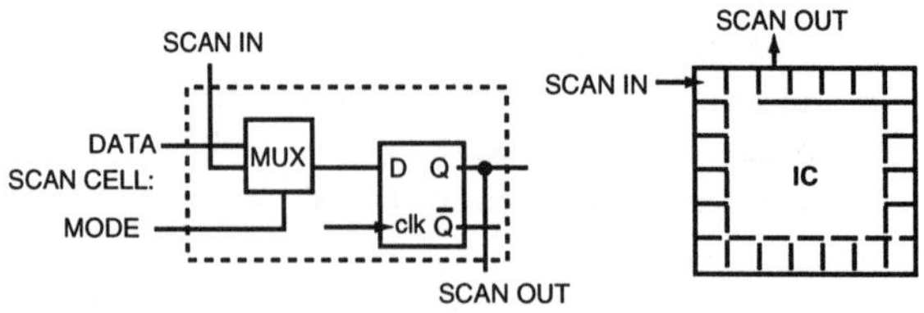

FIGURE 10.11 The use of boundary scan eases fault isolation in an MCM.

Thermal design is often important in an MCM. An MCM will have a higher heat density than the equivalent PCB, sometimes necessitating a more complex thermal solution. If the MCM is dissipating more than 1 W, it is necessary to check if any heat sinks and/or thermal spreaders are necessary.

Sometimes, this higher concentration of heat in an MCM can work to the designer's advantage. If the MCM uses one larger heat sink, as compared with the multiple heat sinks required on the single-chip packaged version, then there is the potential for cost savings.

Generally, electrical design is easier for a MCM than for a PCB; the nets connecting the chips are shorter, and the parasitic inductances and capacitances smaller. With MCM-D technology, it is possible to closely space the power and ground planes, so as to produce an excellent decoupling capacitor. Electrical design, however, can not be ignored in an MCM; 300-MHz MCMs will have the save design complexity as 75-MHz PCBs.

MCMs are often used for mixed signal (mixed analog/RF and digital) designs. The electrical design issues are similar for mixed signal MCMs as for mixed signal PCBs, and there is a definite lack of tools to help the mixed signal designer. Current design practices tend to be qualitative. There is an important fact to remember that is unique to mixed signal MCM design. The on-MCM power and ground supplies are separated, by the package parasitics, from the on-PCB power and ground supplies. Many designers have assumed that the on-MCM and on-PCB references voltages are the same only to find noise problems appearing in their prototypes.

For more information on MCM design techniques, the reader is referred to Doane and Franzon [1992], Tummula and Rymaszewski [1989], and Messner et al. [1992].

10.6 Conclusions and the Future

Multichip module technology is a not just a packaging technology. Its use often compels the design engineer to rethink the design of silicon chips. MCM technology allows designers to integrate multiple chips into a package that looks and performs as if it was a single-chip solution. It adds new capabilities to the system and component designers repertoire.

At this writing, the biggest challenge in MCM design is obtaining the bare die at the required test level and in the correct volume. Fortunately, the infrastructure for obtaining bare die is improving rapidly, making the consideration of MCM products easier.

The second biggest issue in MCM design today is the cost of the substrate. Currently, the cost of an MCM is more than the equivalent cost of single-chip packaging and assembly onto a printed circuit board. The cost issue is changing as well. By using flex circuit and other techniques, MCM-L technology will soon be to the point where the larger per-area cost, as compared with a PCB, will be directly compensated by the smaller area needed. By using large panel processing, less wasteful manufacturing techniques, and integrating the exterior package into the substrate manufacturing, a packaged MCM-D will soon cost the same as a high-performance single-chip package.

Defining Terms

Boundary scan: Technique used to test chips after assembly onto a PCB or MCM. The chips are designed so that test vectors can be scanned into their input/output registers. These vectors are then used to determine if the chips, and the connections between the chips are working.

Ceramic MCM (MCM-C): A MCM built using ceramic packaging techniques.

Deposited MCM (MCM-D): A MCM built using the deposition and thin-film lithography techniques that are similar to those used in integrated circuit manufacturing.

Flip-chip solder-bump: Chip attachment technique in which pads and the surface of the chip have a solder ball placed on them. The chips are then flipped and mated with the MCM or PCB, and the soldered reflowed to create a solder joint. Allows area attachment.

Known good die (KGD): Bare silicon chips (die) tested to some known level.

Laminate MCM (MCM-L): A MCM built using advanced PCB manufacturing techniques.

Multichip module (MCM): A single package containing several chips.

Printed circuit board (PCB): Conventional board found in most electronic assemblies.

Tape automated bonding (TAB): A manufacturing technique in which leads are punched into a metal tape, chips are attached to the inside ends of the leads, and then the chip and lead frame are mounted on the MCM or PCB.

Wire bond: Where a chip is attached to an MCM or PCB by drawing a wire from each pad on the chip to the corresponding pad on the MCM or PCB.

Acknowledgments

The author wishes to thank Andrew Stanaski for his helpful proofreading and for Fig. 10.10 and the comments related to it. He also wishes to thank Jan Vardaman and Daryl A. Doane for providing the tremendous learning experiences that lead to a lot of the knowledge provided here.

References

Dehkordi, P., Ramamurthi, K., Bouldin, D., Davidson, H., and Sandborn, P. 1995. Impact of packaging technology on system partitioning: A case study. In *1995 IEEE MCM Conference*, pp. 144–151.

Doane, D.A. and Franzon, P.D., eds. 1992. *Multichip Module Technologies and Alternatives: The Basics.* Van Nostrand Reinhold, New York.

Franzon, P.D., Stanaski, A., Tekmen, Y., and Banerjia. S. 1996. System design optimization for MCM. *Trans. CPMT.*

Messner, G., Turlik, I., Balde, J.W., and Garrou, P.E., eds. 1992. *Thin Film Multichip Modules.* ISHM.

Sandborn, P.A. and Moreno., H. 1994. *Conceptual Design of Multichip Modules and Systems.* Kluwer, Norwell, MA.

Tummula, R.R. and Rymaszewski, E.J., eds. 1989. *Microelectronics Packaging Handbook.* Van Nostrand Reinhold, Princeton, NJ.

Further Information

Additional information on the topic of multichip module technology is available from the following sources:

The major books in this area are listed in the References.

The primary journals are the *IEEE Transactions on Components, Packaging and Manufacturing Technology, Parts A and B,* and *Advancing Microelectronics*, published by ISHM.

The foremost trade magazine is *Advanced Packaging*, available free of charge to qualified subscribers.

The two main technical conferences are: (1) The IEEE Multichip Module Conference (MCMC), held in early Spring, emphasizes design and applications of multichip modules, and (2) the ISHM/IEEE Multichip Module Conference, held in mid Spring, has a large trade show and emphasizes advances in technology.

11

Software Development Tools for Field Programmable Gate Array Devices[1]

Bradly K. Fawcett
Xilinx Inc.

11.1 Introduction

Since their introduction in 1985, **field programmable gate arrays (FPGAs)** have rapidly emerged as a leading **application-specific integrated circuit** (**ASIC**) technology. These versatile devices can be programmed to execute a wide variety of digital logic functions, allowing designers to customize logic for a particular application without the expense and risk of designing their own integrated circuit (IC) devices. FPGAs already account for more design starts than traditional, mask-programmed gate arrays and continue to be one of the fastest growing segments of the digital logic market.

The logic functions implemented in an FPGA device are determined by a configuration program that is loaded into the device. Just as a software engineer uses a variety of specialized tools, assemblers, compilers, debuggers, etc., to generate programs for a particular computer system, logic designers need powerful development tools to create configuration programs for FPGAs. The mushrooming popularity of FPGAs and the growing complexity of FPGA-based designs has led to a proliferation of FPGA development software from both FPGA device manufacturers and electronic design automation (EDA) tool vendors.

FPGA development tools must be capable of implementing complex logic designs on an engineer's desktop personal computer (PC) or workstation, delivering the ease-of-design and fast time-to-market benefits that have popularized FPGA technology. Recent advances in FPGA architectures have been matched by equally dramatic improvements to FPGA development tools. The latest FPGA development tools offer an increasing variety of design entry strategies, FPGA-specific **logic synthesis**, increased design **portability**, improved design implementation tools and **simulation** support, and framework integration. The result of these continuing improvements is increased productivity for the system designer.

The basic methodology for FPGA design consists of three interrelated steps: entry, implementation, and validation (Fig. 11.1). Design entry is the process of specifying the logic functions needed for the particular application. Traditionally, logic designers compose schematics with graphical symbols representing the desired logic functions and their interconnections. Text-based entry—where in logic functions are described in equations and statements in a manner similar to writing computer programs—is growing in popularity as designs grow in size and complexity. Design implementation involves creating the physical realization of the desired logic in the chosen FPGA architecture, including mapping the design into the logic resources of the FPGA and determining an optimal placement and internal interconnect scheme for those resources. This process is highly automated. Validation refers to the testing of a completed design's operation; validation strategies include in-system testing, static timing analysis, and device-, board- or system-level simulation. The design process invariably is iterative, returning to the design entry phase for correction and optimization.

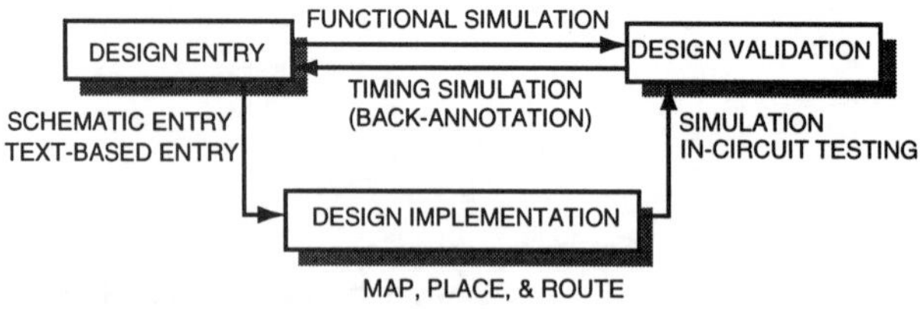

FIGURE 11.1 The basic FPGA design methodology consists of three steps: entry, implementation, and validation.

Today's leading FPGA suppliers support open development environments on PC and workstation platforms that allow the designer to select the design entry and simulation tools of his or her choice. Thus, familiar tools and design techniques can be used to develop FPGA designs. The *place and route* tools that implement a design in a given FPGA architecture, however, are unique to each architecture and tend to be proprietary products of the FPGA vendor. Typically, design descriptions are passed between third-party EDA tools (such as schematic editors and simulators) and an FPGA's implementation tools in a **netlist** format specific to the FPGA vendor. Several FPGA vendors have established formal alliance programs to encourage the development of interfaces to FPGA tools by third-party EDA vendors, and well over 100 different EDA products support FPGA design.

FPGA Architecture

In the 1970s, digital systems were made up of microprocessor and memory ICs and lots of small packages, generally referred to as *glue logic*. As systems became larger and more complex, an economical means was needed for integrating all of the logic into higher density IC devices. But large ICs are affordable only when mass produced, and logic circuits differ from product to product and application to application.

These needs led to the rise of the ASIC in the late 1970s and early 1980s. The most popular type of ASIC is the mask-programmed gate array. As the name implies, a gate array is an IC device composed of an array or matrix of logic gates. These are mass produced up to the point of determining the metal connections that join the gates together into logic circuits. Gate arrays are customized for a particular application by the creation of the metal mask layers that determine the gate interconnection scheme. Typically, a gate array user pays a fee, referred to as nonrecurring engineering charges (NRE), to have metal mask layers created to implement a given application. Prototype devices are fabricated, tested by the customer, and then moved into volume production.

However, gate arrays have their drawbacks. Lead times for prototype development and production ramp-up delay product introduction. The first prototypes often do not function as well as planned; multiple iterations can add months to the design cycle. Test programs need to be generated for each device, a task often as difficult as the initial design of the part. In short, gate array development entails considerable risk and expense.

FPGAs, invented in the mid-1980s, overcome many of the disadvantages of gate arrays. These high-density integrated circuits combine the flexibility and extendibility of a gate array architecture with user programmability, eliminating many of the risks and delays of ASIC development. Like gate arrays FPGAs contain a matrix of logic elements that can be interconnected in any desired configuration to implement a given application. These interconnections, however, are controlled by programmable switches that, using the correct equipment, can be programmed by the user. Thus, application-specific designs can be implemented by the system designer; prototypes can be quickly made, tested, and altered as needed. FPGAs are a standard, mass-produced product from the semiconductor vendor's point of view, but a custom device in the end application from the system designer's point of view.

The general structure of FPGAs, shown in Fig. 11.2, includes three main types of configurable elements: input/output (I/O) blocks, logic blocks, and programmable interconnect. I/O blocks provide an interface between the external package pins and the internal logic. The array of logic blocks provides the functional elements from which the user's logic is constructed; logic blocks typically contain resources for implementing both boolean logic functions and latching data. Programmable interconnect resources provide the routing paths for connecting the I/O and logic blocks into the desired networks. Several architectures are available today, varying in the size, structure, and number of logic and I/O blocks and the amount and connectivity of the routing resources.

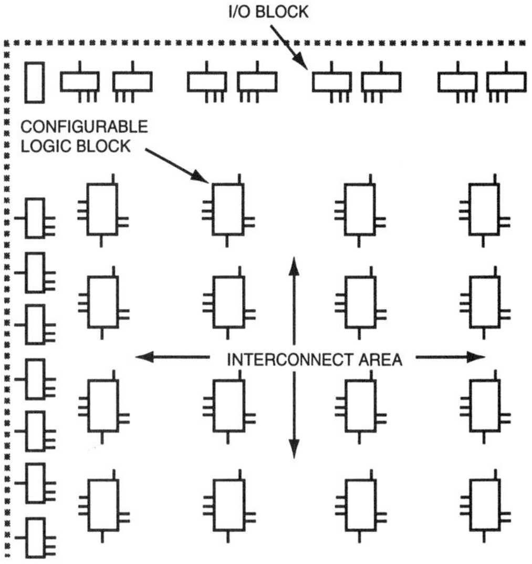

FIGURE 11.2 The main elements of an FPGA architecture are an array of logic blocks, a periphery of input/output blocks, and programmable interconnect.

Like microprocessors, FPGAs are program-driven devices. Two programming technologies are prevalent. Antifuse-based devices can be programmed only once; these devices are programmed by a special piece of equipment prior to insertion in a printed circuit board, just like programmable read-only memories (PROMs). Static-memory-based FPGAs are configured while resident in the target system, much like a programmable microprocessor peripheral; the configuration program is either loaded automatically from external memory on power-up, or is downloaded by a processor during system initialization. Static random access memory- (SRAM-) based FPGAs can be programmed and reprogrammed an unlimited number of times, giving rise to the concept of reconfigurable hardware, that is, logic that can be altered on the fly during the operation of a system.

FPGAs hold less logic and operate at a lower speed than the largest and fastest mask-programmed gate arrays; however, the rapid advance of FPGA technology is quickly closing the gap.

11.2 Design Entry

Design entry is the process of specifying the logic to be implemented within the FPGA device. One result of the proliferation of FPGA development tools is an abundance of design entry methodologies; designers have their choice of several different styles and types of design entry.

Schematic editors remain the most-popular tool for design entry. Use of a schematic editor for FPGA design requires both the appropriate library of logic symbols and a translator to the netlist format used by the FPGA vendor's implementation tools; typically, these two items are supplied as a package by the FPGA or EDA tool vendor. FPGA symbol libraries and netlist interfaces are available for many popular schematic editors, such as Viewdraw™ from Viewlogic Systems, SDT from OrCAD Systems, Design Architect™ from Mentor Graphics, and Composer from Cadence Design Systems.

Schematic libraries for FPGAs contain three types of symbols: primitives, soft macros, and hard macros. Primitives are those symbols that are directly recognized by the implementation software and usually correspond to a logic resource in the target FPGA, such as I/O pads, logic gates, flip-flops, and buffers. Macros are schematics that contain primitives and other macros to implement predefined functions such as counters, shift registers, and decoders. Soft macros have predefined functionality, but their physical implementation (mapping, placement, and routing within the FPGA) is determined completely by the implementation tools. Hard macros, also referred to as relationally placed macros, contain predefined mapping and placement information that optimizes their implementation. FPGA libraries tend to be extensive, reflecting the wide variety of digital functions that can be implemented in these devices. For example, the schematic library for the Xilinx XC4000 FPGA family contains over 300 macros and primitives, ranging from two-input NAND gates to 16-b accumulators. Of course, users can define their own macros based on already defined primitives and macros.

11.3 Hardware Description Languages (HDLs) and Logic Synthesis

The structural approach typical of schematic entry is well suited for describing *glue logic* (that is, the small pieces of interface and control logic found in every system), but is less effective for describing behavioral designs such as state machines. Furthermore, as the density and complexity of FPGA-based designs increases to 10,000 gates and beyond, gate-level schematic entry tools can be cumbersome. These factors have led to the increased use of text-based hardware description languages (HDLs) and logic synthesis tools for FPGA design. (Logic synthesis is the process of generating an optimized gate-level netlist description of a logic design from a higher level structural description.)

Two types of hardware description languages for programmable logic are in use today. Single-module-level HDLs typically are used to describe small modules within a schematic-based hierarchical design; examples include the PALASM®, Data I/O's ABEL™, and Logical Devices' CUPL™ design languages. These languages were developed to support the small, PAL®-like programmable logic devices of the late 1970s. System-level HDLs that support more complex constructs and higher levels of abstraction emerged in the 1980s; the most popular system-level HDLs are VHSIC-HDL (VHDL) and Verilog-HDL. System-level HDLs are modular in that large designs can be created by linking multiple modules constructed using the HDL.

The flexibility of logic synthesis tools and FPGAs complement each other in the design environment. With HDLs, decisions can be tested early in the design cycle through simulation of the HDL description. Design changes are made easily, allowing experimentation and the exploration of tradeoffs in architectural design. FPGAs further enhance this design flexibility by allowing the user to implement and test the design at the workbench. The synthesis compilation time is short enough to allow for exploring design decisions at their gate-level implementations in the FPGA. Thus, tradeoffs in design features, speed, and size are easily explored.

For HDL-based design entry to be useful, however, the synthesis tools must be effective in producing a gate-level design optimized for the target FPGA technology. Early FPGA synthesis tools were modifications of gate array tools that synthesized the design to a collection of gates without regard to the FPGA architecture, often resulting in inefficient implementations. Newer releases perform architecture-specific optimizations, with vastly improved results. Synthesis tools tailored for FPGA architectures are available from vendors such as Synopsys, Viewlogic Systems, Mentor Graphics, Cadence Design Systems, Innovative Synthesis Technologies, and Exemplar Logic.

In general, a synthesis compiler can map logic functions into an FPGA's architectural resources in either of two ways: through inference or through instantiation. Ideally, the compiler should infer the use of resources as much as possible; that is, the compiler should be able to automatically map logic and I/O functions into the appropriate FPGA resource through its synthesis and optimization algorithms, starting with a *generic* HDL code. As logic synthesis tools continue to improve, more FPGA resources are becoming accessible through inference. On the other hand, sometimes specific FPGA resources can be accessed only through instantiation; that is, the user explicitly specifies that these resources are to be used by embedding directives within the HDL code. For example, access to the input/output capabilities of FPGAs—such as I/O registers, internal pull-ups, and slew rate controls—often must be specified explicitly in the HDL code.

Current synthesis algorithms tend to perform well (in terms of the efficiency of the resulting implementation) for random logic elements such as state machines, lookup tables, and decoders. The synthesis of data path and arithmetic structures is more problematic, especially if the target FPGA includes dedicated architectural resources for these functions (e.g., dedicated arithmetic carry logic). The emerging library of parameterized macros (LPM) standard helps alleviate this problem by allowing the synthesis compiler to pass high-level modules (as opposed to gate-level realizations) to the FPGA vendor's implementation tools for optimization in the target device. In other words, rather than passing a collection of gates to the FPGA-specific back-end tools, the synthesis compiler passes a higher level definition of the function (e.g., this is an 8-b binary up–down counter) and allows the FPGA vendor's device-specific LPM compiler to generate the optimized implementation. (Examples of LPM compilers include X-BLOX™ from Xilinx, Scuba from AT&T, and Actgen from Actel.)

11.4 Design Portability and Reuse

Designs typically are more evolutionary than revolutionary in nature, building on previous product development efforts. Ideally, new product development should be able to take advantage of the latest devices and technologies but still reuse proven portions of previous designs. The ability to easily port a design to different device architectures facilitates design reuse in future products.

Design portability is one of the main attractions of high-level design languages. In theory, a design described in an HDL can be technology transparent, relying on synthesis compilers to automatically map the logic into the targeted device architecture. In reality, technology-independent FPGA design with HDLs is difficult; different approaches and coding styles are required for different FPGA architectures, and some FPGA features are accessible only through instantiation. As FPGA synthesis technology continues to improve, generating device-independent HDL code for FPGA design will become easier.

Schematic-based designs also offer some degree of portability. For example, with the unified library in the Xilinx development system, all primitives and macros common to two or more Xilinx device families are consistent in name and appearance. Thus, the symbol for an 8-b binary counter has the same label, size, and pin locations whether targeting the XC2000, XC3000, or XC4000 FPGA family, or the XC7000 electrically erasable PROM-based programmable logic device (EPLD) family. Migration of a design from one family to another requires only a change in the compilation target.

11.5 Mixed-Mode Design Entry

Most FPGA development systems support the use of hierarchical design techniques, with top-level drawings or descriptions defining the major functional blocks and lower-level descriptions defining the logic in each block. Hierarchical design entry has several advantages: it adds structure to the design process, eases debugging, and makes it easier to divide the design effort among members of a design team. Large, complex designs can be handled in smaller pieces. Commonly-used functions can be stored in a library as macros for repeated use.

Ideally, the FPGA implementation software can combine hierarchical elements that are specified with multiple design entry tools, allowing the use of the most convenient entry method for each portion of the design. In this type of mixed-mode design entry, designers can intermix schematic and text, gate-level and behavioral-level design. Mixed-mode design is likely to dominate design entry techniques over the next few years.

11.6 Implementation Tools

Once entered, the physical implementation of an FPGA design involves several steps, as illustrated in Fig. 11.3: translating the input files to the appropriate netlist format, merging together the elements of a hierarchical design, deleting unused logic, mapping the design into the logic resources of the FPGA, determining an optimal placement for the logic and I/O blocks of the FPGA, selecting the routing resources to interconnect the blocks in the array, and generating the actual configuration information used to program the device. A high degree of automation is applied to these tasks. Typically, the entire process can be triggered by invoking a single design compilation utility.

Design implementation begins with the translation and merging of all the hierarchical elements of the design into a single netlist that describes all of the logic of the design. This is followed by logic reduction, the process of deleting any unused logic from the design. Automatic logic deletion permits the liberal use of the macro library without any penalty when some functionality of the macro is not utilized; the unused portions of the macro's logic are deleted from the design automatically.

Sophisticated algorithms are then used to map the design into the FPGA architecture and determine an optimal placement and routing for the design. (FPGA implementation tools are commonly referred to as place and route tools.) The algorithms and techniques used to map, place, and route FPGA designs have been steadily improving since the first automated tools for FPGAs were introduced in 1987. As the implementation algorithms have been refined, this knowledge has influenced the design of new FPGA architectures that are tightly coupled with the tools.

The latest place and route algorithms are timing driven; that is, timing analysis of the signal paths within the application is performed during the placement and routing of the design. The speed at which a design can operate is governed by the speed of the individual paths that make up the design. A path is a set of interconnects and combinational logic blocks that extend from a package pin or clocked register to another package pin or clocked register. Tools such as XACT-Performance™ from Xilinx and TimingWizard™ from NeoCAD allow the user to specify performance requirements along entire paths in an FPGA design (as opposed to the traditional method of assigning net criticality to individual nets). Thus, the user can enter performance specifications that directly reflect the actual performance requirements of the design, and the implementation programs will use this information to guide the placement and routing process.

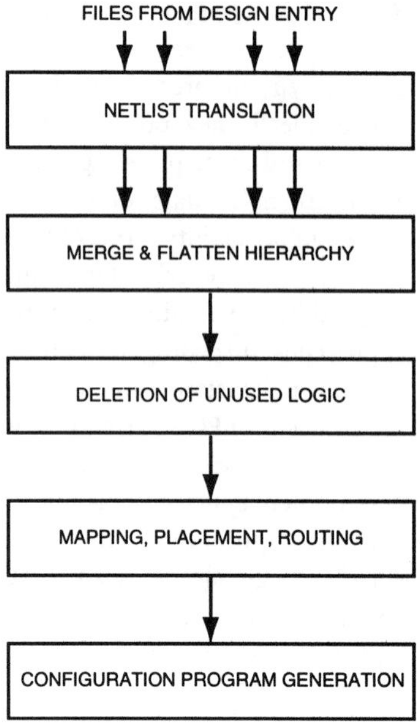

FIGURE 11.3 Steps involved in FPGA design implementation.

For demanding applications, the user may choose to exercise various degrees of control over the automated implementation process. Optionally, user-designated mapping, placement, and routing information can be specified during design entry, such as directives controlling the mapping of selected logic blocks or the assignment of I/O signals to particular pin locations. Available placement controls include *regional constraints*; that is, a selected logic block (or blocks) can have its placement restricted to a certain user-defined region within the array (e.g., place this counter in this corner of the array). The placement algorithms then choose the best location for the block anywhere within that region. This allows the user to perform basic floorplanning to control the overall structure of the implementation; the implementation of highly structured designs can greatly benefit from the floorplanning techniques familiar to designers of large gate arrays.

Incremental design support is another key feature of the latest generation of FPGA implementation tools. Often, minor changes are required near the end of a design cycle to fix last-minute bugs or respond to a change in specifications. The design of large, complex systems can be simplified by first implementing a small portion of the design and then iteratively adding other modules to it. Both these situations require tools that support incremental design techniques.

Incremental design techniques involve using a proven implementation of a design as a guide to the placement and routing of a new version of that design. Where the two designs match, the newer design mimics the old one exactly, preserving its placement and routing and, therefore, its timing characteristics. For example, suppose a large design has been successfully implemented and tested. A small change is then required. By using the guide option, the majority of the placement and routing will remain unchanged, and so only the relatively small, altered portion of the design needs to be reimplemented and retested. With this technique, minor changes can be implemented and verified in a matter of a few minutes. This same option, in concert with placement constraints, supports the incremental construction of a large design. For example, just one small portion of a design could be implemented within an FPGA, perhaps using regional placement constraints to confine it to one portion of the FPGA's logic array. Then additional logic is added to the design, using the previous implementation as its guide. The earlier logic maintains its original placement and routing as the new logic is implemented and then tested. Additional modules are added in a similar manner until the design is complete.

After the design has been fully placed and routed, the final step is the generation of the binary configuration program that is downloaded into the FPGA to implement the specified logic; the configuration program generator is analogous to a microprocessor's assembler.

FPGA Implementation Algorithms

The algorithms used by FPGA place and route programs typically are quite complex. Some have their roots in the algorithms developed to synthesize and place logic into mask-programmed gate arrays. The implementation of FPGA designs involves three interdependent, computationally intensive tasks: mapping, placement, and routing.

Mapping involves partitioning the design's logic into pieces that fit into the logic and I/O blocks of the target FPGA architecture. Mapping algorithms for FPGAs must distinguish between combinational logic gates and other functions such as flip-flops and I/O pads; the latter correspond directly to the equivalent resources within the FPGA, whereas the combinational logic must be mapped into the logic resources internal to the logic blocks. For example, many FPGAs use memory lookup tables to implement combinational logic functions, and a number of algorithms have been developed to map logic into lookup tables. These algorithms attempt to minimize the total number of lookup tables or the number of levels of lookup tables in the final implementation. They differ in the method used to partition the logic, but a key feature of all of these algorithms is a matching step that determines if a given subset of the logic can fit into a single logic block. Some mapping algorithms take into account routability considerations as well as the internal structure of the logic block.

The placement step takes the defined logic and I/O blocks from the technology mapper and assigns them to physical locations in the target FPGA. Many masked-programmed gate array placement algorithms can be applied to FPGAs with some modifications. In these algorithms, placement is optimized using cost functions based on various criteria such as the total net length of resulting block interconnections; net length often corresponds directly to both routability and performance within FPGA architectures. Timing-driven placement tools also incorporate performance requirements into the cost functions of the placement algorithms. Most current FPGA placement tools use some combination of three main algorithmic approaches: **mincut, simulated annealing,** and **general force-directed relaxation** (**GFDR**).

The mincut method begins by dividing the complete design into two clusters, placing half the blocks on each side of a cutline. Then follows a sequence of steps in which elements are swapped, moving across the cutline to reduce cost; the main factor in the cost function is the number of interconnects that must cross the cutline, with the goal of keeping connected blocks close together. The algorithm continues dividing each cluster into smaller clusters until each cluster represents one logic block.

Simulating annealing is a more complex algorithm modeled after the crystallization of a solid from its molten form at high temperature. Starting with a randomly chosen placement, elements are swapped. If the resulting new placement results in a lower overall cost, it is accepted. If the new placement results in a higher cost, there is still a finite probability that the new move will be accepted; the probability of a bad move being accepted is proportional to the temperature during the annealing process. Accepting occasional bad swaps prevents the algorithm from settling into a local minimum when a more optimal solution may exist. Temperatures are high at the beginning of the process and are gradually lowered.

In the GFDR algorithm, nets between blocks are modeled like physical springs, with the strength of the spring being proportional to the criticality of the net and other cost considerations. Placement improvement involves swapping the location of two blocks or cycling the locations of several blocks. The prime candidate destination for a given block is the center of gravity of those blocks having connections to the one being placed.

Once placement is established, the actual routing paths that are used to make the needed interconnections are chosen. Most FPGAs use some variation of a maze router. A maze router attempts to find the shortest path between two or more points; routing is along vertical and horizontal axes only, and the number of routing paths through a given channel between blocks is limited, requiring occasional detours. Most FPGA routers include rip-up and retry capability, wherein an already established route can be removed and rerouted to make room for another signal. Timing-driven routers calculate interconnect delays as they progress, thereby including performance information in the decision process.

FPGA users should have realistic expectations of the capabilities of the implementation tools. The algorithms are complex, and run times ranging from several minutes to several hours can be expected, depending on the size of the FPGA device and the complexity of the design. High-end workstations are recommended for large (>10,000 gates) designs.

11.7 Design Validation

Validation of FPGA designs typically is accomplished through a combination of in-circuit testing and software simulation. The user-programmable nature of FPGAs allows designs to be tested immediately in the target application. As designs increase in density and complexity, however, more circuit paths may have timing problems, and timing simulation becomes an invaluable tool. To support timing simulation, FPGA implementation tools include timing calculators that determine the postlayout timing of implemented designs, including the actual delays of routing paths. This information is annotated into gate level libraries for simulators such as Viewlogic System's Viewsim™, OrCAD's VST, and Mentor Graphics' Quicksim™. Many simulation interfaces support back annotation (the mapping of timing information back into the signal names and symbols of the original schematic) to ease the debugging effort.

To better manage increasing design complexity, a growing number of users employ board-level and system-level simulation spanning multiple device types, in addition to simulating each FPGA on its own. The libraries provided by FPGA vendors typically contain enough information to support board-level simulation. Alternatively, FPGA behavioral models, such as Logic Modeling's SmartModel™ library, can be used with a variety of third-party simulators to support system-level verification.

Another verification technique gaining in popularity is static timing analysis of synchronous designs. Static timing analyzers examine a design's logic and timing to calculate the performance along signal paths, identify possible race conditions, and detect setup and hold-time violations. Unlike simulators, timing analyzers do not require that the user generate input stimulus patterns or test vectors. Most users, however, limit the use of static timing analysis to fully synchronous designs only; the technique is difficult to apply accurately to asynchronous circuits.

FPGA Design Flow Example

The creation of an FPGA design using a schematic editor involves entering the design and testing its functionality, implementing and testing its performance, and programming it. (See Fig. 54.4.) Usually, report files are generated at each step, along with any error or warning messages.

First, then, the symbol library for the target FPGA architecture is used to create the schematic. Some symbols have attributes that are assigned on the schematic. For example, an output buffer may be programmed for a fast or slow slew rate, or an output signal may be assigned to a particular pin location. It is generally a good idea for all symbols and nets on a schematic to be assigned symbolic names. The subsequent preservation of component and net names by the assorted development tools will ease the debugging effort.

Once a design has been entered, its functionality can be tested by being simulated in software, an effective means of identifying logic errors in a design before it is implemented in the FPGA. Because the design elements have not yet been placed and routed, no timing information for the design is available, and so the simulator uses unit delays instead. Most often, a utility program is involved to run the translation routines that create the netlist in a format compatible with the simulator. In more detail, the utility translates the netlists from the schematic editor into the FPGA vendor's netlist format, merges files from the hierarchical blocks that constitute the design, and translates the resulting flattened netlist into the netlist format required by the simulator.

As for implementation, placed and routed FPGA designs can be generated automatically by executing a compilation utility. This software retrieves the design's input files and from them creates the FPGA configuration program. The steps here are: translating the input files to the FPGA's netlist format, merging the elements of a hierarchical design,

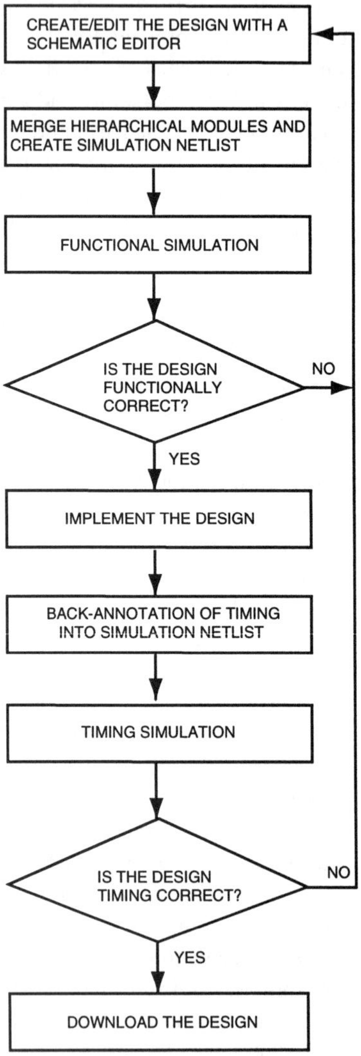

FIGURE 11.4 A typical FPGA design flow proceeds through schematic entry, logic simulation, implementation, and timing simulation. The steps are tightly coupled and iterative, returning to design entry for correction or optimization.

deleting unused logic, mapping the design into the FPGA's logic resources, and routing the logic and I/O blocks, and generating the configuration program that is loaded into the FPGA.

Next, timing simulation verifies the design's functionality by inserting worst-case delay information in a fully implemented FPGA design file. A utility program reads the routed FPGA design file and automatically creates the files and directories needed for the timing simulation. Because the timing delay information for the design is passed through all of the translation programs, actual worst-case timing delays are used during simulation. This reduces the need for hardware debugging as well as determining if the design works under worst-case conditions.

Finally, the user-programmable nature of FPGAs allows designs to be tested immediately in the target application. FPGAs based on static RAMs are usually supported by a cable for downloading configuration files from a PC or workstation to an FPAG residing in the target system. One-time-programmable antifuse-based FPGAs must pass through a dedicated device programmer before being inserted into the target system.

Frameworks and Tool Integration

The typical FPGA development environment includes a mix of generic design tools and architecture-specific implementation tools. Ideally, these tools are molded into an integrated, easy-to-use development environment.

Most FPGA vendors provide their own design management software. For example, the Xilinx Design Manager organizes various implementation tools and utility programs into a set of pull-down menus arranged to reflect the user's normal design flow. These tools can be hooked into other framework environments, but, at some point, the user goes from the host EDA vendor's environment into the Xilinx tool set.

More tightly integrated tool sets, however, are now available. For example, Viewlogic Systems, Mentor Graphics, Cadence Design Systems, and Data I/O are among the EDA vendors that package, sell, and support design kits that provide full front-to-back design capabilities by melding FPGA vendors' implementation tools into their own frameworks.

In summary, FPGA tools are evolving into fully integrated tool sets that support sophisticated design environments. Dramatic advances are being made in the areas of design entry methods, synthesis, placement and routing algorithms, logic simulation, and tool integration, to name but a few. The result is a boon to FPGA users, who have access to increasingly effective EDA tools supporting a wide variety of needs, desires, and budgets.

Author's Note

The vendors listed herein were chosen as representative of the many quality EDA vendors that space prevented our mentioning.

Defining Terms

Application-specific integrated circuit (ASIC): A custom VLSI integrated circuit that is designed for a specific application either through the generation of a mask set used during IC manufacture or programming by the user. Types of ASICs include custom cell ICs, gate arrays, and field programmable gate arrays.

Field programmable gate array (FPGA): A very large-scale integrated (VLSI) circuit consisting of a matrix of programmable logic elements that can be configured and interconnected in any desired configuration through programming by the user.

General force directed relaxation (GFDR): A placement algorithm used in ASIC implementation in which nets connecting logic elements are modeled like physical springs.

Logic synthesis: The automated process of generating a gate-level netlist for a logic design from a higher level structural definition. Typically a register-transfer-level description of a design written in a hardware description language (such as Verilog or VHDL) is translated into a gate-level netlist targeted for implementation in an ASIC device.

Mincut: A placement algorithm used in ASIC implementation based on successively dividing logic elements into clusters based on a cost function.

Netlist: A text file containing information that describes a logic circuit's gates and interconnections.

Portability: The ability to migrate a logic design between different ASIC architectures.

Simulated annealing: A placement algorithm used in ASIC implementation tools modeled after the crystallization of a solid from its molten form.

Simulation: The process of using a software program to model the operation of a logic circuit in order to verify its functionality and examine timing relationships.

References

Chan, P. and Mourad, S. 1994. *Digital Design Using Field Programmable Gate Arrays.* PTR Prentice Hall, Englewood Cliffs, NJ.

Francis, R. et al. 1990. Chortle: A Technology Mapping Program for Table-Based Field Programmable Gate Arrays. *Proceedings of the 27th Annual Design Automation Conference.*

Frankle, J. 1992. Iterative and Adaptive Slack Allocation for Performance-Driven Layout and FPGA Routing. *Proceedings of the 29th Annual Design Automation Conference.*

Goering, R. 1994. FPGA Synthesis: Promise Yet to be Fulfilled. *Electronic Engineering Times*, April 11, 1994.

Oldfield, J. and Dorf, R. 1995. *Field-Programmable Gate Arrays.* John Wiley & Sons Inc., New York.

Trimberger, S. 1994. *Field-Programmable Gate Array Technology.* Kluwer Academic Publishers. Boston, MA.

Further Information

Several good texts discuss FPGA architectures and the algorithms used in FPGA implementation tools, including *Field-Programmable Gate Arrays* by Brown, Francis, Rose, and Vranesic (Kluwer Academic, Norwell, MA, 1992), *Field-Programmable Gate Array Technology* edited by S. Trimberger (Kluwer Academic, Norwell, MA, 1994), and *Digital Design Using Field Programmable Gate Arrays* by Chan and Mourad (PTR Prentice–Hall, Englewood Cliffs, NJ, 1994).

The latest developments are discussed at technical workshops and documented in conference proceedings. The International Workshop on Field Programmable Logic and Applications, held each September in Europe, is organized by the University of Oxford's Continuing Education Dept. The proceedings from two workshops have been incorporated into two books: *FPGAs* and *More FPGAs*, edited by Moore and Luk (Abingdon EE&CS Books, Oxford, England). The ACM International Workshop on Field Programmable Gate Arrays is sponsored by the Association for Computing Machinery Special Interest Group on Design Automation (ACM SIGDA), based in New York City; the second annual workshop was held in February, 1994, in Berkeley, CA. Kingston, Ontario was the site of the second annual Canadian Workshop on Field Programmable Gate Arrays, organized by the Canadian Microelectronics Corporation in June, 1994. The IEEE sponsors several annual conferences on ASIC technology that often include papers on FPGA tools, including EuroASIC (part of the European Design and Test Conference), the Design Automation Conference, and the IEEE International ASIC Conference and Exhibit. Proceedings are available from the IEEE Computer Society Press.

The February, 1994, issue of *ASIC & EDA* magazine includes a report on the programmable logic tool market. A comprehensive chart accompanying the article lists FPGA development tool vendors and summarizes their products offerings.

Lastly, extensive product information is available from the FPGA and EDA tool vendors themselves.

12

Integrated Circuit Packages[1]

Victor Meeldijk
Diagnostic/Retrieval Systems Inc.

12.1 Introduction

In 1958 the first integrated circuit (IC) invented had one transistor on it; today the Motorola Power PC 603 has over 1.6×10^6 transistors, the Intel Pentium has roughly 3.1×10^6 transistors, the DEC Alpha microprocessor version has 9.3×10^6 transistors, and programmable parts are available with 10,000 gate arrays. As semiconductor devices become more complex, the interconnections from the die to the circuit hardware keep evolving. Devices with high clock rates and high-power dissipation, or with multiple die, are leading to various new packages.

The use of bare die in chip on board (COB) and multichip module (MCM) applications is increasing with designers looking for end item size and weight reduction (automotive applications have bond wires going directly from the die to a connector). The use of bare die eliminates the timing delays (caused by stray inductance and capacitance) associated with the leadframes and the device input/outputs (I/Os). For burst mode static RAMs, 1/2–2 ns or a 20% improvement in access time is achieved with bare die product. Using bare die is not without problems, including testing issues and cost. At this time it is more costly for vendors to handle and ship bare die than packaged devices. Using bare parts makes it important to use *known good die* (KGD); otherwise the final assembly has to be scrapped as the device cannot be removed. Industry standards on bare die testing are still evolving. Small portable devices can also use flip-chip bonding to a circuit. Current technology bonding machines can bond IC chips with 1×10^6 bumps of 10-μm size, with a 30-μm pitch.

IC packaging can be divided into the following categories:

- Surface mount packages (plastic or ceramic)
- Chip-scale packaging
- Bare die
- Through-hole packages
- Module assemblies

In addition to bare die and surface mount techniques, there are still the older device packages for through-hole applications, where the lead of the package goes through the printed wiring board (see Figs. 12.1 and 12.2). Modules are also available that are assemblies of either packaged parts or die.

[1] Portions of this chapter were adapted from: Meeldijk, V., 1995. *Electronic Components: Selection and Application Guidelines*, Chap. 12. Wiley-Interscience, New York. With permission.

0-8493-0050-9/00/$0.00+$.50

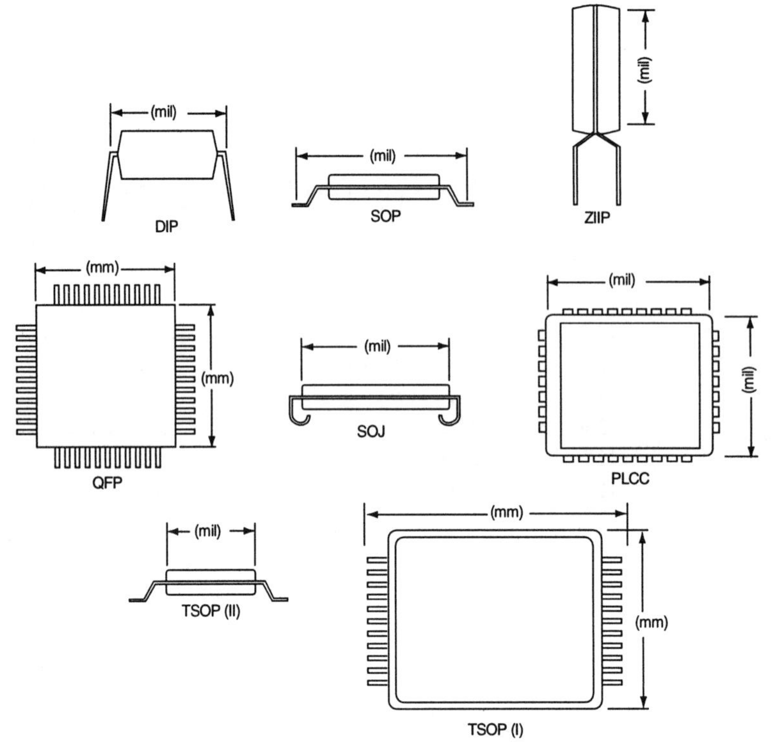

FIGURE 12.1 Some typical IC packages for through-hole applications.

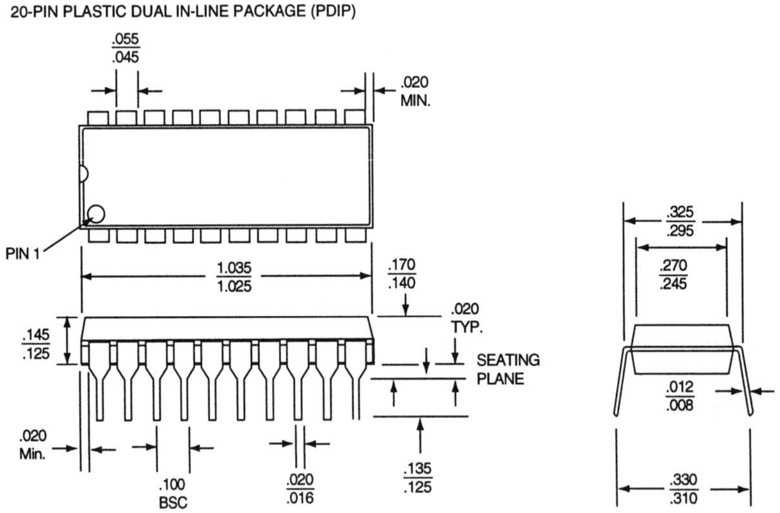

FIGURE 12.2 Common IC package dimensions. (*Source:* Courtesy of Altera Corporation, San Jose, CA.)

Although there are standards for some IC packages, such as the dual-in-line packages and the transistor outline (TO) registered Joint Electronic Device Engineering Council (JEDEC) packages, many IC packages styles are (initially) unique to the vendor that developed them (such as in 1990 when the Intel developed the molded plastic quad flat package).

20-PIN PLASTIC J-LEAD CHIP CARRIER (PLCC)

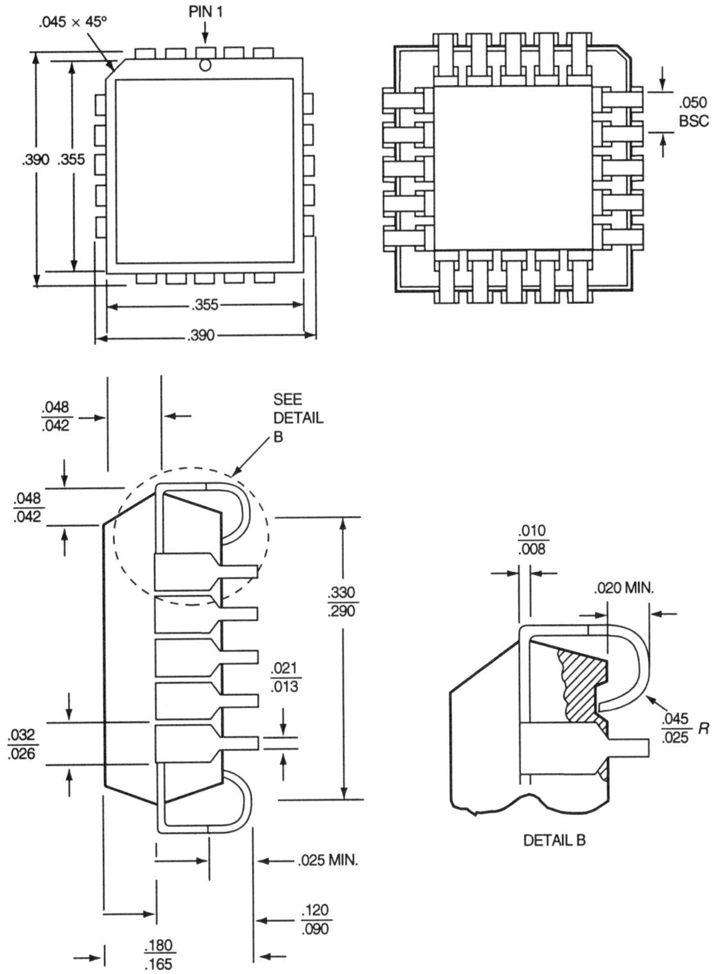

20-PIN PLASTIC SMALL-OUTLINE IC (SOIC)

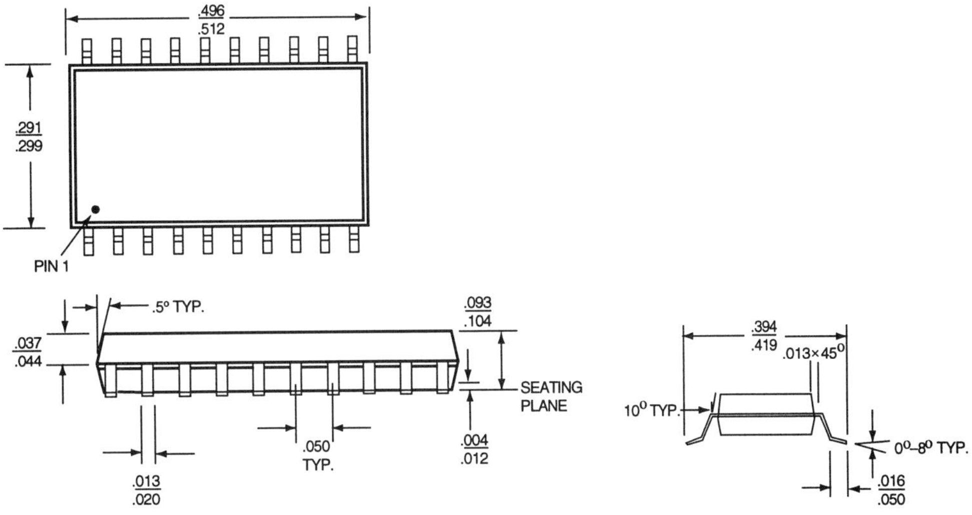

FIGURE 12.2 (Continued) Common IC package dimensions. (*Source:* Courtesy of Altera Corporation, San Jose, CA.)

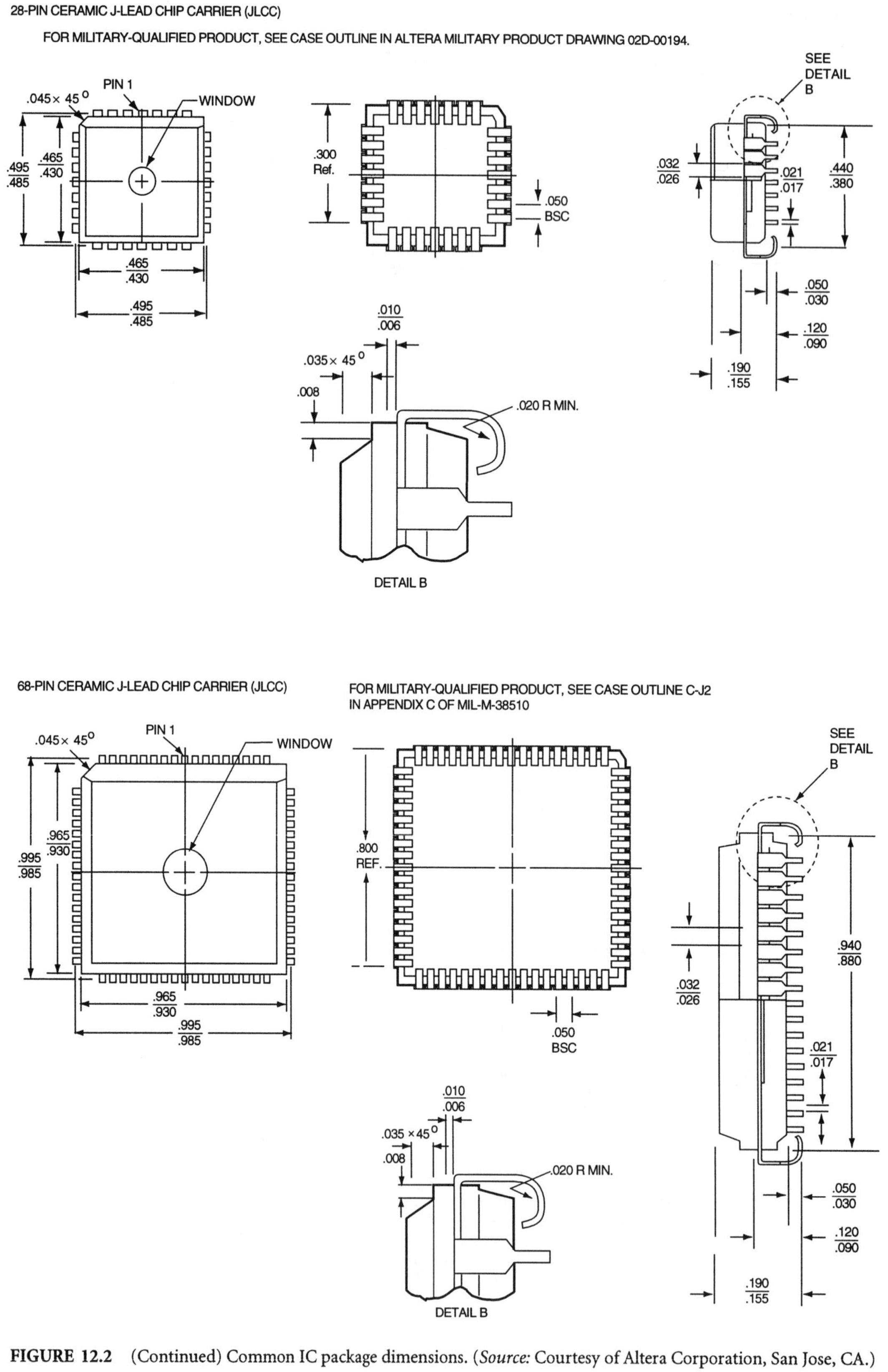

FIGURE 12.2 (Continued) Common IC package dimensions. (*Source:* Courtesy of Altera Corporation, San Jose, CA.)

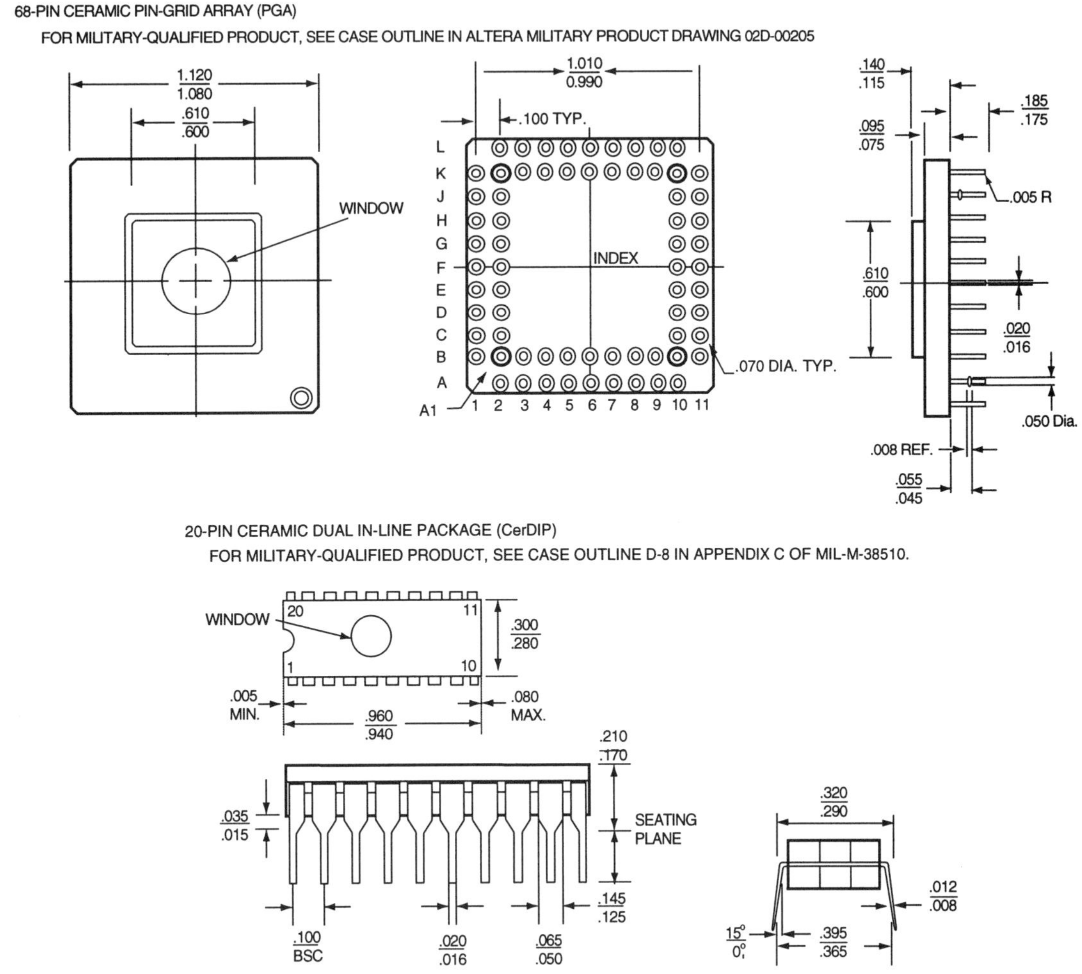

FIGURE 12.2 (Continued) Common IC package dimensions. (*Source:* Courtesy of Altera Corporation, San Jose, CA.)

All JEDEC (a subdivision of the Electronics Industries Association [EIA]) semiconductor outlines are included in their Publication 95 and terms are defined in their Publication 30, *Descriptive Designation System for Semiconductor Device Packages* (note: EIAJ is the Japanese EIA).

12.2 Surface Mount Packages

Plastic surface mount packages result in a device that is light, small, able to withstand physical shock and *g* forces, and inexpensive due to a one-step manufacturing process. The plastic is molded around the lead frame of the device. Work is still continuing on developing coatings for the die, such as polyimide, which may resolve hermeticity problems and coefficient of thermal expansion mismatches between the die and the plastic package.

Plastic parts shipped in sealed bags with desiccant are designed for 12 months storage and should only be opened when the parts are to be used. Parts stored for longer than this time, especially plastic quad flatpack (PQFP) packaged devices, should be baked to remove moisture that has entered the package. Plastic packages are hydroscopic and absorb moisture to a level dependent on the storage environment. This moisture can vaporize during rapid heating, such as in a solder reflow process, and these stresses can cause package cracking (known as the popcorn effect). Subsequent high-temperature and moisture exposures can allow contaminants to enter the IC and cause failure at a later time due to corrosion. Consult the device manufacturer for details on proper device handling.

Hermetic packages, such as the older flat pack design or ceramic leadless chip carriers (CLCCs), are used in harsh applications (such as military and space applications) where water vapor and contaminants can shorten the life of the device. Thus they are used in mission critical communication, and navigation and avionic systems. The reliability of CLCC package connections can be further improved by the attachment of leads to the connection pads, to eliminate differences in the thermal coefficient of expansion between the parts and the circuit board material.

Metal packages, with glass seals, provide the highest level of hermetic sealing followed by glasses and ceramics (Fig. 12.3). These parts have a higher temperature range than plastic encapsulated parts (typically, from −55°C to +125°C vs 0°C to 70°C, or 85°C, for plastic encapsulated devices). Although aluminum oxide, the most commonly used ceramic material, has a thermal conductivity that is an order of magnitude less than a plastic packaged device, in a ceramic sealed device the circuit die does not come into contact with the ceramic packaging material. Thus, temperature cycling will not effect the die, and parts can withstand thousands of temperature cycles without damage (1000 temperature cycles would simulate 20 years use on a commercial airliner). Large die plastic encapsulated parts can fail after only 250 temperature cycles. As mentioned earlier, die coatings being developed may resolve this problem. One of the main drawbacks in specifying ceramic packaged devices is the more complicated (than plastic encapsulated parts) manufacturing process. These devices are made up of three ceramic layers, which result in a higher cost device. Availability is limited, and lead times are often longer than standard plastic devices (this excludes parts that are relatively new and may only be available in a ceramic package, such as a pin grid array [PGA]).

Surface mount packages (and related packaging terms) include the following.

First is ball grid array (BGA) (Figs. 12.4–12.6), a packaging technique developed by IBM (also known as an overmolded pad-array carrier [OMPAC], a plastic version developed by Motorola, see Fig. 12.7). Leads, or pads, are replaced by solder balls that replace high pin count quad flatpacks (QFPs). Several hundred arrayed leads can be accommodated in areas that a few hundred peripherally arranged leads would occupy.

The package transfers heat more efficiently than a QFP. To further assist heat dissipation there is a metal-lid BGA that uses metal ribbons and thermal silicon to transfer heat to the package's metal lid (pioneered by Samsung Electronics). In late 1994, Olin International Technologies announced an aluminum (metal) BGA package using aluminum for both the substrate and IC housing. Thin-film traces as fine as 1 mil can be screened directly on the aluminum. Lead inductance and capacitance are reduced by as much as 50%. The case acts as a heatsink, and popcorning is eliminated as there is only one layer.

Disadvantages of the BGA package are that solder connections cannot be visually inspected and removed parts cannot be reused (the solder balls are melted). Problems generally occur only from defective components, bad boards, setup errors, or worn/damaged machines in the assembly line. BGA package disadvantages according to users (when comparing BGAs to quad flatpacks, tape automated bonding, and PGAs) are that the package has low rated switching performance, heat sinkability, lead compliance, thermal cycling, and hermeticity.

Other BGA versions include the perimeter lead BGA, which does not have any solder balls under the chip, and the SuperBGA (Amkor Electronics), which has a height of 1.1–1.5 mm vs 2.2–2.5 mm for standard BGA package.

Other surface mount packages (and related packaging terms) follow:

- Ceramic ball grid array (CBGA).
- Ceramic column grid array (CCGA), invented by IBM, where solder columns replace the solder balls.
- Cerpack, a flatpack composed of a ceramic base and lid. The leadframe is sealed by a glass frit.
- Cerquad, a ceramic equivalent of plastic leaded chip carriers consisting of a glass sealed ceramic package with J leads and ultraviolet window capability.
- Chip carrier (CC), a rectangular, or square, package with I/O connections on all four sides.
- Ceramic leaded chip carrier (CLDCC).
- Ceramic leadless chip carrier [CLCC (or CLLCC)], see LCCC.
- Ceramic pin grid array (CPGA).
- Ceramic quad flat package with J leads (CQFJ).
- Ceramic quad flatpack (CQFP), an aluminum ceramic integrated circuit package with four sets of leads extending from the sides and parallel to the base of the IC.
- Ceramic small outline with J leads (CSOJ).

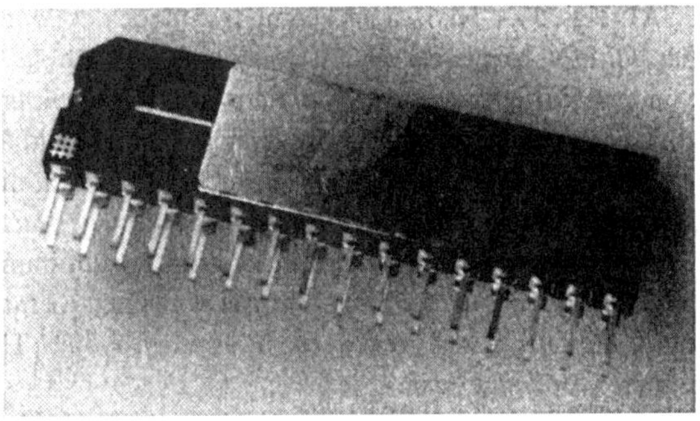

FIGURE 12.3 Typical hermetic IC packages (from the left going clockwise): PGA, flat pack, VIL, LCC and DIP. The bottom photo shows the VIL from a different angle so both rows of pins can be seen.

- Chip scale grid array (CSGA), like a BGA, but with a smaller solder ball pitch (0.5–0.65 vs 1.3–2.5 mm). This is used for devices with 200 or more leads.
- Dual small outline package (DSOP).
- EDQUAD, a trademark of ASAT Inc. for a plastic package that incorporates a heat sink; available as thin QFP (TQFP) and QFP (see Fig. 12.8).
- Flatpack (or quad flatpack), one of the oldest surface mount packages, used mainly on military programs. Typically, flatpacks have 14–50 leads on both sides of the body on 0.050-in centers.
- Flat SIP, a single inline package except the leads have a 90° bend.
- Flatpack (FP).
- Gull winged leadless chip carrier (GCC). Although easier to inspect, it is weaker than the J lead.
- Gull wing, leads which exit the body and bend downward, resembling a seagull in flight. Gull wings are typically used on small outline (SO) package, but are very fragile, easily bent, and difficult to socket for testing or burn-in. J leads solve these problems.
- Hermetic chip carrier (HCC).
- HD-PQFP, originated by Intel for a high density PQFP package with more than 196 leads with a 0.4-mm pitch.

- Heat spreader QFP package (HQFP), package has a copper heat spreader and heat sink.
- Hermetic small outline transistor (H-SOT) packaged device.
- I lead, IC leads that are formed perpendicular to the printed circuit board making a butt solder joint.
- J leads, leads that are rolled under the body of the package in the shape of the letter J. They are typically used on plastic chip carrier packages.
- J bend leads, leaded chip carrier (JCC).
- J leaded ceramic chip carrier (JLCC).
- Leadless chip carrier (LCC), a chip package with I/O pads on the perimeter of the package. Cavity (lid) up packages, with the backside of the die facing the substrate, have heat dissipated through the substrate. These JEDEC B and C ceramic packages are not suitable for air-cooled systems or for the attachment of heat sinks. Cavity (lid) down parts, (JEDEC A and D) with the die facing away from the substrate, are suitable for air-cooled systems. Rectangular, ceramic leadless E and F are intended for memory devices, with direct attachment in the lid-up position. Computing Devices International devised patented C and S leads that can be attached, as shown in Fig. 12.9.
- Leadless ceramic chip carrier (LCCC), or ceramic leadless chip carrier. JEDEC registered type A must be socketed when on a printed circuit board or a ceramic substrate board, and type B must be soldered. The LCC minipack must be soldered onto printed circuit boards.
- Leaded ceramic chip carrier (LDCC), leaded ceramic chip carrier packages include JEDEC types A, B, C, and D. Leaded type B parts are direct soldered to a substrate. Leaded type A parts can be socketed or direct soldered, and include sub-categories: leaded ceramic, premolded plastic, and postmolded plastic (which are not designated as LDCC devices).

FIGURE 12.4 A 256-pin perimeter BGA (no connections are under the die). (*Source:* Courtesy of AMKOR/ANAM-AMKOR Electronics, Chandler, AZ.)

FIGURE 12.5 A 255 bump BGA, 1.5-mm pitch between leads, 27 × 27 mm body size (*Source:* Courtesy of Motorola, Inc., Phoenix, AZ. © Motorola. With permission.)

FIGURE 12.6 A 1.27-mm staggered pitch, 313 bump BGA with distributed vias, 35-mm body size. (*Source:* Courtesy of Motorola, Inc., Phoenix, AZ. © Motorola. With permission.)

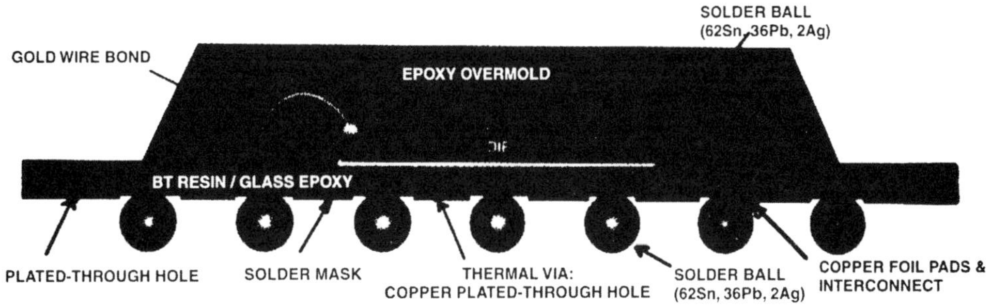

FIGURE 12.7 A plastic (OMPAC) BGA. (*Source:* Courtesy of Motorola Inc. Phoenix, AZ. © Motorola. With permission.)

- Land grid array (LGA), an Intel package used for parts such as the 80386L microprocessor. The package is similar to a PGA except that it has gold-plated pads (called *landing pads*) vs pins.
- Leadless inverted device (LID), a shaped metallized ceramic form used as an intermediate carrier for the semiconductor chip (die). It is especially adapted for attachment to conductor lands of a thick-film or thin-film network by reflow solder bonding.
- Little foot (a trademark of Siliconix), a tiny small outline integrated circuit (SOIC) package.
- Leadless ceramic chip carrier (LLCC).
- Mini-BGA (mBGA), developed by Sandia National Laboratories. Layers of polyimide and metal are applied to the die to redistribute original die pads from a peripheral arrangement into an area array configuration.

- Miniflat, a flat package (MFP) approximately 0.102–0.113 in high (2.6–2.85 mm) that may have leads on two or four sides of the package.
- Multilayer molded (MM) package, an Intel PQFP package with improved high-speed operation due to separate power and ground planes, which reduces device capacitance.
- MM-PQFP, isolated ground and power planes are incorporated within the molded body of the PQFP.
- Metal quad flatpack (MQFP), sometimes used to refer to a plastic metric quad flat pack.
- MQuad, designed by the Olin Corporation, a high thermal dissipation aluminum package available in plastic leaded chip carrier (PLCC) and QFP packages (with 28–300 leads, and clock speeds of 150 MHz) (see Fig. 12.10).

FIGURE 12.8 EDQUAD Package. (*Source:* Courtesy of ASAT, Inc., Palo Alto, CA.)

- OMPAC, see BGA.
- Pad array carrier (PAD), a surface mount equivalent of the PGA package. This package extends surface mount silicon efficiency from 15 to 40%. A disadvantage of this package is that the solder joints cannot be visually inspected, although X-ray inspection can be used to verify the integrity of the blind solder connections (and to check for solder balls and bridging). Individual solder joints, however, cannot be repaired.
- Plastic (encapsulated) ball grid array (PBGA), see OMPAC.
- Plastic leaded chip carrier [(PLCC) also known as PLDCC or a quad pack by some manufacturers, up to 100 pin packages, commonly 0.050-in pin spacing. Depending on the number of pins, pin 1 is in different locations but is confirmed by a dot in the molding, and the progression of the pin count continues counterclockwise. The leads are J shaped and protrude from the package. These parts can be either surface mounted or socketed. (This package was originally known as a *postmolded type A* leaded device.)
- Plastic quad flatpack (PQFP), this JEDEC approved package is used for devices that have 44–256 I/Os, 0.025-in lead spacing (EIA-JEDEC packages), 0.0256- and 0.0316-in spacing (EIAJ), and 0.0135-in pin

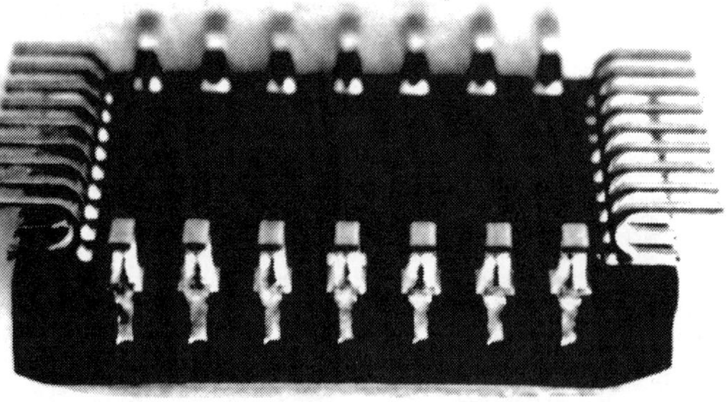

FIGURE 12.9 An LCC package with C leads attached to compensate for the thermal coefficient of expansion mismatch between the part and the circuit board, thus lowering solder joint stresses. (*Source:* Courtesy of Computing Devices International, Bloomington, MN.)

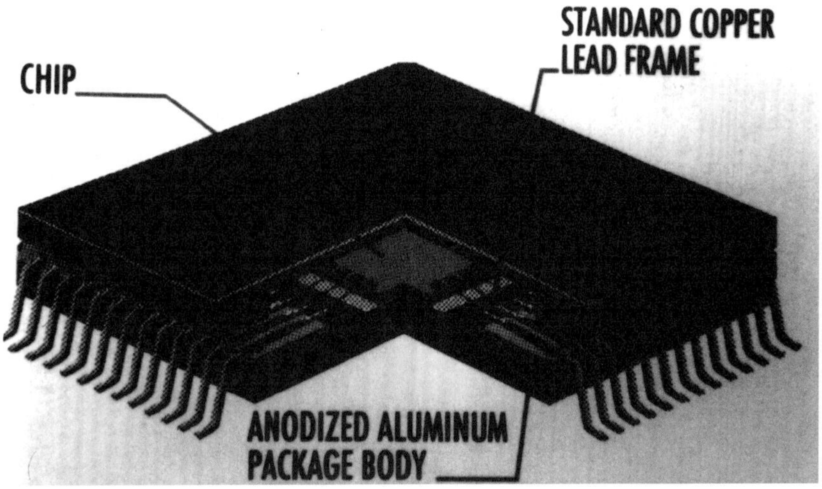

FIGURE 12.10 An MQUAD (trademark OLIN Corporation) package. (*Source:* Courtesy of Swire Technologies, San Jose, CA.)

spacing. This package has gull wing leads on all four sides and is characterized by bumpers on the corners. The body is slightly thicker than the QFP. PQFP packages are susceptible to moisture-induced cracking in applications requiring reflow soldering.

- Plastic surface mount component (PSMC).
- Quad flat J (QFJ) leaded package.
- Quad flatpack (QFP), a flatpack with leads on all four sides. The typical JEDEC approved pinouts are 100, 132, and 196, with lead pitch from 0.040 to 0.016 in. The EIAJ types are slightly thinner than the JEDEC equivalents (Fig. 12.11).
- Quarter quad flatpack (QQFT), see SQFT.
- Quad surface mount (QSM).
- Quarter-size small outline package (QSOP).
- Quad pack, see PLCC.
- Quad very small outline package (QVSOP), introduced by Quality Semiconductor. This package combines the body size of a half-width, 14-pin SOIC, with the 25-mil lead pitch of a PQFP (or a 150 × 390 mil, 48-pin package). The package has designation JEDEC MO-154.
- Repatterned die, see BGA.
- Square (body) chip carrier (SCC).
- Slam pack, a square ceramic package that looks like an LCC and is always used with a socket.
- Small outline (SO) (with versions such as SO wide, SO narrow, or SO large), this design originated with the Swiss watch industry in the 1960s (and was reportedly nicknamed SO for Swiss Outline). It was used in the modern electronics industry by N.V. Philips (formerly known as Signetics in the U.S.) in 1971. This package is also known as a *mini-flat* (which has slightly different dimensions than the JEDEC parts) by the Japanese.
- Small outline IC (SOIC), 8–28 pin packages, commonly 0.050-in pin spacing, with gull wing leads. This package style is about 50–70% of the size of a standard dual inline package (DIP) part (30% as thick).
- Small outline package with J leads on two sides (SOJ). The leads are bent back around the chip carrier.
- Small outline large (SOL), generally refers to a package that is 0.300 mils wide vs 150 mils wide (for an SO package). It is a larger version of the small shrink outline package (SSOP).

- Small outline package (SOP), a package with two rows of narrowly spaced gull-wing leads (same as SOIC).
- Small outline transistor (SOT), a plastic leaded package originally for diodes and transistors but also used for some ICs (such as Hall effect sensors).
- Small outline wide (SOW), see SOL.
- Shrink quad flatpack (or flat package) (SQFP), typically a 64-pin small quad flat package (1/4 height of QFT) with a lead pitch of 0.016 in or less. Also known as quarter quad flat pack (QQFT).
- Square surface mounting (SSM).
- Shrink small outline package (SSOP).
- Shrink small outline IC (SSOIC), a plastic package with gull-wing leads on two sides, with a lead pitch equal to or less than 0.025 in.
- Tape ball grid array (TBGA).

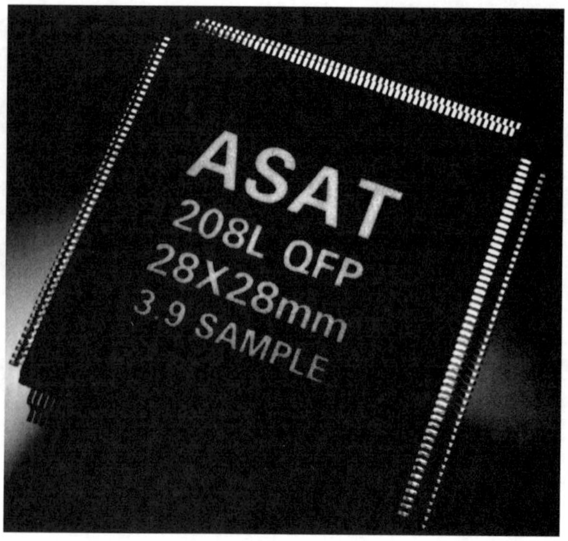

FIGURE 12.11 A QFP package. (*Source:* Courtesy of ASAT, Inc., Palo Alto, CA.)

- Thin QFP (TQFP). Typical sizes are 1.0- and 1.4-mm body thickness, in sizes ranging from 10 × 10 mm to 20 × 20 mm and lead count from 64 to 144 leads.
- Thin small outline package (TSOP). The type-1 plastic surface mount parts have 0.5-mm gull-wing leads on the edges of the parts (the shorter dimension) rather than along the sides (longer dimension). The type-2 parts have the leads on the longer dimension. These packages are generally used for memory ICs.
- Thin shrink (or sometimes called scaled) small outline package (TSSOP), half the height of a standard SOIC.
- Vertical mount package (VPAK), a package conceptually like a zig-zag package (ZIP) except instead of through-hole leads it has surface-mount leads (L shaped leads). This package was introduced by Texas Instruments in late 1991 for 4- and 16-Mb direct random access memories (DRAMs).

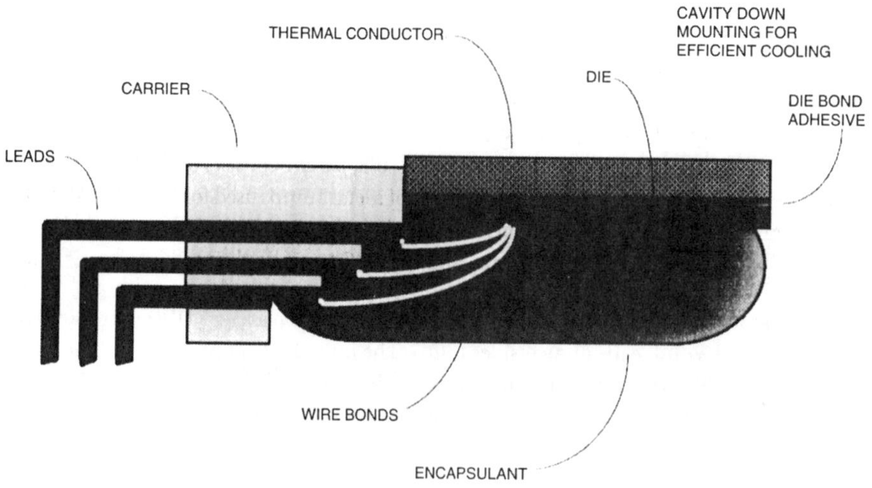

FIGURE 12.12 Very small peripherial array architecture. (*Source:* Courtesy of ARCHISTRAT Technologies, Division of the PANDA Project, Boca Raton, FL.)

- Very small (and thin) quad flat package (VQFP). Lead pitch is 0.5 mm, with 32, 48, 64, 80, 100, 128, and 208 pins per package.
- Very small outline (VSO), usually used to denote 25-mil pitch packages with gull-wing leads.
- Very small outline package (VSOP), with 25-mil spaced leads at the ends of the package (also TSOP and SSOP and SSOIC).
- Very small peripheral array (VSPA^{TM}), a high pin count package, smaller than an equivalent QFP, developed by Archistrat Technologies (Fig. 12.12). The IC die is mounted to a metal plate that is bonded to a high-temperature plastic (Vectra) carrier. The sides of the plastic carrier have rows of holes to accept 4 rows of pins. The pins are connected to the die via wire bonds.

12.3 Chip-Scale Packaging

This form of packaging technique is designed to have the size and performance of bare die parts but with the handling and testability of packaged devices. The package size is no more than 1.2 times the original chip with various techniques to connect the part to the circuit. Flip chips have enlarged solder pads (bumps or balls). Flip-chip packages include the controlled collapse chip connection (C4), developed by IBM, and the direct chip attachment (DCA), developed by Motorola. In 1964, IBM invented flip-chip interconnection technology to enable production-level joining of discrete transistors. General designations are as follows:

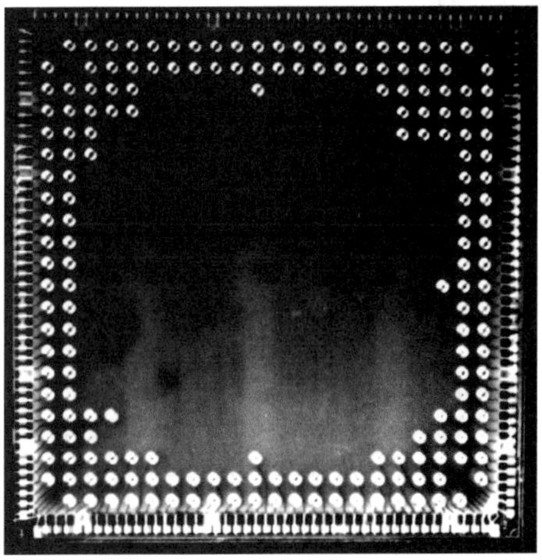

FIGURE 12.13 A μBGA package. (*Source:* Courtesy of Tessera, San Jose, CA.)

- Bumped chip, see flip chip.
- Flip chip—a semiconductor package where the I/O terminations are in the form of bumps on one side of the package (also called *bumped chip*). After the surface of the chip has been passivated, or treated, it is flipped over and attached to a matching substrate.
- MiniBGA, uses a predetermined grid array for the solder bumps and is similar to the flip chip.
- Repatterned die, see MiniBGA.
- Microsurface mount packages (MSMT)—a deposited horizontal metallization, instead of bond wires, goes from the bond area to a connection post. In 1994, Micro MST, Inc. patented this technology where semiconductors are mass packaged in the wafer state and can be tested as individual devices. This package is approximately the same size as the semiconductor die and is used for devices with less than 200 leads. It generally has lower parasitic inductance than BGA or QFP packages.
- Slightly larger than IC carrier (SLICC) package.
- MicroBGA (μBGA), also known as *chip-scale packaging* (Mitsubishi); slightly larger than IC or SLICC packaging (Motorola, Inc). A package designed by Tessera, μBGA consists of a flex circuit with gold traces bonded to the die pads. The die pads fan in to an array of metal bumps used for the second-level assembly. An elastomeric adhesive is used to attach the flexible circuit to the chip. The package compensates for thermal mismatches between the die and the substrate, and the μBGA can be clamped into a socket for tests before attachment in the final assembly (see Fig. 12.13).

12.4 Bare Die

Bare, or unpackaged, parts offer the smallest size, with no signal delays associated with the device package. Current issues related to unpackaged devices are packaging, handling, and testing. The most common bare die and tape packages (bare die mounted on tape) include the following:

- C4PBGA, an IC package developed by Motorola that attaches the IC die to a plastic substrate using the C4 process. It uses a multilayer substrate, unlike the chip-scale or SLICC package.
- Chip scale package (CSP), a Mitsubishi package where the IC is surrounded by a protective covering through which external electrode bumps on the bottom provide electrical contacts. The package, only slightly larger than the chip it houses, has a height about 0.4 mm.
- Chip on board (COB)—the die is mounted directly on the printed circuit substrate (or board). See also tape automated bonding (TAB) and tape carrier packages (TCB).
- Chip on flex (COF), a variation of COB but instead of bonding the bare die to a substrate, it is bonded to a piece of 0.15 in-thick polyimide film (such as FR4) that has a top layer of gold-plated copper. The traces provide bonding ares for wires from the die and a lead pitch similar to QFP packages. It has approximately the same weight and height as COB but failed parts can be removed. It, however, uses about the same board real estate as QFP packages and is more expensive than COB because of the polyimide substrate.
- Demountable tape automated bonding (DTAB), developed by Hewlett–Packard. Mechanical screw and plates align and hold the IC to the board (pressure between the tape and the board provides the contact). The IC can be replaced by removing the screws (vs a standard TAB package, which is soldered in place).
- Sealed chips on tape (SCOT), chips mounted on tape supported leads and sealed (usually with a blob of plastic).
- Tape automated bonding (TAB), see TCP.
- Tape carrier packages (TCP), formerly called TAB packages. The chip is mounted to a dielectric film, which has copper foil connection patterns on it. The chip is sealed with a resin compound. This device assembly is mounted directly to a circuit without a plastic or ceramic package. Some tape carrier packages resemble TSOP packages (with gull-wing leads) but are thinner because the bottom of the IC die is exposed. The TCP design is about one-half the volume and one-third the weight of an equivalent pin count TSOP.
- Tape carrier ring (TCR), or a guard ring package, similar to the TCP package but includes a plastic ring to support the outer rings during test, burn-in, and shipment.
- Ultra-high volume density (UHVD), a process developed by General Electric to interconnect bare ICs on prefabricated laminated polyimide film.

12.5 Through-Hole Packages

These packages, as the name implies, mount in holes (usually plated through with deposited metal) on the printed circuit board. These device packages include the following:

- Batwing, a package (sometimes a DIP type, but can be surface mounted) with two side tabs used for heat dissipation.
- Ceramic dual inline package (CERDIP). The DIP package was developed by Fairchild Semiconductor and Texas Instruments followed with a metal topped ceramic package that resolved problems with the early ceramic packaged parts.
- Dual inline package (DIL).
- Dual-inline package (DIP), a component with two straight parallel rows of pins or lead wires. The number of leads may be from 8–68 pins (although more than 75% of DIP devices have 14–16 pins), 0.100-in pin spacing with width anywhere from 0.300-mil centers to 0.900-mil centers. The skinny (or shrink) DIP (SDIP) has 0.300-mil centers (spacing between the rows) vs 0.600-mil centers. The SDIP usually has 24–28 pins. DIP parts may be ceramic (pins go through a glass frit seal), sidebrazed (pins are brazed onto metal pads on the side of the package), or plastic (where the die is moulded into a plastic package). There is also a shrink DIP, another small package through-hole part.
- Plastic DIP (PDIP).
- Pin grid array (PGA), a plastic or ceramic square package with pins covering the entire bottom surface of the package. Lead pitch is either 0.1 or 0.05 in perpendicular to the plane of the package. Packages have various pin counts (68 or more). The chip die can be placed opposite the pins (cavity up) or nested in the grid array (cavity array).

- Pinned uncommitted memory array (PUMA II), a PGA package application specific integrated circuit (ASIC) memory array with four 32-pad LCC sites on top of a 66 PGA. Each of the four sites can be individually accessed via a chip select signal thus allowing a user definable configuration (i.e., ×8, ×16, ×24, ×32). It provides for an ASIC memory array without tooling. The substrate is a multilayered cofired alumina substrate with three rows of 11 pins. There is a channel between the pins so that the part can be used with a heat sink rail (or ladder). Onboard decoupling capacitors are mounted in a recess in this channel. This type of device is available from various companies including Mosaic Semiconductor (1.12 × 1.12 in square), Cypress Semiconductor (their 66 pin PGA module, the HG01, is 1.09 × 1.09 max), and Dense-Pac Microsystems (Veraspac or VPAC family, 1.09 × 1.09 max.).
- Plastic pin grid array (PPGA).
- Quad inline package (QIP or QUIL).
- Quad inline package (QUIP)—similar to a DIP except the QUIP has a dual row of pins along the package edge. Row to row spacing is 0.100 in, with adjacent rows aligned directly across from each other.
- Skinny (or shrink) DIP (SDIP), see DIP.
- Shrink DIP, a dual inline package with 24–64 pins with 0.070-in lead spacing.
- Single inline (SIL).
- Single inline module (SIM)—electrical connections are made to a row of conductors along one side.
- Single inline package (SIP), a vertically mounted module with a single row of pins along one edge for through-hole mounting. The pins are 0.100 in apart. SIPs may also include heatsinks.
- Skinny DIP (SK-DIP), see SDIP.
- Transistor outline-XX (TO-XX), refers to a package style registered with JEDEC.
- Vertical inline package (VIL).
- Zig-zag inline package (ZIP). This may be either a DIP package that has all the leads on one edge in a staggered zig-zag pattern or a SIP package. Lead spacing is 0.050 in from pin to pin. In modules the leads are on both sides in a staggered zig-zag pattern. Lead spacing is 0.100 in between pins on the same side (or 0.050 in from pin to pin).

12.6 Module Assemblies

We use this terminology to refer to packaging schemes that take either packaged parts or bare die and use them to make an assembly. This may be by either mounting the parts to a substrate or printed circuit board or by stacking parts or die to create dense memory modules. This packaging scheme may be surface mount or through-hole. The circuit board assemblies may be through-hole (having pins) or designed for socket mounting (with conductive traces). Variations include the following:

- Dual-inline memory module (DIMM), pioneered by Hitachi, has memory chips mounted on both sides of a PC board (with 168 pins, 84 pins on each side). The DIMM has keying features for 5-V, 3.3-V, and 2.5-V inputs and for indicating whether it is made with asynchronous DRAMs or flash, static random access memory (SRAM) or DRAM ICs.
- Flexible-rigid-assembly memory module (FRAMM), a memory packaging scheme by Memory X. FRAMM modules use a combination of rigid and flexible PC board assemblies, with the flexible board interconnecting to rigid PC boards. The modules have standard JEDEC 30- and 72-pin SIMM outputs. TSOP DRAMS are mounted on both sides of the rigid PC boards.
- Full stack technology—packing 20–100 dice horizontally, in a loaf-of-bread configuration. This manufacturing method was developed by Irvine Sensors Corporation (see also short stack).
- HDIP module, a hermetic DIP module that has hermetically sealed components mounted on the top and bottom of a ceramic substrate. This package style is generally used in anticipation of a monolithic part (such as a memory device) that will be available at a later time that will fit the same footprint as the module.
- Hermetic vertical DIP (HVDIP) module, a vertically mounted ceramic module with pins along both edges (through-hole mounting). Components used in this module are hermetically sealed. Pins on opposite sides of the module are aligned and are on 0.100-in spacing.

- Leaded multichip module (LMM). An LMMC is an LMM connector.
- Multichip module (MCM), a circuit package with SMT (surface mount) IC chips mounted and interconnected via a substrate similar to a multilayer PC board. MCM-L has a substrate that can be FR-4 dielectric but is often polyimide based laminations (first standardized by JEDEC JC-11 committee). Other dielectric substrates are ceramic (MCM-C), which can have passive devices (resistors, capacitors, and inductors) built into the substrate; thin film (i.e., silicon and ceramic construction), with deposited conductors (which can have capacitors built into the substrate) (MCM-D); laminated film (MCM-LF) made up of layers of modified polyimide film; or multichip on a flexible circuit (MCM-F). In 1994, some manufacturers started manufacturing substrates with MCM-D thin-film deposits onto MCM-L laminate substrates in a technique pioneered by IBM. MCM devices inherently offer higher speed and performance at lower cost than conventional devices. Manufacturing issues associated with MCM design include availability of unpackaged die, how to test/inspect the die, and test/rework of the MCM module. MCM modules for military uses fall under MIL-PRF-38534 (see *memory cube*). There is also an MCM-V, with V standing for a vertical, which is a three-dimensional module. This design is also known as Trimrod, by the developer Thomson-CSF and the European Commission (see *memory cube*).
- Memory cube, a three-dimensional module consisting of stacked memory devices such as DRAMs or SRAMs (see Ribcage and Uniframe Stakpak Module). This design was developed by RTB Technology and was licensed to various other semiconductor companies. The concept is similar to the stacked chip design, by Irvine Sensors, but the cubes are manufactured with standard parts vs IC die that have all I/Os on one edge (which are stacked and electrically connected together).
- Ribcage, a trademark of Staktek Corporation for a three-dimensional memory module design. Also see Uniframe Stakpack Module.
- Short stack, a method of stacking semiconductor dice vertically, where memory ICs (such as 4 SRAMs) are assembled into a three-dimensional thin-film monolithic package with the same footprint as a single SRAM. A variation on the full stack technology also developed by Irvine Sensors Corporation, short stacks hold up to 10 ICs. The stacks may be unpackaged for use in hybrid, multichip, and chip-on-board applications. Interconnect techniques include wire-bond, tab, and flip-chip methods. Another stacked memory method, which uses special silicon wafer segments and vertical interconnects between the wafers in the interior of the silicon segments, was developed and patented by WaferDrive Corporation. This method was developed for PCMCIA card applications.
- Single inline memory module (SIMM), an assembly containing memory chips. The bottom edge of the SIMM, which is part of the substrate material, acts as an edge card connector. SIMM modules are designed to be used with sockets that may hold the SIMM upright or at an angle, that reduces the height of the module on the circuit board. Typical SIMM parts are 4×9 (4 meg memory by 9), 1×9, 1×8, 256×9 and 256×8. Also 9-b data width SIMM modules are produced under license to Wang Laboratories, which developed the SIMM module and socket in the early 1980s as an inexpensive memory expansion for a small workstation.
- Stackable leadless chip carrier (SLCC)—developed by Dense-Pac Microsystems, this multidimensional module consists of stacked chip carriers. Stacking is accomplished by aligning the packages together and tin dipping each of the four sides. SLCC can achieve a density of 40:1 over conventional packages.
- Uniframe Stakpak Module, a tradename three-dimensional stack of TSOP memory ICs by Staktek Corporation (Fig. 12.14).
- VDIP module, vertically mounted modules with plastic encapsulated components and epoxy encapsulated chips on them. VDIP modules have 0.100 in spaced pins along both sides of the substrate, with the pins on the alternate sides aligned.
- Other Definitions:

  Plastic encapsulated microcircuit (PEM).
  Premolded plastic package (PMP).

FIGURE 12.14 An 8-Mb DRAM Uniframe STAKPAK™ memory module. (*Source:* Courtesy of STAKTEK, Austin, TX.)

References

Bindra, A. Nov. 28, 1994. Very small package hopes for big IC impact; May 16, 1994. Sandia shrinks size of BGA packages. *Electronic Engineering Times.*

Costlow, T. Jan. 16, 1995. MCM substrates mixed; Oct. 10, 1994: BGAs honed to meet diverse needs; Sept. 19, 1994. Chip-scale hits the fast track; Sept. 12, 1994. Olin mines metal for ball-grid arrays; Aug. 12,1994. Mitsubishi scales down IC package; July 11, 1994: Amkor trims BGA height; May 9, 1994: MCM design details remain perplexing; April 18, 1994: MCM standard created. *Electronic Engineering Times.*

Crum, S. Sept. 1994. Minimal IC packages to be available in high volume. *Electronic Packaging and Production*, News column.

Derman, G. Feb. 28, 1994. Interconnects and packaging. *Electronic Engineering Times.*

Hutchins, C.L. Nov. 1994. Understanding grid array packages. *SMT* (*Surface Mount Technology.*) Hutchins and Associates, Raleigh, NC.

Karnezos, M. Feb. 28, 1994. DTAB mounts speed and power threat. *Electronic Engineering Times.*

Lineback, J.R. Nov. 1994. Chip makers try to add quality while removing the package. *Electronic Business Buyer.*

Locke, D. Nov. 28, 1994. Plastic packs pare mil IC pricing. *Electronic Engineering Times.* American Microsystems.

Meeldijk, V. 1995. *Electronic Components: Selection and Application Guidelines.* Wiley-Interscience, New York.

Murray, J. June 11, 1994. MCMs pose many production problems. *Electronic Engineering Times.*

O'Brien, K., managing ed. May/June 1994. *Advanced Packaging.* μBGA offers flip chip alternative; copper heat sink QFP, dual use sought in Denver. News column.

Reuning, K. Aug. 1994. Squeezing the most from BGAs. *U.S. Tech.* deHaart Inc.

Woolnough, R. April 25, 1994. Thomson 3-D modules ready for prime time. *Electronic Engineering Times.*

Further Information

The following sources can be referenced for additional data on new IC packaging and related packaging issues (i.e., converting IC packages, lead finish considerations, packaging and failure modes):

Computer Aided Life Cycle Engineering (CALCE), Electronic Packaging Research Center, University of Maryland, College Park, MD, 301-405-5323. CALCE is involved in research and software development associated with the design of high reliability advanced electronic packages.

Electronic components: *Selection and Application Guidelines*, by Victor Meeldijk, John Wiley and Sons Interscience Division, ©1995. The integrated circuits chapter discusses IC packages, lead material, and failure mechanisms related to packaging.

High Density Packaging Users Group (HDP User Group), a nonprofit organization, formed in 1990, serving users and suppliers. Scottsdale, AZ, 602-951-1963; Alvsjo, Sweden, 46 8 86 9868.

IEEE Components, Packaging, and Manufacturing Technology Society, New York. Periodicals, Piscataway, NJ 800-678-4333, 908-981-0060.

International Electronic Packaging Society (IEPS), Wheaton, IL, 708-260-1044.

International Society for Hybrid Microelectronics (also known as the Microelectronics Society) (ISHM), Reston, VA, 800-535-4746, 703-758-1060. This technical society was founded in 1967 and covers various microelectronic issues such as MCMs and advanced packaging.

The Institute for Interconnecting and Packaging Electronic Circuits (IPC), Northbrook, IL, 847-509-9700. This organization issues standards on packaging and printed circuitry.

Microelectronics and Computer Technology Corporation, Austin TX, along with Lehigh University worked toward replacing hermetic packages with more effective lightweight protective coatings. This Reliability without Hermeticity (RwoH) research contract was awarded by the U.S. Air Force Wright Laboratory. The RwoH is an industry working group, initiated in 1988. For more information call 512-338-3740, David Clegg.

Semiconductor Technology Center, Neffs, PA, 610-799-0419 This organization does market research on manufacturing design, holds training seminars, and is involved in the Flip Chip/Tab and Flex Circuit symposiums held each year.

The Surface Mount Technology Association, Triangle Park, NC, 612-920-7682.

The following publications, which offer free subscriptions to qualified readers, cover packaging issues:

Advanced Packaging, IHS Publishing Group, Box 159, 17730 West Paterson Road, Libertyville, IL 60048-0159, 708-362-8711, Fax: 708-362-3484.

This publisher also issues *Surface Mount Technology: Electronic Engineering Times*, a CMP Publication, 600 Community Drive, Manhasset, N.Y. 11030, 516-562-5000, 516-562-5882 (subscription services). Subscription inquiries: 516-733-6800, Fax: 516-562-5409.

Electronic Packaging and Production, a Cahners Publication, Division Reed Publishing, 275 Washington Street, Newton, MA 02158-1630.

Circulation/Subscriptions, Cahners Publishing Co., Box 7541, Highlands Ranch, CO 80163-9341, 303-470-47000.

13

Testing of Integrated Circuits

Wayne Needham
Intel Corporation

13.1 Introduction

The goal of the test process for integrated circuits is to separate good devices from bad devices, and to test good devices as good. Bad devices tested as bad become yield loss, but they also represent opportunity for cost reduction as we reduce number of **defects** in the process. When good devices are tested as bad, overkill occurs and this directly impacts cost and profits. Finally, bad devices tested as good signal a quality problem, usually represented as *defects per million* (DPM) devices delivered.

Unfortunately, the test environment is not the same as the operating environment. The test process may reject good devices and may accept defective devices as good. Here lies one of the basic problems of today's test methods, and it must be understood and addressed as we look at how to generate and test integrated circuits. Figure 13.1 shows this miscorrelation of test to use.

Bad devices are removed from good devices through a series of test techniques that enable the separation, by either voltage- or current-based testing. The match between the test process and the actual system use will not be addressed in this chapter.

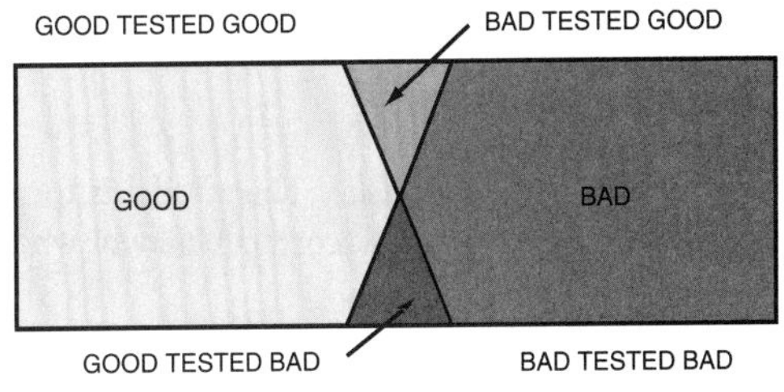

FIGURE 13.1 Test miscorrelations.

13.2 Defect Types

Defects in an integrated circuit process fall into three broad categories. They are:

- Opens or abnormally high-resistance portions of signal lines and power lines that increase the series resistance between connected points. This extra resistance slows the charging of the parasitic capacitance of the line, thus resulting in a time delay. These defects also degrade voltage to a transistor circuit, which reduces levels or causes delays.
- Shorts, bridges, or resistive connections between layers of an integrated circuits, which form a resistance path in a normal insulating area. Given sufficiently low R value, the path may not be able to drive a 1 or a 0.
- Parameter variations such as threshold voltage and saturation current deviations changed by contamination, particles, implant, or process parameter changes. These usually result in speed or drive level changes.

0-8493-0050-9/00/$0.00+$.50

Table 13.1 shows several types of defects and how they are introduced into an integrated circuit. It also describes electrical characteristics.

Defects, although common in integrated circuit processes, are difficult to quantify via test methods. Therefore, the industry has developed abstractions of defects called **faults**. Faults attempt to model the way defects manifest themselves in the test environment.

TABLE 13.1 Typical IC Defects and Failure Modes

Defect	Source	Electrical Characteristic
Extra metal or interconnect	particle	ohmic connection between lines
Constrained geometry	insufficient or under etching	higher than normal resistance
Foreign material induced in the process	contamination	varying resistance or short

Traditional Faults and Fault Models

The most widely used fault model is the *single stuck-at fault* model. It is available in a variety of simulation environments with good support tools. The model represents a defect as a short of a signal line to a power supply line. This causes the signal to remain fixed at either a logic one or a zero. Thus, the logical result of an operation is not propagated to an output. Figure 13.2 shows the equivalent of a single stuck-at one or zero fault in a logic circuit. Although this fault model is the most common one, it was developed over 30 years ago, and does not always reflect today's more advanced processes. Today's very large-scale integrated (VLSI) circuits and submicron processes more commonly have failures that closely approximate bridges between signal lines. These bridges do not map well to the stuck-at fault model.

Other proposed or available fault models include:

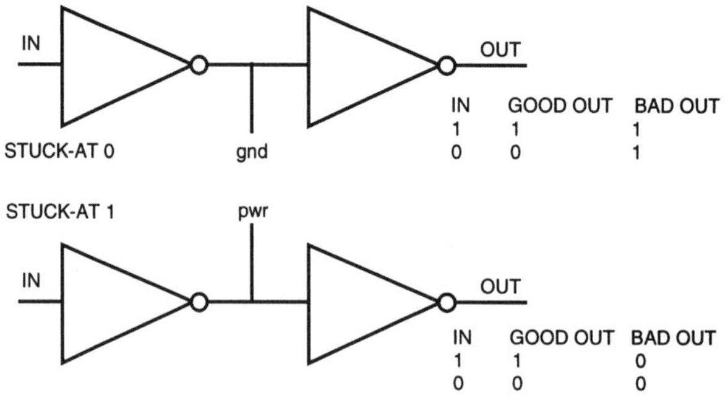

FIGURE 13.2 The single stuck-at model.

- The *open fault model:* This model is for open circuits. Unfortunately in CMOS, missing contacts and transistors can exhibit correct behavior when switching at speed, even if some of the transistors or contacts are missing.
- The *leakage fault model:* This model assumes leakage in transistors and lends itself to an I_{DDQ}[1] test technique (discussed later in this chapter).
- *Timing fault models* (delay, gate and transition faults): All of these models assume a change in the AC characteristics of the circuit. Either a single gate, line, or block of circuitry is slower than needed or simulated, and this slow signal causes a problem with at-speed operation of the integrated circuit.
- *Pseudostuck-at model fault model* for I_{DDQ}: This model assumes nodes look like they are stuck at a one or zero logic level.
- *Voting fault models*, and bridge models: Adjacent lines are assumed connected and the logic result is an X (unknown), or a vote with the highest drive strength circuit dominating the logic level.
- Also for memory circuits:
- *Neighborhood pattern faults*: This fault assumes operations in one memory cell will affect adjacent memory cells (usually limited to the four closest cells).
- *Coupling faults:* This fault model assumes that full rows or columns couple signals into the cells or adjacent rows and columns.
- *Retention faults:* These faults are similar to the open fault for logic (discussed previously). Failures to this fault model do not retain data over an extended period of time.

[1] I_{DDQ} is the I_{IEEE} symbol for quotient current in a CMOS circuit.

There are many other fault models.

It is important to realize that the stuck-at model may not necessarily catch failures and/or defects in the fabrication processes. This problem is being addressed by the computer-aided design (CAD) industry. Today, numerous solutions are available, or on the horizon, that address the issues of fault models. Tools currently available can support delay faults. Bridging fault routines have been demonstrated in university research, and should be available from selected CAD vendors within a few years. Soon, we will not be confined to testing solely through the single stuck-at model.

13.3 Concepts of Test

The main test method of an integrated circuit is to control and observe nodes. Doing this verifies the logical operation that ensures that the circuit is fault free. This process is usually done by generating stimulus patterns for the input and comparing the outputs to a known good state. Figure 13.3 demonstrates a classic case where a small portion of an integrated circuit is tested with a stimulus pattern set for inputs and an expected set for outputs. Note this test set is logic, not layout, dependent and checks only stuck at faults.

Types of Test

There are three basic types of test for integrated circuits. These are parametric tests, voltage tests, and current tests. Each has a place in the testing process.

Parametrics tests ensure the correct operation of the transistors (voltage and current) of the integrated circuit. Parametric tests include timing, input and output voltage, and current tests. Generally, they measure circuit performance. In all of these tests, numbers are associated with the measurements. Typical readings include: $T_x < 334$ ns, $I_{OH} < 4$ mA, $V_{OI} < 0.4$ V, etc.

The next major category of testing is power supply current tests. Current tests include active switching current or power test, power down test, and standby or stable quiescent current tests. This last test, often called I_{DDQ}, will be discussed later.

The most common method of testing integrated circuits is the voltage-based test. Voltage-based test for logic operations includes functions such as placing a one on the input of an integrated circuit and ensuring that the proper one or zero appears on the output of the IC at the right time. Logical values are repeated, in most cases, at high and low operating voltage. The example in Fig. 13.3 is a voltage-based test.

Test Methods

To implement voltage based testing or any of the other types of tests, initialization of the circuit must take place, and control and observation of the internal nodes must occur. The following are four basic categories of tests:

- External stored response testing. This is the most common form for today's integrated circuits and it relies heavily on large IC **automated test equipment** (ATE).

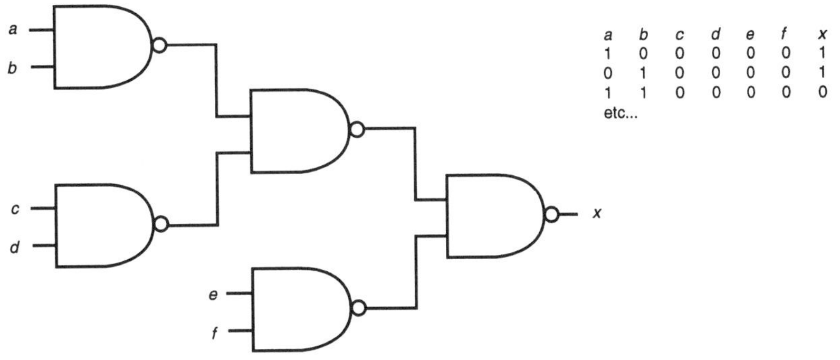

FIGURE 13.3 Logic diagram and test set.

- **Builtin-self-test (BIST).** This is method of defining and building test circuity onto the integrated circuit. This technique provides stimulus patterns and observes the logical output of the integrated circuit.
- **Scan** testing. Some or all of the sequential elements are converted into a shift register for control and observation purposes.
- Parametric tests. The values of circuit parameters are directly measured. This includes the I_{DDQ} test, which is good at detecting certain classes of defects. To implement an I_{DDQ} test, a CMOS integrated circuit clock is stopped and the quiescent current is measured. All active circuitry must be placed in a low-current state, including all analog circuitry.

External Stored Response Testing

Figure 13.4 shows a typical integrated circuit being tested by a stored response tester. Note that the patterns for the input and output of the integrated circuit are applied by primary inputs, and the output is compared to known good response. The process of generating stored response patterns is usually done by simulation. Often these patterns are trace files from the original logic simulation of the circuit. These files may be used during simulation for correct operation of the logic and later transferred to a tester for debug, circuit verification, and test. The patterns may take up considerable storage space, but the test pattern can easily detect structural and/or functional defects in an integrated circuit.

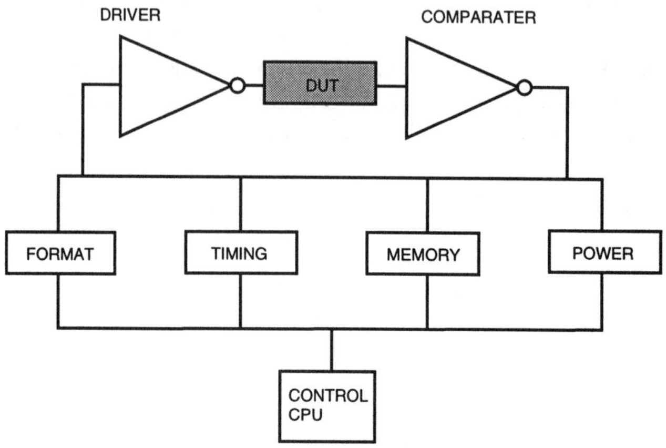

FIGURE 13.4 Stored response tester and DUT.

To exercise the **device under test (DUT)**, a stored response functional tester relies on patterns saved from the logic simulation. These patterns include input and output states and must account for unknown or X states. The set of patterns for functional test can usually be captured as trace files, or change files, and then ported to the test equipment. Trace or change files capture the logic state of the DUT as its logical outputs change over time. Typically, these patterns represent the logical operation of the device under test. As such, they are good for debugging the operation and design of integrated circuits. They are not the best patterns for fault isolation or yield analysis of the device.

BIST

The most common implementation of BIST is to include a **linear feedback shift register (LFSR)** as an input source. An LFSR is constructed from a shift register with the least significant bit fed back to selected stages with exclusive-OR gates. This feedback mechanism creates a polynomial, which successively divides the quotient and adds back the remainder. Only certain polynomials create pseudorandom patterns. The selection of the polynomial must be carefully considered. The LFSR source, when initialized and clocked correctly, generates a pseudorandom pattern that is used for the stimulus patterns for a block of logic in an integrated circuit.

The common method of output compaction is to use a **multiple input shift register (MISR)**. A MISR is an LFSR with the outputs of the logic block connected to the stages of the LFSR with additional exclusive-OR gates. The block of logic, if operating defect free, would give a single correct signature (see Fig. 13.5). If there is a defect in the block of logic, the resultant logic error would be captured in the output MISR. The MISR constantly divides the output with feedback to the appropriate stages such that an error will remain in the MISR until the results are read at the end of the test sequence. Because the logic states are compacted in a MISR, the results may tend to mask errors if the error repeats itself in the output, or if there are multiple errors in the block. Given a number of errors, the output could contain the correct answer even though the logic block contains errors or

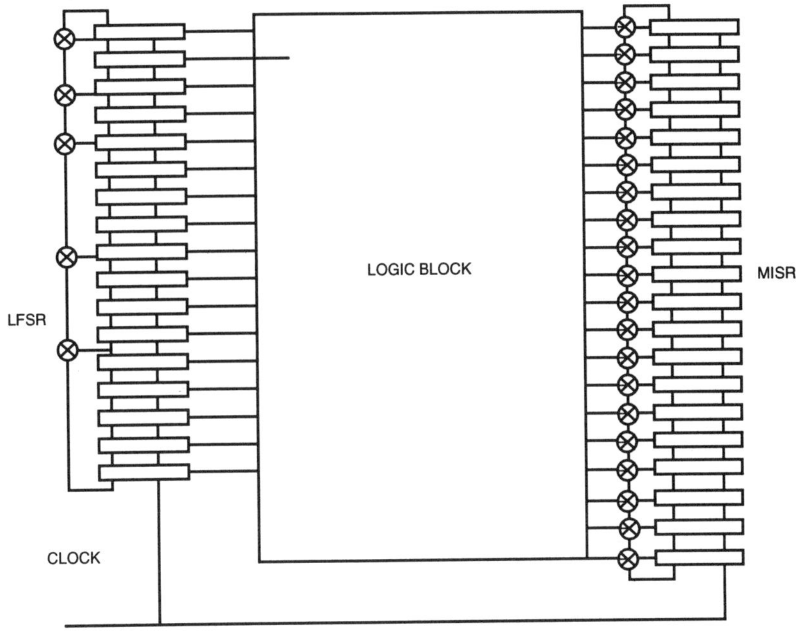

FIGURE 13.5 BIST implementation example with inputs and outputs.

defects. This problem is called *aliasing*. For example, the output register shown in Fig. 13.5 is 20-bits long; 2^{20} is approximately one million, which means that there is a very high chance that without an aliasing defect, the output will be the correct state of one out of a million with a fault free integrated circuit.

It should be noted that the example is shown for combination logic only. Sequential logic becomes more difficult as initialization, illegal states, state machines, global feedback, and many other cases must be accounted for in the generation of BIST design and patterns.

Scan

Scan is a technique where storage elements (latches and flip-flops) are changed to dual mode elements. For instance, Fig. 13.6 shows a normal flip-flop of an integrated circuit. Also shown in Fig. 13.6 is a flip-flop converted to a scan register element. During normal operation, the D input and system clock control the output of the flip-flop. During scan operation, the scan clocks are used to control shifting of data into and out of the shift register. Data is shifted into and out of the master and scan latch. The slave latch can be clocked by the system clock in order to generate a signal for the required stimulation of the logic between this scan latch and the next scan latches. Figure 13.7 shows latches and logic between latches. This method successfully reduces the large sequential test problem to a rather simple combinatorial problem. The unfortunate problem with scan is that the area overhead is, by far, the largest of all **design for test (DFT)** methods. This area increase is due to the fact that each storage element must increase in transistor count to accommodate the added function of scan.

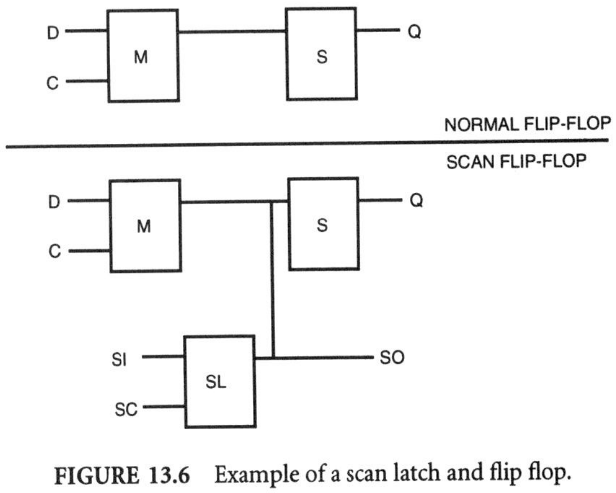

FIGURE 13.6 Example of a scan latch and flip flop.

Note that the scan latch contains extra elements for control and observation. These extra elements include the scan-clock (SC), scan-in (SI), and scan-out (SO). As shown in Fig. 13.7, the scan-clock is used to drive all scan latches. This controls the shift register function of the scan latch, and will allow a known one or zero to be set for each latch in the scan chain. For each scan clock applied, a value of one or zero is placed on the scan-in line, which is then shifted to the scan-out and the next scan-in latch. This shifting stops when the entire chain is initialized. At the end of the scan in sequence, a system clock is triggered, which launches data from the scan portion of the latch of the Q output. Then it goes through the combinational logic to the next stage D input where another system clock stores it in the latch. The next set of scan loads, and shifts out the stored value of the logic operation for comparison. If the logic operation was correct, the scan value will be correct in the shift out sequence.

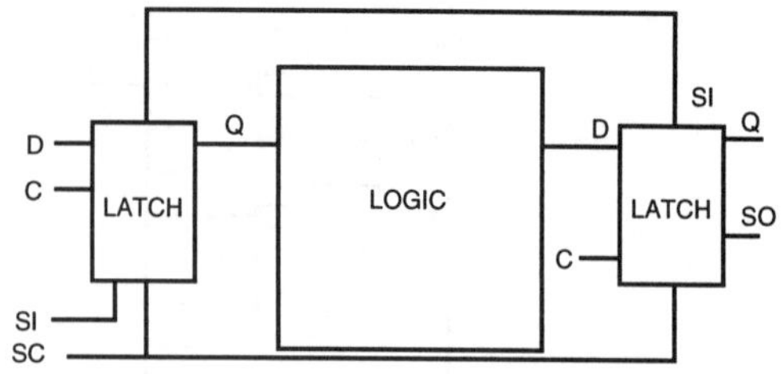

FIGURE 13.7 Use of scan latch for logic.

I_{DDQ}

I_{DDQ} is perhaps the simplest test concept. Figure 13.8 shows a typical dual inverter structure with a potential defect. It also shows the characteristic of I_{DDQ}. The upper circuit is defect free, and the lower circuit has a leakage defect shown as R. I_{DDQ} is shown for each clock cycle. Note the elevated current in the defective circuit during clock cycle A. This elevated current is called I_{DDQ}. It is an easy way to check for most bridging and many other types of defects. The I_{DDQ} test controls and observes approximately half of the transistors in a IC for each test. Therefore, the test is considered massively parallel and is very efficient at detecting bridge defects and partial shorts. Another consideration is the need to design for an I_{DDQ} test. All DC paths, pullups, bus drivers, and contention paths must be designed for zero current when the clock is stopped. One single transistor remaining on, can disable the use of an I_{DDQ} test.

Note that V_{out} and I_{DD} in the defective case may be tolerable if the output voltage is sufficient to drive the next input stage. Not shown here is the timing impact of the degraded voltage. This degraded voltage will cause a time delay in the next stage.

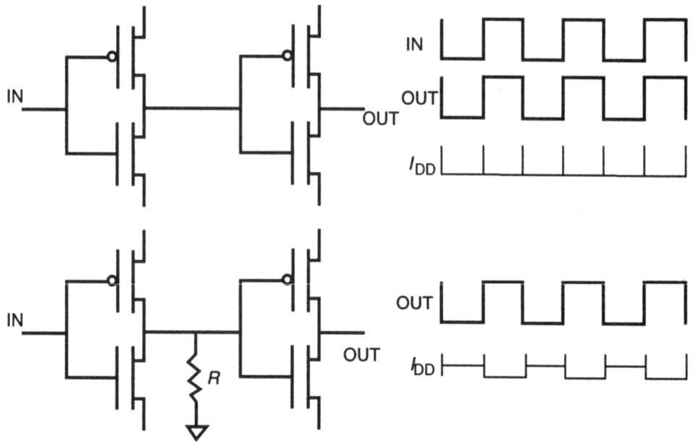

FIGURE 13.8 I_{DDQ} test values.

13.4 Test Tradeoffs

To determine the best type of test, one must understand the defect types, the logical function of the integrated circuit, and test requirements. Examples of test tradeoffs are included in Table 13.2. The rows in the table correspond to the test types as listed earlier: scan, BIST, stored response, and I_{DDQ}. The columns in the table depict certain attributes of test to be described here.

The first column is the test type. Column number two shows pattern complexity. This is the problem of generating the necessary patterns in order to fully stimulate and control the nodes of the integrated circuit. Column number three is illegal states. Often times the logic in integrated circuits, such as bidirectional bus drivers and signals, must be exclusively set one way or another. Examples include decoders, bus drivers and mutually exclusive line drivers. Tests such as scan and BIST, if not designed and implemented correctly, can damage the integrated circuit during the test process by putting the circuit into an internal contention state. Examples include two bus drivers on simultaneously when one drives high and the other drives low.

TABLE 13.2 Attributes of Test

Test Types	Complexity	Illegal States	Aliasing	Silicon Overhead	Test Pattern Generation Time
SCAN	Simple	Controlled	NA	Highest	Shortest
BIST	Complicated with latches	Big problem	Yes	Moderate	Moderate
Stored response	Very complicated	NA	NA	None-little	Longest
I_{DDQ}	Low	None	None	None	Simple

Aliasing is a common problem of BIST techniques. This is shown in column four. For example, consider a memory laid out in rows and columns and assume that one column is a complete failure. If there are multiple failures along the column, let us say 256, and if the signature length is 256, the output could possibly show a correct response even though the entire column or row of the memory is defective.

The next column shows overhead. Overhead is the amount of silicon area necessary to implement the test. Each of the areas of overhead relate to the amount of silicon area needed to implement the specific types of test methods.

The final column is the time for test generation. The unfortunate relationship here is that the techniques with the highest area overhead have the most complete control and are easiest to generate tests. Techniques with the lowest area needs have the most complex test generation problem, and take the longest time. This is a complex trade off.

Tradeoffs of Volume and Testability

It is important to forecast the volume of the integrated circuit to be manufactured prior to selecting DFT and test methods. Engineering effort is nonrecurring and can be averaged over every integrated circuit manufactured. The testing of an integrated circuit can be a simple or a complex process. The criteria used for selecting the proper test technique must include volume, expected quality, and expected cost of the integrated circuit. Time to market and/or time to shipments are key factors, as some of the test techniques for very large integrated circuits take an extremely long time.

Defining Terms

Automated test equipment (ATE): Any computer controlled equipment used to test integrated circuits.

Builtin self-test (BIST): An acronym that generally describes a design technique with input stimulus generation circuitry and output response checking circuitry.

Defect: Any error, particle, or contamination of the circuit structure. Defects may or may not result in faulty circuit operation.

Design for test (DFT): A general term which encompasses all techniques used to improve control and observation of internal nodes.

Device under test (DUT): A term used to describe the device being tested by the ATE.

Fault: The operation of a circuit in error. Usually faulty circuits are the result of defects or damage to the circuit.

Linear feedback shift register (LFSR): A method of construction of a register with feedback to generate pseudo random numbers.

Multiple input shift register (MISR): Usually an LFSR with inputs consisting of the previous stage exclusive OR'ed with input data. This method compacts the response of a circuit into a polynomial.

Scan: A design technique where sequential elements (latches, or flipflops) are chained together in a mode that allows data shifting into and out of the latches. In general, scan allows easy access to the logic between latches and greatly reduces the test generation time and effort.

References

Abramovici, M. et al. 1990. *Digital Systems Testing and Testable Design*. IEEE Press, New York.

Needham, W.M. 1991. *Designer's Guide to Testable ASIC Devices*. Van Nostrand Reinhold, New York.

van de Goor, A.J. 1991. *Testing Semiconductor Memories, Theory and Practice*. John Wiley, New York.

Further Information

IEEE International Test Conference (ITC) held each Fall in Washington DC. This is the largest gathering of test experts, test vendors and researchers in the world.

IEEE Design Automation Conference (DAC) held each summer in various sites throughout the world.

IEEE VLSI Test Symposium held each spring in Cherry Hill, New Jersey.

Test vendors of various types of equipment provide training for operation, programming, and maintenance of ATE. Visit their booths at the ITC.

CAD vendors supply a variety of simulation tools for test analysis and debugging. See them at the DAC and ITC.

II

Conversion Factors, Standards and Constants

Jerry C. Whitaker
Editor-in-Chief

ONE OF THE GOALS OF THIS BOOK is to provide easy reference to a wide variety od data that engineere need on the job. Section II represents the culmination of that effort. Extensive tables and other data are given on a wide range of topics. as listed above. Particular attention should be given to Chapter 15, **Frequency Bands and Assignments,** where a detailed listing of frequency allocation and communications services is given. Other chapters in this section provide important reference data on the properties of materials, international standards, conversion factors, general mathematical tables, and abbrevations for common terms in the communication industry.

14

Properties of Materials

James F. Shackelford
University of California, Davis

14.1 Introduction

The term *materials science and engineering* refers to that branch of engineering dealing with the processing, selection, and evaluation of solid-state materials [Shackelford 1996]. As such, this is a highly interdisciplinary field. This chapter reflects the fact that engineers outside of the specialized field of materials science and engineering largely need guidance in the selection of materials for their specific applications. A comprehensive source of property data for engineering materials is available in *The CRC Materials Science and Engineering Handbook* [Shackelford, Alexander, and Park 1994]. This brief chapter will be devoted to defining key terms associated with the properties of engineering materials and providing representative tables of such properties. Because the underlying principle of the fundamental understanding of solid-state materials is the fact that atomic- or microscopic-scale structure is responsible for the nature of materials properties, we shall begin with a discussion of structure. This will be followed by a discussion of the importance of specifying the chemical composition of commercial materials. These discussions will be followed by the definition of the main categories of materials properties.

14.2 Structure

A central tenet of materials science is that the behavior of materials (represented by their **properties**) is determined by their structure on the atomic and microscopic scales [Shackelford 1996]. Perhaps the most fundamental aspect of the structure–property relationship is to appreciate the basic skeletal arrangement of atoms in **crystalline** solids. Table 14.1 illustrates the fundamental possibilities, known as the 14 **Bravais lattices**. All crystalline structures of real materials can be produced by decorating the unit cell patterns of Table 14.1 with one or more atoms and repetitively stacking the unit cell structure through three-dimensional space.

14.3 Composition

The properties of commercially available materials are determined by chemical composition as well as structure [Shackelford 1996]. As a result, extensive numbering systems have been developed to label materials, especially metal **alloys**. Table 14.2 gives an example for gray cast irons.

0-8493-0050-9/00/$0.00+$.50

TABLE 14.1 The 14 Bravais Lattices

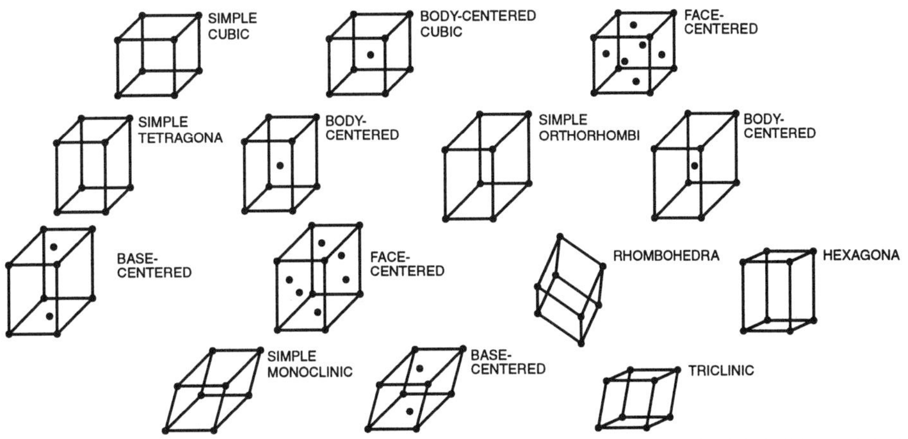

Source: Shackelford, J.F. 1996. *Introduction to Materials Science for Engineers*, 4th ed., p. 58. Prentice–Hall, Upper Saddle River, NJ.

14.4 Physical Properties

Among the most basic and practical characteristics of engineering materials are their physical properties. Table 14.3 gives the **density** of a wide range of materials in units of mega gram per cubic meter (= gram per cubic centimeter), whereas Table 14.4 gives the **melting points** for several common metals and ceramics.

14.5 Mechanical Properties

Central to the selection of materials for structural applications is their behavior in response to mechanical loads. A wide variety of mechanical properties are available to help guide materials selection [Shackelford, Alexander, and Park 1994]. The most basic of the mechanical properties are defined in terms of the **engineering stress** and the **engineering strain**. The engineering stress σ is defined as

$$\sigma = \frac{P}{A_0} \tag{14.1}$$

where P is the load on the sample with an original (zero stress) cross-sectional area A_0. The engineering strain ε is defined as

$$\varepsilon = \frac{[l - l_0]}{l_0} = \frac{\Delta l}{l_0} \tag{14.2}$$

where l is the sample length at a given load and l_0 is the original (zero stress) length. The maximum engineering stress that can be withstood by the material during its load history is termed the *ultimate tensile strength*, or simply **tensile strength** (TS). An example of the tensile strength for selected wrought (meaning worked, as opposed to cast) aluminum alloys is given in Table 14.5. The stiffness of a material is indicated by the linear relationship between engineering stress and engineering strain for relatively small levels of load application. The **modulus of elasticity** E, also known as **Young's modulus**, is

TABLE 14.2 Composition Limits of Selected Gray Cast Irons (%)

UNS	SAE Grade	C	Mn	Si	P	S
F10004	G1800	3.40 to 3.70	0.50 to 0.80	2.80 to 2.30	0.15	0.15
F10005	G2500	3.20 to 3.50	0.60 to 0.90	2.40 to 2.00	0.12	0.15
F10009	G2500	3.40 min	0.60 to 0.90	1.60 to 2.10	0.12	0.12
F10006	G3000	3.10 to 3.40	0.60 to 0.90	2.30 to 1.90	0.10	0.16
F10007	G3500	3.00 to 3.30	0.60 to 0.90	2.20 to 1.80	0.08	0.16
F10010	G3500	3.40 min	0.60 to 0.90	1.30 to 1.80	0.08	0.12
F10011	G3500	3.50 min	0.60 to 0.90	1.30 to 1.80	0.08	0.12
F10008	G4000	3.00 to 3.30	0.70 to 1.00	2.10 to 1.80	0.07	0.16
F10012	G4000	3.10 to 3.60	0.60 to 0.90	1.95 to 2.40	0.07	0.12

Source: Data from ASM. 1984. *ASM Metals Reference Book*, 2nd ed., p. 166. American Society for Metals, Metals Park, OH.

given by the ratio

$$E = \frac{\sigma}{\varepsilon} \tag{14.3}$$

Table 14.6 gives values of Young's modulus for selected compositions of glass materials. The ductility of a material is indicated by the percent elongation at failure ($= 100 \times \varepsilon_{\text{failure}}$), representing the general ability of the material to be plastically (i.e., permanently) deformed. The percent elongation at failure for selected polymers is given in Table 14.7.

TABLE 14.3 Density of Selected Materials, Mg/m^3

Metal		Ceramic		Glass		Polymer	
Ag	10.50	Al_2O_3	3.97–3.986	SiO_2	2.20	ABS	1.05–1.07
Al	2.7	BN (cub)	3.49	SiO_2 10 wt% Na_2O	2.291	Acrylic	1.17–1.19
Au	19.28	BeO	3.01–3.03	SiO_2 19.55 wt% Na_2O	2.383	Epoxy	1.80–2.00
Co	8.8	MgO	3.581	SiO_2 29.20 wt% Na_2O	2.459	HDPE	0.96
Cr	7.19	SiC(hex)	3.217	SiO_2 39.66 wt% Na_2O	2.521	Nylon, type 6	1.12–1.14
Cu	8.93	Si_3N_4 (α)	3.184	SiO_2 39.0 wt% CaO	2.746	Nylon 6/6	1.13–1.15
Fe	7.87	Si_3N_4 (β)	3.187			Phenolic	1.32–1.46
Ni	8.91	TiO_2 (rutile)	4.25			Polyacetal	1.425
Pb	11.34	UO_2	10.949–10.97			Polycarbonate	1.2
Pt	21.44	ZrO_2 (CaO)	5.5			Polyester	1.31
Ti	4.51	Al_2O_3 MgO	3.580			Polystyrene	1.04
W	19.25	$3Al_2O_3$ $2SiO_2$	2.6–3.26			PTFE	2.1–2.3

TABLE 14.4 Melting Point of Selected Metals and Ceramics

Metal	M.P. (°C)	Ceramic	M.P. (°C)
Ag	962	Al_2O_3	2049
Al	660	BN	2727
Au	1064	B_2O_3	450
Co	1495	BeO	2452
Cr	1857	NiO	1984
Cu	1083	PbO	886
Fe	1535	SiC	2697
Ni	1453	Si_3N_4	2442
Pb	328	SiO_2	1723
Pt	1772	WC	2627
Ti	1660	ZnO	1975
W	3410	ZrO_2	2850

TABLE 14.5 Tensile Strength of Selected Wrought Aluminum Alloys

Alloy	Temper	TS (MPa)
1050	0	76
1050	H16	130
2024	0	185
2024	T361	495
3003	0	110
3003	H16	180
5050	0	145
5050	H34	195
6061	0	125
6061	T6, T651	310
7075	0	230
7075	T6, T651	570

TABLE 14.6 Young's Modulus of Selected Glasses, GPa

Type	E
SiO_2	72.76–74.15
SiO_2 20 mol % Na_2O	62.0
SiO_2 30 mol % Na_2O	60.5
SiO_2 35 mol % Na_2O	60.2
SiO_2 24.6 mol % PbO	47.1
SiO_2 50.0 mol % PbO	44.1
SiO_2 65.0 mol % PbO	41.2
SiO_2 60 mol % B_2O_3	23.3
SiO_2 90 mol % B_2O_3	20.9
B_2O_3	17.2–17.7
B_2O_3 10 mol % Na_2O	31.4
B_2O_3 20 mol % Na_2O	43.2

TABLE 14.7 Total Elongation at Failure of Selected Polymers

Polymer	Elongation
ABS	5–20
Acrylic	2–7
Epoxy	4.4
HDPE	700–1000
Nylon, type 6	30–100
Nylon 6/6	15–300
Phenolic	0.4–0.8
Polyacetal	25
Polycarbonate	110
Polyester	300
Polypropylene	100–600
PTFE	250–350

[a]% in 50 mm section.

Source: Selected data for Tables 14.3–14.7 from Shackelford, J.F., Alexander, W., and Park, J.S., eds. 1994. *CRC Materials Science and Engineering Handbook*, 2nd ed. CRC Press, Boca Raton, FL.

14.6 Thermal Properties

Many applications of engineering materials depend on their response to a thermal environment. The **thermal conductivity** k is defined by **Fourier's law**

$$k = \frac{-\left[\frac{dQ}{dt}\right]}{\left[A\left(\frac{dT}{dx}\right)\right]} \tag{14.4}$$

where dQ/dt is the rate of heat transfer across an area A due to a temperature gradient dT/dx. It is also important to note that the dimensions of a material will, in general, increase with temperature. Increases in temperature lead to greater thermal vibration of the atoms and an increase in average separation distance of adjacent atoms. The **linear coefficient of thermal expansion** α is given by

TABLE 14.8 Thermal Conductivity and Thermal Expansion of Alloy Cast Irons

Alloy	Thermal Conductivity, W/(m K)	Thermal Expansion Coefficient, 10^{-6} mm/(mm °C)
Low-C white iron	22[a]	12[b]
Martensitic nickel-chromium iron	30[a]	8–9[b]
High-nickel gray iron	38–40	8.1–19.3
High-nickel ductile iron	13.4	12.6–18.7
Medium-silicon iron	37	10.8
High-chromium iron	20	9.3–9.9
High-nickel iron	37–40	8.1–19.3
Nickel-chromium-silicon iron	30	12.6–16.2
High-nickel (20%) ductile iron	13	18.7

[a]Estimated.
[b]10–260°C. (*Source:* Data from ASM. 1984. *ASM Metals Reference Book*, 2nd ed., p. 172. American Society for Metals, Metals Park, OH.)

$$\alpha = \frac{dl}{l dT} \tag{14.5}$$

with α having units of millimeter per millimeter degree Celsius. Examples of thermal conductivity and thermal expansion coefficient for alloy cast irons are given in Table 14.8.

14.7 Chemical Properties

A wide variety of data is available to characterize the nature of the reaction between engineering materials and their chemical environments [Shackelford, Alexander, and Park 1994]. Perhaps no such data are more fundamental and practical than the **electromotive force series** of metals shown in Table 14.9. The voltage associated with various half-cell reactions in standard aqueous environments are arranged in order, with the materials associated with more anodic reactions tending to be corroded in the presence of a metal associated with a more cathodic reaction.

TABLE 14.9 Electromotive Force Series of Metals

Metal	Potential, V	Metal	Potential, V	Metal	Potential, V
Anodic or Corroded End					
Li	−3.04	Al	−1.70	Pb	−0.13
Rb	−2.93	Mn	−1.04	H	0.00
K	−2.92	Zn	−0.76	Cu	0.52
Ba	−2.90	Cr	−0.60	Ag	0.80
Sr	−2.89	Cd	−0.40	Hg	0.85
Ca	−2.80	Ti	−0.33	Pd	1.0
Na	−2.71	Co	−0.28	Pt	1.2
Mg	−2.37	Ni	−0.23	Au	1.5
Be	−1.70	Sn	−0.14		
			Cathodic or Noble Metal End		

Source: Data compiled by J.S. Park from Bolz, R.E. and Tuve, G.L., eds. 1973. *CRC Handbook of Tables for Applied Engineering Science.* CRC Press, Boca Raton, FL.

14.8 Electrical and Optical Properties

To this point, we have concentrated on various properties dealing largely with the structural applications of engineering materials. In many cases the electromagnetic nature of the materials may determine their engineering

TABLE 14.10 Electrical Resistivity of Selected Materials

Metal (Alloy Cast Iron)	$\rho, \Omega \cdot m$	Ceramic	$\rho, \Omega \cdot m$	Polymer	$\rho, \Omega \cdot m$
Low-C white cast iron	$0.53 \cdot 10^{-6}$	Al_2O_3	$>10^{13}$	ABS	$2\text{–}4 \cdot 10^{13}$
Martensitic Ni-Cr iron	$0.80 \cdot 10^{-6}$	B_4C	$0.3\text{–}0.8 \cdot 10^{-2}$	Acrylic	$>10^{13}$
High-Si iron	$0.50 \cdot 10^{-6}$	BN	$1.7 \cdot 10^{11}$	HDPE	$>10^{15}$
High-Ni iron	$1.4\text{–}1.7 \cdot 10^{-6}$	BeO	$>10^{15}$	Nylon 6/6	$10^{12}\text{–}10^{13}$
Ni-Cr-Si iron	$1.5\text{–}1.7 \cdot 10^{-6}$	MgO	$1.3 \cdot 10^{13}$	Phenolic	$10^{7}\text{–}10^{11}$
High-Al iron	$2.4 \cdot 10^{-6}$	SiC	$1\text{–}1 \cdot 10^{10}$	Polyacetal	10^{13}
Medium-Si ductile iron	$0.58\text{-}0.87 \cdot 10^{-6}$	Si_3N_4	10^{11}	Polypro- pylene	$>10^{15}$
High-Ni (20%) ductile iron	$1.02 \cdot 10^{-6}$	SiO_2	10^{16}	PTFE	$>10^{16}$

Source: Selected data from Shackelford, J.F., Alexander, W., and Park, J.S., eds. 1994. *CRC Materials Science and Engineering Handbook,* 2nd ed. CRC Press, Boca Raton, FL.

TABLE 14.11 Refractive Index of Selected Polymers

Polymer	n
Acrylic	1.485–1.500
Cellulose Acetate	1.46–1.50
Epoxy	1.61
HDPE	1.54
Polycarbonate	1.586
PTFE	1.35
Polyester	1.50–1.58
Polystyrene	1.6
SAN	1.565–1.569
Vinylidene Chloride	1.60–1.63

Source: Data compiled by J.S. Park from Lynch, C.T., ed. 1975. *CRC Handbook of Materials Science, Vol. 3.* CRC Press, Inc., Boca Raton, FL; and ASM. 1988. *Engineered Materials Handbook, Vol. 2, Engineering Plastics.* ASM International, Metals Park, OH.

applications. Perhaps the most fundamental relationship in this regard is **Ohm's law,** which states that the magnitude of current flow I through a circuit with a given resistance R and voltage V is related by

$$V = IR \tag{14.6}$$

where V is in units of volts, I is in amperes, and R is in ohms. The resistance value depends on the specific sample geometry. In general, R increases with sample length l and decreases with sample area A. As a result, the property more characteristic of a given material and independent of its geometry is **resistivity** ρ defined as

$$\rho = \frac{[RA]}{l} \tag{14.7}$$

The units for resistivity are ohm meter ($\Omega \cdot m$). Table 14.10 gives the values of electrical resistivity for various materials, indicating that metals typically have low resistivities (and correspondingly high electrical conductivities) and ceramics and polymers typically have high resistivities (and correspondingly low conductivities).

An important aspect of the electromagnetic nature of materials is their optical properties. Among the most fundamental optical characteristics of a light transmitting material is the **index of refraction** n defined as

$$n = \frac{v_{\text{vac}}}{v} \tag{14.8}$$

where v_{vac} is the speed of light in vacuum (essentially equal to that in air) and v is the speed of light in a transparent material. The index of refraction for a variety of polymers is given in Table 14.11.

14.9 Additional Data

The following, more detailed tabular data are reprinted from: Dorf, R.C. 1993. *The Electrical Engineering Handbook.* CRC Press, Boca Raton, FL, pp. 2527–2544.

Electrical Resistivity

Electrical Resistivity of Pure Metals

The first part of this table gives the electrical resistivity, in units of $10^{-8}\ \Omega \cdot m$, for 28 common metallic elements as a function of temperature. The data refer to polycrystalline samples. The number of significant figures indicates the accuracy of the values. However, at low temperatures (especially below 50 K) the electrical resistivity is extremely sensitive to sample purity. Thus the low-temperature values refer to samples of specified purity and treatment.

The second part of the table gives resistivity values in the neighborhood of room temperature for other metallic elements that have not been studied over an extended temperature range.

Electrical Resistivity in $10^{-8}\ \Omega \cdot m$

T/K	Aluminum	Barium	Beryllium	Calcium	Cesium	Chromium	Copper
1	0.000100	0.081	0.0332	0.045	0.0026		0.00200
10	0.000193	0.189	0.0332	0.047	0.243		0.00202
20	0.000755	0.94	0.0336	0.060	0.86		0.00280
40	0.0181	2.91	0.0367	0.175	1.99		0.0239
60	0.0959	4.86	0.067	0.40	3.07		0.0971
80	0.245	6.83	0.075	0.65	4.16		0.215
100	0.442	8.85	0.133	0.91	5.28	1.6	0.348
150	1.006	14.3	0.510	1.56	8.43	4.5	0.699
200	1.587	20.2	1.29	2.19	12.2	7.7	1.046
273	2.417	30.2	3.02	3.11	18.7	11.8	1.543
293	2.650	33.2	3.56	3.36	20.5	12.5	1.678
298	2.709	34.0	3.70	3.42	20.8	12.6	1.712
300	2.733	34.3	3.76	3.45	21.0	12.7	1.725
400	3.87	51.4	6.76	4.7		15.8	2.402
500	4.99	72.4	9.9	6.0		20.1	3.090
600	6.13	98.2	13.2	7.3		24.7	3.792
700	7.35	130	16.5	8.7		29.5	4.514
800	8.70	168	20.0	10.0		34.6	5.262
900	10.18	216	23.7	11.4		39.9	6.041

T/K	Gold	Hafnium	Iron	Lead	Lithium	Magnesium	Manganese
1	0.0220	1.00	0.0225		0.007	0.0062	7.02
10	0.0226	1.00	0.0238		0.008	0.0069	18.9
20	0.035	1.11	0.0287		0.012	0.0123	54
40	0.141	2.52	0.0758		0.074	0.074	116
60	0.308	4.53	0.271		0.345	0.261	131
80	0.481	6.75	0.693	4.9	1.00	0.557	132
100	0.650	9.12	1.28	6.4	1.73	0.91	132
150	1.061	15.0	3.15	9.9	3.72	1.84	136
200	1.462	21.0	5.20	13.6	5.71	2.75	139
273	2.051	30.4	8.57	19.2	8.53	4.05	143
293	2.214	33.1	9.61	20.8	9.28	4.39	144
298	2.255	33.7	9.87	21.1	9.47	4.48	144
300	2.271	34.0	9.98	21.3	9.55	4.51	144
400	3.107	48.1	16.1	29.6	13.4	6.19	147
500	3.97	63.1	23.7	38.3		7.86	149
600	4.87	78.5	32.9			9.52	151
700	5.82		44.0			11.2	152
800	6.81		57.1			12.8	
900	7.86					14.4	

T/K	Molybdenum	Nickel	Palladium	Platinum	Potassium	Rubidium	Silver
1	0.00070	0.0032	0.0200	0.002	0.0008	0.0131	0.00100
10	0.00089	0.0057	0.0242	0.0154	0.0160	0.109	0.00115
20	0.00261	0.0140	0.0563	0.0484	0.117	0.444	0.0042
40	0.0457	0.068	0.334	0.409	0.480	1.21	0.0539
60	0.206	0.242	0.938	1.107	0.90	1.94	0.162
80	0.482	0.545	1.75	1.922	1.34	2.65	0.289
100	0.858	0.96	2.62	2.755	1.79	3.36	0.418
150	1.99	2.21	4.80	4.76	2.99	5.27	0.726
200	3.13	3.67	6.88	6.77	4.26	7.49	1.029
273	4.85	6.16	9.78	9.6	6.49	11.5	1.467
293	5.34	6.93	10.54	10.5	7.20	12.8	1.587
298	5.47	7.12	10.73	10.7	7.39	13.1	1.617
300	5.52	7.20	10.80	10.8	7.47	13.3	1.629
400	8.02	11.8	14.48	14.6			2.241

Electrical Resistivity in $10^{-8}\ \Omega \cdot m$ (*continued*)

T/K	Molybdenum	Nickel	Palladium	Platinum	Potassium	Rubidium	Silver
500	10.6	17.7	17.94	18.3			2.87
600	13.1	25.5	21.2	21.9			3.53
700	15.8	32.1	24.2	25.4			4.21
800	18.4	35.5	27.1	28.7			4.91
900	21.2	38.6	29.4	32.0			5.64

T/K	Sodium	Strontium	Tantalum	Tungsten	Vanadium	Zinc	Zirconium
1	0.0009	0.80	0.10	0.000016		0.0100	0.250
10	0.0015	0.80	0.102	0.000137	0.0145	0.0112	0.253
20	0.016	0.92	0.146	0.00196	0.039	0.0387	0.357
40	0.172	1.70	0.751	0.0544	0.304	0.306	1.44
60	0.447	2.68	1.65	0.266	1.11	0.715	3.75
80	0.80	3.64	2.62	0.606	2.41	1.15	6.64
100	1.16	4.58	3.64	1.02	4.01	1.60	9.79
150	2.03	6.84	6.19	2.09	8.2	2.71	17.8
200	2.89	9.04	8.66	3.18	12.4	3.83	26.3
273	4.33	12.3	12.2	4.82	18.1	5.46	38.8
293	4.77	13.2	13.1	5.28	19.7	5.90	42.1
298	4.88	13.4	13.4	5.39	20.1	6.01	42.9
300	4.93	13.5	13.5	5.44	20.2	6.06	43.3
400		17.8	18.2	7.83	28.0	8.37	60.3
500		22.2	22.9	10.3	34.8	10.82	76.5
600		26.7	27.4	13.0	41.1	13.49	91.5
700		31.2	31.8	15.7	47.2		104.2
800		35.6	35.9	18.6	53.1		114.9
900			40.1	21.5	58.7		123.1

Electrical Resistivity of Other Metallic Elements in the Neighborhood of Room Temperature

Element	*T*/K	Electrical Resistivity $10^{-8}\ \Omega \cdot m$	Element	*T*/K	Electrical Resistivity $10^{-8}\ \Omega \cdot m$
Antimony	273	39	Polonium	273	40
Bismuth	273	107	Praseodymium	290–300	70.0
Cadmium	273	6.8	Promethium	290–300	75
Cerium	290–300	82.8	Protactinium	273	17.7
Cobalt	273	5.6	Rhenium	273	17.2
Dysprosium	290–300	92.6	Rhodium	273	4.3
Erbium	290–300	86.0	Ruthenium	273	7.1
Europium	290–300	90.0	Samarium	290–300	94.0
Gadolinium	290–300	131	Scandium	290–300	56.2
Gallium	273	13.6	Terbium	290–300	115
Holmium	290–300	81.4	Thallium	273	15
Indium	273	8.0	Thorium	273	14.7
Iridium	273	4.7	Thulium	290–300	67.6
Lanthanum	290–300	61.5	Tin	273	11.5
Lutetium	290–300	58.2	Titanium	273	39
Mercury	273	94.1	Uranium	273	28
Neodymium	290–300	64.3	Ytterbium	290–300	25.0
Niobium	273	15.2	Yttrium	290–300	59.6
Osmium	273	8.1			

Electrical Resistivity of Selected Alloys

Values of the resistivity are given in units of 10^{-8} Ω· m. General comments in the preceding table for pure metals also apply here.

Alloy—Aluminum-Copper

Wt % Al	273 K	293 K	300 K	350 K	400 K
99[a]	2.51	2.74	2.82	3.38	3.95
95[a]	2.88	3.10	3.18	3.75	4.33
90[b]	3.36	3.59	3.67	4.25	4.86
85[b]	3.87	4.10	4.19	4.79	5.42
80[b]	4.33	4.58	4.67	5.31	5.99
70[b]	5.03	5.31	5.41	6.16	6.94
60[b]	5.56	5.88	5.99	6.77	7.63
50[b]	6.22	6.55	6.67	7.55	8.52
40[c]	7.57	7.96	8.10	9.12	10.2
30[c]	11.2	11.8	12.0	13.5	15.2
25[f]	16.3[aa]	17.2	17.6	19.8	22.2
15[h]	—	12.3	—	—	—
19[g]	1.8[aa]	11.0	11.1	11.7	12.3
5[e]	9.43	9.61	9.68	10.2	10.7
1[b]	4.46	4.60	4.65	5.00	5.37

Alloy—Aluminum-Magnesium

	273 K	293 K	300 K	350 K	400 K
99[c]	2.96	3.18	3.26	3.82	4.39
95[c]	5.05	5.28	5.36	5.93	6.51
90[c]	7.52	7.76	7.85	8.43	9.02
85	—	—	—	—	—
80	—	—	—	—	—
70	—	—	—	—	—
60	—	—	—	—	—
50	—	—	—	—	—
40	—	—	—	—	—
30	—	—	—	—	—
25	—	—	—	—	—
15	—	—	—	—	—
10[b]	17.1	17.4	17.6	18.4	19.2
5[b]	13.1	13.4	13.5	14.3	15.2
1[a]	5.92	6.25	6.37	7.20	8.03

Alloy—Copper-Gold

Wt % Cu	273 K	293 K	300 K	350 K	400 K
99[c]	1.73	1.86[aa]	1.91[aa]	2.24[aa]	2.58[aa]
95[c]	2.41	2.54[aa]	2.59[aa]	2.92[aa]	3.26[aa]
90[c]	3.29	4.42[aa]	3.46[aa]	3.79[aa]	4.12[aa]
85[c]	4.20	4.33	4.38[aa]	4.71[aa]	5.05[aa]
80[c]	5.15	5.28	5.32	5.65	5.99
70[c]	7.12	7.25	7.30	7.64	7.99
60[c]	9.18	9.13	9.36	9.70	10.05
50[c]	11.07	11.20	11.25	11.60	11.94
40[c]	12.70	12.85	12.90[aa]	13.27[aa]	13.65[aa]
30[c]	13.77	13.93	13.99[aa]	14.38[aa]	14.78[aa]
25[c]	13.93	14.09	14.14	14.54	14.94
15[c]	12.75	12.91	12.96[aa]	13.36[aa]	13.77
10[c]	10.70	10.86	10.91	11.31	11.72
5[c]	7.25	7.41[aa]	7.46	7.87	8.28
1[c]	3.40	3.57	3.62	4.03	4.45

Alloy—Copper-Nickel

Wt % Cu	273 K	293 K	300 K	350 K	400 K
99[c]	2.71	2.85	2.91	3.27	3.62
95[c]	7.60	7.71	7.82	8.22	8.62
90[c]	13.69	13.89	13.96	14.40	14.81
85[c]	19.63	19.83	19.90	2032	20.70
80[c]	25.46	25.66	25.72	26.12[aa]	26.44[aa]
70[i]	36.67	36.72	36.76	36.85	36.89
60[i]	45.43	45.38	45.35	45.20	45.01
50[i]	50.19	50.05	50.01	49.73	49.50
40[c]	47.42	47.73	47.82	48.28	48.49
30[i]	40.19	41.79	42.34	44.51	45.40
25[c]	33.46	35.11	35.69	39.67[aa]	42.81[aa]
15[c]	22.00	23.35	23.85	27.60	31.38
10[c]	16.65	17.82	18.26	21.51	25.19
5[c]	11.49	12.50	12.90	15.69	18.78
1[c]	7.23	8.08	8.37	10.63[aa]	13.18[aa]

Alloy—Copper-Palladium

Wt % Cu	273 K	293 K	300 K	350 K	400 K
99[c]	2.10	2.23	2.27	2.59	2.92
95[c]	4.21	4.35	4.40	4.74	5.08
90[c]	6.89	7.03	7.08	7.41	7.74
85[c]	9.48	9.61	9.66	10.01	10.36
80[c]	11.99	12.12	12.16	12.51[aa]	12.87
70[c]	16.87	17.01	17.06	17.41	17.78
60[c]	21.73	21.87	21.92	22.30	22.69
50[c]	27.62	27.79	27.86	28.25	28.64
40[c]	35.31	35.51	35.57	36.03	36.47
30[c]	46.50	46.66	46.71	47.11	47.47
25[c]	46.25	46.45	46.52	46.99[aa]	47.43[aa]
15[c]	36.52	36.99	37.16	38.28	39.35
10[c]	28.90	29.51	29.73	31.19[aa]	32.56[aa]
5[c]	20.00	20.75	21.02	22.84[aa]	24.54[aa]
1[c]	11.90	12.67	12.93[aa]	14.82[aa]	16.68[aa]

Alloy—Copper-Zinc

Wt % Cu	273 K	293 K	300 K	350 K	400 K
99[b]	1.84	1.97	2.02	2.36	2.71
95[b]	2.78	2.92	2.97	3.33	3.69
90[b]	3.66	3.81	3.86	4.25	4.63
85[b]	4.37	4.54	4.60	5.02	5.44
80[b]	5.01	5.19	5.26	5.71	6.17
70[b]	5.87	6.08	6.15	6.67	7.19
60	—	—	—	—	—
50	—	—	—	—	—
40	—	—	—	—	—
30	—	—	—	—	—
25	—	—	—	—	—
15	—	—	—	—	—
10	—	—	—	—	—
5	—	—	—	—	—
1	—	—	—	—	—

Electrical Resistivity of Selected Alloys (*continued*)

	273 K	293 K	300 K	350 K	400 K
Alloy—Gold-Palladium					
Wt % Au					
99[c]	2.69	2.86	2.91	3.32	3.73
95[c]	5.21	5.35	5.41	5.79	6.17
90[i]	8.01	8.17	8.22	8.56	8.93
85[b]	10.50[aa]	10.66	10.72[aa]	11.100[aa]	11.48[aa]
80[b]	12.75	12.93	12.99	13.45	13.93
70[c]	18.23	18.46	18.54	19.10	19.67
60[b]	26.70	26.94	27.02	27.63[aa]	28.23[aa]
50[a]	27.23	27.63	27.76	28.64[aa]	29.42[aa]
40[a]	24.65	25.23	25.42	26.74	27.95
30[b]	20.82	21.49	21.72	23.35	24.92
25[b]	18.86	19.53	19.77	21.51	23.19
15[a]	15.08	15.77	16.01	17.80	19.61
10[a]	13.25	13.95	14.20[aa]	16.00[aa]	17.81[aa]
5[a]	11.49[aa]	12.21	12.46[aa]	14.26[aa]	16.07[aa]
1[a]	10.07	10.85[aa]	11.12[aa]	12.99[aa]	14.80[aa]
Alloy—Gold-Silver					
Wt % Au					
99[b]	2.58	2.75	2.80[aa]	3.22[aa]	3.63[aa]
95[a]	4.58	4.74	4.79	5.19	5.59
90[i]	6.57	6.73	6.78	7.19	7.58
85[j]	8.14	8.30	8.36[aa]	8.75	9.15
80[j]	9.34	9.50	9.55	9.94	10.33
70[j]	10.70	10.86	10.91	11.29	11.68[aa]
60[j]	10.92	11.07	11.12	11.50	11.87
50[j]	10.23	10.37	10.42	10.78	11.14
40[j]	8.92	9.06	9.11	9.46[aa]	9.81
30[a]	7.34	7.47	7.52	7.85	8.19
25[a]	6.46	6.59	6.63	6.96	7.30[aa]
15[a]	4.55	4.67	4.72	5.03	5.34
10[a]	3.54	3.66	3.71	4.00	4.31
5[i]	2.52	2.64[aa]	2.68[aa]	2.96[aa]	3.25[aa]
1[b]	1.69	1.80	1.84[aa]	2.12[aa]	2.42[aa]
Alloy—Iron-Nickel					
Wt % Fe					
99[a]	10.9	12.0	12.4	—	18.7
95[c]	18.7	19.9	20.2	—	26.8
90[c]	24.2	25.5	25.9	—	33.2
85[c]	27.8	29.2	29.7	—	37.3
80[c]	30.1	31.6	32.2	—	40.0
70[b]	32.3	33.9	34.4	—	42.4
60[c]	53.8	57.1	58.2	—	73.9
50[d]	28.4	30.6	31.4	—	43.7
40[d]	19.6	21.6	22.5	—	34.0
30[c]	15.3	17.1	17.7	—	27.4
25[b]	14.3	15.9	16.4	—	25.1
15[c]	12.6	13.8	14.2	—	21.1
10[c]	11.4	12.5	12.9	—	18.9
5[c]	9.66	10.6	10.9	—	16.1[aa]
1[b]	7.17	7.94	8.12	—	12.8
Alloy—Silver-Palladium					
Wt % Ag					
99[b]	1.891	2.007	2.049	2.35	2.66
95[b]	3.58	3.70	3.74	4.04	4.34
90[b]	5.82	5.94	5.98	6.28	6.59
85[k]	7.92[aa]	8.04[aa]	8.08	8.38[aa]	8.68[aa]
80[k]	10.01	10.13	10.17	10.47	10.78
70[k]	14.53	14.65	14.69	14.99	15.30
60[i]	20.9	21.1	21.2	21.6	22.0
50[k]	31.2	31.4	31.5	32.0	32.4
40[m]	42.2	42.2	42.2	42.3	42.3
30[b]	40.4	40.6	40.7	41.3	41.7
25[k]	36.67[aa]	37.06	37.19	38.1[aa]	38.8[aa]
15[i]	27.08[aa]	26.68[aa]	27.89[aa]	29.3[aa]	30.6[aa]
10[i]	21.69	22.39	22.63	24.3	25.9
5[b]	15.98	16.72	16.98	18.8[aa]	20.5[aa]
1[a]	11.06	11.82	12.08[aa]	13.92[aa]	15.70[aa]

[a] Uncertainty in resistivity is ±2%.
[b] Uncertainty in resistivity is ±3%.
[c] Uncertainty in resistivity is ±5%.
[d] Uncertainty in resistivity is ±7% below 300 K and ±5% at 300 and 400 K.
[e] Uncertainty in resistivity is ±7%.
[f] Uncertainty in resistivity is ±8%.
[g] Uncertainty in resistivity is ±10%.
[h] Uncertainty in resistivity is ±12%.
[i] Uncertainty in resistivity is ±4%.
[j] Uncertainty in resistivity is ±1%.
[k] Uncertainty in resistivity is ±3% up to 300 K and ± 4% above 300 K.
[m] Uncertainty in resistivity is ±2% up to 300 K and ± 4% above 300 K.
[a] Crystal usually a mixture of α-hcp and fcc lattice.
[aa] In temperature range where no experimental data are available.

Electrical Resistivity of Selected Alloy Cast Irons

Description	Electrical resistivity, $\mu\Omega \cdot m$
Abrasion-resistant white irons	
Low-C white iron	0.53
Martensitic nickel–chromium iron	0.80
Corrosion-resistant irons	
High-silicon iron	0.50
High-chromium iron	
High-nickel gray iron	1.0[a]
High-nickel ductile iron	1.0[a]
Heat-resistant gray irons	
Medium-silicon iron	
High-chromium iron	
High-nickel iron	1.4–1.7
Nickel-chromium–silicon iron	1.5–1.7
High-aluminum iron	2.4
Heat-resistant ductile irons	
Medium-silicon ductile iron	0.58–0.87
High-nickel ductile (20 Ni)	1.02
High-nickel ductile (23 Ni)	1.0[a]

[a] Estimated. (*Source:* Data from 1984. *ASM Metals Reference Book*, 2nd ed. American Society for Metals, Metals Park, OH.)

Resistivity of Selected Ceramics (Listed by Ceramic)

Ceramic	Resistivity, $\Omega \cdot$ cm
Borides	
Chromium diboride (CrB_2)	21×10^{-6}
Hafnium diboride (HfB_2)	10–12 $\times 10^{-6}$ at room temp.
Tantalum diboride (TaB_2)	68×10^{-6}
Titanium diboride (TiB_2) (polycrystalline)	
85% dense	26.5–28.4 $\times 10^{-6}$ at room temp.
85% dense	9.0×10^{-6} at room temp.
100% dense, extrapolated values	8.7–14.1 $\times 10^{-6}$ at room temp.
	3.7×10^{-6} at liquid air temp.
Titanium diboride (TiB_2) (monocrystalline)	
Crystal length 5 cm, 39 deg. and 59 deg. orientation with respect to growth axis	$6.6 \pm 0.2 \times 10^{-6}$ at room temp.
Crystal length 1.5 cm, 16.5 deg. and 90 deg. orientation with respect to growth axis	$6.7 \pm 0.2 \times 10^{-6}$ at room temp.
Zirconium diboride (ZrB_2)	9.2×10^{-6} at 20°C
	1.8×10^{-6} at liquid air temp.
Carbides: boron carbide (B_4C)	0.3–0.8

Dielectric Constants

Dielectric Constants of Solids: Temperatures in the Range 17–22°C.

Material	Freq., Hz	Dielectric constant	Material	Freq., Hz	Dielectric Constant
Acetamide	4×10^8	4.0	Phenanthrene	4×10^8	2.80
Acetanilide	–	2.9	Phenol (10°C)	4×10^8	4.3
Acetic acid (2°C)	4×10^8	4.1	Phosphorus, red	10^8	4.1
Aluminum oleate	4×10^8	2.40	Phosphorus, yellow	10^8	3.6
Ammonium bromide	10^8	7.1	Potassium aluminum sulfate	10^6	3.8
Ammonium chloride	10^8	7.0	Potassium carbonate (15°C)	10^8	5.6
Antimony trichloride	10^8	5.34	Potassium chlorate	6×10^7	5.1
Apatite ⊥ optic axis	3×10^8	9.50	Potassium chloride	10^4	5.03
Apatite ‖ optic axis	3×10^8	7.41	Potassium chromate	6×10^7	7.3
Asphalt	$<3 \times 10^6$	2.68	Potassium iodide	6×10^7	5.6
Barium chloride (anhyd.)	6×10^7	11.4	Potassium nitrate	6×10^7	5.0
Barium chloride ($2H_2O$)	6×10^7	9.4	Potassium sulfate	6×10^7	5.9
Barium nitrate	6×10^7	5.9	Quartz ⊥ optic axis	3×10^7	4.34
Barium sulfate(15°C)	10^8	11.4	Quartz ‖ optic axis	3×10^7	4.27
Beryl ⊥ optic axis	10^4	7.02	Resorcinol	4×10^8	3.2
Beryl ‖ optic axis	10^4	6.08	Ruby⊥ optic axis	10^4	13.27
Calcite ⊥ optic axis	10^4	8.5	Ruby ‖ optic axis	10^4	11.28
Calcite ‖ optic axis	10^4	8.0	Rutile ⊥ optic axis	10^8	86
Calcium carbonate	10^6	6.14	Rutile ‖ optic axis	10^8	170
Calcium fluoride	10^4	7.36	Selenium	10^8	6.6
Calcium sulfate ($2H_2O$)	10^4	5.66	Silver bromide	10^6	12.2
Cassiterite ⊥ optic axis	10^{12}	23.4	Silver chloride	10^6	11.2
Cassiterite ‖ optic axis	10^{12}	24	Silver cyanide	10^6	5.6
d-Cocaine	5×10^8	3.10	Smithsonite ⊥ optic axis	10^{12}	9.3
Cupric oleate	4×10^8	2.80	Smithsonite ‖ optic axis	10^{10}	9.4
Cupric oxide (15°C)	10^8	18.1	Sodium carbonate (anhyd.)	6×10^7	8.4
Cupric sulfate (anhyd.)	6×10^7	10.3			
Cupric sulfate ($5H_2O$)	6×10^7	7.8			
Diamond	10^8	5.5			
Diphenylymethane	4×10^8	2.7			

Dielectric Constants of Solids: Temperatures in the Range 17–22°C (*continued*)

Material	Freq., Hz	Dielectric constant	Material	Freq., Hz	Dielectric Constant
Dolomite ⊥ optic axis	10^8	8.0			
Dolomite ∥ optic axis	10^8	6.8	Sodium carbonate ($10H_2O$)	6×10^7	5.3
Ferrous oxide (15°C)	10^8	14.2			
Iodine	10^8	4	Sodium chloride	10^4	6.12
Lead acetate	10^6	2.6	Sodium nitrate	–	5.2
Lead carbonate (15°C)	10^8	18.6	Sodium oleate	4×10^8	2.75
Lead chloride	10^6	4.2	Sodium perchlorate	6×10^7	5.4
Lead monoxide (15°C)	10^8	25.9	Sucrose (mean)	3×10^8	3.32
Lead nitrate	6×10^7	37.7	Sulfur (mean)	–	4.0
Lead oleate	4×10^8	3.27	Thallium chloride	10^6	46.9
Lead sulfate	10^6	14.3	*p*-Toluidine	4×10^8	3.0
Lead sulfide (15°)	10^6	17.9	Tourmaline ⊥ optic axis	10^4	7.10
Malachite (mean)	10^{12}	7.2			
Mercuric chloride	10^6	3.2	Tourmaline ∥ optic axis	10^4	6.3
Mercurous chloride	10^6	9.4			
Naphthalene	4×10^8	2.52	Urea	4×10^8	3.5
			Zircon ⊥, ∥	10^8	12

Dielectric Constants of Ceramics

Material	Dielectric constant, 10^6 Hz	Dielectric strength V/mil	Volume resistivity $\Omega \cdot$ cm (23°C)	Loss factor[a]
Alumina	4.5–8.4	40–160	10^{11}–10^{14}	0.0002–0.01
Corderite	4.5–5.4	40–250	10^{12}–10^{14}	0.004–0.012
Forsterite	6.2	240	10^{14}	0.0004
Porcelain (dry process)	6.0–8.0	40–240	10^{12}–10^{14}	0.0003–0.02
Porcelain (wet process)	6.0–7.0	90–400	10^{12}–10^{14}	0.006–0.01
Porcelain, zircon	7.1–10.5	250–400	10^{13}–10^{15}	0.0002–0.008
Steatite	5.5–7.5	200–400	10^{13}–10^{15}	0.0002–0.004
Titanates (Ba, Sr, Ca, Mg, and Pb)	15–12.000	50–300	10^8–10^{15}	0.0001–0.02
Titanium dioxide	14–110	100–210	10^{13}–10^{18}	0.0002–0.005

[a]Power factor × dielectric constant equals loss factor.

Dielectric Constants of Glasses

Type	Dielectric constant at 100 MHz (20°C)	Volume resistivity (350°C M $\Omega \cdot$ cm)	Loss factor[a]
Corning 0010	6.32	10	0.015
Corning 0080	6.75	0.13	0.058
Corning 0120	6.65	100	0.012
Pyrex 1710	6.00	2,500	0.025
Pyrex 3320	4.71	–	0.019
Pyrex 7040	4.65	80	0.013
Pyrex 7050	4.77	16	0.017
Pyrex 7052	5.07	25	0.019
Pyrex 7060	4.70	13	0.018
Pyrex 7070	4.00	1,300	0.0048
Vycor 7230	3.83	–	0.0061
Pyrex 7720	4.50	16	0.014
Pyrex 7740	5.00	4	0.040
Pyrex 7750	4.28	50	0.011
Pyrex 7760	4.50	50	0.0081

Dielectric Constants of Glasses (*continued*)

Type	Dielectric constant at 100 MHz (20°C)	Volume resistivity (350°C M Ω · cm)	Loss factor[a]
Vycor 7900	3.9	130	0.0023
Vycor 7910	3.8	1,600	0.00091
Vycor 7911	3.8	4,000	0.00072
Corning 8870	9.5	5,000	0.0085
G.E. Clear (silica glass)	3.81	4,000–30,000	0.00038
Quartz (fused)	3.75–4.1 (1 MHz)	–	0.0002 (1 MHz)

[a]Power factor × dielectric constant equals loss factor.

Properties of Semiconductors

Semiconducting Properties of Selected Materials

Substance	Minimum energy gap, eV R.T.	0 K	$\frac{dE_g}{dT}$ $\times 10^4$, eV/°C	$\frac{dE_g}{dP}$ $\times 10^6$, eV · cm²/kg	Density of states electron effective mass m_{d_n} (m_o)	Electron mobility and temperature dependence μ_n, cm²/V·s	$-x$	Density of states hole effective mass m_{d_p}, (m_o)	Hole mobility and temperature dependence μ_p, cm²/V · s	$-x$
Si	1.107	1.153	−2.3	−2.0	1.1	1,900	2.6	0.56	500	2.3
Ge	0.67	0.744	−3.7	±7.3	0.55	3,800	1.66	0.3	1,820	2.33
αSn	0.08	0.094	−0.5		0.02	2,500	1.65	0.3	2,400	2.0
Te	0.33				0.68	1,100		0.19	560	
III–V Compounds										
AlAs	2.2	2.3				1,200			420	
AlSb	1.6	1.7	−3.5	−1.6	0.09	200	1.5	0.4	500	1.8
GaP	2.24	2.40	−5.4	−1.7	0.35	300	1.5	0.5	150	1.5
GaAs	1.35	1.53	−5.0	+9.4	0.068	9,000	1.0	0.5	500	2.1
GaSb	0.67	0.78	−3.5	+12	0.050	5,000	2.0	0.23	1,400	0.9
InP	1.27	1.41	−4.6	+4.6	0.067	5,000	2.0		200	2.4
InAs	0.36	0.43	−2.8	+8	0.022	33,000	1.2	0.41	460	2.3
InSb	0.165	0.23	−2.8	+15	0.014	78,000	1.6	0.4	750	2.1
II–VI Compounds										
ZnO	3.2		−9.5	+0.6	0.38	180	1.5			
ZnS	3.54		−5.3	+5.7		180			5(400°C)	
ZnSe	2.58	2.80	−7.2	+6		540			28	
ZnTe	2.26			+6		340			100	
CdO	2.5 ± 0.1		−6		0.1	120				
CdS	2.42		−5	+3.3	0.165	400		0.8		
CdSe	1.74	1.85	−4.6		0.13	650	1.0	0.6		
CdTe	1.44	1.56	−4.1	+8	0.14	1,200		0.35	50	
HgSe	0.30				0.030	20,000	2.0			
HgTe	0.15		−1		0.017	25,000		0.5	350	
Halite Structure Compounds										
PbS	0.37	0.28	+4		0.16	800		0.1	1,000	2.2
PbSe	0.26	0.16	+4		0.3	1,500		0.34	1,500	2.2
PbTe	0.25	0.19	+4	−7	0.21	1,600		0.14	750	2.2
Others										
ZnSb	0.50	0.56			0.15	10				1.5
CdSb	0.45	0.57	−5.4		0.15	300			2,000	1.5
Bi_2S_3	1.3					200			1,100	
Bi_2Se_3	0.27					600			675	
Bi_2Te_3	0.13		−0.95		0.58	1,200	1.68	1.07	510	1.95

Semiconducting Properties of Selected Materials (*continued*)

Substance	Minimum energy gap, eV R.T.	0 K	$\frac{dE_g}{dT}$ $\times 10^4$, eV/°C	$\frac{dE_g}{dP}$ $\times 10^6$, eV · cm²/kg	Density of states electron effective mass m_{d_n} (m_o)	Electron mobility and temperature dependence μ_n, cm²/V·s	$-x$	Density of states hole effective mass m_{d_p}, (m_o)	Hole mobility and temperature dependence μ_p, cm²/V · s	$-x$
Mg_2Si		0.77	−6.4		0.46	400	2.5		70	
Mg_2Ge		0.74	−9			280	2		110	
Mg_2Sn	0.21	0.33	−3.5		0.37	320			260	
Mg_3Sb_2		0.32				20			82	
Zn_3As_2	0.93					10	1.1		10	
Cd_3As_2	0.55				0.046	100,000	0.88			
GaSe	2.05		3.8						20	
GaTe	1.66	1.80	−3.6			14	−5			
InSe	1.8					9000				
TlSe	0.57		−3.9		0.3	30		0.6	20	1.5
$CdSnAs_2$	0.23				0.05	25,000	1.7			
Ga_2Te_3	1.1	1.55	−4.8							
α-In_2Te_3	1.1	1.2			0.7				50	1.1
β-In_2Te_3	1.0								5	
$Hg_5In_2Te_8$	0.5								11,000	
SnO_2									78	

Band Properties of Semiconductors

Part A. Data on Valence Bands of Semiconductors (Room Temperature)

Substance	Band curvature effective mass (expressed as fraction of free electron mass): Heavy holes	Light holes	Split-off band holes	Energy separation of split-off band, eV	Measured (light) hole mobility cm²/V · s
	Semiconductors with Valence Band Maximum at the Center of the Brillouin Zone ("F")				
Si	0.52	0.16	0.25	0.044	500
Ge	0.34	0.043	0.08	0.3	1,820
Sn	0.3				2,400
AlAs					
AlSb	0.4			0.7	550
GaP				0.13	100
GaAs	0.8	0.12	0.20	0.34	400
GaSb	0.23	0.06		0.7	1,400
InP				0.21	150
InAs	0.41	0.025	0.083	0.43	460
InSb	0.4	0.015		0.85	750
CbTe	0.35				50
HgTe	0.5				350

Semiconductors with Multiple Valence Band Maxima

Substance	Number of equivalent valleys and direction	Band curvature effective masses: Longitudinal m_L	Transverse m_T	Anisotropy, $K = m_L/m_T$	Measured (light) hole mobility, cm²/V · s
PbSe	4 "L" [111]	0.095	0.047	2.0	1,500
PbTe	4 "L" [111]	0.27	0.02	10	750
Bi_2Te_3	6	0.207	~0.045	4.5	515

Part B. Data on conduction Bands of Semiconductors (Room Temperature Data)

Single Valley Semiconductors

Substance	Energy gap, eV	Effective mass (m_o)	Mobility $cm^2/V \cdot s$	Comments
GaAs	1.35	0.067	8,500	3 (or 6?) equivalent [100] valleys 0.36 eV above this maximum with a mobility of ~50
InP	1.27	0.067	5,000	3 (or 6?) equivalent [100] valleys 0.4 eV above this minimum
InAs	0.36	0.022	33,000	Equivalent valleys ~1.0 eV above this minimum
InSb	0.165	0.014	78,000	
CdTe	1.44	0.11	1,000	4 (or 8?) equivalent [111] valleys 0.51 eV above this minimum

Multivalley Semiconductors

Substance	Energy gap	Number of equivalent valleys and direction	Band curvature effective mass: Longitudinal m_L	Band curvature effective mass: Transverse m_T	Anisotropy $K = m_L/m_T$
Si	1.107	6 in [100] Δ	0.90	0.192	4.7
Ge	0.67	4 in [111] at L	1.588	0.0815	19.5
GaSb	0.67	as Ge (?)	~1.0	~0.2	~5
PbSe	0.26	4 in [111] at L	0.085	0.05	1.7
PbTe	0.25	4 in [111] at L	0.21	0.029	5.5
Bi_2Te_3	0.13	6			~0.05

Resistivity of Semiconducting Minerals

Mineral	ρ, $\Omega \cdot m$	Mineral	ρ, $\Omega \cdot m$
Diamond (C)	2.7	Gersdorffite, NiAsS	1 to 160 $\times 10^{-6}$
Sulfides		Glaucodote, (Co, Fe)AsS	5 to 100 $\times 10^{-6}$
Argentite, Ag_2S	1.5 to 2.0 $\times 10^{-3}$	Antimonide	
Bismuthinite, Bi_2S_3	3 to 570	Dyscrasite, Ag_3Sb	0.12 to 1.2 $\times 10^{-6}$
Bornite, $Fe_2S_3 \cdot nCu_2S$	1.6 to 6000 $\times 10^{-6}$	Arsenides	
Chalcocite, Cu_2S	80 to 100 $\times 10^{-6}$	Allemonite, $SbAs_2$	70 to 60,000
Chalcopyrite, $Fe_2S_3 \cdot Cu_2S$	150 to 9000 $\times 10^{-6}$	Lollingite, $FeAs_2$	2 to 270 $\times 10^{-6}$
Covellite, CuS	0.30 to 83 $\times 10^{-6}$	Nicollite, NiAs	0.1 to 2 $\times 10^{-6}$
Galena, PbS	6.8 $\times 10^{-6}$ to 9.0 $\times 10^{-2}$	Skutterudite, $CoAs_3$	1 to 400 $\times 10^{-6}$
Haverite, MnS_2	10 to 20	Smaltite, $CoAs_2$	1 to 12 $\times 10^{-6}$
Marcasite, FeS_2	1 to 150 $\times 10^{-3}$	Tellurides	
Metacinnabarite, 4HgS	2 $\times 10^{-6}$ to 1 $\times 10^{-3}$	Altaite, PbTe	20 to 200 $\times 10^{-6}$
Millerite, NiS	2 to 4 $\times 10^{-7}$	Calavarite, $AuTe_2$	6 to 12 $\times 10^{-6}$
Molybdenite, MoS_2	0.12 to 7.5	Coloradoite, HgTe	4 to 100 $\times 10^{-6}$
Pentlandite, $(Fe, Ni)_9S_8$	1 to 11 $\times 10^{-6}$	Hessite, Ag_2Te	4 to 100 $\times 10^{-6}$
Pyrrhotite, Fe_7S_8	2 to 160 $\times 10^{-6}$	Nagyagite, $Pb_6Au(S, Te)_{14}$	20 to 80 $\times 10^{-6}$
Pyrite, FeS_2	1.2 to 600 $\times 10^{-3}$	Sylvanite, $AgAuTe_4$	4 to 20 $\times 10^{-6}$
Sphalerite, ZnS	2.7 $\times 10^{-3}$ to 1.2 $\times 10^{4}$	Oxides	
Antimony-sulfur compounds		Braunite, Mn_2O_3	0.16 to 1.0
Berthierite, $FeSb_2S_4$	0.0083 to 2.0	Cassiterite, SnO_2	4.5 $\times 10^{-4}$ to 10,000
Boulangerite, $Pb_5Sb_4S_{11}$	2 $\times 10^{3}$ to 4 $\times 10^{4}$	Cuprite, Cu_2O	10 to 50
Cylindrite, $Pb_3Sn_4Sb_2S_{14}$	2.5 to 60	Hollandite, $(Ba, Na, K)Mn_8O_{16}$	2 to 100 $\times 10^{-3}$
Franckeite, $Pb_5Sn_3Sb_2S_{14}$	1.2 to 4	Ilmenite, $FeTiO_3$	0.001 to 4
Hauchecornite, $Ni_9(Bi, Sb)_2S_8$	1 to 83 $\times 10^{-6}$	Magnetite, Fe_3O_4	52 $\times 10^{-6}$
Jamesonite, $Pb_4FeSb_6S_{14}$	0.020 to 0.15	Manganite, $MnO \cdot OH$	0.018 to 0.5
Tetrahedrite, Cu_3SbS_3	0.30 to 30,000	Melaconite, CuO	6000

Resistivity of Semiconducting Minerals (*continued*)

Mineral	$\rho, \Omega \cdot m$	Mineral	$\rho, \Omega \cdot m$
Arsenic-sulfur compounds		Psilomelane, $KMnO \cdot MnO_2 \cdot nH_2O$	0.04 to 6000
Arsenopyrite, FeAsS	20 to 300×10^{-6}	Pyrolusite, MnO_2	0.007 to 30
Cobaltite, CoAsS	6.5 to 130×10^{-3}	Rutile, TiO_2	29 to 910
Enargite, Cu_3AsS_4	0.2 to 40×10^{-3}	Uraninite, UO	1.5 to 200

Source: Carmichael, R.S., ed. 1982. *Handbook of Physical Properties of Rocks*, Vol. I. CRC Press, Boca Raton, FL.

Properties of Magnetic Alloys

Name	Composition,[a] weight percent					Remanence, B_r, G	Coercive force, H_e, O	Maximum energy product, $(BH)_{max}$, G–O $\times 10^{-6}$
	Al	Ni	Co	Cu	Other			
U.S.								
Alnico I	12	20–22	5			7,100	440	1.4
Alnico II	10	17	12.5	6		7,200	540	1.6
Alnico III	12	24–26		3		6,900	470	1.35
Alnico IV	12	27–28	5			5,500	700	1.3
Alnico V[b]	8	14	24	3		12,500	600	5.0
Alnico V DG[b]	8	14	24	3		13,100	640	6.0
Alnico VI[b]	8	15	24	3	1.25 Ti	10,500	750	3.75
Alnico VII[b]	8.5	18	24	3	5 Ti	7,200	1,050	2.75
Alnico XII	6	18	35		8 Ti	5,800	950	1.6
					1 Mn	10,000	50	0.2
					0.9 C			
Chromium steel					3.5 Cr	9,700	65	0.3
					0.9 C			
					0.3 Mn			
Cobalt steel			17		2.5 Cr	9,500	150	0.65
					8 W			
					0.75 C			
Cunico		21	29	50		3,400	660	0.80
Cunife		20		60		5,400	550	1.5
Ferroxdur 1		$BaFe_{12}O_{19}$				2,200	1,800	1.0
Ferroxdur 2		$BaF_{12}O_{19}$ (oriented)				3,840	2,000	3.5
Platinum-cobalt			23		77 Pt	6,000	4,300	7.5
Remalloy			12		17 Mo	10,500	250	1.1
Silmanol	4.4				86.6 Ag	550	6,000	0.075
					8.8 Mn			
Tungsten steel					5 W	10,000	70	0.32
					0.3 Mn			
					0.7 C			
Vicalloy I			52		10 V	8,800	300	1.0
Vicalloy II (wire)			52		14 V	10,000	510	3.5
Germany								
Alni 90	12	21				8,000	350	1.2
Alni 120	13	27				6,000	570	1.2
Alnico 130	12	23	5			6,300	620	1.4
Alnico 160	11	24	12	4		6,200	700	1.6
Alnico 190	12	21	15	4		7,000	700	1.8
Alnico 250	8	19	23	4	6 Ti	6,500	1,000	2.2
Alnico 400[b]	9	15	23	4		12,000	650	4.8
Alnico 580[b] (semicolumnar)	9	15	23	4		13,000	700	6.0

Properties of Magnetic Alloys (*continued*)

Name	Composition,[a] weight percent Al	Ni	Co	Cu	Other	Remanence, B_r, G	Coercive force, H_e, O	Maximum energy product, $(BH)_{max}$, G–O $\times 10^{-6}$
Oerstit 800	9	18	19	4	4 Ti	6,600	750	1.95
Great Britain								
Alcomax I	7.5	11	25	3	1.5 Ti	12,000	475	3.5
Alcomax II	8	11.5	24	4.5		12,400	575	4.7
Alcomax IISC (semicolumnar)	8	11	22	4.5		12,800	600	5.15
Alcomax III	8	13.5	24	3	0.8 Nb	12,500	670	5.10
Alcomax IIISC (semicolumnar)	8	13.5	24	3	0.8 Nb	13,000	700	5.80
Alcomax IV	8	13.5	24	3	2.5 Nb	11,200	750	4.30
Alcomax IVSC (semicolumnar)	8	13.5	24	3	2.5 Nb	11,700	780	5.10
Alni, high B_r	13	24		3.5		6,200	480	1.25
Alni, normal						5,600	580	1.25
Alni, high H_e	12	32			0–0.5Ti	5,000	680	1.25
Alnico, high B_r	10	17	12	6		8,000	500	1.70
Alnico, normal						7,250	560	1.70
Alnico, high H_e	10	20	13.5	6	0.25 Ti	6,600	620	1.70
Columax (columnar)	similar to Alcomax III or IV					13,000–14,000	700–800	7.0–8.5
Hycomax	9	21	20	1.6		9,500	830	3.3

[a]Remainder of unlisted composition is either iron or iron plus trace impurities.
[b]Cast anisotropic. Unmarked are cast isotropic.

Properties of Magnetic Alloys: High-Permeability Magnetic Alloys

Name	Composition[a], weight percent	Sp. gr., g/cm³	Tensile Strength kg/mm[2b]	Form	Remark	Use
Silicon iron AISI M 15	Si 4	7.68–7.64	44.3	—	Annealed 4 h 802–1093°C	Low core losses
Silicon iron AISI M 8	Si 3	7.68–7.64	44.2	Grain oriented	Annealed 4 h 802–1204°C	
45 Permalloy	Ni 45; Mn 0.3	8.17	—	—	—	Audio transformer, coils, relays
Monimax	Ni 47; Mo 3	8.27	—	—	—	High-frequency coils
4–79 Permalloy	Ni 79; Mo 4; Mn 0.3	8.74	55.4	—	H_2 annealed 1121°C	Audio coils, transformers, magnetic shields
Sinimax	Ni 43; Si 3	7.70	—	—	—	High-frequency coils
Nu-metal	Ni 75; Cr 2; Cu 5	8.58	44.8	—	H_2 annealed 1221°C	Audio coils, magnetic shields, transformers
Supermalloy	Ni 79; Mo 5; Mn 0.3	8.77	—	—	—	Pulse transformers, magnetic amplifiers, coils
2-V Permendur	Co 40; V 2	8.15	46.3	—	—	DC electromagnets, pole tips

[a]Iron is additional alloying metal.
[b]$kg/mm^2 \times 1422.33 = lbs/in^2$.

Properties of Magnetic Alloys: Cast Permanent Magnetic Alloys

Alloy name (country of manufacture[a])	Composition[b] weight percent	Sp.gr., g/cm^3	Thermal expansion Cm×10^{-6} / cm×°C	Thermal expansion Between °C	Tensile strength kg/mm^{2c}	Form	Remark[d]	Use
Alnico I (US)	Al 12; Ni 20–22; Co 5	6.9	12.6	20–300	2.9	Cast	i	Permanent magnets
Alnico II (US)	Al 10; Ni 17; Cu 6; Co 12.5	7.1	12.4	20–300	2.1 45.7	Cast Sintered	i	Temperature controls magnetic toys and novelties
Alnico III (US)	Al 12; Ni 24–26; Cu 3	6.9	12	20–300	8.5	Cast	i	Tractor magnetos
Alnico IV (US)	Al 12; Ni 27–28; Co 5	7.0	13.1	20–300	6.3 42.1	Cast Sintered	i	Application requiring high coercive force
Alnico V (US)	Al 8; Ni 14; Co 24; Cu 3	7.3	11.3	—	3.8 35	Cast Sintered	 a	Application requiring high energy
Alnico V DG (US)	Al 8; Ni 14; Co 24; Cu 3	7.3	11.3	—	—	—	a, c	
Alnico VI (US)	Al 8; Ni 15; Co 24; Cu 3; Ti 1.25	7.3	11.4	—	16.1	Cast	a	Application requiring high energy
Alnico VII (US)	Al 8.5; Ni 18; Cu 3; Co 24; Ti 5	7.17	11.4	—	—	—	a	
Alnico XII (US)	Al 6; Ni 18; Co 35; Ti 8	7.2	11	20–300	—	—	—	Permanent magnets
Comol (US)	Co 12; Mo 17	8.16	9.3	20–300	88.6	—	—	Permanent magnets
Cunife (US)	Cu 60; Ni 20	8.52	—	—	70.3	—	—	Permanent magnets
Cunico (US)	Cu 50; Ni 21	8.31	—	—	70.3	—	—	Permanent magnets
Barium ferrite Feroxdur (US)	$BaFe_{12}O_{19}$	4.7	10	—	70.3	—	—	Ceramics
Alcomax I (GB)	Al 7.5; Ni 11; Co 25; Cu 3; Ti 1.5	—	—	—	—	—	a	Permanent magnets
Alcomax II (GB)	Al 8; Ni 11.5; Co 24; Cu 4.5	—	—	—	—	—	a	Permanent magnets
Alcomax II SC (GB)	Al 8; Ni 11; Co 22; Cu 4.5	7.3	—	—	—	—	a, sc	
Alcomax III (GB)	Al 8; Ni 13.5; Co 24; Nb 0.8	7.3	—	—	—	—	a	Magnets for motors, loudspeakers
Alcomax IV (GB)	Al 8; Ni 13.5; Cu 3; Co 24; Nb 2.5	—	—	—	—	—	—	Magnets for cycle-dynamos
Columax (GB)	Similar to Alcomax III or IV	—	—	—	—	—	a, sc	Permanent magnets, heat treatable
Hycomax (GB)	Al 9; Ni 21; Co 20; Cu 1.6	—	—	—	—	—	a	Permanent magnets
Alnico (high H_c) (GB)	Al 10; Ni 20; Co 13.5; Cu 6; Ti 0.25	7.3	—	—	—	—	i	
Alnico (high B_r) (GB)	Al 10; Ni 17; Co 12; Cu 6	7.3	—	—	—	—	i	
Alni (high H_c) (GB)	Al 12; Ni 32; Ti 0–0.5	6.9	—	—	—	—	i	
Alni (high B_r) (GB)	Al 13; Ni 24; Cu 3.5	—	—	—	—	—	i	
Alnico 580 (Ger)	Al 9; Ni 15; Co 23; Cu 4	—	—	—	—	—	i	
Alnico 400 (Ger)	Al 9; Ni 15; Co 23; Cu 4	—	—	—	—	—	a	

Properties of Magnetic Alloys: Cast Permanent Magnetic Alloys (*continued*)

Alloy name (country of manufacture[a]	Composition[b] weight percent	Sp.gr., g/cm^3	Thermal expansion Cm$\times 10^{-6}$ cm$\times$° C	Thermal expansion Between °C	Tensile strength kg/ mm^{2c}	Form	Remark[d]	Use
Oerstit 800 (Ger)	Al 9; Ni 18; Co 19; Cu 4; Ti 4	—	—	—	—	—	i	Permanent magnets
Alnico 250 (Ger)	Al 8; Ni 19; Co 23; Cu 4; Ti 6	—	—	—	—	—	i	
Alnico 190 (Ger)	Al 12; Ni 21; Cu 4; Co 15	—	—	—	—	—	i	
Alnico 160 (Austria)	Al 11; Ni 24; Co 12; Cu 4	—	—	—	—	—	i	Permanent magnets, sintered
Alnico 130 (Ger)	Al 12; Ni 23; Co 5	—	—	—	—	—	i	
Alni 120 (Ger)	Al 13; Ni 27	—	—	—	—	—	i	
Alni 90 (Ger)	Al 12; Ni 21	—	—	—	—	—	i	

[a]US, United States; GB, Great Britain; Ger, Germany.
[b]Iron is the additional alloying metal for each of the magnets listed.
[c] kg/mm^2 × 1422.33 = lb/in^2.
[d]i. = isotropic; a. = anisotropic; c. = columnar; sc. = semicolumnar.

Properties of Antiferromagnetic Compounds

Compound	Crystal Symmetry	θ_N[a] K	θ_P[b] K	$(P_A)_{\text{eff}}$[c] μ_B	P_A[d] μ_B
$CoCl_2$	Rhombohedral	25	−38.1	5.18	3.1 ± 0.6
CoF_2	Tetragonal	38	50	5.15	3.0
CoO	Tetragonal	291	330	5.1	3.8
Cr	Cubic	475			
Cr_2O_3	Rhombohedral	307	485	3.73	3.0
CrSb	Hexagonal	723	550	4.92	2.7
$CuBr_2$	Monoclinic	189	246	1.9	
$CuCl_2 \cdot 2H_2O$	Orthorhombic	4.3	4–5	1.9	
$CuCl_2$	Monoclinic	∼70	109	2.08	
$FeCl_2$	Hexagonal	24	−48	5.38	4.4 ± 0.7
FeF_2	Tetragonal	79–90	117	5.56	4.64
FeO	Rhombohedral	198	507	7.06	3.32
α-Fe_2O_3	Rhombohedral	953	2940	6.4	5.0
α-Mn	Cubic	95			
$MnBr_2 \cdot 4H_2O$	Monoclinic	2.1	$\left\{\begin{matrix}2.5\\1.3\end{matrix}\right\}$	5.93	
$MnCl_2 \cdot 4H_2O$	Monoclinic	1.66	1.8	5.94	
MnF_2	Tetragonal	72–75	113.2	5.71	5
MnO	Rhombohedral	122	610	5.95	5.0
β-MnS	Cubic	160	982	5.82	5.0
MnSe	Cubic	∼173	361	5.67	
MnTe	Hexagonal	310–323	690	6.07	5.0
$NiCl_2$	Hexagonal	50	−68	3.32	
NiF_2	Tetragonal	78.5–83	115.6	3.5	2.0

Properties of Antiferromagnetic Compounds (*continued*)

Compound	Crystal Symmetry	θ_N[a] K	θ_P[b] K	$(P_A)_{eff}$[c] μ_B	P_A[d] μ_B
NiO	Rhombohedral	533–650	~2000	4.6	2.0
$TiCl_3$		100			
V_2O_3		170			

[a]θ_N = Néel temperature, determined from susceptibility maxima or from the disappearance of magnetic scattering.
[b]θ_P = a constant in the Curie-Weiss law written in the form $\chi_A = C_A/(T+\theta_P)$, which is valid for antiferromagnetic material for $T > \theta_N$.
[c]$(P_A)_{eff}$ = effective moment per atom, derived from the atomic Curie constant $C_A = (P_A)^2_{eff}(N^2/3R)$ and expressed in units of the Bohr magneton, $\mu_B = 0.9273 \times 10^{-20}$ erg G^{-1}.
[d]P_A = magnetic moment per atom, obtained from neutron diffraction measurements in the ordered state.

Properties of Magnetic Alloys: Saturation Constants and Curie Points of Ferromagnetic Elements

Element	σ_s[a] (20°C)	M_s[b] (20°C)	σ_s (0 K)	n_B[c]	Curie point, °C
Fe	218.0	1,714	221.9	2.219	770
Co	161	1,422	162.5	1.715	1,131
Ni	54.39	484.1	57.50	0.604	358
Gd	0	0	253.5	7.12	16

[a]σ_s = saturation magnetic moment/gram.
[b]M_s = saturation magnetic moment/cm^3, in cgs units.
[c]n_B = magnetic moment per atom in Bohr magnetons.
(*Source:* 1963. *American Institute of Physics Handbook.* McGraw–Hill, New York.)

Magnetic Properties of Transformer Steels

Ordinary Transformer Steel		
B(Gauss)	H(Oersted)	Permeability = B/H
2,000	0.60	3,340
4,000	0.87	4,600
6,000	1.10	5,450
8,000	1.48	5,400
10,000	2.28	4,380
12,000	3.85	3,120
14,000	10.9	1,280
16,000	43.0	372
18,000	149	121

High Silicon Transformer Steels

B	H	Permeability
2,000	0.50	4,000
4,000	0.70	5,720
6,000	0.90	6,670
8,000	1.28	6,250
10,000	1.99	5,020
12,000	3.60	3,340
14,000	9.80	1,430
16,000	47.4	338
18,000	165	109

Initial Permeability of High Purity Iron for Various Temperatures

L. Alberts and B.J. Shepstone	
Temperature °C	Permeability (Gauss/oersted)
0	920
200	1,040
400	1,440
600	2,550
700	3,900
770	12,580

Saturation Constants for Magnetic Substances

Substance	Field Intensity (For Saturation)	Induced Magnetization (For Saturation)	Substance	Field Intensity (For Saturation)	Induced Magnetization (For Saturation)
Cobalt	9,000	1,300	Nickel, hard	8,000	400
Iron, wrought	2,000	1,700	annealed	7,000	515
cast	4,000	1,200	Vicker's steel	15,000	1,600
Manganese steel	7,000	200			

Magnetic Materials: High-Permeability Materials

Material	Form	Approximate % Composition					Typical Heat Treatment, °C	Permeability at $B = 20$, G	Maximum Permeability	Saturation flux density B, G	Hysteresis[a] loss, W_h, ergs/cm^2	Coercive[a] force H_a O	Resistivity $\mu \cdot \Omega$cm	Density, g/cm^3
		Fe	Ni	Co	Mo	Other								
Cold rolled steel	Sheet	98.5	–	–	–	–	950 Anneal	180	2,000	21,000	–	1.8	10	7.88
Iron	Sheet	99.91	–	–	–	–	950 Anneal	200	5,000	21,500	5,000	1.0	10	7.88
Purified iron	Sheet	99.95	–	–	–	–	1480 H_2 + 880	5,000	180,000	21,500	300	0.05	10	7.88
4% Silicon-iron	Sheet	96	–	–	–	4 Si	800 Anneal	500	7,000	19,700	3,500	0.5	60	7.65
Grain oriented[b]	Sheet	97	–	–	–	3 Si	800 Anneal	1,500	30,000	20,000	–	0.15	47	7.67
45 Permalloy	Sheet	54.7	45	–	–	0.3 Mn	1050 Anneal	2,500	25,000	16,000	1,200	0.3	45	8.17
45 permalloy[c]	Sheet	54.7	45	–	–	0.3 Mn	1200 H_2 Anneal	4,000	50,000	16,000	–	0.07	45	8.17
Hipernik	Sheet	50	50	–	–	–	1200 H_2 Anneal	4,500	70,000	16,000	220	0.05	50	8.25
Monimax	Sheet	–	–	–	–	–	1125 H_2 Anneal	2,000	35,000	15,000	–	0.1	80	8.27
Sinimax	Sheet	–	–	–	–	–	1125 H_2 Anneal	3,000	35,000	11,000	–	–	90	–
78 Permalloy	Sheet	21.2	78.5	–	–	0.3 Mn	1050 + 600 Q^d	8,000	100,000	10,700	200	0.05	16	8.60
4–79 Permalloy	Sheet	16.7	79	–	4	0.3 Mn	1100 + Q	20,000	100,000	8,700	200	0.05	55	8.72
Mu metal	Sheet	18	75	–	–	2 Cr, 5 Cu	1175 H_2	20,000	100,000	6,500	–	0.05	62	8.58
Supermalloy	Sheet	15.7	79	–	5	0.3 Mn	1300 H_2+ Q	100,000	800,000	8,000	–	0.002	60	8.77
Permendur	Sheet	49.7	–	50	–	0.3 Mn	800 Anneal	800	5,000	24,500	12,000	2.0	7	8.3
2 V Permendur	Sheet	49	–	49	–	2 V	800 Anneal	800	4,500	24,000	6,000	2.0	26	8.2
Hiperco	Sheet	64	–	34	–	Cr	850 Anneal	650	10,000	24,200	–	1.0	25	8.0
2–81 Permalloy	Insulated powder	17	81	–	2	–	650 Anneal	125	130	8,000	–	<1.0	10^6	7.8
Carbonyl iron	Insulated powder	99.9	–	–	–	–	–	55	132	–	–	–	–	7.86
Ferroxcube III	Sintered powder	$MnFe_2O_4 + ZnFe_2O_4$						1,000	1,500	2,500	–	0.1	10^8	5.0

[a]At saturation.
[b]Properties in direction of rolling.
[c]Similar properties for Nicaloi, 4750 alloy, Carpenter 49, Armco 48.
[d]Q, quench or controlled cooling.

Magnetic Materials: Permanent Magnet Alloys

Material	% composition (remainder Fe)	Heat treatment[a] (temperature, °C)	Magnetizing force H_{max}, O	Coercive force H_c, O	Residual induction B_r, G	Energy product BH_{max} $\times 10^{-6}$	Method of fabrication[b]	Mechanical properties[c]	Weight lb/In.^{3}
Carbon steel	1 Mn, 0.9 C	Q 800	300	50	10,000	0.20	HR, M, P	H, S	0.280
Tungsten steel	5 W, 0.3 Mn, 0.7 C	Q 850	300	70	10,300	0.32	HR, M, P	H, S	0.292
Chromium steel	3.5 Cr, 0.9 C, 0.3 Mn	Q 830	300	65	9,700	0.30	HR, M, P	H, S	0.280
17% Cobalt steel	17 Co, 0.75 C, 2.5 Cr, 8 W	–	1,000	150	9,500	0.65	HR, M, P	H, S	–
36% Cobalt steel	36 Co, 0.7 C, 4 Cr, 5 W	Q 950	1,000	240	9,500	0.97	HR, M, P	H, S	0.296
Remalloy or Comol	17 Mo, 12 Co	Q 1200, B 700	1,000	250	10,500	1.1	HR, M, P	H	0.295
Alnico I	12 Al, 20 Ni, 5 Co	A 1200, B 700	2,000	440	7,200	1.4	C, G	H, B	0.249
Alnico II	10 Al, 17 Ni, 2.5 Co, 6 Cu	A 1200, B 600	2,000	550	7,200	1.6	C, G	H, B	0.256
Alnico II (sintered)	10 Al, 17 Ni, 2.5 Co, 6 Cu	A 1300	2,000	520	6,900	1.4	Sn, G	H	0.249
Alnico IV	12 Al, 28 Ni, 5 Co	Q 1200, B 650	3,000	700	5,500	1.3	Sn, C, G	H	0.253
Alnico V	8 Al, 14 Ni, 24 Co, 3 Cu	AF 1300, B 600	2,000	550	12,500	4.5	C, G	H, B	0.264
Alnico VI	8 Al, 15 Ni, 24 Co, 3 Cu, 1 Ti	–	3,000	750	10,000	3.5	C, G	H, B	0.268
Alnico XII	6 Al, 18 Ni, 35 Co, 8 Ti	–	3,000	950	5,800	1.5	C, G	H, B	0.26
Vicalloy I	52 Co, 10 V	B 600	1,000	300	8,800	1.0	C, CR, M, P	D	0.295
Vicalloy II (wire)	52 Co, 14 V	CW + B 600	2,000	510	10,000	3.5	C, CR, M, P	D	0.292
Cunife (wire)	60 Cu, 20 Ni	CW + B 600	2,400	550	5,400	1.5	C, CR, M, P	D, M	0.311
Cunico	50 Cu, 21 Ni, 29 Co	–	3,200	660	3,400	0.80	C, CR, M, P	D, M	0.300
Vectolite	$30Fe_2O_3$, $44Fe_3O_4$, $26C_2O_3$	–	3,000	1,000	1,600	0.60	Sn, G	W	0.113
Silmanal	86.8Ag, 8.8Mn, 4.4Al	–	20,000	6,000[d]	550	0.075	C, CR, M, P	D, M	0.325
Platinum-cobalt	77 Pt, 23 Co	Q 1200, B 650	15,000	3,600	5,900	6.5	C, CR, M	D	–
Hyflux	Fine powder	–	2,000	390	6,600	0.97	–	–	0.176

[a] Q, quenched in oil or water; A, air colled; B, baked; F, cooled in magnetic field; CW, cold worked.

[b] HR, hot rolled or froged; CR, cold rolled or drawn; M, machined; G, must be ground; P, punched; C, cast; Sn, sintered.

[c] H, hard; B, brittle; S, strong; D, ductile; M, malleable; W, weak.

[d] Value given is intrinsic H_c.

Resistance of Wires

The following table gives the approximate resistance of various metallic conductors. The values have been computed from the resistivities at 20°C, except as otherwise stated, and for the dimensions of wire indicated. Owing to differences in purity in the case of elements and of composition in alloys, the values can be considered only as approximations.

The following dimensions have been adopted in the computation.

B. & S. gauge	Diameter mm	Diameter mils 1 mil = 0.001 in	B. & S. gauge	Diameter mm	Diameter mils 1 mil = 0.001 in
10	2.588	101.9	26	0.4049	15.94
12	2.053	80.81	27	0.3606	14.20
14	1.628	64.08	28	0.3211	12.64
16	1.291	50.82	30	0.2546	10.03
18	1.024	40.30	32	0.2019	7.950
20	0.8118	31.96	34	0.1601	6.305
22	0.6438	25.35	36	0.1270	5.000
24	0.5106	20.10	40	0.07987	3.145

B. & S. No.	Ω/cm	Ω/ft	B. & S. No.	Ω/cm	Ω/ft	B.& S. No.	Ω/cm	Ω/ft
Advance[a] (0°C) $\varrho = 48. \times 10^{-6}$ Ω · cm			Aluminum $\varrho = 2.828 \times 10^{-6}$ Ω · cm			Brass $\varrho = 7.00 \times 10^{-6}$ Ω · cm		
10	0.000912	0.0278	10	0.0000538	0.00164	10	0.000133	0.00406
12	0.00145	0.0442	12	0.0000855	0.00260	12	0.000212	0.00645
14	0.00231	0.0703	14	0.000136	0.00414	14	0.000336	0.0103
16	0.00367	0.112	16	0.000216	0.00658	16	0.000535	0.0163
18	0.00583	0.178	18	0.000344	0.0105	18	0.000850	0.0259
20	0.00927	0.283	20	0.000546	0.0167	20	0.00135	0.0412
22	0.0147	0.449	22	0.000869	0.0265	22	0.00215	0.655
24	0.0234	0.715	24	0.00138	0.0421	24	0.00342	0.104
26	0.0373	1.14	26	0.00220	0.0669	26	0.00543	0.166
27	0.0470	1.43	27	0.00277	0.0844	27	0.00686	0.209
28	0.0593	1.81	28	0.00349	0.106	28	0.00864	0.263
30	0.0942	2.87	30	0.00555	0.169	30	0.0137	0.419
32	0.150	4.57	32	0.00883	0.269	32	0.0219	0.666
34	0.238	7.26	34	0.0140	0.428	34	0.0348	1.06
36	0.379	11.5	36	0.0223	0.680	36	0.0552	1.68
40	0.958	29.2	40	0.0564	1.72	40	0.140	4.26
Climax $\varrho = 87 \times 10^{-6}$ Ω · cm			Constantan (0°C) $\varrho = 44.1 \times 10^{-6}$ Ω · cm			Copper, annealed $\varrho = 1.724 \times 10^{-6}$ Ω · cm		
10	0.00165	0.0504	10	0.000838	0.0255	10	0.0000328	0.000999
12	0.00263	0.0801	12	0.00133	0.0406	12	0.0000521	0.00159
14	0.00418	0.127	14	0.00212	0.0646	14	0.0000828	0.00253
16	0.00665	0.203	16	0.00337	0.103	16	0.000132	0.00401
18	0.0106	0.322	18	0.00536	0.163	18	0.000209	0.00638
20	0.0168	0.512	20	0.00852	0.260	20	0.000333	0.0102
22	0.0267	0.815	22	0.0135	0.413	22	0.000530	0.0161
24	0.0425	1.30	24	0.0215	0.657	24	0.000842	0.0257
26	0.0675	2.06	26	0.0342	1.04	26	0.00134	0.0408
27	0.0852	2.60	27	0.0432	1.32	27	0.00169	0.0515
28	0.107	3.27	28	0.0545	1.66	28	0.00213	0.0649
30	0.171	5.21	30	0.0866	2.64	30	0.00339	0.103
32	0.272	8.28	32	0.138	4.20	32	0.00538	0.164
34	0.432	13.2	34	0.219	6.67	34	0.00856	0.261
36	0.687	20.9	36	0.348	10.6	36	0.0136	0.415
40	1.74	52.9	40	0.880	26.8	40	0.0344	1.05

[a]Trademark.

Resistance of Wires (*continued*)

Eureka[a] (0°C) $\varrho = 47. \times 10^{-6}$ Ω · cm

B. & S. No.	Ω/cm	Ω/ft
10	0.000893	0.0272
12	0.00142	0.0433
14	0.00226	0.0688
16	0.00359	0.109
18	0.00571	0.174
20	0.00908	0.277
22	0.0144	0.440
24	0.0230	0.700
26	0.0365	1.11
27	0.0460	1.40
28	0.0580	1.77
30	0.0923	2.81
32	0.147	4.47
34	0.233	7.11
36	0.371	11.3
40	0.938	28.6

Excello $\varrho = 92. \times 10^{-6}$ Ω · cm

B. & S. No.	Ω/cm	Ω/ft
10	0.00175	0.0533
12	0.00278	0.0847
14	0.00442	0.135
16	0.00703	0.214
18	0.0112	0.341
20	0.0178	0.542
22	0.0283	0.861
24	0.0449	1.37
26	0.0714	2.18
27	0.0901	2.75
28	0.114	3.46
30	0.181	5.51
32	0.287	8.75
34	0.457	13.9
36	0.726	22.1
40	1.84	56.0

German silver $\varrho = 33 \times 10^{-6}$ Ω · cm

B. & S. No.	Ω/cm	Ω/ft
10	0.000627	0.0191
12	0.000997	0.304
14	0.00159	0.0483
16	0.00252	0.0768
18	0.00401	0.122
20	0.00638	0.194
22	0.0101	0.309
24	0.0161	0.491
26	0.0256	0.781
27	0.0323	0.985
28	0.0408	1.24
30	0.0648	1.97
32	0.103	3.14
34	0.164	4.99
36	0.260	0.794
40	0.659	20.1

Gold $\varrho = 2.44 \times 10^{-6}$ Ω · cm

B. & S. No.	Ω/cm	Ω/ft
10	0.0000464	0.00141
12	0.0000737	0.00225
14	0.000117	0.00357
16	0.000186	0.00568
18	0.000296	0.00904
20	0.000471	0.0144
22	0.000750	0.0228
24	0.00119	0.0363
26	0.00189	0.0577
27	0.00239	0.728
28	0.00301	0.0918
30	0.00479	0.146
32	0.00762	0.232
34	0.0121	0.369
36	0.0193	0.587
40	0.0487	1.48

Iron $\varrho = 10 \times 10^{-6}$ Ω · cm

B. & S. No.	Ω/cm	Ω/ft
10	0.000190	0.00579
12	0.000302	0.00921
14	0.000481	0.0146
16	0.000764	0.0233
18	0.00121	0.0370
20	0.00193	0.0589
22	0.00307	0.0936
24	0.00489	0.149
26	0.00776	0.237
27	0.00979	0.299
28	0.0123	0.376
30	0.0196	0.598
32	0.0312	0.952
34	0.0497	1.51
36	0.789	2.41
40	0.200	6.08

Lead $\varrho = 22 \times 10^{-6}$ Ω · cm

B. & S. No.	Ω/cm	Ω/ft
10	0.000418	0.0127
12	0.000665	0.0203
14	0.00106	0.0322
16	0.00168	0.0512
18	0.00267	0.0815
20	0.00425	0.130
22	0.00676	0.206
24	0.0107	0.328
26	0.0171	0.521
27	0.0215	0.657
28	0.0272	0.828
30	0.0432	1.32
32	0.0687	2.09
34	0.109	3.33
36	0.174	5.29
40	0.439	13.4

Magnesium $\varrho = 4.6 \times 10^{-6}$ Ω · cm

B. & S. No.	Ω/cm	Ω/ft
10	0.0000874	0.00267
12	0.000139	0.00424
14	0.000221	0.00674
16	0.000351	0.0107
18	0.000559	0.0170
20	0.000889	0.0271
22	0.00141	0.0431
24	0.00225	0.0685
26	0.00357	0.109
27	0.00451	0.137
28	0.00568	0.173
30	0.00903	0.275
32	0.0144	0.438
34	0.0228	0.696
36	0.0363	1.11
40	0.0918	2.80

Manganin $\varrho = 44 \times 10^{-6}$ Ω · cm

B. & S. No.	Ω/cm	Ω/ft
10	0.000836	0.0255
12	0.00133	0.0405
14	0.00211	0.0644
16	0.00336	0.102
18	0.00535	0.163
20	0.00850	0.259
22	0.0135	0.412
24	0.0215	0.655
26	0.0342	1.04
27	0.0431	1.31
28	0.0543	1.66
30	0.0864	2.63
32	0.137	4.19
34	0.218	6.66
36	0.347	10.6
40	0.878	26.8

Molybdenum $\varrho = 5.7 \times 10^{-6}$ Ω · cm

B. & S. No.	Ω/cm	Ω/ft
10	0.000108	0.00330
12	0.000172	0.00525
14	0.000274	0.00835
16	0.000435	0.0133
18	0.000693	0.0211
20	0.00110	0.0336
22	0.00175	0.0534
24	0.00278	0.0849
26	0.00443	0.135
27	0.00558	0.170
28	0.00704	0.215
30	0.0112	0.341
32	0.0178	0.542
34	0.0283	0.863
36	0.0450	1.37
40	0.114	3.47

[a]Trademark.

Resistance of Wires (*continued*)

Monel Metal $\varrho = 42 \times 10^{-6}$ Ω · cm

B. & S. No.	Ω/cm	Ω/ft
10	0.000798	0.0243
12	0.00127	0.0387
14	0.00202	0.0615
16	0.00321	0.0978
18	0.00510	0.156
20	0.00811	0.247
22	0.0129	0.393
24	0.0205	0.625
26	0.0326	0.994
27	0.0411	1.25
28	0.0519	1.58
30	0.0825	2.51
32	0.131	4.00
34	0.209	6.36
36	0.331	10.1
40	0.838	25.6

[a]Nichrome $\varrho = 150 \times 10^{-6}$ Ω · cm

B. & S. No.	Ω/cm	Ω/ft
10	0.0021281	0.06488
12	0.0033751	0.1029
14	0.0054054	0.1648
16	0.0085116	0.2595
18	0.0138383	0.4219
20	0.0216218	0.6592
22	0.0346040	1.055
24	0.0548088	1.671
26	0.0875760	2.670
28	0.1394328	4.251
30	0.2214000	6.750
32	0.346040	10.55
34	0.557600	17.00
36	0.885600	27.00
38	1.383832	42.19
40	2.303872	70.24

Nickel $\varrho = 7.8 \times 10^{-6}$ Ω · cm

B. & S. No.	Ω/cm	Ω/ft
10	0.000148	0.00452
12	0.000236	0.00718
14	0.000375	0.0114
16	0.000596	0.0182
18	0.000948	0.0289
20	0.00151	0.0459
22	0.00240	0.0730
24	0.00381	0.116
26	0.00606	0.185
27	0.00764	0.233
28	0.00963	0.294
30	0.0153	0.467
32	0.0244	0.742
34	0.0387	1.18
36	0.0616	1.88
40	0.156	4.75

Platinum $\varrho = 10 \times 10^{-6}$ Ω · cm

10	0.000190	0.00579
12	0.000302	0.00921
14	0.000481	0.0146
16	0.000764	0.0233
18	0.00121	0.0370
20	0.00193	0.0589
22	0.00307	0.0936
24	0.00489	0.149
26	0.00776	0.237
27	0.00979	0.299
28	0.0123	0.376
30	0.0196	0.598
32	0.0312	0.952
34	0.0497	1.51
36	0.0789	2.41
40	0.200	6.08

Silver (18°C) $\varrho = 1.629 \times 10^{-6}$ Ω · cm

10	0.0000310	0.000944
12	0.0000492	0.00150
14	0.0000783	0.00239
16	0.000124	0.00379
18	0.000198	0.00603
20	0.000315	0.00959
22	0.000500	0.0153
24	0.000796	0.0243
26	0.00126	0.0386
27	0.00160	0.0486
28	0.00201	0.0613
30	0.00320	0.0975
32	0.00509	0.155
34	0.00809	0.247
36	0.0129	0.392
40	0.325	0.991

Steel, piano wire (0°C) $\varrho = 11.8 \times 10^{-6}$ Ω · cm

10	0.000224	0.00684
12	0.000357	0.0109
14	0.000567	0.0173
16	0.000901	0.0275
18	0.00143	0.0437
20	0.00228	0.0695
22	0.00363	0.110
24	0.00576	0.176
26	0.00916	0.279
27	0.0116	0.352
28	0.0146	0.444
30	0.0232	0.706
32	0.0368	1.12
34	0.0586	1.79
36	0.0931	2.84
40	0.236	7.18

Steel, Invar (35% Ni) $\varrho = 81 \times 10^{-6}$ Ω · cm

10	0.00154	0.0469
12	0.00245	0.0746
14	0.00389	0.119
16	0.00619	0.189
18	0.00984	0.300
20	0.0156	0.477
22	0.0249	0.758
24	0.0396	1.21
26	0.0629	1.92
27	0.0793	2.42
28	0.100	3.05
30	0.159	4.85
32	0.253	7.71
34	0.402	12.3
36	0.639	19.5
40	1.62	49.3

Tantalum $\varrho = 15.5 \times 10^{-6}$ Ω · cm

10	0.000295	0.00898
12	0.000468	0.0143
14	0.000745	0.0227
16	0.00118	0.0361
18	0.00188	0.0574
20	0.00299	0.0913
22	0.00476	0.145
24	0.00757	0.231
26	0.0120	0.367
27	0.0152	0.463
28	0.0191	0.583
30	0.0304	0.928
32	0.0484	1.47
34	0.0770	2.35
36	0.122	3.73
40	0.309	9.43

Tin $\varrho = 11.5 \times 10^{-6}$ Ω · cm

10	0.000219	0.00666
12	0.000348	0.0106
14	0.000553	0.0168
16	0.000879	0.0268
18	0.00140	0.0426
20	0.00222	0.0677
22	0.00353	0.108
24	0.00562	0.171
26	0.00893	0.272
27	0.0113	0.343
28	0.0142	0.433
30	0.0226	0.688
32	0.0359	1.09
34	0.0571	1.74
36	0.0908	2.77
40	0.230	7.00

[a]Trademark.

Resistance of Wires (*continued*)

B & S. No.	Ω/cm	Ω/ft	B. & S. No.	Ω/cm	Ω/ft
Tungsten $\varrho = 5.51 \times 10^{-6} \Omega \cdot cm$			Zinc (0°C) $\varrho = 5.75 \times 10^{-6} \Omega \cdot cm$		
10	0.000105	0.00319	10	0.000109	0.00333
12	0.000167	0.00508	12	0.000174	0.00530
14	0.000265	0.00807	14	0.000276	0.00842
16	0.000421	0.0128	16	0.000439	0.0134
18	0.000669	0.0204	18	0.000699	0.0213
20	0.00106	0.0324	20	0.00111	0.0339
22	0.00169	0.0516	22	0.00177	0.0538
24	0.00269	0.0820	24	0.00281	0.0856
26	0.00428	0.130	26	0.00446	0.136
27	0.00540	0.164	27	0.00563	0.172
28	0.00680	0.207	28	0.00710	0.216
30	0.0108	0.330	30	0.0113	0.344
32	0.0172	0.524	32	0.0180	0.547
34	0.0274	0.834	34	0.0286	0.870
36	0.0435	1.33	36	0.0454	1.38
40	0.110	3.35	40	0.115	3.50

Defining Terms

Alloy: Metal composed of more than one element.

Bravais lattice: One of the 14 possible arrangements of points in three-dimensional space.

Crystalline: Having constituent atoms stacked together in a regular, repeating pattern.

Density: Mass per unit volume.

Electromotive force series: Systematic listing of half-cell reaction voltages.

Engineering strain: Increase in sample length at a given load divided by the original (stress-free) length.

Engineering stress: Load on a sample divided by the original (stress-free) area.

Fourier's law: Relationship between rate of heat transfer and temperature gradient.

Index of refraction: Ratio of speed of light in vacuum to that in a transparent material.

Linear coefficient of thermal expansion: Material parameter indicating dimensional change as a function of increasing temperature.

Melting point: Temperature of transformation from solid to liquid upon heating.

Ohm's law: Relationship between voltage, current, and resistance in an electrical circuit.

Property: Observable characteristic of a material.

Resistivity: Electrical resistance normalized for sample geometry.

Tensile strength: Maximum engineering stress during a tensile test.

Thermal conductivity: Proportionality constant in Fourier's law.

Young's modulus (modulus of elasticity): Ratio of engineering stress to engineering strain for relatively small levels of load application.

References

ASM. 1984. *ASM Metals Reference Book*, 2nd ed. American Society for Metals, Metals Park, OH.

ASM. 1988. *Engineered Materials Handbook*, Vol. 2, *Engineering Plastics.* ASM International, Metals Park, OH.

Bolz, R.E. and Tuve, G.L., eds. 1973. *CRC Handbook of Tables for Applied Engineering Science.* CRC Press, Boca Raton, FL.

Lynch, C.T., ed. 1975. *CRC Handbook of Materials Science*, Vol. 3. CRC Press, Boca Raton, FL.

Shackelford, J.F. 1996. *Introduction to Materials Science for Engineers*, 4th ed. Prentice–Hall, Upper Saddle River, NJ.

Shackelford, J.F., Alexander, W., and Park, J.S., eds. 1994. *The CRC Materials Science and Engineering Handbook*, 2nd ed. CRC Press, Boca Raton, FL.

Further Information

A general introduction to the field of materials science and engineering is available from a variety of introductory textbooks. In addition to Shackelford [1996], readily available references include:

Askeland, D.R. 1994. *The Science and Engineering of Materials*, 3rd ed. PWS–Kent, Boston.

Callister, W.D. 1994. *Materials Science and Engineering-An Introduction*, 3rd ed. Wiley, New York.

Flinn, R.A. and Trojan, P.K. 1990. *Engineering Materials and Their Applications*, 4th ed. Houghton Mifflin, Boston.

Smith, W.F. 1990. *Principles of Materials Science and Engineering*, 2nd ed. McGraw–Hill New York.

Van Vlack, L.H. 1989. *Elements of Materials Science and Engineering*, 6th ed. Addison–Wesley Reading, MA.

As noted earlier, *The CRC Materials Science and Engineering Handbook* [Shackelford, Alexander, and Park 1994] is available as a comprehensive source of property data for engineering materials. In addition, ASM International has published between 1982 and 1996 the *ASM Handbook*, a 19-volume set concentrating on metals and alloys. ASM International has also published a 4-volume set entitled the *Engineered Materials Handbook*, covering composites, engineering plastics, adhesives and sealants, and ceramics and glasses.

15

Frequency Bands and Assignments[1]

Robert D. Greenberg
Federal Communications Commission

15.1 U.S. Table of Frequency Allocations

The U.S. Table of Frequency Allocations (columns 4–7 of Table 15.1) is based on the International plan for Region 2 because the relevant area of jurisdiction is located primarily in Region 2 (i.e., the 50 States, the District of Columbia, the Caribbean insular areas[2] and some of the Pacific insular areas [3]).[4] Because there is a need to provide radio spectrum for both Federal government and non-Federal government operations, the U.S. Table is divided into the Government Table of Frequency Allocation and the Nongovernment Table of Frequency Allocations. The Government plan, as shown in column 4, is administered by the National Telecommunications and Information Administration (NTIA),[5] whereas the non-Government plan, as shown in column 5, is administered by the Federal Communications Commission (FCC).[6]

In the U.S., radio spectrum may be allocated to either government or nongovernment use exclusively, or for shared use. In the case of shared use, the type of service(s) permitted need not be the same (e.g., Government fixed, nongovernment mobile). The terms used to designate categories of service in columns 4 and 5, correspond to the terms employed by the International Telecommunication Union (ITU) in the international *Radio Regulations* [RR 1982].

Categories of Services

Any segment of the radio spectrum may be allocated to the government and/or nongovernment sectors either on an exclusive or shared basis for use by one or more radio services. In the case where an allocation has been made to more than one service, such services are listed in the following order:

[1]Chapter adapted from the Federal Communications Commission, 47 Code of Federal Regulations, Part 2 (10-1-93 edition).

[2]The Caribbean insular areas are: the Commonwealth of Puerto Rico; the unincorporated territory of the U.S. Virgin Islands; and Navassa Island, Quita Sueno Bank, Roncador Bank, Serrana Bank, and Serranilla Bank.

[3]The Pacific insular areas located in region 2 are: Johnston Island and Midway Island.

[4]The operation of stations in the Pacific insular areas located in region 3 are generally governed by the International plan for region 3 (i.e., column 3 of Table 15.1. The Pacific insular areas located in region 3 are: the Commonwealth of the Northern Mariana Islands; the unincorporated territory of American Samoa; the unincorporated territory of Guam; and Baker Island. Howland Island, Jarvis Island, Kingman Reef, Palmyra Island, and Wake Island.

[5]Section 305(a) of the Communications Act of 1934, as amended; Executive Order 12046 (26 March 1978) and Department of Commerce Organization Order 10–10 (9 May 1979).

[6]The Communications Act of 1934, as amended.

0-8493-0050-9/00/$0.00+$.50

- Services, the names of which are printed in upper case (example: FIXED): these are called *primary* services.
- Services, the names of which are printed in upper case between oblique strokes (example: /RADIOLOCATION): these are called *permitted services.*
- Services, the names of which are printed in lower case (example: Mobile): these are called *secondary* services.

Permitted and primary services have equal rights, except that, in the preparation of frequency plans, the primary services, as compared with the permitted services, shall have prior choice of frequencies.

Stations of a secondary service have limitations. They shall not cause harmful interference to stations of primary or permitted services to which frequencies are already assigned or to which frequencies may be assigned at a later date. They cannot claim protection from harmful interference from stations of a primary or permitted service to which frequencies are already assigned or may be assigned at a later date. They can claim protection, however, from harmful interference from stations of the same or other secondary service(s) to which frequencies may be assigned at a later date.

Format of the U.S. Table

The frequency band referred to in each allocation, column 4 for government and column 5 for nongovernment, is indicated in the left-hand top corner of the column. If there is no service or footnote indicated for a band of frequencies in either column 4 or 5, then the government or the nongovernment sector, respectively, has no access to that band except as provided for by FCC rules. The government allocation plan, given in column 4, is included for informational purposes only. In the case where there is a parenthetical addition to an allocation in the U.S. Table [example: FIXED-SATELLITE (space-to-earth)], that service allocation is restricted to the type of operation so indicated.

The following symbols are used to designate footnotes in the U.S. Table:

Any footnote not prefixed by a letter, denotes an international footnote. Where such a footnote is applicable, without modification, to the U.S. Table, the symbol appears in the U.S. Table (column 4 or 5) and denotes a stipulation affecting both the government and nongovernment plans.

Any footnote consisting of the letters US followed by one or more digits denotes a stipulation affecting both the government and nongovernment plans.

Any footnote consisting of the letters NG followed by one or more digits, for example, NG1, denotes a stipulation applicable only to the nongovernment plan (column 5).

Any footnote consisting of the letter G following by one or more digits, for example, G1, denotes a stipulation applicable only to the government plan (column 4).

Column 6 provides a reference to indicate which Rule part(s) (e.g., private land mobile radio services, domestic public land mobile radio services, etc.) are given assignments within the allocation plan specified in column 5 for any given band of frequencies. The exact use that can be made of any given frequency or frequencies (e.g., channelling plans, allowable emissions, etc.) is given in the Rule part(s) so indicated. The Rule parts in this column are not allocations. They are provided for informational purposes only.

Column 7 is used to denote certain frequencies that have national and/or international significance.

Nomenclature of Frequencies (§ 2.101)

Band No.	Frequency Subdivision	Frequency Range
4	VLF (very low frequency)	Below 30 kHz
5	LF (low frequency)	30–300 kHz
6	MF (medium frequency)	300–3000 kHz
7	HF (high frequency)	3–30 MHz
8	VHF (very high frequency)	30–300 MHz
9	UHF (ultra high frequency)	300–3000 MHz
10	SHF (super high frequency)	3–30 GHz
11	EHF (extremely high frequency)	30–300 GHz
12		300–3000 GHz

TABLE 15.1 Table of Frequency Allocations

International Table			U.S. Table		FCC Use Designators	
			Government	Nongovernment		
Region 1—allocation, kHz (1)	Region 2—allocation, kHz (2)	Region 3—allocation, kHz (3)	Allocation, kHz (4)	Allocation, kHz (5)	Rule part(s) (6)	Special-use frequencies (7)
Below 9	(Not allocated.) 444 445		Below 9. (Not allocated.) 444 445	Below 9. (Not allocated.) 444 445		
9–14	RADIONAVIGATION.		9–14 RADIONAVIGATION US18 US294	9–14 RADIONAVIGATION. US18 US294		
14–19.95	FIXED. MARITIME MOBILE. 448 446 447		14–19.95 FIXED. MARITIME MOBILE. 448 US294	14–19.95 Fixed 448 US294	INTERNATIONAL FIXED PUBLIC (23).	
19.95–20.05	STANDARD FREQUENCY AND TIME SIGNAL (20 kHz).		19.95–20.05 STANDARD FREQUENCY AND TIME SIGNAL. US294	19.95–20.05 STANDARD FREQUENCY AND TIME SIGNAL. US294		20 kHz Standard Frequency.
20.05–70	FIXED. MARITIME MOBILE. 448 447 449		20.05–59 FIXED. MARITIME MOBILE. 446 US294	20.05–59 FIXED. 448 US294	INTERNATIONAL FIXED PUBLIC (23).	
			59–61 STANDARD FREQUENCY AND TIME SIGNAL. US294	59–61 STANDARD FREQUENCY AND TIME SIGNAL. US294		60 kHz Standard Frequency.
			61–70 FIXED. MARITIME MOBILE. 448 US294	61–70 FIXED. 448 US294	INTERNATIONAL FIXED PUBLIC (23).	
70–72 RADIONAVIGATION 451	70–90 FIXED. MARITIME MOBILE 448	70–72 RADIONAVIGATION 451 Fixed.	70–90 FIXED. MARITIME MOBILE. Radiolocation.	70–90 FIXED. Radiolocation.	INTERNATIONAL FIXED PUBLIC (23). Private Land Mobile	

	MARITIME RADIONAVIGATION 451 Radiolocation. 452	Maritime Mobile 448. 450	US288 US294	US288 US294	(90).	
72–84 FIXED. MARITIME MOBILE 448 RADIONAVIGATION 451 447	452	72–84 FIXED. MARITIME MOBILE 448 RADIONAVIGATION 451				
84–86 RADIONAVIGATION 451		84–86 RADIONAVIGATION 451 Fixed. Maritime Mobile 448 450				
86–90 FIXED. MARITIME MOBILE 448 RADIONAVIGATION 447		86–90 FIXED. MARITIME MOBILE 448 RADIONAVIGATION 451	448 451 US294	448 451 US294		
90–110	RADIONAVIGATION 453 Fixed. 453A 454		90–110 RADIONAVIGATION 453 US18 US104 US294	90–110 RADIONAVIGATION 453 US18 US104 US294	Private Land Mobile (90).	
110–112 FIXED. MARITIME MOBILE. RADIONAVIGATION. 454	110–130 FIXED. MARITIME MOBILE. MARITIME RADIO-NAVIGATION 451 Radiolocation. 452 454	110–112 FIXED. MARITIME MOBILE. RADIONAVIGATION 451 454	110–130 FIXED. MARITIME MOBILE. Radiolocation. 454 US294	110–130 FIXED. MARITIME MOBILE. Radiolocation. 454 US294	INTERNATIONAL FIXED PUBLIC (23). MARITIME (80). Private Land Mobile (90).	
112–115 RADIONAVIGATION 451		112–117.6 RADIONAVIGATION 451 Fixed. Maritime Mobile.				
115–117.6 RADIONAVIGATION 451						

TABLE 15.1 Table of Frequency Allocations (*Continued*)

International Table			U.S. Table		FCC Use Designators	
			Government	Nongovernment		
Region 1—allocation, kHz (1)	Region 2—allocation, kHz (2)	Region 3—allocation, kHz (3)	Allocation, kHz (4)	Allocation, kHz (5)	Rule part(s) (6)	Special-use frequencies (7)
Fixed. Maritime Mobile. 454 456		454 455				
117.6–126.0 FIXED. MARITIME MOBILE. RADIONAVIGATION 451 454		117.6–126.0 FIXED. MARITIME MOBILE. RADIONAVIGATION 451 454				
126–129 RADIONAVIGATION 451		126–129 RADIONAVIGATION 451 Fixed. Maritime Mobile. 454 455				
129–130 FIXED. MARITIME MOBILE. RADIONAVIGATION 451 454		129–130 FIXED. MARITIME MOBILE. RADIONAVIGATION 451 454	451 454 US294	451 454 US294		
130–148.5 MARITIME MOBILE. /FIXED/. 454 457	130–160 FIXED. MARITIME MOBILE. 454	130–160 FIXED. MARITIME MOBILE. RADIONAVIGATION 454	130–160 FIXED. MARITIME MOBILE. 454 US294	130–160 FIXED. MARITIME MOBILE. 454 US294	INTERNATIONAL FIXED PUBLIC (23). MARITIME (80).	
148.5–255 BROADCASTING	160–190 FIXED. 459	160–190 FIXED. Aeronautical Radionavigation.	160–190 FIXED. MARITIME MOBILE. 459 US294	160–190 FIXED. 459 US294	INTERNATIONAL FIXED PUBLIC (23). MARITIME (80).	
	190–200 AERONAUTICAL RADIONAVIGATION.		190–200 AERONAUTICAL RADIONAVIGATION. US18 US226 US294	190–200 AERONAUTICAL RADIONAVIGATION. US18 US226 US294		

460 461 462	200–275 AERONAUTICAL RADIONAVIGATION. Aeronautical Mobile.	200–285 AERONAUTICAL RADIONAVIGATION. Aeronautical Mobile.	200–275 AERONAUTICAL RADIONAVIGATION. Aeronautical Mobile. US18 US294	200–275 AERONAUTICAL RADIONAVIGATION. Aeronautical Mobile. US18 US294	AVIATION (87).	
255–283.5 BROADCASTING. /AERONAUTICAL RADIONAVIGATION/ 463 462 464 464A	275–285 AERONAUTICAL RADIONAVIGATION. Aeronautical Mobile. Maritime Radionavigation (radiobeacons).		275–285 AERONAUTICAL RADIONAVIGATION. Aeronautical Mobile. Maritime Radionavigation (radiobeacons). US18 US294	275–285 AERONAUTICAL RADIONAVIGATION. Aeronautical Mobile. Maritime Radionavigation (radiobeacons). US18 US294		
283.5–315 MARITIME RADIO- NAVIGATION (radiobeacons) 466 /AERONAUTICAL RADIONAVIGATION/. 464A 465 466A	285–315 MARITIME RADIONAVIGATION (radiobeacons) 466 /AERONAUTICAL RADIONAVIGATION/.		285–325 MARITIME RADIONAVIGATION (radiobeacons) 466 Aeronautical Radionavigation (radiobeacons). US18 US294	285–325 MARITIME RADIONAVIGATION (radiobeacons) 466 Aeronautical Radionavigation (radiobeacons). US18 US294	AVIATION (87).	
315–325 AERONAUTICAL RADIONAVIGATION. Maritime Radionavigation (radiobeacons) 466 465 467	315–325 MARITIME RADIONAVIGATION (radiobeacons) 466 Aeronautical Radionavigation.	315–325 AERONAUTICAL RADIONAVIGATION. MARITIME RADIONAVIGATION (radiobeacons) 466				
325–405 AERONAUTICAL RADIONAVIGATION.	325–335 AERONAUTICAL RADIONAVIGATION. Aeronautical Mobile Maritime Radionavigation (radiobeacons).	325–405 AERONAUTICAL RADIONAVIGATION. Aeronautical Mobile.	325–335 AERONAUTICAL RADIONAVIGATION (radiobeacons). Aeronautical Mobile. Maritime Radionavigation (radiobeacons). US18 US294	325–335 AERONAUTICAL RADIONAVIGATION (radiobeacons). Aeronautical Mobile. Maritime Radionavigation (radiobeacons). US18 US294	AVIATION (87).	

TABLE 15.1 Table of Frequency Allocations (*Continued*)

International Table			U.S. Table		FCC Use Designators	
			Government	Nongovernment		
Region 1—allocation, kHz (1)	Region 2—allocation, kHz (2)	Region 3—allocation, kHz (3)	Allocation, kHz (4)	Allocation, kHz (5)	Rule part(s) (6)	Special-use frequencies (7)
465	335–405 AERONAUTICAL RADIONAVIGATION. Aeronautical Mobile.		335–405 AERONAUTICAL RADIONAVIGATION (radiobeacons). Aeronautical Mobile. US18 US294	335–405 AERONAUTICAL RADIONAVIGATION (radiobeacons). Aeronautical Mobile. US18 US294	AVIATION (87).	
405–415 RADIONAVIGATION 468 465	405–415 RADIONAVIGATION 468 Aeronautical Mobile.		405–415 RADIONAVIGATION 468 Aeronautical Mobile. US18 US294	405–415 RADIONAVIGATION 468 Aeronautical Mobile. US18 US294	AVIATION (87).	
415–435 AERONAUTICAL RADIO-NAVIGATION. /MARITIME MOBILE/ 470 465	415–495 MARITIME MOBILE 470 AERONAUTICAL RADIONAVIGATION. 470A	415–435 MARITIME MOBILE 470 AERONAUTICAL RADIONAVIGATION 470A	415–435 AERONAUTICAL RADIONAVIGATION. MARITIME MOBILE 470 469A US294	415–435 AERONAUTICAL RADIONAVIGATION. MARITIME MOBILE. 470 469A US294	AVIATION (87) MARITIME (80)	
435–495 MARITIME MOBILE 470 Aeronautical Radionavigation. 465 471 472A	469 469A 471 472A	469 469A 471 472A	435–495 MARITIME MOBILE 470 AERONAUTICAL RADIO-NAVIGATION. 471 US231 472A US294	435–495 MARITIME MOBILE 470 471 472A US231 US294	MARITIME (80).	
495–505	MOBILE (distress and calling). 472		495–505 MOBILE (distress and calling). 472	495–505 MOBILE (distress and calling). 472	MARITIME (80).	500 kHz: Distress and calling frequency.
505–526.5 MARITIME MOBILE 470 /AERONAUTICAL RADIONAVIGATION/	505–510 MARITIME MOBILE 470	505–526.5 MARITIME MOBILE 470 474 /AERONAUTICAL RADIONAVIGATION/. Aeronautical Mobile Land Mobile.	505–510 MARITIME MOBILE 470	505–510 MARITIME MOBILE 470	MARITIME (80).	

	471		471	471		
	510–525 MOBILE. 474 AERONAUTICAL RADIONAVIGATION.		510–525 AERONAUTICAL RADIONAVIGATION (radiobeacons). MARITIME MOBILE (Ships only). 474 US14 US18 US225	510–525 AERONAUTICAL RADIONAVIGATION (radiobeacons). MARITIME MOBILE (Ships only). 474 US14 US18 US225	AVIATION (87). MARITIME (80).	518 kHz: International NAVTEX in the Maritime Mobile Service.
465 471 474 475 476	525–535 BROADCASTING 477 AERONAUTICAL RADIONAVIGATION.	471	525–535 MOBILE. AERONAUTICAL RADIONAVIGATION (radiobeacons). US18 US221 US239	525–535 MOBILE. AERONAUTICAL RADIONAVIGATION (radiobeacons). US18 US221 US239	AVIATION (87) PRIVATE LAND MOBILE (90)	530 kHz: Travelers information.
526.5–1606.5 BROADCASTING. 478		526.5–535 BROADCASTING. Mobile. 479				
535–1705 480 US238, US299, US321, 478	535–1705 BROAD- CASTING 480 US238, US299, US321, NG128	RADIO BROADCASTING (AM) (73). Alaska Fixed (80). Auxiliary Broadcasting (74). Private Land Mobile (90)	535–1705 kHz: Travelers Information.			
	1605–1625		1605–1615 MOBILE. 480 US221	1605–1615 MOBILE. 480 US221	AUXILIARY BROADCASTING (74). PRIVATE LAND MOBILE (90).	1610 kHz: Travelers information.
1606.5–1625 MARITIME MOBILE. 480A/FIXED/. /LAND MOBILE/. 483 484	BROADCASTING 480 480A 481	1606.5–1800 FIXED. MOBILE. RADIOLOCATION. RADIONAVIGATION. 482				
			1615–1625 480 US237	1615–1625 BROADCASTING 480 US237 US299	ALASKA FIXED (80). AUXILIARY BROADCASTING (74). Private Land Mobile (90).	
1625–1635	1625–1705		1625–1705	1625–1705	ALASKA FIXED (80).	

TABLE 15.1 Table of Frequency Allocations (*Continued*)

International Table			U.S. Table		FCC Use Designators	
			Government	Nongovernment		
Region 1—allocation, kHz (1)	Region 2—allocation, kHz (2)	Region 3—allocation, kHz (3)	Allocation, kHz (4)	Allocation, kHz (5)	Rule part(s) (6)	Special-use frequencies (7)
RADIOLOCATION 487 485 486	BROADCASTING 480 /FIXED/. /MOBILE/. Radiolocation 480A 481		Radiolocation. 480 US238	BROADCASTING 480 Radiolocation US238 US299	AUXILIARY BROADCASTING (74). Private Land Mobile (90).	
1635–1800 MARITIME MOBILE /FIXED/. /LAND MOBILE/. 480A 483 484 488	1705–1800 FIXED. MOBILE. RADIOLOCATION. AERONAUTICAL RADIONAVIGATION.		1705–1800 FIXED. MOBILE. RADIOLOCATION. US240	1705–1800 FIXED. MOBILE. RADIOLOCATION. US240	DISASTER (99). INTERNATIONAL FIXED PUBLIC (23). MARITIME (80). PRIVATE LAND MOBILE (90).	
1800–1810 RADIOLOCATION 487 485 486 1810–1850 AMATEUR. 490 491 492 493	1800–1850 AMATEUR.	1800–2000 AMATEUR. FIXED. MOBILE except aeronautical mobile. RADIONAVIGATION. Radiolocation.	1800–1900	1800–1900 AMATEUR.	AMATEUR (97).	
1850–2000 FIXED. MOBILE except aeronautical mobile.	1850–2000 AMATEUR. FIXED. MOBILE expect aeronautical mobile. RADIOLOCATION. RADIONAVIGATION. 494		1900–2000 RADIOLOCATION.	1900–2000 RADIOLOCATION.	PRIVATE LAND MOBILE (90). Amateur (97).	

484 488 495		489	US290	US290		
2000–2025 FIXED. MOBILE except aeronautical mobile (R). 484 495 2025–2045 FIXED. MOBILE except aeronautical mobile (R). Meteorological Aids 496 484 495	2000–2065 FIXED. MOBILE.		2000–2065 FIXED. MOBILE.	2000–2065 MORITIME MOBILE. NG19	MARITIME (80).	
2045–2160 MARITIME MOBILE. /FIXED/. /LAND MOBILE/. 483 484	2065–2107 MARITIME MOBILE 497 498		2065–2107 MARITIME MOBILE 497	2065–2107 MARITIME MOBILE 497	MARITIME (80).	
2160–2170 RADIOLOCATION 487 485 486 499	2107–2170 FIXED. MOBILE.		2107–2170 FIXED. MOBILE.	2107–2170 FIXED. MARITIME MOBILE. LAND MOBILE. NG19	AVIATION (87). INTERNATIONAL FIXED PUBLIC (23). MARITIME (80). PRIVATE LAND MOBILE (90).	
2170.0–2173.5	MARITIME MOBILE.		2170–2173.5 MARITIME MOBILE.	2170–2173.5 MARITIME MOBILE.	MARITIME (80).	
2173.5–2190.5	MOBILE (distress and calling). 500 501 500A 500B	MOBILE (distress and calling). 500 501 500A 500B	2173.5–2190.5 MOBILE (distress and calling). 500 501 US279 500A 500B	2173.5–2190.5 MOBILE (distress and calling). 500 501 US279 500A 500B	AVIATION (87). MARITIME (80).	2182 kHz: Distress and calling.
2190.5–2194.0	MARITIME MOBILE.		2190.5–2194 MARITIME MOBILE.	2190.5–2194 MARITIME MOBILE.	MARITIME (80).	
2194–2300 FIXED. MOBILE except	2194–2300 FIXED. MOBILE.		2194–2495 FIXED. MOBILE.	2194–2495 FIXED. LAND MOBILE.	AVIATION (87). INTERNATIONAL	

TABLE 15.1 Table of Frequency Allocations (*Continued*)

International Table			U.S. Table		FCC Use Designators	
Region 1—allocation, kHz (1)	Region 2—allocation, kHz (2)	Region 3—allocation, kHz (3)	Government Allocation, kHz (4)	Nongovernment Allocation, kHz (5)	Rule part(s) (6)	Special-use frequencies (7)
aeronautical mobile (R). 484 495 502	502			MARITIME MOBILE. NG19	FIXED PUBLIC (23). MARITIME (80) PRIVATE LAND MOBILE (90).	
2300–2498 FIXED. MOBILE except aeronautical mobile (R). BROADCASTING 503 495	2300–2495 FIXED. MOBILE. BROADCASTING 503.					
2498–2501 STANDARD FREQUENCY AND TIME SIGNAL (2500 kHz).	2495–2501 STANDARD FREQUENCY AND TIME SIGNAL (2500 kHz).		2495–2505 STANDARD FREQUENCY AND TIME SIGNAL. G106	2495–2505 STANDARD FREQUENCY AND TIME SIGNAL.		2500 kHz: Standard frequency
2501–2502	STANDARD FREQUENCY AND TIME SIGNAL. Space Research.					
2502–2625 FIXED. MOBILE except aeronautical mobile (R). 484 495 504	2502–2505 STANDARD FREQUENCY AND TIME SIGNAL.					
	2505–2850 FIXED. MOBILE.		2505–2850 FIXED. MOBILE.	2505–2850 FIXED. LAND MOBILE. MARITIME MOBILE.	AVIATION (87). INTERNATIONAL FIXED PUBLIC (23). MARITIME (80). PRIVATE LAND MOBILE (90).	
2625–2650 MARITIME MOBILE.						

MARITIME RADIO-NAVIGATION. 484						
2650–2850 FIXED. MOBILE except aeronautical mobile (R). 484 495			US285	US285		
2850–3025	AERONAUTICAL MOBILE (R). 501 505		2850–3025 AERONAUTICAL MOBILE (R). 501 505 US283	2850–3025 AERONAUTICAL MOBILE (R). 501 505 US283	AVIATION (87).	
3025–3155	AERONAUTICAL MOBILE (OR).		3025–3155 AERONAUTICAL MOBILE (OR).	3025–3155 AERONAUTICAL MOBILE (OR).		
3155–3200	FIXED. MOBILE except aeronautical mobile (R). 506 507		3155–3230 FIXED. MOBILE except aeronautical mobile (R).	3155–3230 FIXED. MOBILE except aeronautical mobile (R).	AVIATION (87). INTERNATIONAL FIXED PUBLIC (23). MARITIME (80). PRIVATE LAND MOBILE (90).	
3200–3230	FIXED MOBILE except aeronautical mobile (R). BROADCASTING 503. 506					
3230–3400	FIXED. MOBILE except aeronautical mobile. BROADCASTING 503 506 508		3230–3400 FIXED. MOBILE except aeronautical mobile. Radiolocation.	3230–3400 FIXED. MOBILE except aeronautical mobile. Radiolocation.	AVIATION (87). INTERNATIONAL FIXED PUBLIC (23). MARITIME (80). PRIVATE LAND MOBILE (90).	
3400–3500	AERONAUTICAL MOBILE (R).		3400–3500 AERONAUTICAL MOBILE (R). US283	3400–3500 AERONAUTICAL MOBILE (R). US283	AVIATION (87).	
3500–3800	3500–3750	3500–3900	3500–4000	3500–4000		

TABLE 15.1 Table of Frequency Allocations (*Continued*)

International Table			U.S. Table		FCC Use Designators	
			Government	Nongovernment		
Region 1—allocation, kHz (1)	Region 2—allocation, kHz (2)	Region 3—allocation, kHz (3)	Allocation, kHz (4)	Allocation, kHz (5)	Rule part(s) (6)	Special-use frequencies (7)
AMATEUR 510 FIXED. MOBILE except aeronautical mobile. 484	AMATEUR 510 509 511	AMATEUR 510 FIXED. MOBILE.	510	AMATEUR 510	AMATEUR (97).	
3800–3900 FIXED. AERONAUTICAL MOBILE (OR). LAND MOBILE.	3750–4000 AMATEUR 510 FIXED. MOBILE except aeronautical mobile (R). 511 512 514 515					
3900–3950 AERONAUTICAL MOBILE (OR). 513		3900–3950 AERONAUTICAL MOBILE. BROADCASTING.				
3950–4000 FIXED. BROADCASTING.		3950–4000 FIXED. BROADCASTING. 516				
4000–4063	FIXED. MARITIME MOBILE 517 516		4000–4438 MARITIME MOBILE 500A 500B 520 520B US82 US236 US296	4000–4438 MARITIME MOBILE 500A 500B 520 520B US82 US236 US296	INTERNATIONAL FIXED PUBLIC (23). MARITIME (80).	
4063–4438	MARITIME MOBILE 500A 500B 520 520A 520B 518 519					
4438–4650 FIXED. MOBILE except		4438–4650 FIXED. MOBILE except	4438–4650 FIXED. MOBILE except	4438–4650 FIXED. MOBILE except	AVIATION (87). INTERNATIONAL	

aeronautical mobile (R)		aeronautical mobile (R)	aeronautical mobile (R)	aeronautical mobile.	FIXED PUBLIC (23). MARITIME (80). PRIVATE LAND MOBILE (90).	
4650–4700	AERONAUTICAL MOBILE (R).		4650–4700 AERONAUTICAL MOBILE (R). US282 US283	4650–4700 AERONAUTICAL MOBILE (R). US282 US283	AVIATION (87).	
4700–4750	AERONAUTICAL MOBILE (OR).		4700–4750 AERONAUTICAL MOBILE (OR).	4700–4750 AERONAUTICAL MOBILE (OR).		
4750–4850 FIXED. AERONAUTICAL MOBILE (OR). LAND MOBILE BROADCASTING 503	4750–4850 FIXED. MOBILE except aeronautical. BROADCASTING 503	4750–4850 FIXED. BROADCASTING 503 Land Mobile.	4750–4850 FIXED. MOBILE except aeronautical mobile.	4750–4850 FIXED. MOBILE except aeronautical mobile (R).	AVIATION (87). INTERNATIONAL FIXED PUBLIC (23). MARITIME (80).	
4850–4995	FIXED. LAND MOBILE. BROADCASTING 503.		4850–4995 FIXED. MOBILE.	4850–4995 FIXED.	AVIATION (87). INTERNATIONAL FIXED PUBLIC (23). MARITIME (80).	
4995–5003	STANDARD FREQUENCY AND TIME SIGNAL (5000 kHz).		4995–5005 STANDARD FREQUENCY AND TIME SIGNAL	4995–5005 STANDARD FREQUENCY AND TIME SIGNAL.		5000 kHz: Standard frequency.
5003–5005	STANDARD FREQUENCY AND TIME SIGNAL. Space Research.		G106			
5005–5060	FIXED. BROADCASTING 503		5005–5060 FIXED.	5005–5060 FIXED.	AVIATION (87). INTERNATIONAL FIXED PUBLIC (23). MARITIME (80). PRIVATE LAND MOBILE (90).	
5060–5250	FIXED. Mobile except aeronautical mobile.		5060–5450 FIXED. MOBILE except aeronautical mobile.	5060–5450 FIXED. MOBILE except aeronautical mobile.	AVIATION (87). INTERNATIONAL FIXED PUBLIC (23).	

TABLE 15.1 Table of Frequency Allocations (*Continued*)

	International Table		U.S. Table		FCC Use Designators	
			Government	Nongovernment		
Region 1—allocation, kHz (1)	Region 2—allocation, kHz (2)	Region 3—allocation, kHz (3)	Allocation, kHz (4)	Allocation, kHz (5)	Rule part(s) (6)	Special-use frequencies (7)
	521				MARITIME (80). PRIVATE LAND MOBILE (90).	
5250–5450	FIXED. MOBILE except aeronautical mobile.		US212	US212		
5450–5480 FIXED. AERONAUTICAL MOBILE (OR). LAND MOBILE.	5450–5480 AERONAUTICAL MOBILE (R).	5450–5480 FIXED. AERONAUTICAL MOBILE (OR). LAND MOBILE.	5450–5480 AERONAUTICAL MOBILE (R).	5450–5480 AERONAUTICAL MOBILE (R).	AVIATION (87).	
5480–5680	AERONAUTICAL MOBILE (R). 501 505		501 505 US283	501 505 US283		
5680–5730	AERONAUTICAL MOBILE (OR). 501 505		5680–5730 AERONAUTICAL MOBILE (OR). 501 505	5680–5730 AERONAUTICAL MOBILE (OR). 501 505		
5730–5950 FIXED. LAND MOBILE	5730–5950 FIXED. MOBILE except aeronautical mobile (R).	5730–5950 FIXED. MOBILE except aeronautical mobile (R)	5730–5950 FIXED. MOBILE except aeronautical mobile (R).	5730–5950 FIXED. MOBILE except aeronautical mobile (R).	AVIATION (87). INTERNATIONAL FIXED PUBLIC (23). MARITIME (80).	
5950–6200	BROADCASTING		5950–6200 BROADCASTING US280	5950–6200 BROADCASTING US280	RADIO BROADCAST (HF)(73).	
6200–6525	MARITIME MOBILE 500A 500B 520 520B. 522		6200–6525 MARITIME MOBILE 500A 500B 520 520B. US82 US296	6200–6525 MARITIME MOBILE 500A 500B 520 520B. US82 US296	MARITIME (80).	
6525–6685	AERONAUTICAL		6525–6685 AERONAUTICAL	6525–6685 AERONAUTICAL	AVIATION (87).	

	MOBILE (R).		MOBILE (R). US283	MOBILE (R). US283		
6685–6765	AERONAUTICAL MOBILE (OR).		6765–7000 AERONAUTICAL MOBILE (OR).	6765–7000 AERONAUTICAL MOBILE (OR).		
6765–7000	FIXED. Land Mobile 525 524		6765–7000 FIXED. Mobile. 524	6765–7000 FIXED. Mobile. 524	AVIATION (87). INTERNATIONAL FIXED PUBLIC (23).	6780 + 15 kHz: Industrial, scientific, and medical frequency.
7000–7100	AMATEUR 510 AMATEUR-SATELLITE. 526 527		7000–7300 510 528	7000–7100 AMATEUR 510 AMATEUR-SATELLITE.	AMATEUR (97).	
7100–7300 BROADCASTING.	7100–7300 AMATEUR 510 528	7100–7300 BROADCASTING.		7100–7300 AMATEUR 510 528	AMATEUR (97).	
7300–8100	FIXED. Land Mobile. 529		7300–8100 FIXED. Mobile.	7300–8100 FIXED. Mobile.	AVIATION (87). INTERNATIONAL FIXED PUBLIC (23). MARITIME (80). PRIVATE LAND MOBILE (90).	
8100–8195	FIXED. MARITIME MOBILE.		8100–8815 MARITIME MOBILE. 500A 500B 520B 529A 501 US82 US236 US296	8100–8815 MARITIME mobile. 500A 500B 520B 529A 501 US82 US236 US296	MARITIME (80).	
8195–8815	MARITIME MOBILE. 500A 500B 520B 529A 501					
8815–8965	AERONAUTICAL MOBILE (R).		8815–8965 AERONAUTICAL MOBILE (R).	8815–8965 AERONAUTICAL MOBILE (R).	Aviation (87).	
8965–9040	AERONAUTICAL MOBILE (OR).		8965–9040 AERONAUTICAL MOBILE (OR).	8965–9040 AERONAUTICAL MOBILE (OR).		

TABLE 15.1 Table of Frequency Allocations (*Continued*)

	International Table		U.S. Table		FCC Use Designators	
			Government	Nongovernment		
Region 1—allocation, kHz (1)	Region 2—allocation, kHz (2)	Region 3—allocation, kHz (3)	Allocation, kHz (4)	Allocation, kHz (5)	Rule part(s) (6)	Special-use frequencies (7)
9040–9500	FIXED.		9040–9500 FIXED.	9040–9050 FIXED.	Aviation (87). INTERNATIONAL FIXED PUBLIC (23). MARITIME (80).	
9500–9900	BROADCASTING. 530 531		9500–9900 BROADCASTING. US235	9500–9900 BROADCASTING. US235	RADIO BROADCAST (HF) (73). INTERNATIONAL FIXED PUBLIC (23).	
9900–9995	FIXED.		9900–9995 FIXED.	9900–9995 FIXED.	AVIATION (87). INTERNATIONAL FIXED PUBLIC (23).	
9995–10003	STANDARD FREQUENCY AND TIME SIGNAL (10000 kHz). 501		9995–10005 STANDARD FREQUENCY AND TIME SIGNAL. 501 G106	9995–10005 STANDARD FREQUENCY AND TIME SIGNAL. 501		10000 kHz: Standard frequency.
10003–10005	STANDARD FREQUENCY AND TIME SIGNAL Space Research. 501					
10005–10100	AERONAUTICAL MOBILE (R). 501		10005–10100 AERONAUTICAL MOBILE (R). 501 US283	10005–10100 AERONAUTICAL MOBILE (R). 501 US283	AVIATION (87).	
10100–10150	FIXED. Amateur 510.		10100–10150 510 US247	10100–10150 AMATEUR 510. US247	AMATEUR (97).	

10150–11175	FIXED. MOBILE except aeronautical mobile (R).		10150–11175 FIXED. MOBILE except aeronautical mobile (R).	10150–11175 FIXED. MOBILE except aeronautical mobile (R).	10150–11175 AVIATION (87). INTERNATIONAL FIXED PUBLIC (23).	
11175–11275	AERONAUTICAL MOBILE (OR).		11175–11275 AERONAUTICAL MOBILE (OR).	11175–11275 AERONAUTICAL MOBILE (OR).		
11275–11400	AERONAUTICAL MOBILE (R).		11275–11400 AERONAUTICAL MOBILE (R). US283	11275–11400 AERONAUTICAL MOBILE (R). US283	AVIATION (87).	
11400–11650	FIXED.		11400–11650 FIXED.	11400–11650 FIXED.	AVIATION (87). INTERNATIONAL FIXED PUBLIC (23).	
11650–12050	BROADCASTING. 530 531		11650–12050 BROADCASTING. US235	11650–12050 BROADCASTING. US235	RADIO BROADCAST (HF) (73). INTERNATIONAL FIXED PUBLIC (23).	
12050–12230	FIXED.		12050–12230 FIXED.	12050–12230 FIXED.	AVIATION (87). INTERNATIONAL FIXED PUBLIC (23).	
12230–13200	MARITIME MOBILE. 500A 500B 520B 529A 532		12230–13200 MARITIME MOBILE. 500A 500B 520B 529A US82 US296	12230–13200 MARITIME MOBILE. 500A 500B 520B 529A US82 US296	INTERNATIONAL FIXED PUBLIC (23). MARITIME (80).	
13200–13260	AERONAUTICAL MOBILE (OR).		13200–13260 AERONAUTICAL MOBILE (OR).	13200–13260 AERONAUTICAL MOBILE (OR).		
13260–13360	AERONAUTICAL MOBILE (R).		13260–13360 AERONAUTICAL MOBILE (R). US283	13260–13360 AERONAUTICAL MOBILE (R). US283	AVIATION (87).	
13360–13410	FIXED.		13360–13410 RADIO ASTRONOMY.	13360–13410 RADIO ASTRONOMY.		

TABLE 15.1 Table of Frequency Allocations (*Continued*)

	International Table		U.S. Table		FCC Use Designators	
			Government	Nongovernment		
Region 1—allocation, kHz (1)	Region 2—allocation, kHz (2)	Region 3—allocation, kHz (3)	Allocation, kHz (4)	Allocation, kHz (5)	Rule part(s) (6)	Special-use frequencies (7)
	RADIO ASTRONOMY. 533		533 G115	533		
13410–13600	FIXED. Mobile except aeronautical mobile (R). 534		13410–13600 FIXED. Mobile except aeronautical mobile (R). 534	13410–13600 FIXED. 534	AVIATION (87). INTERNATIONAL FIXED PUBLIC (23).	13560 ± 7 kHz: Industrial, scientific, and medical frequency.
13600–13800	BROADCASTING. 531		13600–13800 BROADCASTING. US235	13600–13800 BROADCASTING. US235	RADIO BROADCAST (HF) (73). INTERNATIONAL FIXED PUBLIC (23).	
13800–14000	FIXED. Mobile except aeronautical mobile (R)		13800–14000 FIXED. Mobile except aeronautical mobile (R).	13800–14000 FIXED.	AVIATION (87). INTERNATIONAL FIXED PUBLIC (23).	
14000–14250	AMATEUR 510 AMATEUR-SATELLITE.		14000–14350	14000–14250 AMATEUR 510 AMATEUR-SATELLITE.	AMATEUR (97).	
14250–14350	AMATEUR 510 535		510	14250–14350 AMATEUR 510	AMATEUR (97).	
14350–14990	FIXED. Mobile except aeronautical mobile (R).		14350–14990 FIXED. Mobile except aeronautical mobile (R).	14350–14990 FIXED.	AVIATION (87). INTERNATIONAL FIXED PUBLIC (23).	
14990–15005	STANDARD FREQUENCY AND TIME SIGNAL (15000 kHz). 501		14990–15010 STANDARD FREQUENCY AND TIME SIGNAL.	14990–15010 STANDARD FREQUENCY AND TIME SIGNAL.		1500 kHz: Standard frequency.
15005–15010	STANDARD FREQUENCY AND TIME					

	SIGNAL. Space Research.		501 G106	501		
15010–15100	AERONAUTICAL MOBILE (OR).		15010–15100 AERONAUTICAL MOBILE (OR).	15010–15100 AERONAUTICAL MOBILE (OR).		
15100–15600	BROADCASTING. 531		15100–15600 BROADCASTING. US235	15100–15600 BROADCASTING. US235	RADIO BROADCAST (HF) (73). INTERNATIONAL FIXED PUBLIC (23).	
15600–16360	FIXED. 536		15600–16360 FIXED.	15600–16360 FIXED.	AVIATION (87). INTERNATIONAL FIXED PUBLIC (23).	
16360–17410	MARITIME MOBILE. 500A 500B 520B 529A 532		16360–17410 MARITIME MOBILE. 500A 500B 520B 529A US82 US296	16360–17410 MARITIME MOBILE. 500A 500B 520B 529A US82 US296	MARITIME (80)	
17410–17550	FIXED.		17410–17550 FIXED.	17410–17550 FIXED.	AVIATION (87). INTERNATIONAL FIXED PUBLIC (23).	
17550–17900	BROADCASTING. 531		17550–17900 BROADCASTING. US235	17550–17900 BROADCASTING. US235	RADIO BROADCAST (HF) (73). INTERNATIONAL FIXED PUBLIC (23).	
17900–17970	AERONAUTICAL MOBILE (R).		17900–17970 AERONAUTICAL MOBILE (R). US283	17900–17970 AERONAUTICAL MOBILE (R). US283	AVIATION (87).	
17970–18030	AERONAUTICAL MOBILE (OR).		17970–18030 AERONAUTICAL MOBILE (OR).	17970–18030 AERONAUTICAL MOBILE (OR).		

TABLE 15.1 Table of Frequency Allocations (*Continued*)

	International Table		U.S. Table		FCC Use Designators	
			Government	Nongovernment		
Region 1—allocation, kHz (1)	Region 2—allocation, kHz (2)	Region 3—allocation, kHz (3)	Allocation, kHz (4)	Allocation, kHz (5)	Rule part(s) (6)	Special-use frequencies (7)
18030–18052	FIXED.		18030–18068 FIXED.	18030–18068 FIXED.	INTERNATIONAL FIXED PUBLIC (23). MARITIME (80).	
18052–18068	FIXED. Space Research.					
18068–18168	AMATEUR 510 AMATEUR-SATELLITE. 537 538		18068–18168 510 US248	18068–18168 AMATEUR 510 AMATEUR-SATELLITE. US248	AMATEUR (97). INTERNATIONAL FIXED PUBLIC (23). MARITIME (80).	
18168–18780	FIXED. Mobile except aeronautical mobile		18168–18780 FIXED. Mobile	18168–18780 FIXED. Mobile	AVIATION (87). INTERNATIONAL FIXED PUBLIC (23). MARITIME (80).	
18780–18900	MARITIME MOBILE. 532		18780–18900 MARITIME MOBILE. US82 US296	18780–18900 MARITIME MOBILE. US82 US296	INTERNATIONAL FIXED PUBLIC (23). MARITIME (80).	
18900–19680	FIXED.		18900–19680 FIXED.	18900–19680 FIXED.	AVIATION (87). INTERNATIONAL FIXED PUBLIC (23).	
19680–19800	MARITIME MOBILE. 520B 532		19680–19800 MARITIME MOBILE. 520B	19680–19800 MARITIME MOBILE. 520B	MARITIME (80).	
19800–19990	FIXED.		19800–19990 FIXED.	19800–19990 FIXED.	AVIATION (87). INTERNATIONAL FIXED PUBLIC (23).	

19990–19995	STANDARD FREQUENCY AND TIME SIGNAL Space Research. 501		19990–20010 STANDARD FREQUENCY AND TIME SIGNAL 501 G106	19990–20010 STANDARD FREQUENCY AND TIME SIGNAL 501		20000 kHz: Standard frequency.
19995–20010	STANDARD FREQUENCY AND TIME SIGNAL (20000 kHz). 501					
20010–21000	FIXED. Mobile.		20010–21000 FIXED. Mobile.	20010–21000 FIXED.		
21000–21450	AMATEUR 510 AMATEUR-SATELLITE.		21000–21450 510	21000–21450 AMATEUR 510 AMATEUR-SATELLITE.	AMATEUR (97).	
21450–21850	BROADCASTING. 531		21450–21850 BROADCASTING. US235	21450–21850 BROADCASTING. US235	INTERNATIONAL FIXED PUBLIC (23) RADIO BROADCAST (HF) (73).	
21850–21870	FIXED. 539		21850–21924 FIXED.	21850–21924 FIXED.	AVIATION (87). INTERNATIONAL FIXED PUBLIC (23).	
21870–21924	AERONAUTICAL FIXED.					
	AERONAUTICAL MOBILE (R).		AERONAUTICAL MOBILE (R).	AERONAUTICAL MOBILE (R).	AVIATION (87).	
22000–22855	MARITIME MOBILE. 520B 532 540		22000–22855 MARITIME MOBILE. 520B US82 US296	22000–22855 MARITIME MOBILE. 520B US82 US296	INTERNATIONAL FIXED PUBLIC (23). MARITIME (80).	
22855–23000			22855–23000	22855–23000		

TABLE 15.1 Table of Frequency Allocations (*Continued*)

	International Table		U.S. Table		FCC Use Designators	
			Government	Nongovernment		
Region 1—allocation, kHz (1)	Region 2—allocation, kHz (2)	Region 3—allocation, kHz (3)	Allocation, kHz (4)	Allocation, kHz (5)	Rule part(s) (6)	Special-use frequencies (7)
	FIXED. 540		FIXED.	FIXED.	AVIATION (87). INTERNATIONAL FIXED PUBLIC (23).	
23000–23200	FIXED. Mobile except aeronautical mobile (R). 540		23000–23200 FIXED. Mobile except aeronautical mobile (R).	23000–23200 FIXED.	AVIATION (87). INTERNATIONAL FIXED PUBLIC (23).	
23200–23350	AERONAUTICAL FIXED. AERONAUTICAL MOBILE (OR).		23200–23350 AERONAUTICAL MOBILE (OR).	23200–23350 AERONAUTICAL MOBILE (OR).		
23350–24000	FIXED. Mobile except aeronautical mobile 541. 542		23350–24890 FIXED. Mobile except aeronautical mobile.	23350–24890 FIXED.	AVIATION (87). INTERNATIONAL FIXED PUBLIC (23).	
24000–24890	FIXED. LAND MOBILE. 542					
24890–24990	AMATEUR 510 AMATEUR-SATELLITE. 542 543		24890–24990 510 US248	24890–24990 AMATEUR 510 AMATEUR-SATELLITE. US248	AMATEUR (97).	
24990–25005	STANDARD FREQUENCY AND TIME SIGNAL (25000 kHz).		24990–25010 STANDARD FREQUENCY AND TIME SIGNAL G106	24990–25010 STANDARD FREQUENCY AND TIME SIGNAL		25000 kHz: Standard frequency.
25005–25010	STANDARD FREQUENCY AND TIME SIGNAL Space Research.					

25010–25070	FIXED. MOBILE expect aeronautical mobile.		25010–25070	25010–25070 LAND MOBILE. NG112	PRIVATE LAND MOBILE (90).	
25070–25210	MARITIME MOBILE. 544		25070–25210 MARITIME MOBILE. US82 US281 US296	25070–25210 MARITIME MOBILE. US82 US281 US296 NG112	MARITIME (80). PRIVATE LAND MOBILE (90).	
25210–25550	FIXED. MOBILE except aeronautical mobile.		25210–25330	25210–25330 LAND MOBILE.	PRIVATE LAND MOBILE (90).	
			25330–25550 FIXED. MOBILE except aeronautical mobile.	25330–25550		
25550–25670	RADIO ASTRONOMY. 545		25550–25670 RADIO ASTRONOMY. 545 US74	25550–25670 RADIO ASTRONOMY. 545 US74		
25670–26100	BROADCASTING.		25670–26100 BROADCASTING. US25	25670–26100 BROADCASTING. US25	AUXILIARY BRODCASTING (74). RADIO BROADCAST (HF) (73).	
26100–26175	MARITIME MOBILE. 520B 544		26100–26175 MARITIME MOBILE. 520B	26100–26175 MARITIME MOBILE. 520B	AUXILIARY BROADCASTING (74). MARITIME (80).	
26175–27500	FIXED. MOBILE except aeronautical mobile.		26175–26480	26175–26480 LAND MOBILE.	AUXILIARY BROADCASTING (74).	
			26480–26950 FIXED. MOBILE except aeronautical mobile.	26480–26950		

TABLE 15.1 Table of Frequency Allocations (*Continued*)

	International Table		U.S. Table		FCC Use Designators	
			Government	Nongovernment		
Region 1—allocation, kHz (1)	Region 2—allocation, kHz (2)	Region 3—allocation, kHz (3)	Allocation, kHz (4)	Allocation, kHz (5)	Rule part(s) (6)	Special-use frequencies (7)
			US10	US10		
			26950–27540	26950–26960 FIXED. 546	INTERNATIONAL FIXED PUBLIC (23).	
				26960–27230 MOBILE except aeronautical mobile. 546	PERSONAL (95).	27120 ± 160 kHz: Industrial, scientific, and medical frequency.
	546			27230–27410 FIXED. MOBILE except aeronautical mobile. 546	PERSONAL (95). PRIVATE LAND MOBILE (90).	
27500–28000	METEOROLOGICAL AIDS. FIXED. MOBILE.		546	27410–27540 LAND MOBILE.	PRIVATE LAND MOBILE (90).	
			27540–28000 FIXED. MOBILE.	27540–28000		

	International Table		U.S. Table		FCC Use Designators	
			Government	Nongovernment		
Region 1—allocation, MHz (1)	Region 2—allocation, MHz (2)	Region 3—allocation, MHz (3)	Allocation, MHz (4)	Allocation, MHz (5)	Rule part(s) (6)	Special-use frequencies (7)
28.0–29.7	AMATEUR. AMATEUR-SATELLITE.		28.0–29.7	28.0–29.7 AMATEUR. AMATEUR-SATELLITE.	AMATEUR (97).	

29.7–30.005	FIXED. MOBILE.		29.7–29.89	29.7–29.8 LAND MOBILE.	PRIVATE LAND MOBILE (90).	
				29.8–29.89 FIXED.	AVIATION (87). INTERNATIONAL FIXED PUBLIC (23).	
			29.89–29.91 FIXED. MOBILE.	29.89–29.91		
			29.91–30.0	29.91–30.0 FIXED.	AVIATION (87). INTERNATIONAL FIXED PUBLIC (23).	
			30.0–30.56 MOBILE. Fixed.	30.0–30.56		
30.005–30.01	FIXED. MOBILE. SPACE RESEARCH. SPACE OPERATIONS (Satellite identification)					
30.01–37.5	FIXED. MOBILE.					
			30.56–32.0	30.56–32.0 LAND MOBILE. NG124	PRIVATE LAND MOBILE (90).	
			32.0–33.0 FIXED MOBILE.	32.0–33.0		
			33.0–34.0	33.0–34.0 LAND MOBILE. NG124	PRIVATE LAND MOBILE (90).	
			34.0–35.0 FIXED. MOBILE.	34.0–35.0		
			35.0–36.0	35.0–35.19 LAND MOBILE. NG124	PRIVATE LAND MOBILE (90).	
				35.19–35.69 LAND MOBILE.	DOMESTIC PUBLIC	

TABLE 15.1 Table of Frequency Allocations (*Continued*)

International Table			U.S. Table		FCC Use Designators	
			Government	Nongovernment		
Region 1—allocation, MHz (1)	Region 2—allocation, MHz (2)	Region 3—allocation, MHz (3)	Allocation, MHz (4)	Allocation, MHz (5)	Rule part(s) (6)	Special-use frequencies (7)
				NG124	LAND MOBILE (22). PRIVATE LAND MOBILE (90).	
				35.69–36.0 LAND MOBILE. NG124	PRIVATE LAND MOBILE (90).	
			36.0–37.0 FIXED. MOBILE. US220	36.0–37.0 US220		
			37.0–37.5	37.0–37.5 LAND MOBILE. NG124	PRIVATE LAND MOBILE (90).	
37.5–38.25	FIXED. MOBILE. Radio Astronomy. 547		37.5–38.0 Radio Astronomy. 547	37.5–38.0 LAND MOBILE. Radio Astronomy. 547 NG59 NG124	PRIVATE LAND MOBILE (90).	
			38.0–38.25 FIXED. MOBILE. RADIO ASTRONOMY. 547 US81	38.0–38.25 RADIO ASTRONOMY. 547 US81		
38.25–39.986	FIXED. MOBILE.		38.25–39.0 FIXED. MOBILE.	38.25–39.0		
			39.0–40.0	39.0–40.0 LAND MOBILE. NG124	PRIVATE LAND MOBILE (90).	
39.986–40.02	FIXED. MOBILE.					

	Space Research.					
40.02–40.98	FIXED. MOBILE. 548		40.0–42.0 FIXED MOBILE. 548 US210 US220	40.0–42.0 548 US210 US220		40.68 MHz ± 0.02 MHz: Industrial, scientific and medical frequency.
40.98–41.015	FIXED. MOBILE. Space Research. 549 550 551					
41.015–44.0	FIXED. MOBILE. 549 550 551	LAND MOBILE.	42.0–46.6 PRIVATE LAND MOBILE (90).	42.0–43.19 NG124 NG141		
				43.19–43.69 LAND MOBILE. NG124	DOMESTIC PUBLIC LAND MOBILE (22). PRIVATE LAND MOBILE (90).	
44.0–47.0	FIXED. MOBILE. 551 552			43.69–46.6 LAND MOBILE. NG124 NG141	PRIVATE LAND MOBILE (90).	
			46.6–47.0 FIXED. MOBILE.	46.6–45.0		
47.0–68.0 BROADCASTING.	47.0–50.0 FIXED. MOBILE.	47.0–50.0 FIXED. MOBILE. BROADCASTING.	47.0–49.6	47.0–49.6 LAND MOBILE. NG124	PRIVATE LAND MOBILE (90).	
			49.6–50.0 FIXED.	49.6–50.0		

TABLE 15.1 Table of Frequency Allocations (*Continued*)

	International Table		U.S. Table		FCC Use Designators	
			Government	Nongovernment		
Region 1—allocation, MHz (1)	Region 2—allocation, MHz (2)	Region 3—allocation, MHz (3)	Allocation, MHz (4)	Allocation, MHz (5)	Rule part(s) (6)	Special-use frequencies (7)
			MOBILE.			
553 554 555 559 561	50.0–54.0 AMATEUR. 556 557 558 560		50.0–54.0	50.0–54.0 AMATEUR.	AMATEUR (97).	
	54.0–68.0 BROADCASTING. Fixed. Mobile. 562	54.0–68.0 FIXED. MOBILE. BROADCASTING.	54.0–72.0	540–72.0 BROADCASTING. NG128 NG149	RADIO BROADCAST (TT) (73). Auxiliary Broadcasting (74).	
68.0–74.8 FIXED. MOBILE except aeronautical mobile.	68.0–72.0 BROADCASTING. Fixed. Mobile. 563	68.0–74.8 FIXED. MOBILE,				
564 565 567 568 571 572	72.0–73.0 FIXED. MOBILE.	566 568 551 572	72.0–73.0	72.0–73.0 FIXED. MOBILE. NG3 NG49 NG56	DOMESTIC PUBLIC LAND MOBILE (22). PERSONAL (96). PRIVATE LAND MOBILE (90).	
	73.0–74.6 RADIO ASTRONOMY. 570		73.0–74.6 RADIO ASTRONOMY. US74	73.0–74.6 RADIO ASTRONOMY. US74		
	74.6–74.8 FIXED. MOBILE. 572		74.6–74.8 FIXED. MOBILE. 572 US273	74.6–74.8 FIXED. MOBILE. 572 US273	PRIVATE LAND MOBILE (90).	
74.8–75.2	AERONAUTICAL RADIONAVIGATION. 572 572A		74.8–75.2 AERONAUTICAL RADIONAVIGATION. 572	74.8–75.2 AERONAUTICAL RADIONAVIGATION. 572	AVIATION (87).	75 MHz Marker Beacon.
75.2–87.5 FIXED. MOBILE except aeronautical	75.2–75.4 FIXED. MOBILE.		75.2–75.4 FIXED. MOBILE.	75.2–75.4 FIXED. MOBILE.	PRIVATE LAND	

mobile	571 572		572 US273	572 US273	MOBILE (90)	
565 571 572 575 578	75.4–76 FIXED. MOBILE.	75.4–87 FIXED. MOBILE. 573 574 577 579	75.4–76	75.4–76 FIXED. MOBILE NG3 NG49 NG56	DOMESTIC PUBLIC LAND MOBILE (22). PERSONAL (95). PRIVATE LAND MOBILE (90)	
	76.0–88.0 BROADCASTING. Fixed. Mobile. 576	87–100 FIXED. MOBILE. BROADCASTING.	76.0–88.0 NG128	76.0–88.0 BROADCASTING. NG129 NG149	RADIO BROADCAST (TV) (73). Auxiliary Broadcasting (74).	
87.5–100 BROADCASTING. 581 582	88–100 BROADCASTING.	580	88–108 US93	88–108 BROADCASTING. US93 NG2 NG128 NG129	RADIO BROADCAST (FM) (73). Auxiliary Broadcasting (74).	
100–108	BROADCASTING. 582 584 585 586 587 588 589					
108–117.975	AERONAUTICAL RADIONAVIGATION. 590A		108–117.975 AERONAUTICAL RADIONAVIGATION. US93	108–117.975 AERONAUTICAL RADIONAVIGATION. US93		
117.975–136.0	AERONAUTICAL MOBILE (R).		117.975–121.9375 AERONAUTICAL MOBILE (R). 501 591 592 593 US26 US28	117.975–121.9375 AERONAUTICAL MOBILE (R). 501 591 592 593 US26 US28	AVIATION (87).	

TABLE 15.1 Table of Frequency Allocations (*Continued*)

International Table			U.S. Table		FCC Use Designators	
			Government	Nongovernment		
Region 1—allocation, MHz (1)	Region 2—allocation, MHz (2)	Region 3—allocation, MHz (3)	Allocation, MHz (4)	Allocation, MHz (5)	Rule part(s) (6)	Special-use frequencies (7)
	501 591 592 593 594		121.9375–123.0875 591 US30 US31 US33 US80 US102 US213	121.9375–123.0875 AERONAUTICAL MOBILE 591 US30 US31 US33 US80 US102 US213	AVIATION (87).	
			123.0875–123.5875 AERONAUTICAL MOBILE 591 593 US32 US33 US112	123.0875–123.5875 AERONAUTICAL MOBILE 591 593 US32 US33 US112	AVIATION (87).	123.1 MHz for Scene-of-Action Communication (SAR).
			123.5875–128.8125 AERONAUTICAL MOBILE (R). 591 US26	123.5875–128.8125 AERONAUTICAL MOBILE (R). 591 US26	AVIATION (87).	
			128.8125–132.0125 591	128.8125–132.0125 AERONAUTICAL MOBILE (R). 591	AVIATION (87).	
			132.0125–136.0 AERONAUTICAL MOBILE (R). 591 US26	132.0125–136.0 AERONAUTICAL MOBILE (R). 591 US26	AVIATION (87).	
136–137	AERONAUTICAL MOBILE (R). FIXED. Mobile except aeronautical mobile (R). 591 595 594A		136.0–137.0 591 US244	136.0–137.0 AERONAUTICAL MOBILE (R). 591 US244	AVIATION (87). SATELLITE COMMUNICATIONS (25).	
137.0–138.0	SPACE OPERATION (space-to-Earth).		137.0–138.0 SPACE OPERATION (space-to-Earth).	137.0–138.0 SPACE OPERATION (space-to-Earth).	SATELLITE COMMUNICATIONS (25)	

	METEOROLOGICAL- SATELLITE (space- to-Earth). SPACE RESEARCH (space-to-Earth) Fixed. Mobile except aeronautical mobile (R). 596 597 598 599		METEOROLOGICAL- SATELLITE (space- to-Earth). SPACE RESEARCH (space-to-Earth)	METEOROLOGICAL- SATELLITE (space- to-Earth). SPACE RESEARCH (space-to-Earth)		
138.0–143.6 AERONAUTICAL MOBILE (OR). 600 601 601 604	138.0–143.6 FIXED. MOBILE. /RADIOLOCATION/. Space Research (space-to-Earth).	138.0–143.6 FIXED. MOBILE. Space Research (space-to-Earth)./ 599 603	138.0–144.0 FIXED. MOBILE.	138.0–144.0		
143.6–143.65 AERONAUTICAL MOBILE (OR). SPACE RESEARCH (space-to-Earth). 501 602 604	143.6–143.65 FIXED. MOBILE. SPACE RESEARCH (space-to-Earth). /RADIOLOCATION/.	143.6–143.65 FIXED. MOBILE. SPACE RESEARCH (space-to-Earth). 599 603				
143.65–144.0 AERONAUTICAL MOBILE (OR). 600 601 602 604	143.65–144.0 FIXED. MOBILE. /RADIOLOCATION/. Space Research (space-to-Earth).	143.65–144.0 FIXED. MOBILE. Space Research (space-to-Earth)./ 599 603	US10 G30	US10		
144.0–146.0	AMATEUR 510 AMATEUR-SATELLITE. 605 606		144.0–146.0 510	144.0–146.0 AMATEUR 510 AMATEUR-SATELLITE.	AMATEUR (97).	
146.0–149.9 FIXED. MOBILE except aeronautical Mobile (R).	146.0–148.0 AMATEUR. 607	146.0–148.0 AMATEUR. FIXED. MOBILE. 607	146.0–148.0	146.0–148.0 AMATEUR.	AMATEUR (97).	
608	148.0–149.9 FIXED. MOBILE. 608		148.0–149.9 FIXED. MOBILE. 606 US10 G30	148.0–149.9 606 US10	SATELLITE COMMUNICATIONS (25).	

TABLE 15.1 Table of Frequency Allocations (*Continued*)

International Table			U.S. Table		FCC Use Designators	
			Government	Nongovernment		
Region 1—allocation, MHz (1)	Region 2—allocation, MHz (2)	Region 3—allocation, MHz (3)	Allocation, MHz (4)	Allocation, MHz (5)	Rule part(s) (6)	Special-use frequencies (7)
149.9–150.05	RADIONAVIGATION-SATELLITE. 609 609A		149.9–150.05 RADIONAVIGATION SATELLITE. 609A	149.9–150.05 RADIONAVIGATION SATELLITE. 609A		
150.05–153 FIXED. MOBILE except aeronautical mobile. RADIO ASTRONOMY. 610 612	150.05–156.7625 FIXED. MOBILE. 611 613 613A		150.05–150.8 FIXED. MOBILE. US216 G30	150.05–150.8 US216		
			150.8–156.2475 613 US216	150.8–152 LAND MOBILE. NG51 NG112 NG124	PRIVATE LAND MOBILE (90).	
				152.0–152.255 LAND MOBILE. US216	DOMESTIC PUBLIC LAND MOBILE (22).	
				152.255–152.495 LAND MOBILE. NG124	PRIVATE LAND MOBILE (90).	
				152.495–152.855 LAND MOBILE. NG4	DOMESTIC PUBLIC LAND MOBILE (22).	
				152.855–156.2475 LAND MOBILE.	PRIVATE LAND MOBILE (90). AUXILIARY BROADCASTING (74) MARITIME (80).	
153–154 FIXED. MOBILE except aeronautical mobile (R).						

Meteorological Aids. 154.0–156.7625 FIXED. MOBILE except aeronautical mobile (R). 613 613A				613 613A NG4 NG112 NG117 NG124 NG148		
156.7625–156.8375	MARITIME MOBILE (distress and calling). 501 613		156.2475–157.0375 613 613A US77 US106 US107 US266	156.2475–157.0375 MARITIME MOBILE. NG117 613 613A US77 US106 US107 US266		
156.8375–174 FIXED. MOBILE except aeronautical mobile. 613 613B 614 615	156.8375–174 FIXED. MOBILE. 613 616 617 618		157.0375–157.1875 MARITIME MOBILE. 613 US214 US266 G109	157.0375–157.1875 613 US214 US266	Private Land Mobile (90).	
			157.1875–157.45 613 US223 US266	157.1875–157.45 MARITIME MOBILE. 613 US223 US266 NG111	MARITIME (80).	
			157.45–161.575	157.45–157.755 LAND MOBILE. 613 US266 NG111 NG124	PRIVATE LAND MOBILE (90).	
				157.755–158.115 LAND MOBILE. 613	DOMESTIC PUBLIC LAND MOBILE (22).	
				158.115–161.575		

TABLE 15.1 Table of Frequency Allocations (*Continued*)

	International Table		U.S. Table		FCC Use Designators	
			Government	Nongovernment		
Region 1—allocation, MHz (1)	Region 2—allocation, MHz (2)	Region 3—allocation, MHz (3)	Allocation, MHz (4)	Allocation, MHz (5)	Rule part(s) (6)	Special-use frequencies (7)
			613 US266	LAND MOBILE NG6 613 NG6 NG28 NG70 NG112 NG124 NG148	DOMESTIC PUBLIC LAND MOBILE (22). PRIVATE LAND MOBILE (90) MARITIME (80).	
			161.575–161.625 613 US77	161.575–161.625 MARITIME MOBILE. 613 US77 NG6 NG17	DOMESTIC PUBLIC LAND MOBILE (22). MARITIME (80)	
			161.625–161.775 613	161.625–161.775 LAND MOBILE. 613 NG6	AUXILIARY BROADCASTING (74). DOMESTIC PUBLIC LAND MOBILE (22).	
			161.775–162.0125 613 US266	161.775–162.0125 MARITIME MOBILE. 613 US266 NG6	DOMESTIC PUBLIC LAND MOBILE (22). MARITIME (80).	
			162.0125–173.2 FIXED. MOBILE. 613 US8 US11 US13 US216 US223 US300 US312 G5	162.0125–173.2 613 US8 US11 US13 US216 US223 US300 US312	Auxillary Broadcasting (74). Private Land Mobile (90).	

			173.2–173.4	173.2–173.4 FIXED. Land Mobile. NG124	Private Land Mobile (90).	
			173.4–174.0 FIXED MOBILE. G5	173.4–174.0		
174–223 BROADCASTING.	174–216 BROADCASTING. Fixed. Mobile. 620	174–233 FIXED. MOBILE. BROADCASTING.	174–216	174–216 BROADCASTING. NG115 NG128 NG149	RADIO BROADCAST (TV) (73). Auxiliary Broadcasting (74).	
	216–220–FIXED. MARITIME MOBILE. Radiolocation 627, 627A,		216–220 MARITIME MOBILE. Aeronautical-Mobile. Fixed. Land Mobile, Radiolocation 627, US210 US229 US274 US317 G2.	216–220 MARITIME MOBILE. Aeronautical Mobile. Fixed. Land Mobile, 627, US210 US229 US274, US317 NG121.	MARITIME (80). Private Land Mobile (90). Personal Radio Service (95).	
	220–225 AMATEUR. FIXED. MOBILE. Radiolocation 627		220–222 Land mobile Radiolocation: 627 US243, G2	220–222 Land mobile 627, US243	Private land mobile (90).	
223–230 BROADCASTING. Fixed. Mobile. 622 628 629 631 632 633 634 635.		223–230 FIXED. 636 637	222–225 Radiolocation: 627 US243, G2	222–225 Amateur 627, US243	Amateur (97)	
	225–235 FIXED. MOBILE.		225.0–328.6 FIXED. MOBILE. 501 592 642 644	225.0–328.6 501 592 642 644.		
230–235 FIXED.		230–235 FIXED.				

TABLE 15.1 Table of Frequency Allocations (*Continued*)

International Table			U.S. Table		FCC Use Designators	
			Government	Nongovernment		
Region 1—allocation, MHz (1)	Region 2—allocation, MHz (2)	Region 3—allocation, MHz (3)	Allocation, MHz (4)	Allocation, MHz (5)	Rule part(s) (6)	Special-use frequencies (7)
MOBILE. 629 632 633 634 635 638 639.		MOBILE. AERONAUTICAL RADIONAVIGATION. 637	G27 G100			
235–267	FIXED. MOBILE. 501 592 635 640 641 642					
267–272	FIXED. MOBILE. Space Operation (space-to-Earth). 641 643					
272–273	SPACE OPERATION (space-to-Earth). FIXED. MOBILE. 641					
273–322	FIXED. MOBILE. 641					
322.0–328.6	FIXED. MOBILE. RADIO ASTRONOMY. 644					
328.6–335.4	AERONAUTICAL RADIONAVIGATION. 645 645A		328.6–335.4 AERONAUTICAL RADIONAVIGATION. 645	328.6–335.4 AERONAUTICAL RADIONAVIGATION. 645		
335.4–399.9	FIXED. MOBILE. 641		335.4–399.9 FIXED. MOBILE. G27 G100	335.4–399.9		

399.9–400.05	RADIONAVIGATION. SATELLITE. 609 645B		399.9–400.05 RADIONAVIGATION. SATELLITE. 645B	399.9–400.05 RADIONAVIGATION. SATELLITE. 645B		
400.05–400.15	STANDARD FREQUENCY AND TIME SIGNAL-SATELLITE. (400.1 MHz) 646 647		400.05–400.15 STANDARD FREQUENCY AND TIME SIGNAL-SATELLITE. 646	400.05–400.15 STANDARD FREQUENCY AND TIME SIGNAL-SATELLITE. 646		400.1 MHz: Standard frequency.
400.15–401.0	METEOROLOGICAL AIDS. METEOROLOGICAL- SATELLITE (space- to-Earth). SPACE RESEARCH (space-to-Earth). Space Operation (space-to-Earth). 647		400.15–401.0 METEOROLOGICAL AIDS (radiosonde). METEOROLOGICAL- SATELLITE (space- to-Earth) SPACE RESEARCH (space-to-Earth). Space Operation (space-to-Earth). US70	400.15–401.0 METEOROLOGICAL AIDS (radiosonde). SPACE RESEARCH (space-to-Earth). Space Operation (space-to-Earth). US70	SATELLITE COMMUNICATION (25).	
401–402	METEOROLOGICAL AIDS. SPACE OPERATION (space-to-Earth). Earth Exploration- Satellite (Earth-to-space). Fixed. Meteorological-Satellite (Earth-to-space). Mobile except aeronautical mobile.		401–402 METEOROLOGICAL AIDS (radiosonde). SPACE OPERATION (space-to-Earth). Earth Exploration- Satellite (Earth-to-space). Meteorological-Satellite (Earth-to-space). US70	401–402 METEOROLOGICAL AIDS (radiosonde). SPACE OPERATION (space-to-Earth). Earth Exploration- Satellite (Earth-to-space). Meteorological-Satellite (Earth-to-space). US70	SATELLITE COMMUNICATIONS (25).	
402–403	METEOROLOGICAL AIDS. Earth Exploration- Satellite (Earth-to-space). Fixed. Meteorological-Satellite		402–403 METEOROLOGICAL AIDS (radiosonde) Earth Exploration- Satellite (Earth-to-space). Meteorological-Satellite (Earth-to-space).	402–403 METEOROLOGICAL AIDS (radiosonde). Earth Exploration- Satellite (Earth-to-space). Meteorological-Satellite (Earth-to-space).		

TABLE 15.1 Table of Frequency Allocations (*Continued*)

Region 1—allocation, MHz (1)	International Table: Region 2—allocation, MHz (2)	Region 3—allocation, MHz (3)	U.S. Table: Government: Allocation, MHz (4)	U.S. Table: Nongovernment: Allocation, MHz (5)	FCC Use Designators: Rule part(s) (6)	FCC Use Designators: Special-use frequencies (7)
	(Earth-to-space). Mobile except aeronautical mobile.		US70	US70		
403–406	METEOROLOGICAL AIDS. Fixed. Mobile except aeronautical mobile. 648		403–406 METEOROLOGICAL AIDS (radiosonde). US70 G6	403–406 METEOROLOGICAL AIDS (radiosonde). US70		
406.0–406.1	MOBILE-SATELLITE (Earth-to-space). 649 649A		406.0–406.1 MOBILE-SATELLITE (Earth-to-space). 649 649A	406.0–406.1 MOBILE-SATELLITE (Earth-to-space). 649 649A		
406.1–410.0	FIXED. MOBILE except aeronautical mobile. RADIO ASTRONOMY. 648 650		406.1–410.0 FIXED. MOBILE. RADIO ASTRONOMY. US13 US74 US117 G5 G6	406.1–410.0 RADIO ASTRONOMY. US13 US74 US117		
410–420	FIXED. MOBILE except aeronautical mobile.		410–420 FIXED. MOBILE. US13 G5	410–420 US13		
420–430	FIXED. MOBILE except aeronautical mobile. Radiolocation. 651 652 653		420–450 RADIOLOCATION. 664 668 US7 US87 US217 US228 US230 G2 G8	420–450 Amateur. 664 668 US7 US87 US217 US228 US230 NG135	LAND MOBILE (90). Amateur (97).	
430–440	430–440					

AMATEUR. RADIOLOCATION. 653 654 655 656 657 658 659 661 662 663 664 665	RADIOLOCATION. Amateur. 653 658 659 660 660A 663 664					
440–450	FIXED. MOBILE except aeronautical mobile. Radiolocation. 651 652 653 666 667 668					
450–460	FIXED. MOBILE. 653 668 669 670		450–460 668 669 670 US87	450–451 LAND MOBILE. 668 US87	AUXILIARY BROADCASTING (74). SATELLITE COMMUNICATION (25).	
				451–454 LAND MOBILE. NG112 NG124	PRIVATE LAND MOBILE (90).	
				454–455 LAND MOBILE. NG12 NG112 NG148	DOMESTIC PUBLIC LAND MOBILE (22). MARITIME (80).	
				455–456 LAND MOBILE.	AUXILIARY BROADCASTING (74).	
				456–459 LAND MOBILE. 669 670 NG112 NG124	PRIVATE LAND MOBILE (90).	
				459–460 LAND MOBILE. NG12 NG112 NG148	DOMESTIC PUBLIC LAND MOBILE (22). MARITIME (80)	

TABLE 15.1 Table of Frequency Allocations (*Continued*)

International Table			U.S. Table		FCC Use Designators	
			Government	Nongovernment		
Region 1—allocation, MHz (1)	Region 2—allocation, MHz (2)	Region 3—allocation, MHz (3)	Allocation, MHz (4)	Allocation, MHz (5)	Rule part(s) (6)	Special-use frequencies (7)
460–470	FIXED. MOBILE. Meterological-Satellite (space-to-Earth). 669 670 671 672		460–470 Meterological Satellite (space-to-Earth). 669 670 671 US201 US209 US216	460–462.5375 LAND MOBILE. 671 US201 US209 NG124	PRIVATE LAND MOBILE (90).	
				462.5375–462.7375 LAND MOBILE. 671 US201	PERSONAL (95).	
				462.7375–467.5375 LAND MOBILE. 669 671 US201 US209 US216 NG124	PRIVATE LAND MOBILE (90).	
				467.5375–467.7375 LAND MOBILE. 669 671 US201	PERSONAL (95).	
				467.7375–470.0 LAND MOBILE. 669 670 671 US201 US216 NG124	PRIVATE LAND MOBILE (90).	
470–790 BROADCASTING.	470–512 BROADCASTING. Fixed. Mobile.	470–585 FIXED. MOBILE. BROADCASTING.	470–512	470–512 BROADCASTING. LAND MOBILE.	RADIO BROADCAST (TV) (73). DOMESTIC PUBLIC LAND MOBILE (22). PRIVATE LAND MOBILE (90).	

	674 675			NG66 NG114 NG127 NG128 NG149	Auxiliary Broadcasting (74).	
667A 676 682 683 684 685 686 686A 687 689 693 694	512–608 BROADCASTING. 678	673 677 679 585–610 FIXED. MOBILE. BROADCASTING. RADIONAVIGATION. 688 689 690	512–608	512–608 BROADCASTING. NG128 NG149	RADIO BROADCAST (TV) (73). Auxiliary Broadcasting (74).	
	608–614 RADIO ASTRONOMY. Mobile-Satellite except aeronautical mobile-satellite (Earth-to-space).	610–890 FIXED. MOBILE. BROADCASTING.	608–614 RADIO ASTRONOMY. US74 US246	608–614 RADIO ASTRONOMY. US74 US246		
790–862 FIXED. BROADCASTING. 694 695 695A 696 697 700B 702	614–806 BROADCASTING. Fixed. Mobile. 675 692 692A 693		614–806	614–806 BROADCASTING NG30 NG43· NG63 NG149	RADIO BROADCAST (TV) (73). Auxiliary Broadcasting (74).	
862–890 FIXED. MOBILE except aeronautical mobile. BROADCASTING 700B 703.	806–890 FIXED. MOBILE. BROADCASTING.		806–902	806–821 LAND MOBILE. NG30 NG43 NG63 NG31	PRIVATE LAND MOBILE (90).	
				821–824		

TABLE 15.1 Table of Frequency Allocations (*Continued*)

	International Table		U.S. Table		FCC Use Designators	
			Government	Nongovernment		
Region 1—allocation, MHz (1)	Region 2—allocation, MHz (2)	Region 3—allocation, MHz (3)	Allocation, MHz (4)	Allocation, MHz (5)	Rule part(s) (6)	Special-use frequencies (7)
704	692A 700 700A	677 688 689 690 691 693 701	US116 US268 G2	LAND MOBILE. NG30 NG43 NG63	PRIVATE LAND MOBILE (90)	
				824–849 LAND MOBILE. NG30 NG43 NG63 NG151	DOMESTIC PUBLIC LAND MOBILE (22).	
				849–851 AERONAUTICAL MOBILE. NG30 NG63 NG153	PUBLIC MOBILE (22).	
				851–866 LAND MOBILE. NG30 NG63 NG31	PRIVATE LAND MOBILE(90)	
				866–869 LAND MOBILE. NG30 NG63	PRIVATE LAND MOBILE (90)	
				869–894 LAND MOBILE. NG30 NG63 US116 US268 NG151	DOMESTIC PUBLIC LAND MOBILE (22).	
890–942 FIXED. MOBILE except aeronautical mobile. BROADCASTING 703. Radiolocation. 704	890–902 FIXED. MOBILE except aeronautical mobile. Radiolocation. 700A 704A 705	890–942 FIXED. MOBILE. BROADCASTING. Radiolocation. 706		894–896 AERONAUTICAL MOBILE. US116 US268 NG153	PUBLIC MOBILE (22).	
				896–901 LAND MOBILE US116 US268	PRIVATE LAND MOBILE (90)	

				901–902 FIXED MOBILE US116 US268 US330	PERSONAL COMMUNICATIONS SERVICES (99)	
	902–928 FIXED. Amateur. Mobile except aeronautical mobile. Radiolocation. 705 707 707A		902–928 RADIOLOCATION. 707 US215 US218 US267 US275 G11 G59	902–928 707 US215 US21 US267 US275	Amateur (97).	915 ± 13 MHz: Industrial, scientific and medical frequency.
	928–942 FIXED. MOBILE except aeronautical mobile. Radiolocation. 705		928–932 US116 US215 US268 G2	928–929 FIXED. US116 US215 US268	DOMESTIC PUBLIC LAND MOBILE (22). PRIVATE LAND MOBILE (90). PRIVATE OPERATIONAL FIXED MICROWAVE (94).	
				929–930 LAND MOBILE US116 US215 US268	DOMESTIC PUBLIC LAND MOBILE (22). PRIVATE LAND MOBILE (90).	
				930–931 FIXED. MOBILE. US116 US215 US268 US330	PERSONAL COMMUNICATIONS SERVICES (99)	
				931–932 LAND MOBILE. US116 US215 US268	DOMESTIC PUBLIC LAND MOBILE (22). PRIVATE LAND MOBILE (90).	

TABLE 15.1 Table of Frequency Allocations (*Continued*)

International Table			U.S. Table		FCC Use Designators	
			Government	Nongovernment		
Region 1—allocation, MHz (1)	Region 2—allocation, MHz (2)	Region 3—allocation, MHz (3)	Allocation, MHz (4)	Allocation, MHz (5)	Rule part(s) (6)	Special-use frequencies (7)
			932–935 FIXED US215 US268 G2	932–935 FIXED US215 US268		
942–960 FIXED. MOBILE except aeronautical mobile. BROADCASTING 703	942–960 FIXED. MOBILE.	942–960 FIXED. MOBILE. BROADCASTING.	935–941 US116 US215 US268 G2	935–940 LAND MOBILE US116 US215 US268	PRIVATE LAND MOBILE (90)	
				940–941 FIXED MOBILE US116 US268 US330	PERSONAL COMMUNICATIONS SERVICES (99)	
			941–944 FIXED US268 US301 US302	941–944 FIXED US268 US301 US302		
704	708	701	944–960	944–960 FIXED NG120	AUXILIARY BROADCASTING (74). DOMESTIC PUBLIC FIXED (22). INTERNATIONAL FIXED PUBLIC (23). PRIVATE OPERATIONAL FIXED MICROWAVE (94).	
960–1215	AERONAUTICAL RADIONAVIGATION. 709		960–1215 AERONAUTICAL RADIONAVIGATION. 709 US224	960–1215 AERONAUTICAL RADIONAVIGATION. 709 US224	AVIATION (87).	

1215–1240	RADIOLOCATION. RADIONAVIGATION-SATELLITE (space-to-Earth) 710. 711 712 712A 713		1215–1240 RADIOLOCATION. RADIONAVIGATION-SATELLITE (space-to-Earth). 713 G56	1215–1240 713		
1240–1260	RADIOLOCATION. RADIONAVIGATION-SATELLITE (space-to-Earth) 710. Amateur. 711 712 712A 713 714		1240–1300 RADIOLOCATION. 664 713 714 G56	1240–1300 Amateur. 664 713 714	Amateur (97).	
1260–1300	RADIOLOCATION. Amateur. 664 711 712 712A 713 714					
1300–1350	AERONAUTICAL RADIONAVIGATION 717 Radiolocation 715 716 718		1300–1350 AERONAUTICAL RADIONAVIGATION 717 Radiolocation 718 G2	1300–1350 AERONAUTICAL RADIONAVIGATION 717 718	AVIATION (87).	
1350–1400 FIXED. MOBILE. RADIOLOCATION. 718 719 720	1350–1400 RADIOLOCATION. 714 718 720		1350–1400 RADIOLOCATION. Fixed. Mobile. 714 718 720 G2 G27 G114 US311	1350–1400 714 718 720		
1400–1427	EARTH EXPLORATION-SATELLITE (passive). RADIO ASTRONOMY. SPACE RESEARCH (passive). 721 722		1400–1427 EARTH EXPLORATION-SATELLITE (passive). RADIO ASTRONOMY. SPACE RESEARCH (passive). 722 US74 US246	1400–1427 EARTH EXPLORATION-SATELLITE (passive). RADIO ASTRONOMY. SPACE RESEARCH (passive). 722 US74 US246		
1427–1429	SPACE OPERATION (Earth-to-space).		1427–1429 SPACE OPERATION (Earth-to-space).	1427–1429 SPACE OPERATION (Earth-to-space).	Private Land Mobile (90).	

TABLE 15.1 Table of Frequency Allocations (*Continued*)

International Table			U.S. Table		FCC Use Designators	
			Government	Nongovernment		
Region 1—allocation, MHz (1)	Region 2—allocation, MHz (2)	Region 3—allocation, MHz (3)	Allocation, MHz (4)	Allocation, MHz (5)	Rule part(s) (6)	Special-use frequencies (7)
	FIXED. MOBILE except aero-nautical mobile. 722		FIXED. MOBILE except aero-nautical mobile. 722 G30	Fixed (telemetering). Land Mobile (telemetering and telecommand). 722	Satellite Communica-tions (25).	
1429–1525 FIXED. MOBILE except aeronauti-cal mobile. 722	1429–1525 FIXED. MOBILE 723 722		1429–1435 FIXED. MOBILE. 722 G30	1429–1435 Land Mobile (telemetering and telecommand). Fixed (telemetering) 722	Private Land Mobile (90).	
1525–1530 SPACE OPERATION (space-to-Earth). FIXED. MARITIME MOBILE-SATELLITE (space-to-Earth). Land Mobile-Satellite (space-to-Earth) 726B. Earth Exploration-Satellite Mobile except aeronautical mobile 724. 722 723B 725 726A 726D	1525–1530 SPACE OPERATION (space-to-Earth). MOBILE-SATELLITE (space-to-Earth). Earth Exploration-satellite. Fixed. Mobile 723. 722 723A 726A 726D	1525–1530 SPACE OPERATION (space-to-Earth). FIXED. MOBILE-SATELLITE (space-to-Earth). Earth Exploration-satellite. Mobile 723 724. 722 726A 726D	1435–1530 MOBILE (aeronautical telemetering) 722 US78	1435–1530 MOBILE (aeronautical telemetering). 722 US78	AVIATION (87).	
1530–1533 SPACE OPERATION (space-to-Earth). MARITIME MOBILE-SATELLITE (space-to-Earth). LAND MOBILE-SATELLITE (space-to-Earth).	1530–1533 SPACE OPERATION (space-to-Earth). MARITIME MOBILE-SATELLITE (space-to-Earth). LAND MOBILE-SATELLITE (space-to-Earth).	1530–1533 SPACE OPERATION (space-to-Earth). MARITIME MOBILE-SATELLITE (space-to-Earth). LAND MOBILE-SATELLITE (space-to-Earth).	1530–1533 MARITIME MOBILE-SATELLITE (space-to-Earth). MOBILE-SATELLITE (space-to-Earth). Mobile (aeronautical telemetering).	1530–1533 MARITIME MOBILE-SATELLITE (space-to-Earth). MOBILE-SATELLITE (space-to-Earth). Mobile (aeronautical telemetering).	SATELLITE COMMUNICATION (25). Aviation (87).	

Earth Exploration-Satellite. Fixed. Mobile except aeronautical mobile. 722 723B 726A 726D	Earth Exploration-Satellite. Fixed. Mobile 723. 722 726A 726C 726D	Earth Exploration-Satellite. Fixed. Mobile 723. 722 726A 726C 726D				
1533–1535 SPACE OPERATION (space-to-Earth). MARITIME MOBILE-SATELLITE (space-to-Earth). Earth Exploration-Satellite. Fixed. Mobile except aeronautical mobile. Land Mobile-Satellite (space-to-Earth) 726B. 722 723B 726A 726D	1533–1535 SPACE OPERATION (space-to-Earth). MARITIME MOBILE-SATELLITE (space-to-Earth). Earth Exploration-Satellite. Fixed. Mobile 723. Land Mobile-Satellite (space-to-Earth) 726B. 722 726A 726C 726D	1533–1535 SPACE OPERATION (space-to-Earth). MARITIME MOBILE-SATELLITE (space-to-Earth). Earth Exploration-Satellite. Fixed. Mobile 723. Land Mobile-Satellite (space-to-Earth) 726B. 722 726A 726C 726D	722 726A US78 US315	722 726A US78 US315		
1544–1545 MOBILE-SATELLITE (space-to-Earth). 722 726D 727 727A	1544–1545 MOBILE-SATELLITE (space-to-Earth). 722 726D 727 727A	1544–1545 MOBILE-SATELLITE (space-to-Earth). 722 726D 727 727A	1544–1545 MOBILE-SATELLITE (space-to-Earth). 722 727A	1544–1545 MOBILE-SATELLITE (space-to-Earth). 722 727A	MARITIME (80). SATELLITE COMMUNICATION (25).	
1545–1555 AERONAUTICAL MOBILE-SATELLITE (R) (space-to-Earth). 722 726A 726D 727 729 729A 730	1545–1555 AERONAUTICAL MOBILE-SATELLITE (R) (space-to-Earth). 722 726A 726D 727 729 729A 730	1545–1555 AERONAUTICAL MOBILE-SATELLITE (R) (space-to-Earth). 722 726A 726D 727 729 729A 730	1545–1549.5 AERONAUTICAL MOBILE-SATELLITE (R) (space-to-Earth). Mobile-satellite (space-to-Earth). 722 726A US308 US309	1545–1549.5 AERONAUTICAL MOBILE-SATELLITE (R) (space-to-Earth). Mobile-satellite (space-to-Earth). 722 726A US308 US309	AVIATION (87)	
			1549.5–1558.5 AERONAUTICAL MOBILE-SATELLITE (R) (space-to-Earth). MOBILE-SATELLITE (space-to-Earth).	1549.5–1558.5 AERONAUTICAL MOBILE-SATELLITE (R) (space-to-Earth). MOBILE-SATELLITE (space-to-Earth).	AVIATION (87).	

TABLE 15.1 Table of Frequency Allocations (*Continued*)

International Table			U.S. Table		FCC Use Designators	
			Government	Non-government		
Region 1—allocation, MHz (1)	Region 2—allocation, MHz (2)	Region 3—allocation, MHz (3)	Allocation, MHz (4)	Allocation, MHz (5)	Rule part(s) (6)	Special-use frequencies (7)
			722 726A US308 US309	722 726A US308 US309		
1555–1559 LAND MOBILE-SATELLITE (space-to-Earth). 722 726A 726D 727 730 730A 730B 730C	1555–1559 LAND MOBILE-SATELLITE (space-to-Earth). 722 726A 726D 727 730 730A 730B 730C	1555–1559 LAND MOBILE-SATELLITE (space-to-Earth). 722 726A 726D 727 730 730A 730B 730C	1558.5–1559 AERONAUTICAL MOBILE-SATELLITE (R) (space-to-Earth). 722 726A US308 US309	1558.5–1559 AERONAUTICAL MOBILE-SATELLITE (R) (space-to-Earth). 722 726A US308 US309	AVIATION (87).	
1559–1610 722 727 730	AERONAUTICAL RADIONAVIGATION. RADIONAVIGATION-SATELLITE (space-to-Earth). 722 727 730 731 731A 731B 731C 731D	722 US28	1559–1610 AERONAUTICAL RADIONAVIGATION-SATELLITE (space-to-Earth). 722 US208 US260 US280	1559–1610 AERONAUTICAL RADIONAVIGATION-SATELLITE (space-to-Earth)	AVIATION (87).	
1610–1626.5 AERONAUTICAL RADIO-NAVIGATION. 722 727 730 731 731A 731B 731D 732 733 733A 733B 733E 733F 734	1610–1626.5 AERONAUTICAL RADIONAVIGATION. Radiodetermination-satellite (Earth-to-space) 733A 733E 722 731B 731C 732 733 733C 733D 734	1610–1626.5 AERONAUTICAL RADIONAVIGATION. Radiodetermination-satellite (Earth-to-space) 733A 733E 722 727 730 731B 731C 732 733 733B 734	1610–1626.5 AERONAUTICAL RADIONAVIGATION. 722 732 733 734 US208 US260 US306	AERONAUTICAL RADIONAVIGATION. 722 732 733 734 US208 US260 US306	AVIATION (87). Satellite communication (25).	
1626.5–1631.5 MARITIME MOBILE-SATELLITE (Earth-to-space). Land Mobile-Satellite (Earth-to-space) 726B. 722 726A 726D 727 730	1625.5–1631.5 MOBILE-SATELLITE (Earth-to-space). 722 726A 726C 726D 727 730	1626.5–1631.5 MOBILE-SATELLITE (Earth-to-space). 722 726A 726C 726D 727 730	1626.5–1645.5 MARITIME MOBILE-SATELLITE (Earth-to-space). MOBILE-SATELLITE (Earth-to-space).	1626.5–1645.5 MARITIME MOBILE-SATELLITE (Earth-to-space). MOBILE-SATELLITE (Earth-to-space).	MARITIME (80). SATELLITE COMMUNICATION (25).	
1631.5–1634.5 MARITIME MOBILE-	1631.5–1634.5 MARITIME MOBILE-	1631.5–1634.5 MARITIME MOBILE-				

SATELLITE (Earth-to-space). LAND MOBILE-SATELLITE (Earth-to-space). 722 726A 726C 726D 727 730 734A	SATELLITE (Earth-to-space). LAND MOBILE-SATELLITE (Earth-to-space). 722 726A 726C 726D 727 730 734A	SATELLITE (Earth-to-space). LAND MOBILE-SATELLITE (Earth-to-space). 722 726A 726C 726D 727 730 734A				
1634.5–1645.5 MARITIME MOBILE-SATELLITE (Earth-to-space). Land Mobile-Satellite (Earth-to-space) 726B. 722 726A 726C 726D 727 730	1634.5–1645.5 MARITIME MOBILE-SATELLITE (Earth-to-space). Land Mobile-Satellite (Earth-to-space). 726B. 722 726A 726C 726D 727 730	1634.5–1645.5 MARITIME MOBILE-SATELLITE (Earth-to-space). Land Mobile-Satellite (Earth-to-space). 726B. 722 726A 726C 726D 727 730	722 726A US315	722 726A US315		
1645.5–1646.5 MOBILE-SATELLITE (Earth-to-space). 722 726D 734B	1645.5–1646.5 MOBILE-SATELLITE (Earth-to-space). 722 726D 734B	1645.5–1646.5 MOBILE-SATELLITE (Earth-to-space). 722 726D 734B	1645.5–1646.5 MOBILE-SATELLITE (Earth-to-space). 722 734B	1645.5–1646.5 MOBILE-SATELLITE (Earth-to-space). 722 734B	MARITIME (80). SATELLITE COMMUNICATION (25).	
1646.5–1656.5 AERONAUTICAL MOBILE-SATELLITE (R) (Earth-to-space). 722 726A 726D 727 729A 730 735	1646.5–1656.5 AERONAUTICAL MOBILE-SATELLITE (R) (Earth-to-space). 722 726A 726D 727 729A 730 735	1646.5–1656.5 AERONAUTICAL MOBILE-SATELLITE (R) (Earth-to-space). 722 726A 726D 727 729A 730 735	1646.5–1651 AERONAUTICAL MOBILE-SATELLITE (R) (Earth-to-space). Mobile-Satellite (Earth-to-space).	1646.5–1651 AERONAUTICAL MOBILE-SATELLITE (R) (Earth-to-space). Mobile-Satellite (Earth-to-space). 722 726A US308 US309	AVIATION (87). 722 726A US308 US309	
1656.5–1660 LAND MOBILE-SATELLITE (Earth-to-space). 722 726A 726D 727 730 730A 730B 730C 734A	1656.5–1660 LAND MOBILE-SATELLITE (Earth-to-space). 722 726A 726D 727 730 730A 730B 730C 734A	1656.5–1660 LAND MOBILE-SATELLITE (Earth-to-space). 722 726A 726D 727 730 730A 730B 730C 734A	1651–1660 AERONAUTICAL MOBILE-SATELLITE (R) (Earth-to-space). MOBILE-SATELLITE (Earth-to-space). 722 726A US308 US39	1651–1660 AERONAUTICAL MOBILE-SATELLITE (R) (Earth-to-space). MOBILE-SATELLITE (Earth-to-space). 722 726A US308 US39	AVIATION (87).	
1660–1660.5 RADIO ASTRONOMY LAND MOBILE-SATELLITE (Earth-to-space).	1660–1660.5 RADIO ASTRONOMY LAND MOBILE-SATELLITE (Earth-to-space).	1660–1660.5 RADIO ASTRONOMY LAND MOBILE-SATELLITE (Earth-to-space).	1660–1660.5 AERONAUTICAL MOBILE-SATELLITE (R) (Earth-to-space). RADIO ASTRONOMY.	1660–1660.5 AERONAUTICAL MOBILE-SATELLITE (R) (Earth-to-space). RADIO ASTRONOMY.	AVIATION (87).	

TABLE 15.1 Table of Frequency Allocations (*Continued*)

International Table			U.S. Table		FCC Use Designators	
			Government	Nongovernment		
Region 1—allocation, MHz (1)	Region 2—allocation, MHz (2)	Region 3—allocation, MHz (3)	Allocation, MHz (4)	Allocation, MHz (5)	Rule part(s) (6)	Special-use frequencies (7)
722 726A 726D 730A 730B 730C 736	722 726A 726D 730A 730B 730C 736	722 726A 726D 730A 730B 730C 736	722 726A 736 US309	722 726A 736 US309		
1660.5–1668.4	RADIO ASTRONOMY SPACE RESEARCH (passive). Fixed. Mobile except aeronautical mobile. 722 736 737 738 739		1660.5–1668.4 RADIO ASTRONOMY SPACE RESEARCH (passive). 722 US74 US246	1660.5–1668.4 RADIO ASTRONOMY SPACE RESEARCH (passive). 722 US74 US246		
1668.4–1670.0	METEOROLOGICAL AIDS. FIXED. MOBILE except aeronautical mobile. RADIO ASTRONOMY 722 736		1668.4–1670.0 METEOROLOGICAL AIDS (radiosonde). RADIO ASTRONOMY. 722 736 US74 US99	1668.4–1670.0 METEOROLOGICAL AIDS (radiosonde). RADIO ASTRONOMY. 722 736 US74 US99		
1670–1690	METEOROLOGICAL AIDS. FIXED. METEOROLOGICAL-SATELLITE (space-to-Earth). MOBILE except aeronautical mobile. 722		1670–1690 METEOROLOGICAL AIDS (radiosonde). METEOROLOGICAL-SATELLITE (space-to-Earth). 722 US211	1670–1690 METEOROLOGICAL AIDS (radiosonde). METEOROLOGICAL-SATELLITE (space-to-Earth). 722 US211		
1690–1700 METEOROLOGICAL AIDS METEOROLOGICAL-SATELLITE (space-to-	1690–1700 METEOROLOGICAL AIDS. METEOROLOGICAL-		1690–1700 METEOROLOGICAL-AIDS (radiosonde). METEOROLOGICAL-	1690–1700 METEOROLOGICAL-AIDS (radiosonde). METEOROLOGICAL-		

Earth). Mobile except aeronautical mobile. 671 722 742	SATELLITE (space-to-Earth). 671 722 740 742		SATELLITE (space-to-Earth). 671 722	SATELLITE (space-to-Earth). 671 722		
1700–1710 FIXED. METEOROLOGICAL-SATELLITE (space-to-Earth). Mobile except aeronautical mobile. 671 722 743A	1700–1710 FIXED. METEOROLOGICAL SATELLITE (space-to-Earth). MOBILE except aeronautical mobile. 671 722 743		1700–1710 FIXED. METEOROLOGICAL-SATELLITE (space-to-Earth). 671 722 G118	1700–1710 METEOROLOGICAL-SATELLITE (space-to-Earth). Fixed. 671 722		
1710–2290 FIXED. Mobile. 722 743A 744 746 747 748 750	1710–2290 FIXED. MOBILE. 722 744 745 746 747 748 749 750		1710–1850 FIXED. MOBILE. 722 US256 G42	1710–1850 722 US256		
			1850–1990	1850–1990 FIXED. NG153	1850–1990 PRIVATE OPERATIONAL-FIXED MICROWAVE (94).	EMERGING TECHNOLOGIES
			1990–2110 US90 US111 US219 US222	1990–2100 Fixed MOBILE. US90 US111 US219 US222 NG23 NG118	AUXILIARY BROAD-CAST (74) CABLE TELEVISION (78) DOMESTIC PUBLIC FIXED (21)	
			2110–2200 US111 US252			
			2200–2290 FIXED. MOBILE. SPACE RESEARCH (space-to-Earth) (space-to-Earth) US303 G101	2100–2290 US303		

TABLE 15.1 Table of Frequency Allocations (*Continued*)

	International Table		U.S. Table		FCC Use Designators	
			Government	Nongovernment		
Region 1—allocation, MHz (1)	Region 2—allocation, MHz (2)	Region 3—allocation, MHz (3)	Allocation, MHz (4)	Allocation, MHz (5)	Rule part(s) (6)	Special-use frequencies (7)
2290–2300 FIXED. SPACE RESEARCH (space-to-Earth) (deep space). MOBILE except aeronautical mobile. 743A	2290–2300 FIXED. MOBILE except aeronautical mobile. SPACE RESEARCH (space-to-Earth) (deep space).		2290–2300 FIXED. MOBILE except aeronautical mobile. SPACE RESEARCH (space-to-Earth) (deep space only).	2290–2300 SPACE RESEARCH (space-to-Earth) (deep space only).		
2300–2450 FIXED. Amateur. Mobile. Radiolocation.	2300–2450 FIXED. MOBILE. RADIOLOCATION. Amateur.		2300–2310 RADIOLOCATION Fixed. Mobile. US253 G2	2300–2310 Amateur. US253	Amateur (97).	
			2310–2390 MOBILE. RADIOLOCATION. Fixed. US276 G2	2310–2390 MOBILE US276		
664 743A 752	664 751 752		2390–2450 RADIOLOCATION. 664 752 G2	2390–2450 Amateur. 664 752	Amateur (97).	
2450–2483.5 FIXED. MOBILE. Radiolocation. 752 753	2450–2483.5 FIXED. MOBILE. RADIOLOCATION. 752		2450–2483.5 752 US41	2450–2483.5 FIXED. MOBILE. Radiolocation. 752 US41		2450 ± 50 MHz: Industrial, scientific and medical frequency.
2483.5–2500 FIXED MOBILE. Radiolocation. 733F 752 753A 753B 753C 763E	2483.5–2500 FIXED-MOBILE RADIODETERMINATION SATELLITE. (space-to-Earth). Radiolocation. 752 753D	2483.5–2500 FIXED-MOBILE RADIOLOCATION RADIODETERMINATION SATELLITE (space-to-earth) 752 753A 753C	2483.5–2500 752 US41	2483.5–2500 RADIODETERMINATION SATELLITE (space-to-Earth). 752 US41 NG147	Satellite communication (25).	

2500–2655 FIXED 762 763 764 MOBILE except aeronautical mobile. BROADCASTING-SATELLITE 757 760 720 753 756 758 759	2500–2655 FIXED 762 764 FIXED-SATELLITE (space-to-Earth) 761 MOBILE except aeronautical mobile. BROADCASTING-SATELLITE 757 760 720 755	2500–2535 FIXED 762 764 FIXED SATELLITE (space-to-Earth) 761 MOBILE except aeronautical mobile. BROADCASTING-SATELLITE 757 760 754 754A 2536–2655 FIXED 762 764 MOBILE except aeronautical mobile. BROADCASTING-SATELLITE 757 760 720	2500–2655 720 US205 US269	2500–2655 FIXED. BROADCASTING-SATELLITE. 720 US205 US269 NG101 NG102	AUXILIARY BROADCASTING (74) DOMESTIC PUBLIC FIXED RADIO (21)	
2655–2690 FIXED 762 763 764 MOBILE except aeronautical mobile. BROADCASTING-SATELLITE 757 760 Earth Exploration-Satellite (passive). Radio Astronomy. Space Research passive). 758 759 765	2655–2690 FIXED 762 764 FIXED-SATELLITE (Earth-to-space) (space-to-Earth) 761 MOBILE except aeronautical mobile. BROADCASTING-SATELLITE 757 760 Earth Exploration-Satellite (passive). Radio Astronomy. Space Research (passive). 765	2655–2690 FIXED 762 764 FIXED-SATELLITE (Earth-to-space) 761 MOBILE except aeronautical mobile. BROADCASTING-SATELLITE 757 760 Earth Exploration-Satellite (passive). Radio Astronomy. Space Research passive). 765 766	2655–2690 Earth Exploration-Satellite (passive) Radio Astronomy Space Research passive). US205 US269	2655–2690 FIXED. BROADCASTING-SATELLITE. Earth Exploration-Satellite (passive). Radio Astronomy. Space Research (passive) US205 US269 NG47 NG101 NG102	AUXILIARY BROADCASTING (74). PRIVATE OPERATIONAL-FIXED MICROWAVE (94).	
2690–2700	EARTH EXPLORATION-SATELLITE (passive). RADIO ASTRONOMY. SPACE RESEARCH (passive). 767 768 769		2690–2700 EARTH EXPLORATION-SATELLITE(passive). RADIO ASTRONOMY. SPACE RESEARCH (passive). US74 US246	2690–2700 EARTH EXPLORATION SATELLITE (passive). RADIO ASTRONOMY. SPACE RESEARCH (passive). US74 US246		
2700–2900			2700–2900	2700–2900		

TABLE 15.1 Table of Frequency Allocations (*Continued*)

International Table			U.S. Table		FCC Use Designators	
			Government	Nongovernment		
Region 1—allocation, MHz (1)	Region 2—allocation, MHz (2)	Region 3—allocation, MHz (3)	Allocation, MHz (4)	Allocation, MHz (5)	Rule part(s) (6)	Special-use frequencies (7)
	AERONAUTICAL RADIONAVIGATION 717. Radiolocation. 770 771		AERONAUTICAL RADIONAVIGATION 717 METEOROLOGICAL AIDS. Radiolocation. 770 US18 G2 G15	717 770 US18		
2900–3100	RADIONAVIGATION. Radiolocation. 772 773 775A		2900–3100 MARITIME RADIO-NAVIGATION 775A Radiolocation. US44 US316 G56	2900–3100 MARITIME RADIO-NAVIGATION 775A Radiolocation. US44 US316	MARITIME (80).	
3100–3300	RADIOLOCATION. 713 777 778		3100–3300 RADIOLOCATION. 713 778 US110 G59	3100–3300 Radiolocation. 713 778 US110		
3300–3400 RADIOLOCATION 778 779 780	3300–3400 RADIOLOCATION. Amateur. Fixed. Mobile. 778 780	3300–3400 RADIOLOCATION. Amateur. 778 779	3300–3500 RADIOLOCATION. 664 778 US108 G31	3300–3500 Amateur. Radiolocation. 664 778 US108	Amateur (97).	
3400–3600 FIXED. FIXED-SATELLITE (space-to-Earth). Mobile. Radiolocation.	3400–3500 FIXED. FIXED-SATELLITE (space-to-Earth). Amateur. Mobile. Radiolocation 784 664 783					
	3500–3700 FIXED. FIXED-SATELLITE (space-to-Earth) MOBILE except		3500–3600 AERONAUTICAL RADIONAVIGATION (ground-based). RADIOLOCATION	3500–3600 Radiolocation.		

781 782 785	aeronautical mobile Radiolocation 784		US110 G59 G110	US110		
3600–4200 FIXED. FIXED-SATELLITE (space-to-Earth). Mobile.	786		3600–3700 AERONAUTICAL RADIONAVIGATION (ground-based). RADIOLOCATION. US110 US245 G59 G110	3600–3700 FIXED-SATELLITE (space-to-Earth). Radiolocation. US110 US245		
	3700–4200 FIXED. FIXED-SATELLITE (space-to-Earth). MOBILE except aeronautical mobile. 787		3700–4200	3700–4200 FIXED. FIXED-SATELLITE (space-to-Earth). NG41	DOMESTIC PUBLIC FIXED (21). SATELLITE COMMUNICATIONS (25) PRIVATE OPERATIONAL FIXED MICROWAVE (94).	
4200–4400	AERONAUTICAL RADIONAVIGATION 789 788 790 791		4200–4400 AERONAUTICAL RADIONAVIGATION. 791 US261	4200–4400 AERONAUTICAL RADIONAVIGATION. 791 US261	AVIATION (87).	
4400–4500	FIXED. MOBILE.		4400–4500 FIXED. MOBILE.	4400–4500		
4500–4800	FIXED. FIXED-SATELLITE (space-to-Earth). MOBILE. 792A		4500–4800 FIXED. MOBILE. US245	4500–4800 FIXED-SATELLITE (space-to-Earth). 792A US245		
4800–4990	FIXED. MOBILE 793 Radio Astronomy. 720 778 794		4800–4990 FIXED. MOBILE. 720 778 US203 US257	4800–4990 720 778 US203 US257		
4990–5000	FIXED. MOBILE except		4990–5000 RADIO ASTRONOMY. Space Research	4990–5000 RADIO ASTRONOMY. Space Research		

TABLE 15.1 Table of Frequency Allocations (*Continued*)

	International Table		U.S. Table		FCC Use Designators	
			Government	Nongovernment		
Region 1—allocation, MHz (1)	Region 2—allocation, MHz (2)	Region 3—allocation, MHz (3)	Allocation, MHz (4)	Allocation, MHz (5)	Rule part(s) (6)	Special-use frequencies (7)
	aeronautical mobile. RADIO ASTRONOMY. Space Research (passive). 795		(passive). US74 US246	(passive). US74 US246		
5000–5250	AERONAUTICAL RADIONAVIGATION. 733 796 797 797A 797B		5000–5250 AERONAUTICAL RADIONAVIGATION. 733 796 797 797A US211 US260 US307	5000–5250 AERONAUTICAL RADIONAVIGATION. 733 796 797 797A US211 US260 US307	AVIATION (87). Satellite communication (25).	
5250–5255	RADIOLOCATION. Space Research. 713 798		5250–5350 RADIOLOCATION. 713 US110 G59	5250–5350 Radiolocation. 713 US110		
5255–5350	RADIOLOCATION. 713 798					
5350–5460	AERONAUTICAL RADIONAVIGATION 799 Radiolocation.		5350–5460 AERONAUTICAL RADIONAVIGATION 799 RADIOLOCATION. US48 G56	5350–5460 AERONAUTICAL RADIONAVIGATION 799 Radiolocation. US48	AVIATION (87).	
5460–5470	RADIONAVIGATION 799 Radiolocation.		5460–5470 RADIONAVIGATION 799 Radiolocation. US49 US65 G56	5460–5470 RADIONAVIGATION 799 Radiolocation. US49 US65		
5470–5650	MARITIME RADIO-NAVIGATION Radiolocation.		5470–5650 MARITIME RADIO-NAVIGATION. Radiolocation. US50 US65 G56	5470–5650 MARITIME RADIO-NAVIGATION Radiolocation. US50 US65	MARITIME (80).	
			5600–5650 MARITIME RADIO-NAVIGATION.	5600–5650 MARITIME RADIO-NAVIGATION.	MARITIME (80).	

	800 801 802		METEOROLOGICAL AIDS. Radiolocation 772 802 US51 US65 G56	METEOROLOGICAL AIDS. Radiolocation. 772 802 US51 US65		
5650–5725	RADIOLOCATION. Amateur. Space Research (deep space). 664 801 803 804 805		5650–5850 RADIOLOCATION. 664 806 808 G2	5650–5850 Amateur. 644 806 808	Amateur (97). Industrial,Scientific, and Medical Equipment (18).	58000 ± 75 MHz Industrial, scientific and medical frequency.
5725–5850 FIXED-SATELLITE (Earth-to-space). RADIOLOCATION. Amateur. 801 803 805 806 807 808	5725–5850 RADIOLOCATION. Amateur. 803 805 806 808					
5850–5925 FIXED. FIXED-SATELLITE (Earth-to-space MOBILE. 806	5850–5925 FIXED. FIXED-SATELLITE (Earth-to-space) MOBILE. Amateur. Radiolocation. 806	5850–5925 FIXED. FIXED-SATELLITE (Earth-to-space) MOBILE. Radiolocation. 806	5850–5925 RADIOLOCATION. 806 US245 G2	5850–5925 FIXED-SATELLITE (Earth-to-space). Amateur. 806 US245	Amateur (97).	
5925–7075	FIXED. FIXED-SATELLITE (Earth-to-space). MOBILE. 791 792A 809		5925–7125 791 809	5925–6425 FIXED. FIXED SATELLITE (Earth-to-space). NG41	DOMESTIC PUBLIC FIXED (21) SATELLITE COMMUNICATIONS (25). PRIVATE OPERATIONAL FIXED MICROWAVE (94).	
				6425–6525 FIXED-SATELLITE (Earth-to-space). MOBILE.	AUXILIARY BROADCAST (74) CABLE TELEVISION (78) DOMESTIC PUBLIC	

TABLE 15.1 Table of Frequency Allocations (*Continued*)

International Table			U.S. Table		FCC Use Designators	
Region 1—allocation, MHz (1)	Region 2—allocation, MHz (2)	Region 3—allocation, MHz (3)	Government Allocation, MHz (4)	Nongovernment Allocation, MHz (5)	Rule part(s) (6)	Special-use frequencies (7)
				791 809	FIXED (21) PRIVATE OPERATIONAL-FIXED MICROWAVE (94).	
				6525–6875 FIXED. FIXED-SATELLITE (Earth-to-space). 792A 809	DOMESTIC PUBLIC FIXED (21) SATELLITE COMMUNICATIONS (25) PRIVATE OPERATION-FIXED MICROWAVE (94).	
				6875–7075 FIXED. FIXED-SATELLITE (Earth-to-space). MOBILE 792A 809 NG118	AUXILIARY BROADCAST (74) CABLE TELEVISION (78) DOMESTIC PUBLIC FIXED (21).	
7075–7250	FIXED. MOBILE. 809 810 811			7075–7125 FIXED MOBILE. 809 NG118	AUXILIARY BROADCAST (74) CABLE TELEVISION (78) DOMESTIC PUBLIC FIXED (21).	
			7125–7190 FIXED. 809 US252 G116	7125–8450		
			7190–7235 FIXED. SPACE RESEARCH (Earth-to-space).			

			809			
			7235–7250 FIXED. 809	809 US252 US258		
7250–7300	FIXED. FIXED-SATELLITE (space-to-Earth). MOBILE. 812		7250–7300 FIXED-SATELLITE (space-to-Earth). MOBILE-SATELLITE (space-to-Earth). Fixed. G117			
7300–7450	FIXED. FIXED-SATELLITE (space-to-Earth) MOBILE except aeronautical mobile. 812		7300–7450 FIXED. FIXED-SATELLITE (space-to-Earth) Mobile-Satellite (space- to-Earth). G117			
7450–7550	FIXED. FIXED-SATELLITE (space-to-Earth). METEOROLOGICAL- SATELLITE (space- to-Earth). MOBILE except aeronautical mobile.		7450–7550 FIXED. FIXED-SATELLITE (space-to-Earth). METEOROLOGICAL- SATELLITE (space- to-Earth). Mobile-Satellite (space- to-Earth). G104 G117			
7550–7750	FIXED. FIXED-SATELLITE (space-to-Earth). MOBILE except aeronautical mobile.		7550–7750 FIXED. FIXED-SATELLITE (space-to-Earth). Mobile-Satellite (space- to-Earth). G117			
7750–7900	FIXED. MOBILE except aeronautical mobile.		7750–7900 FIXED.			
7900–7975	FIXED. FIXED-SATELLITE		7900–8025 FIXED-SATELLITE (Earth-to-space).			

TABLE 15.1 Table of Frequency Allocations (*Continued*)

International Table			U.S. Table		FCC Use Designators	
			Government	Nongovernment		
Region 1—allocation, MHz (1)	Region 2—allocation, MHz (2)	Region 3—allocation, MHz (3)	Allocation, MHz (4)	Allocation, MHz (5)	Rule part(s) (6)	Special-use frequencies (7)
	(Earth-to-space). MOBILE. 812		MOBILE-SATELLITE (Earth-to-space). Fixed. G117			
7975–8025	FIXED. FIXED-SATELLITE (Earth-to-space). MOBILE. 812					
8025–8175 FIXED. FIXED-SATELLITE (Earth-to-space). MOBILE. Earth Exploration-Satellite (space-to-Earth) 813 815	8025–8175 EARTH EXPLORATION-SATELLITE (space-to-Earth) FIXED. FIXED-SATELLITE (Earth-to-space). MOBILE 814	8025–8175 FIXED. FIXED-SATELLITE (Earth-to-space). MOBILE. Earth Exploration-Satellite (space-to-Earth) 813 815	8025–8175 EARTH EXPLORATION-SATELLITE (space-to-Earth). FIXED. FIXED-SATELLITE (Earth-to-space). Mobile-Satellite (Earth-to-space) (no airborne transmission). US258 G117			
8175–8215 FIXED. FIXED-SATELLITE (Earth-to-space). METEOROLOGICAL-SATELLITE (Earth-to-space). MOBILE. Earth Exploration-Satellite (space-to-Earth) 813 815	8175–8215 EARTH EXPLORATION-SATELLITE (space-to-Earth). FIXED. FIXED-SATELLITE (Earth-to-space). METEOROLOGICAL-SATELLITE (Earth-to-space). MOBILE 814	8175–8215 FIXED. FIXED-SATELLITE (Earth-to-space). METEOROLOGICAL-SATELLITE (Earth-to-space). MOBILE. Earth Exploration-Satellite (space-to-Earth) 813 815	8175–8215 EARTH EXPLORATION SATELLITE (space-to-Earth). FIXED. FIXED-SATELLITE (Earth-to-space). METEOROLOGICAL-SATELLITE (Earth-to-space). Mobile-Satellite (Earth-to-space) (no airborne transmissions). US258 G104 G117			
8215–8400 FIXED.	8125–8400 EARTH EXPLORATION	8215–8400 FIXED	8215–8400 EARTH EXPLORATION			

FIXED-SATELLITE (Earth-to-space). MOBILE. Earth Exploration-Satellite (space-to-Earth) 813 815	SATELLITE (space-to-Earth). FIXED. FIXED-SATELLITE. (Earth-to-space). MOBILE 814	FIXED-SATELLITE (Earth-to-space). MOBILE. Earth Exploration-Satellite (space-to-Earth). 813 815	SATELLITE (space-to-Earth). FIXED. FIXED-SATELLITE (Earth-to-space). Mobile-Satellite (Earth-to-space) (no airborne transmissions). US258 G117			
8400–8500	FIXED. MOBILE except aeronautical mobile. SPACE RESEARCH (space-to-Earth) 816 817 818		8400–8450 FIXED. SPACE RESEARCH (space-to-Earth) (deep space only). FIXED. SPACE RESEARCH (space-to-Earth).	SPACE RESEARCH (space-to-Earth).		
8500–8750	RADIOLOCATION. 713 819 820		8500–9000 RADIOLOCATION.	8500–9000 Radiolocation.		
8750–8850	RADIOLOCATION. AERONAUTICAL RADIONAVIGATION 821 822					
8850–9000	RADIOLOCATION. MARITIME RADIO-NAVIGATION 823 824		713 US53 US110 G59	713 US53 US110		
9000–9200	AERONAUTICAL RADIONAVIGATION. 717 Radiolocation. 822		9000–9200 AERONAUTICAL RADIONAVIGATION 717 Radiolocation. US48 US54	9000–9200 AERONAUTICAL RADIONAVIGATION 717 Radiolocation. US48 US54	AVIATION (87).	

TABLE 15.1 Table of Frequency Allocations (*Continued*)

International Table			U.S. Table		FCC Use Designators	
			Government	Nongovernment		
Region 1—allocation, MHz (1)	Region 2—allocation, MHz (2)	Region 3—allocation, MHz (3)	Allocation, MHz (4)	Allocation, MHz (5)	Rule part(s) (6)	Special-use frequencies (7)
			G2 G19			
9200–9300	RADIOLOCATION. MARITIME RADIO-NAVIGATION 823 824 824A		9200–9300 MARITIME RADIO-NAVIGATION. RADIOLOCATION. US110 G59 823 824A	9200–9300 MARITIME RADIO-NAVIGATION Radiolocation. US110 823 824A		
9300–9500	RADIONAVIGATION RADIOLOCATION. 825A 775A 824A 825		9300–9500 RADIONAVIGATION 825A Meterological Aids. Radiolocation. 775A 824A US51 US56 US67 US71 G56	9300–9500 RADIONAVIGATION 825A Meterological Aids. Radiolocation. 775A 824A US51 US66 US67 US71		
9500–9800	RADIOLOCATION. RADIONAVIGATION. 713		9500–10,000 RADIOLOCATION. 713 828 US110	9500–10,000 Radiolocation. 713 828 US110		
9800–10,000	RADIOLOCATION. Fixed. 826 827 828					

International Table			U.S. Table		FCC Use Designators	
			Government	Nongovernment		
Region 1—allocation, GHz (1)	Region 2—allocation, GHz (2)	Region 3—allocation, GHz (3)	Allocation, GHz (4)	Allocation, GHz (5)	Rule part(s) (6)	Special-use frequencies (7)
10.0–10.45 FIXED. MOBILE. RADIOLOCATION. Amateur. 828	10.00–10.45 RADIOLOCATION. Amateur. 828 829	10.00–10.45 FIXED. MOBILE. RADIOLOCATION. Amateur. 828	10.00–10.45 RADIOLOCATION. 828 US58 US108 G32	10.00–10.45 Amateur. Radiolocation. 828 US58 US108 NG42	Amateur (97). Private Land Mobile (90).	
10.45–10.50	RADIOLOCATION.		10.45–10.50 RADIOLOCATION.	10.45–10.50 Amateur.	Amateur (97).	

	Amateur Amateur-Satellite. 830		US58 US108 G32	Amateur-Satellite. Radiolocation. US58 US108 NG42 NG134	Private Land Mobile (90).	
10.50–10.55 FIXED. MOBILE. Radiolocation.	10.50–10.55 FIXED. MOBILE. RADIOLOCATION.		10.50–10.55 RADIOLOCATION. US59	10.50–10.55 RADIOLOCATION. US59	PRIVATE LAND MOBILE (90).	
10.55–10.60	FIXED. MOBILE except aeronautical mobile. Radiolocation.		10.55–10.60	10.55–10.60 FIXED.	DOMESTIC PUBLIC FIXED (21). PRIVATE OPERATONAL-FIXED MICROWAVE (94)	
10.60–10.68	EARTH EXPLORATION- SATELLITE (passive). FIXED. MOBILE except aeronautical mobile. RADIO ASTRONOMY. SPACE RESEARCH (passive). Radiolocation. 831 832		10.60–10.68 EARTH-EXPLORATION- SATELLITE (passive). SPACE RESEARCH (passive). US265 US277	10.60–10.68 EARTH EXPLORATION- SATELLITE (passive). FIXED. SPACE RESEARCH (passive). US265 US277	DOMESTIC PUBLIC FIXED (21). PRIVATE OPERATIONAL FIXED. MICROWAVE (94).	
10.68–10.70	EARTH EXPLORATION- SATELLITE (passive). RADIO ASTRONOMY. SPACE RESEARCH (passive). 833 834		10.68–10.70 EARTH-EXPLORATION SATELLITE (passive). RADIO ASTRONOMY. SPACE RESEARCH (passive). US74 US246	10.68–10.70 EARTH EXPLORATION SATELLITE (passive). RADIO ASTRONOMY. SPACE RESEARCH (passive). US74 US246		
10.7–11.7 FIXED. FIXED-SATELLITE (space-to-Earth). (Earth-to-space) 835 MOBILE except aeronautical mobile.	10.7–11.7 FIXED. FIXED-SATELLITE (space-to-Earth). MOBILE except aeronautical mobile. 792A		10.7–11.7	10.7–11.7 FIXED. FIXED-SATELLITE (space-to-Earth).	DOMESTIC PUBLIC FIXED (21). PRIVATE OPERATIONAL- FIXED MICROWAVE (94).	

TABLE 15.1 Table of Frequency Allocations (*Continued*)

International Table			U.S. Table		FCC Use Designators	
			Government	Nongovernment		
Region 1—allocation, GHz (1)	Region 2—allocation, GHz (2)	Region 3—allocation, GHz (3)	Allocation, GHz (4)	Allocation, GHz (5)	Rule part(s) (6)	Special-use frequencies (7)
792A			US211	792A US211 NG41 NG104		
11.7–12.5 FIXED. BROADCASTING. BROADCASTING-SATELLITE. Mobile except aeronautical mobile.	11.7–12.1 FIXED 837 FIXED-SATELLITE (space-to-Earth). Mobile except aeronautical mobile. 836 839	11.7–12.2 FIXED. MOBILE except aeronautical mobile. BROADCASTING. BROADCASTING-SATELLITE.	11.7–12.2	11.7–12.2 FIXED.SATELLITE (space-to-Earth). Mobile except aeronautical mobile.	DOMESTIC PUBLIC FIXED (21) SATELLITE COMMUNICATION (25).	
838	12.1–12.2 FIXED-SATELLITE (space-to-Earth). 836 839 842.	838	839	837 839 NG143 NG145		
		12.2–12.5 FIXED. MOBILE except aeronautical mobile. BROADCASTING. 838 845	12.2–12.7	12.2–12.7 FIXED. BROADCASTING-SATELLITE.	INTERNATIONAL PUBLIC (23). PRIVATE OPERATIONAL-FIXED MICROWAVE (94). DIRECT BROADCAST SATELLITE SERVICE (100).	
12.5–12.75 FIXED-SATELLITE (space-to-Earth) (Earth-to-space).	12.2–12.7 FIXED. MOBILE except aeronautical mobile. BROADCASTING. BROADCASTING-SATELLITE. 839 844 846	12.5–12.75 FIXED. FIXED-SATELLITE (space-to-Earth). MOBILE except aeronautical mobile.	839 843 844	839 843 844 NG139		

848 849 850	12.7–12.75 FIXED. FIXED-SATELLITE. MOBILE except aeronautical mobile.	BROADCASTING-SATELLITE 847	12.7–12.75	12.7–12.75 FIXED. FIXED-SATELLITE (Earth-to-space). MOBILE. NG53 NG118	AUXILIARY BROADCASTING (74). CABLE TELEVISION RELAY (78). PRIVATE OPERATIONAL FIXED MICROWAVE (94).	
12.75–13.25	FIXED. FIXED-SATELLITE (Earth-to-space). MOBILE Space Research (deep space) (space-to-Earth). 792A		12.75–13.25 US251	12.75–13.25 FIXED. FIXED-SATELLITE (Earth-to-space). MOBILE. 792A US251 NG53 NG104 NG118	AUXILIARY BROADCASTING (74). CABLE TELEVISION RELAY (78). DOMESTIC PUBLIC FIXED (21). PRIVATE OPERATIONAL-FIXED MICROWAVE (94).	
13.25–13.40	AERONAUTICAL RADIONAVIGATION. 851 852 853		13.25–13.40 AERONAUTICAL RADIONAVIGATION 851 Space Research (Earth-to-space).	13.25–13.40 AERONAUTICAL RADIONAVIGATION 851 Space Research (Earth-to-space).	AVIATION (87).	
13.4–14.0	RADIOLOCATION. Standard Frequency and Time Signal-Satellite (Earth-to-space). Space Research. 713 853 854 855		13.4–14.0 RADIOLOCATION. Standard Frequency and Time Signal-Satellite (Earth-to-space). Space Research. 713 US110 G59	13.4–14.0 Radiolocation. Standard Frequency and Time Signal-Satellite (Earth-to-space). Space Research 713 US110	PRIVATE LAND MOBILE (90).	
14.00–14.25	FIXED-SATELLITE (Earth-to-Space) 858 RADIONAVIGATION 856 Space Research.		14.0–14.2 RADIONAVIGATION. Space Research.	14.0–14.2 FIXED-SATELLITE (Earth-to-space). RADIONAVIGATION. Space Research	Aviation (87). MARITIME (80). SATELLITE COMMUNICATION (25).	

TABLE 15.1 Table of Frequency Allocations (*Continued*)

	International Table		U.S. Table		FCC Use Designators	
			Government	Nongovernment		
Region 1—allocation, GHz (1)	Region 2—allocation, GHz (2)	Region 3—allocation, GHz (3)	Allocation, GHz (4)	Allocation, GHz (5)	Rule part(s) (6)	Special-use frequencies (7)
	857 859		14.2–14.3 US287	14.2–14.3 Fixed-satellite (Earth-to-space) Mobile except aeronautical mobile. US287	Satellite communications (25). Domestic public fixed (21).	
14.25–14.30	FIXED-SATELLITE (Earth-to-space) 858 RADIONAVIGATION 856 Space Research. 857 859 860 861					
14.3–14.4 FIXED. FIXED-SATELLITE (Earth-to-space) 858 MOBILE except aeronautical mobile. Radionavigation-Satellite. 859	14.3–14.4 FIXED-SATELLITE (Earth-to-space) 858 Radionavigation-Satellite. 859	14.3–14.4 FIXED. FIXED-SATELLITE (Earth-to-space) 858 MOBILE except aeronautical mobile. Radionavigation-Satellite. 859	14.3–14.4 US287	14.3–14.4 Fixed-satellite (Earth-to-space) Mobile except aeronautical mobile. US287	Satellite communication (25). Domestic public fixed (21).	
14.40–14.47	FIXED. FIXED-SATELLITE (Earth-to-space) 858 MOBILE except aeronautical mobile. Space Research (space-to-Earth). 859		14.4–14.5 Fixed. Mobile. 862 US203 US287	14.4–14.5 FIXED-SATELLITE (Earth-to-space). 862 US203 US287	SATELLITE COMMUNICATION (25).	
14.47–14.50	FIXED. FIXED-SATELLITE (Earth-to-space) 858					

	MOBILE except aeronautical mobile. Radio Astronomy. 859 862					
14.5–14.8	FIXED. FIXED-SATELLITE (Earth-to-space) 863 MOBILE. Space Research.		14.5000–14.7145 FIXED. Mobile. Space Research.	14.50–15.35 720 US211 US310		
			14.7145–15.1365 MOBILE. Fixed. Space Research. US310 G119			
14.80–15.35	FIXED. MOBILE. Space Research. 720		15.1365–15.35 FIXED. Mobile. Space Research. 720 US211			
15.35–15.40	EARTH EXPLORATION SATELLITE (passive). RADIO ASTRONOMY. SPACE RESEARCH (passive). 864 865		15.35–15.40 EARTH EXPLORATION SATELLITE (passive). RADIO ASTRONOMY. SPACE RESEARCH (passive). US74 US246	15.35–15.40 EARTH EXPLORATION SATELLITE (passive). RADIO ASTRONOMY. SPACE RESEARCH (passive). US74 US246		
15.4–15.7	AERONAUTICAL RADIONAVIGATION. 733 797		15.4–15.7 AERONAUTICAL RADIONAVIGATION. 733 797 US211 US260	15.4–15.7 AERONAUTICAL RADIONAVIGATION. 733 797 US211 US260	AVIATION (87).	
15.7–16.6	RADIOLOCATION. 866 867		15.7–16.6 RADIOLOCATION. US110 G59	15.7–17.2 Radiolocation.	Private Land Mobile (90).	
16.6–17.1	RADIOLOCATION. Space Research (deep space) (Earth-to-space).		16.6–17.1 RADIOLOCATION. Space Research (deep space) (Earth-to-space).			

TABLE 15.1 Table of Frequency Allocations (*Continued*)

Region 1—allocation, GHz (1)	International Table: Region 2—allocation, GHz (2)	Region 3—allocation, GHz (3)	U.S. Table: Government Allocation, GHz (4)	U.S. Table: Nongovernment Allocation, GHz (5)	FCC Use Designators: Rule part(s) (6)	FCC Use Designators: Special-use frequencies (7)
	866 867		US110 G59	US110		
17.1–17.2	RADIOLOCATION. 866 867		17.1–17.2 RADIOLOCATION. US110 G59			
17.2–17.3	RADIOLOCATION. Earth Exploration-Satellite (active). Space Research (active). 866 867		17.2–17.3 RADIOLOCATION. Earth Exploration-Satellite (active). Space Research (active). US110 G59	17.2–17.3 Radiolocation. Earth Exploration-Satellite (active). Space Research (active). US110	Private Land Mobile (90).	
17.3–17.7	FIXED-SATELLITE (Earth-to-space) 869 Radiolocation. 868		17.3–17.7 Radiolocation. US259 US271 G59	17.3–17.7 FIXED-SATELLITE (Earth-to-space). US259 US271 NG140		
17.7–18.1	FIXED. FIXED-SATELLITE (space-to-Earth) (Earth-to-space) 869 MOBILE.		17.7–17.8 US271	17.7–17.8 FIXED. FIXED-SATELLITE (space-to-Earth) (Earth-to-space). MOBILE. US271 NG140 NG144	AUXILIARY BROADCASTING (74). CABLE TELEVISION RELAY (78). DOMESTIC PUBLIC FIXED (21). PRIVATE OPERATIONAL-FIXED MICROWAVE (94).	
18.1–18.6	FIXED. FIXED-SATELLITE (space-to-Earth). MOBILE.		17.8–18.6	17.8–18.6 FIXED. FIXED-SATELLITE (space-to-Earth). MOBILE.	AUXILIARY BROADCASTING (74). CABLE TELEVISION RELAY (78). DOMESTIC PUBLIC FIXED (21). PRIVATE OPERATIONAL	

	870		870	870 NG144	FIXED MICROWAVE (94).	
18.6–18.8 FIXED. FIXED-SATELLITE (space-to-Earth) 872 MOBILE except aeronautical mobile. Earth Exploration- Satellite (passive). Space Research (passive). 871	18.6–18.8 EARTH EXPLORATION- SATELLITE (passive). FIXED. FIXED-SATELLITE (space-to-Earth) 872 MOBILE except aeronautical mobile. SPACE RESEARCH (passive). 871	18.6–18.8 FIXED. FIXED-SATELLITE (space-to-Earth) 872 MOBILE except aeronautical mobile. Earth-Exploration- Satellite (passive). Space Research (passive). 871	18.6–18.8 EARTH EXPLORATION- SATELLITE (passive). SPACE RESEARCH (passive). US254 US255	18.6–18.8 EARTH EXPLORATION- SATELLITE (passive). FIXED. FIXED-SATELLITE (space-to-Earth). MOBILE except aeronautical mobile SPACE RESEARCH (passive). US254 US255	AUXILIARY BROADCASTING (74). CABLE TELEVISION RELAY (78). DOMESTIC PUBLIC FIXED (21). PRIVATE OPERATIONAL FIXED MICROWAVE (94).	
18.8–19.7	FIXED. FIXED-SATELLITE (space-to-Earth). MOBILE.		18.8–19.7	18.8–19.7 FIXED. FIXED-SATELLITE (space-to-Earth). MOBILE.	AUXILIARY BROADCASTING (74). CABLE TELEVISION RELAY (78). DOMESTIC PUBLIC FIXED (21). PRIVATE OPERATIONAL FIXED MICROWAVE (94).	
19.7–20.2	FIXED-SATELLITE (space-to-Earth). MOBILE-SATELLITE (space-to-Earth). 873		19.7–20.2	19.7–20.2 FIXED SATELLITE (space-to-Earth). Mobile-Satellite (space- to-Earth).		
20.2–21.2	FIXED-SATELLITE (space-to-Earth). MOBILE-SATELLITE (space-to-Earth). Standard Frequency and Time Signal- Satellite (space- to-Earth). 873		20.2–21.2 FIXED-SATELLITE (space-to-Earth). MOBILE-SATELLITE (space-to-Earth). Standard Frequency and Time Signal- Satellite (space- to-Earth). G117	20.2–21.2 Standard Frequency and Time Signal- Satellite (space-to-Earth).		
21.2–21.4			21.2–21.4	21.2–21.4	21.2–21.4	

TABLE 15.1 Table of Frequency Allocations (*Continued*)

International Table			U.S. Table		FCC Use Designators	
			Government	Nongovernment		
Region 1—allocation, GHz (1)	Region 2—allocation, GHz (2)	Region 3—allocation, GHz (3)	Allocation, GHz (4)	Allocation, GHz (5)	Rule part(s) (6)	Special-use frequencies (7)
	EARTH EXPLORATION-SATELLITE (passive). FIXED. MOBILE. SPACE RESEARCH (passive).		EARTH EXPLORATION-SATELLITE (passive). FIXED. MOBILE. SPACE RESEARCH (passive). US263	EARTH EXPLORATION-SATELLITE (passive). FIXED. MOBILE. SPACE RESEARCH (passive). US263	DOMESTIC PUBLIC FIXED (21). PRIVATE OPER-ATIONAL-FIXED MICROWAVE (94).	
21.4–22.0	FIXED. MOBILE.		21.4–22.0 FIXED. MOBILE.	21.4–22.0 FIXED. MOBILE.	DOMESTIC PUBLIC FIXED (21). PRIVATE OPERATIONAL-FIXED MICROWAVE (94).	
22.00–22.21	FIXED. MOBILE except aeronautical mobile. 874		22.00–22.21 FIXED. MOBILE except aeronautical mobile. 874	22.00–22.21 FIXED. MOBILE except aeronautical mobile. 874	DOMESTIC PUBLIC MOBILE (22). PRIVATE OPERATIONAL-FIXED MICROWAVE (94).	
22.21–22.50	EARTH EXPLORATION-SATELLITE (passive). FIXED. MOBILE except aeronautical mobile. RADIO ASTRONOMY. SPACE RESEARCH (passive). 875 876		22.21–22.50 EARTH EXPLORATION-SATELLITE (passive). FIXED. MOBILE except aeronautical mobile. RADIO ASTRONOMY. SPACE RESEARCH (passive). 875 US263	22.21–22.50 EARTH EXPLORATION-SATELLITE (passive). FIXED MOBILE except aeronautical mobile. RADIO ASTRONOMY. SPACE RESEARCH (passive). 875 US263	DOMESTIC PUBLIC FIXED (21). PRIVATE OPERATIONAL-FIXED. MICROWAVE (94).	
22.50–22.55 FIXED. MOBILE.	22.50–22.55 BROADCASTING-SATELLITE 877 FIXED. MOBILE.		22.50–22.55 FIXED. MOBILE.	22.50–22.55 BROADCASTING-SATELLITE. FIXED. MOBILE.	DOMESTIC PUBLIC FIXED (21). PRIVATE OPERATIONAL-FIXED	

	878		US211	US211	MICROWAVE (94).	
22.55–23.00 FIXED. INTER-SATELLITE. MOBILE. 879	22.55–23.00 BROADCASTING- SATELLITE 877 FIXED. INTER-SATELLITE. MOBILE. 878 879		22.55–23.00 FIXED. INTER-SATELLITE. MOBILE. 879 US278	22.55–23.00 BROADCASTING- SATELLITE. FIXED. INTER-SATELLITE. MOBILE. 879 US278	DOMESTIC PUBLIC FIXED (21). PRIVATE OPERATIONAL-FIXED MICROWAVE (94).	
23.00–23.55	FIXED. INTER-SATELLITE. MOBILE. 879		23.00–23.55 FIXED. INTER-SATELLITE. MOBILE. 879 US278	23.00–23.55 FIXED. INTER-SATELLITE. MOBILE. 879 US278	DOMESTIC PUBLIC FIXED (21). PRIVATE OPERATIONAL FIXED MICROWAVE (94).	
23.55–23.60	FIXED. MOBILE.		23.55–23.60 FIXED. MOBILE.	23.55–23.60 FIXED. MOBILE.	DOMESTIC PUBLIC FIXED (21). PRIVATE OPERATIONAL-FIXED MICROWAVE (94).	
23.6–24.0	EARTH EXPLORATION SATELLITE (passive). RADIO ASTRONOMY. SPACE RESEARCH (passive). 880		23.6–24.0 EARTH EXPLORATION SATELLITE (passive). RADIO ASTRONOMY. SPACE RESEARCH (passive). US74 US246	23.6–24.0 EARTH EXPLORATION SATELLITE (passive). RADIO ASTRONOMY. SPACE RESEARCH (passive). US74 US246		
24.00–24.05	AMATEUR. AMATEUR-SATELLITE. 881		24.00–24.05 881 US211	24.00–24.05 AMATEUR. AMATEUR-SATELLITE. 881 US211	AMATEUR (97).	
24.05–24.25	RADIOLOCATION. Amateur. Earth Exploration- Satellite (active). 881		24.05–24.25 RADIOLOCATION. Earth Exploration- Satellite (active). 881 US110 G59	24.05–24.25 Amateur. Radiolocation. Earth Exploration- Satellite (active). 881 US110	Amateur (97). Private Land Mobile (90).	24.125 ± 125 GHz: Industrial scientific and medical frequency.
24.25–25.25	RADIONAVIGATION.		24.25–25.25 RADIONAVIGATION.	24.24–25.25 RADIONAVIGATION.	AVIATION (87).	
25.25–27.00			25.25–27.00	25.25–27.00		

TABLE 15.1 Table of Frequency Allocations (*Continued*)

International Table			U.S. Table		FCC Use Designators	
			Government	Nongovernment		
Region 1—allocation, GHz (1)	Region 2—allocation, GHz (2)	Region 3—allocation, GHz (3)	Allocation, GHz (4)	Allocation, GHz (5)	Rule part(s) (6)	Special-use frequencies (7)
	FIXED. MOBILE. Earth Exploration-Satellite (space-to-space). Standard Frequency and Time Signal-Satellite (Earth-to-space).		FIXED. MOBILE. Earth Exploration-Satellite (space-to-space). Standard Frequency and Time Signal-Satellite (Earth-to-space).	Earth Exploration-Satellite (space-to-space). Standard Frequency and Time Signal-Satellite (Earth-to-space).		
27.0–27.5 FIXED. MOBILE. Earth Exploration-Satellite (space-to-space).	27.0–27.5 FIXED. FIXED-SATELLITE (Earth-to-space). MOBILE. Earth Exploration-Satellite (space-to-space).		27.0–27.5 FIXED. MOBILE. Earth-Exploration-Satellite (space-to-space).	27.0–27.5 Earth Exploration Satellite (space-to-space).		
27.5–29.5	FIXED. FIXED-SATELLITE (Earth-to-space). MOBILE.		27.5–29.5	27.5–29.5 FIXED. FIXED-SATELLITE (Earth-to-space). MOBILE.	DOMESTIC PUBLIC FIXED (21).	
29.5–30.0	FIXED-SATELLITE (Earth-to-space). Mobile-Satellite (Earth-to-space). 882 883		29.5–30.0 882	29.5–30.0 FIXED-SATELLITE (Earth-to-space). Mobile-Satellite (Earth-to-space). 882		
30.0–31.0	FIXED-SATELLITE (Earth-to-space). MOBILE-SATELLITE (Earth-to-space). Standard Frequency and Time Signal-Satellite (space-to-Earth). 883		30.0–31.0 FIXED-SATELLITE (Earth-to-space). MOBILE-SATELLITE (Earth-to-space). Standard Frequency and Time Signal-Satellite (space-to-Earth). G117	30.0–31.0 Standard Frequency and Time Signal-Satellite (space-to-Earth).		
31.0–31.3			31.0–31.3	31.0–31.3	AUXILIARY	

	FIXED. MOBILE. Standard Frequency and Time Signal- Satellite (space-to-Earth). Space Research. 884 885 886		Standard Frequency and Time Signal- Satellite (space-to-Earth). 886 US211	FIXED. MOBILE. Standard Frequency and Time Signal- Satellite (space-to-Earth). 884 886 US211	BROADCASTING (74). DOMESTIC PUBLIC FIXED (21). CABLE TELEVISION RELAY (78). GENERAL MOBILE RADIO (95). PRIVATE OPERATIONAL- FIXED MICROWAVE (94).	
31.3–31.5	EARTH EXPLORATION- SATELLITE (passive). RADIO ASTRONOMY. SPACE RESEARCH (passive). 887		31.3–31.8 EARTH EXPLORATION- SATELLITE (passive). RADIO ASTRONOMY. SPACE RESEARCH (passive). US74 US246	31.3–31.8 EARTH EXPLORATION- SATELLITE (passive). RADIO ASTRONOMY. SPACE RESEARCH (passive). US74 US246		
31.5–31.8 EARTH EXPLORATION- SATELLITE (passive). RADIO ASTRONOMY. SPACE RESEARCH (passive) Fixed. Mobile except aeronautical mobile. 888 889	31.5–31.8 EARTH EXPLORATION- SATELLITE (passive). RADIO ASTRONOMY. SPACE RESEARCH (passive). 888	31.5–31.8 EARTH EXPLORATION- SATELLITE (passive). RADIO ASTRONOMY. SPACE RESEARCH (passive). Fixed. Mobile except aeronautical mobile. 888				
31.8–32.0	RADIONAVIGATION. Space Research. 890 891 892		31.8–32.0 RADIONAVIGATION. US69 US211 US262	31.8–32.0 RADIONAVIGATION. US69 US211 US262		
32.0–32.3	INTERSATELLITE RADIONAVIGATION. Space Research. 890 891 892 893		32.0–33.0 INTERSATELLITE RADIONAVIGATION. 893 US69 US262	32.0–33.0 INTERSATELLITE RADIONAVIGATION. 893 US69 US262		
32.3–33.0	INTERSATELLITE RADIONAVIGATION. 892 893					

TABLE 15.1 Table of Frequency Allocations (*Continued*)

International Table			U.S. Table		FCC Use Designators	
			Government	Nongovernment		
Region 1—allocation, GHz (1)	Region 2—allocation, GHz (2)	Region 3—allocation, GHz (3)	Allocation, GHz (4)	Allocation, GHz (5)	Rule part(s) (6)	Special-use frequencies (7)
			US278	US278		
33.0–33.4	RADIONAVIGATION. 892		33.0–33.4 RADIONAVIGATION. US69	33.0–33.4 RADIONAVIGATION. US69		
33.4–34.2	RADIOLOCATION. 892 894		33.4–36.0 RADIOLOCATION. 897 US110 US252 G34	33.4–36.0 Radiolocation. 897 US110 US252	Private Land Mobile (90).	
34.2–35.2	RADIOLOCATION. Space Research. 895 896 894					
35.2–36.0	METEOROLOGICAL AIDS. RADIOLOCATION. 894 897					
36.0–37.0	EARTH EXPLORATION-SATELLITE (passive). FIXED. MOBILE. SPACE RESEARCH (passive). 898		36.0–37.0 EARTH EXPLORATION-SATELLITE (passive). FIXED. MOBILE. SPACE RESEARCH (passive). 898 US263	36.0–37.0 EARTH EXPLORATION SATELLITE (passive). FIXED. MOBILE. SPACE RESEARCH (passive). 898 US263		
37.0–37.5	FIXED. MOBILE. 899		37.0–38.6 FIXED. MOBILE.	37.0–38.6 FIXED. MOBILE.	DOMESTIC PUBLIC FIXED (21). PRIVATE OPERATIONAL-FIXED MICROWAVE (94).	
37.5–39.5	FIXED. FIXED-SATELLITE					

	(space-to-Earth). MOBILE. 899		38.6–39.5 US291	38.6–39.5 FIXED. MOBILE. FIXED-SATELLITE (space-to-Earth) US291	DOMESTIC PUBLIC FIXED (21). PRIVATE OPERATIONAL-FIXED MICROWAVE (90). Auxiliary Broadcasting (74).	
39.5–40.5	FIXED. FIXED-SATELLITE (space-to-Earth). MOBILE. MOBILE-SATELLITE (space-to-Earth).		39.5–40.5 FIXED-SATELLITE (space-to-Earth). MOBILE-SATELLITE (space-to-Earth). US291 G117	39.5–40.5 FIXED. FIXED-SATELLITE (space-to-Earth). MOBILE. MOBILE-SATELLITE (space-to-Earth). US291	DOMESTIC PUBLIC FIXED (21). PRIVATE OPERATIONAL-FIXED MICROWAVE (94). Auxiliary Broadcasting (74).	
			40.0–40.5 FIXED-SATELLITE (space-to-Earth). MOBILE-SATELLITE (space-to-Earth). G117	40.0–40.5 FIXED-SATELLITE (space-to-Earth). MOBILE-SATELLITE (space-to-Earth).		
40.5–42.5	BROADCASTING-SATELLITE. /BROADCASTING/. Fixed. Mobile.		40.5–42.5 US211	40.5–42.5 BROADCASTING-SATELLITE. /BROADCASTING/. Fixed. Mobile. US211		
42.5–43.5	FIXED. FIXED-SATELLITE (Earth-to-space) 901 MOBILE except aeronautical mobile. RADIO ASTRONOMY. 900		42.5–43.5 FIXED. FIXED-SATELLITE (Earth-to-space). MOBILE except aeronautical mobile. RADIO ASTRONOMY. 900	42.5–43.5 FIXED. FIXED-SATELLITE (Earth-to-space). MOBILE except aeronautical mobile. RADIO ASTRONOMY. 900		
43.5–47.0	MOBILE 902		43.5–45.5 FIXED-SATELLITE	43.5–45.5		

TABLE 15.1 Table of Frequency Allocations (*Continued*)

	International Table		U.S. Table		FCC Use Designators	
			Government	Nongovernment		
Region 1—allocation, GHz (1)	Region 2—allocation, GHz (2)	Region 3—allocation, GHz (3)	Allocation, GHz (4)	Allocation, GHz (5)	Rule part(s) (6)	Special-use frequencies (7)
	MOBILE-SATELLITE. RADIONAVIGATION. RADIONAVIGATION-SATELLITE. 903		(Earth-to-space). MOBILE-SATELLITE (Earth-to-space). G117			
			45.5–47.0 MOBILE. MOBILE-SATELLITE (Earth-to-space). RADIONAVIGATION. RADIONAVIGATION-SATELLITE. 903	45.5–47.0 MOBILE. MOBILE-SATELLITE (Earth-to-space). RADIONAVIGATION. RADIONAVIGATION-SATELLITE. 903		
47.0–47.2	AMATEUR. AMATEUR-SATELLITE.		47.0–47.2	47.0–47.2 AMATEUR. AMATEUR-SATELLITE.	AMATEUR (97).	
47.2–50.2	FIXED. FIXED-SATELLITE (Earth-to-space) 901. MOBILE 905. 904		47.2–50.2 FIXED. FIXED-SATELLITE (Earth-to-space). MOBILE. 904 US264 US297	47.2–50.2 FIXED. FIXED-SATELLITE (Earth-to-space). MOBILE. 904 US264 US297		
50.2–50.4	EARTH EXPLORATION-SATELLITE (passive). FIXED. MOBILE. SPACE RESEARCH (passive).		50.2–50.4 EARTH EXPLORATION-SATELLITE (passive). FIXED. MOBILE. SPACE RESEARCH (passive). US263	50.2–50.4 EARTH EXPLORATION-SATELLITE (passive). FIXED. MOBILE. SPACE RESEARCH (passive). US263		
50.4–51.4	FIXED. FIXED-SATELLITE (Earth-to-space). MOBILE		50.4–51.4 FIXED. FIXED-SATELLITE (Earth-to-space). MOBILE.	50.4–51.4 FIXED. FIXED-SATELLITE (Earth-to-space). MOBILE.		

	Mobile-Satellite (Earth-to-space).		MOBILE-SATELLITE (Earth-to-space). G117	MOBILE-SATELLITE (Earth-to-space).		
51.4–54.25	EARTH EXPLORATION-SATELLITE (passive). SPACE RESEARCH (passive). 906 907		51.4–54.25 EARTH EXPLORATION-SATELLITE (passive). RADIO ASTRONOMY. SPACE RESEARCH (passive). US246	51.4–54.25 EARTH EXPLORATION-SATELLITE (passive). RADIO ASTRONOMY. SPACE RESEARCH (passive). US246		
54.25–58.2	EARTH EXPLORATION-SATELLITE (passive). FIXED. INTERSATELLITE. MOBILE 909. SPACE RESEARCH (passive). 908		54.25–58.2 EARTH EXPLORATION-SATELLITE (passive). FIXED. INTERSATELLITE. MOBILE 909. SPACE RESEARCH (passive). US263	54.25–58.2 EARTH EXPLORATION SATELLITE (passive). FIXED. INTERSATELLITE. MOBILE 909. SPACE RESEARCH (passive). US263		
58.2–59.0	EARTH EXPLORATION-SATELLITE (passive). SPACE RESEARCH (passive). 906 907		58.2–59.0 EARTH EXPLORATION-SATELLITE (passive). RADIO ASTRONOMY. SPACE RESEARCH (passive). US246	58.2–59.0 EARTH EXPLORATION SATELLITE (passive). RADIO ASTRONOMY. SPACE RESEARCH (passive). US246		
59–64	FIXED. INTERSATELLITE. MOBILE 909. RADIOLOCATION 910 911		59–64 FIXED. INTERSATELLITE. MOBILE 909. RADIOLOCATION 910 911	59–64 FIXED. INTERSATELLITE. MOBILE 909. RADIOLOCATION 910 911		61.25 GHz ± 250 MHz: Industrial, scientific and medical frequency
64–65	EARTH EXPLORATION SATELLITE (passive). SPACE RESEARCH (passive). 906 907		64–65 EARTH EXPLORATION SATELLITE (passive). RADIO ASTRONOMY. SPACE RESEARCH (passive). US246	64–65 EARTH EXPLORATION SATELLITE (passive). RADIO ASTRONOMY. SPACE RESEARCH (passive). US246		
65–66	EARTH EXPLORATION-SATELLITE.		65–66 EARTH EXPLORATION-SATELLITE.	65–66 EARTH EXPLORATION SATELLITE.		

TABLE 15.1 Table of Frequency Allocations (*Continued*)

	International Table		U.S. Table		FCC Use Designators	
			Government	Nongovernment		
Region 1—allocation, GHz (1)	Region 2—allocation, GHz (2)	Region 3—allocation, GHz (3)	Allocation, GHz (4)	Allocation, GHz (5)	Rule part(s) (6)	Special-use frequencies (7)
	SPACE RESEARCH. Fixed. Mobile.		SPACE RESEARCH. Fixed. Mobile.	SPACE RESEARCH. Fixed. Mobile.		
66–71	MOBILE 902 MOBILE-SATELLITE. RADIONAVIGATION. RADIONAVIGATION-SATELLITE. 903		66–71 MOBILE 902 MOBILE-SATELLITE. RADIONAVIGATION. RADIONAVIGATION-SATELLITE. 903	66–71 MOBILE 902 MOBILE-SATELLITE. RADIONAVIGATION. RADIONAVIGATION-SATELLITE. 903		
71–74	FIXED. FIXED-SATELLITE (Earth-to-space). MOBILE. MOBILE-SATELLITE (Earth-to-space). 906		71–74 FIXED. FIXED-SATELLITE (Earth-to-space). MOBILE. MOBILE-SATELLITE (Earth-to-space). US270	71–74 FIXED. FIXED-SATELLITE (Earth-to-space). MOBILE. MOBILE-SATELLITE (Earth-to-space). US270		
74.0–75.5	FIXED. FIXED-SATELLITE (Earth-to-space). MOBILE.		74.0–75.5 FIXED. FIXED-SATELLITE (Earth-to-space). MOBILE. US297	74.0–75.5 FIXED. FIXED-SATELLITE (Earth-to-space). MOBILE. US297		
75.5–76.0	AMATEUR. AMATEUR-SATELLITE.		75.5–76.0	75.5–76.0 AMATEUR. AMATEUR-SATELLITE.	AMATEUR (97).	
76–81	RADIOLOCATION. Amateur. Amateur-Satellite. 912		76–81 RADIOLOCATION. 912	76–81 RADIOLOCATION. Amateur. Amateur-Satellite. 912	Amateur (97).	
81–84	FIXED. FIXED-SATELLITE (space-to-Earth). MOBILE.		81–84 FIXED. FIXED-SATELLITE (space-to-Earth). MOBILE.	81–84 FIXED. FIXED-SATELLITE (space-to-Earth). MOBILE.		

	MOBILE-SATELLITE (space-to-Earth).		MOBILE-SATELLITE (space-to-Earth).	MOBILE-SATELLITE (space-to-Earth).		
84–86	FIXED. MOBILE. BROADCASTING. BROADCASTING-SATELLITE. 913		84–86 FIXED. MOBILE. 913 US211	84–86 FIXED. MOBILE. BROADCASTING. BROADCASTING-SATELLITE. 913 US211		
86–92	EARTH EXPLORATION-SATELLITE (passive). RADIO ASTRONOMY. SPACE RESEARCH (passive). 907		86–92 EARTH EXPLORATION-SATELLITE (passive). RADIO ASTRONOMY. SPACE RESEARCH (passive). US74 US246	86–92 EARTH EXPLORATION SATELLITE (passive). RADIO ASTRONOMY. SPACE RESEARCH (passive). US74 US246		
92–95	FIXED. FIXED-SATELLITE (Earth-to-space) MOBILE. RADIOLOCATION. 914		92–95 FIXED. FIXED-SATELLITE (Earth-to-space). MOBILE. RADIOLOCATION. 914	92–95 FIXED. FIXED-SATELLITE (Earth-to-space). MOBILE. RADIOLOCATION. 914		
95–100	MOBILE 902 MOBILE-SATELLITE. RADIONAVIGATION. RADIONAVIGATION-SATELLITE. Radiolocation. 903 904		95–100 MOBILE 902 MOBILE-SATELLITE. RADIONAVIGATION. RADIONAVIGATION-SATELLITE. Radiolocation. 903 904	95–100 MOBILE 902 MOBILE-SATELLITE. RADIONAVIGATION. RADIONAVIGATION-SATELLITE. Radiolocation. 903 904		
100–102	EARTH EXPLORATION SATELLITE (passive). FIXED. MOBILE. SPACE RESEARCH (passive). 722		100–102 EARTH EXPLORATION SATELLITE (passive). SPACE RESEARCH (passive). 722 US246	100–102 EARTH EXPLORATION SATELLITE (passive). SPACE RESEARCH (passive). 722 US246		
102–105	FIXED. FIXED-SATELLITE		102–105 FIXED. FIXED-SATELLITE	102–105 FIXED. FIXED-SATELLITE		

TABLE 15.1 Table of Frequency Allocations (*Continued*)

	International Table		U.S. Table		FCC Use Designators	
			Government	Nongovernment		
Region 1—allocation, GHz (1)	Region 2—allocation, GHz (2)	Region 3—allocation, GHz (3)	Allocation, GHz (4)	Allocation, GHz (5)	Rule part(s) (6)	Special-use frequencies (7)
	(space-to-Earth). 722		(space-to-Earth). 722 US211	(space-to-Earth). 722		
105–116	EARTH EXPLORATION-SATELLITE (passive). RADIO ASTRONOMY. SPACE RESEARCH (passive). 722 907		105–116 EARTH EXPLORATION-SATELLITE (passive). RADIO ASTRONOMY. SPACE RESEARCH (passive). 722 US74 US246	105–116 EARTH EXPLORATION-SATELLITE (passive). RADIO ASTRONOMY. SPACE RESEARCH (passive). 722 US74 US246		
116–126	EARTH EXPLORATION-SATELLITE (passive). FIXED. INTERSATELLITE. MOBILE 909. 722 915 916		116–126 EARTH EXPLORATION-SATELLITE (passive). FIXED. INTERSATELLITE. MOBILE 909. SPACE RESEARCH (passive). 722 915 916 US211 US263	116–126 EARTH EXPLORATION-SATELLITE (passive). FIXED. INTERSATELLITE. MOBILE 909. SPACE RESEARCH (passive). 722 915 916 US211 US263		122.5 ± 5GHz Industrial scientific and medical frequency
126–134	FIXED. INTERSATELLITE. MOBILE 909. RADIOLOCATION. 910		126–134 FIXED. INTERSATELLITE. MOBILE 909. RADIOLOCATION. 910	126–134 FIXED. INTERSATELLITE. MOBILE 909. RADIOLOCATION. 910		
134–142	MOBILE 902 MOBILE-SATELLITE. RADIONAVIGATION. RADIONAVIGATION-SATELLITE. Radiolocation. 903 917 918		134–142 MOBILE 902 MOBILE-SATELLITE. RADIONAVIGATION. RADIONAVIGATION-SATELLITE. Radiolocation. 903 917 918	134–142 MOBILE 902 MOBILE-SATELLITE. RADIONAVIGATION. RADIONAVIGATION-SATELLITE. Radiolocation. 903 917 818		
142–144	AMATEUR. AMATEUR-SATELLITE.		142–144	142–144 AMATEUR. AMATEUR-SATELLITE.	AMATEUR (97).	

144–149	RADIOLOCATION. Amateur. Amateur-Satellite. 918		144–149 RADIOLOCATION. 918	144–149 RADIOLOCATION. Amateur. Amateur-Satellite. 918	Amateur (97).	
149–150	FIXED. FIXED-SATELLITE (space-to-Earth). MOBILE.		149–150 FIXED. FIXED-SATELLITE (space-to-Earth). MOBILE.	149–150 FIXED. FIXED-SATELLITE (space-to-Earth). MOBILE.		
150–151	EARTH EXPLORATION. SATELLITE (passive). FIXED. FIXED-SATELLITE (space-to-Earth). MOBILE. SPACE RESEARCH (passive). 919		150–151 EARTH EXPLORATION. SATELLITE (passive). FIXED. FIXED-SATELLITE (space-to-Earth). MOBILE. SPACE RESEARCH (passive). 919 US263	150–151 EARTH EXPLORATION. SATELLITE (passive). FIXED. FIXED-SATELLITE (space-to-Earth). MOBILE. SPACE RESEARCH (passive). 919 US263		
151–164	FIXED. FIXED-SATELLITE. (space-to-Earth).		151–164 FIXED. FIXED-SATELLITE. 211	151–164 FIXED. FIXED-SATELLITE. 211		
164–168	EARTH EXPLORATION-SATELLITE (passive). RADIO ASTRONOMY. SPACE RESEARCH (passive).		164–168 EARTH EXPLORATION-SATELLITE (passive). RADIO ASTRONOMY. SPACE RESEARCH (passive). US246	164–168 EARTH EXPLORATION-SATELLITE (passive). RADIO ASTRONOMY. SPACE RESEARCH (passive). US246		
168–170	FIXED. MOBILE.		168–170 FIXED. MOBILE.	168–170 FIXED. MOBILE.		
170.0–174.5	FIXED. INTERSATELLITE. MOBILE. 909. 919		170.0–174.5 FIXED. INTERSATELLITE. MOBILE. 909. 919	170.0–174.5 FIXED. INTERSATELLITE. MOBILE. 909. 919		
174.5–176.5	EARTH EXPLORATION-		174.5–176.5 EARTH EXPLORATION-	174.5–176.5 EARTH EXPLORATION-		

TABLE 152.1 Table of Frequency Allocations (*Continued*)

	International Table		U.S. Table		FCC Use Designators	
			Government	Nongovernment		
Region 1—allocation, GHz (1)	Region 2—allocation, GHz (2)	Region 3—allocation, GHz (3)	Allocation, GHz (4)	Allocation, GHz (5)	Rule part(s) (6)	Special-use frequencies (7)
	SATELLITE (passive). FIXED. INTERSATELLITE. MOBILE. 909. SPACE RESEARCH (passive). 919		SATELLITE (passive). FIXED. INTERSATELLITE. MOBILE 909. SPACE RESEARCH (passive). 919 US263	SATELLITE (passive). FIXED. INTERSATELLITE. MOBILE 909. SPACE RESEARCH (passive). 919 US263		
176.5–182.0	FIXED. INTERSATELLITE. MOBILE. 909. 919		176.5–182.0 FIXED. INTERSATELLITE. MOBILE. 909. 919 US211	176.5–182.0 FIXED. INTERSATELLITE. MOBILE. 909. 919 US211		
182–185	EARTH EXPLORATION-SATELLITE (passive). RADIO ASTRONOMY. SPACE RESEARCH (passive). 920 921		182–185 EARTH EXPLORATION-SATELLITE (passive). RADIO ASTRONOMY. SPACE RESEARCH (passive). US246	182–185 EARTH EXPLORATION-SATELLITE (passive). RADIO ASTRONOMY. SPACE RESEARCH (passive). US246		
185–190	FIXED. INTERSATELLITE. MOBILE 909 919		185–190 FIXED. INTERSATELLITE. MOBILE 909 919 US211	185–190 FIXED. INTERSATELLITE. MOBILE 909 919 US211		
190–200	MOBILE 902 MOBILE-SATELLITE. RADIONAVIGATION. RADIONAVIGATION-SATELLITE. 722 903		190–200 MOBILE 902 MOBILE-SATELLITE. RADIONAVIGATION. RADIONAVIGATION-SATELLITE. 722 903	190–200 MOBILE 902 MOBILE-SATELLITE. RADIONAVIGATION. RADIONAVIGATION-SATELLITE. 722 903		
200–202	EARTH EXPLORATION-SATELLITE (passive). FIXED. MOBILE. SPACE RESEARCH		200–202 EARTH EXPLORATION-SATELLITE (passive). FIXED. MOBILE. SPACE RESEARCH	200–202 EARTH EXPLORATION SATELLITE (passive). FIXED. MOBILE. SPACE RESEARCH		

	(passive). 722		(passive). 722 US263	(passive). 722 US263		
202–217	FIXED. FIXED-SATELLITE (Earth-to-space). MOBILE. 722		202–217 FIXED. FIXED-SATELLITE (Earth-to-space). MOBILE. 722	202–217 FIXED. FIXED-SATELLITE (Earth-to-space). MOBILE. 722		
217–231	EARTH EXPLORATION-SATELLITE (passive). RADIO ASTRONOMY. SPACE RESEARCH (passive). 722 907		217–231 EARTH EXPLORATION-SATELLITE (passive). RADIO ASTRONOMY. SPACE RESEARCH (passive). 722 US74 US246	217–231 EARTH EXPLORATION-SATELLITE (passive). RADIO ASTRONOMY. SPACE RESEARCH (passive). 722 US74 US246		
231–235	FIXED. FIXED-SATELLITE (space-to-Earth). MOBILE. Radiolocation.		213–235 FIXED. FIXED-SATELLITE (space-to-Earth). MOBILE. Radiolocation. US211	231–235 FIXED. FIXED-SATELLITE (space-to-Earth). MOBILE. Radiolocation. US211		
235–238	EARTH EXPLORATION-SATELLITE (passive). FIXED. FIXED-SATELLITE (space-to-Earth). MOBILE. SPACE RESEARCH (passive).		235–238 EARTH EXPLORATION-SATELLITE (passive). FIXED. FIXED-SATELLITE (space-to-Earth). MOBILE. SPACE RESEARCH (passive). US263	235–238 EARTH EXPLORATION-SATELLITE (passive). FIXED. FIXED-SATELLITE (space-to-Earth). MOBILE. SPACE RESEARCH (passive). US263		
238–241	FIXED. FIXED-SATELLITE (space-to-Earth). MOBILE. Radiolocation.		238–241 FIXED. FIXED-SATELLITE (space-to-Earth). MOBILE. Radiolocation.	238–241 FIXED. FIXED-SATELLITE (space-to-Earth). MOBILE. Radiolocation.		
241–248	RADIOLOCATION.		241–248 RADIOLOCATION.	241–248 RADIOLOCATION.	Amateur (97).	245 ± 1 GHz:

TABLE 15.1 Table of Frequency Allocations (*Continued*)

	International Table		U.S. Table		FCC Use Designators	
			Government	Nongovernment		
Region 1—allocation, GHz (1)	Region 2—allocation, GHz (2)	Region 3—allocation, GHz (3)	Allocation, GHz (4)	Allocation, GHz (5)	Rule part(s) (6)	Special-use frequencies (7)
	Amateur. Amateur-Satellite. 922		922	Amateur. Amateur-Satellite. 922		Industrial, scientific and medical frequency.
248–250	AMATEUR. AMATEUR-SATELLITE.		248–250	248–250 AMATEUR. AMATEUR-SATELLITE.	AMATEUR (97).	
250–252	EARTH EXPLORATION-SATELLITE (passive). SPACE RESEARCH (passive). 923		250–252 EARTH EXPLORATION-SATELLITE (passive). SPACE RESEARCH (passive). 923	250–252 EARTH EXPLORATION-SATELLITE (passive). SPACE RESEARCH (passive). 923		
252–265	MOBILE 902 MOBILE-SATELLITE. RADIONAVIGATION. RADIONAVIGATION-SATELLITE. 903 923 924 925		252–265 MOBILE 902 MOBILE-SATELLITE. RADIONAVIGATION. RADIONAVIGATION-SATELLITE. 903 923 924 US211	252–265 MOBILE 902 MOBILE-SATELLITE. RADIONAVIGATION. RADIONAVIGATION-SATELLITE. 903 923 924 US211		
265–275	FIXED. FIXED-SATELLITE (Earth-to-space). MOBILE. RADIO ASTRONOMY. 926		265–275 FIXED. FIXED-SATELLITE (Earth-to-space). MOBILE. RADIO ASTRONOMY. 926	265–275 FIXED. FIXED-SATELLITE (Earth-to-space). MOBILE. RADIO ASTRONOMY. 926		
275–400	(Not allocated) 927		275–300 FIXED. MOBILE. 927	275–300 FIXED. MOBILE. 927		
			Above 300 (Not allocated). 927	Above 300. (Not allocated). 927	Amateur (97).	

Defining Terms

Accepted interference:[7] Interference at a higher level than defined as permissible interference and which has been agreed upon between two or more administrations without prejudice to other administrations [RR 1982].

Active satellite: A satellite carrying a station intended to transmit or retransmit radiocommunication signals [RR 1982].

Active sensor: A measuring instrument in the Earth exploration-satellite service or in the space research service by means of which information is obtained by transmission and reception of radio waves [RR 1982].

Administration: Any governmental department or service responsible for discharging the obligations undertaken in the Convention of the International Telecommunication Union and the Regulations [CONV 1973].

Aeronautical Earth station: An Earth station in the fixed-satellite service, or, in some cases, in the aeronautical mobile-satellite service, located at a specified fixed point on land to provide a feeder link for the aeronautical mobile-satellite service [RR 1982].

Aeronautical fixed service: A radiocommunication service between specified fixed points provided primarily for the safety of air navigation and for the regular, efficient and economical operation of air transport [RR 1982].

Aeronautical fixed station: A station in the aeronautical fixed service [RR 1982].

Aeronautical mobile off-route (OR) service: An aeronautical mobile service intended for communications, including those relating to flight coordination, primarily outside national or international civil air routes [RR 1982].

Aeronautical mobile route (R) service: An aeronautical mobile service reserved for communications relating to safety and regularity of flight, primarily along national or international civil air routes [RR 1982].

Aeronautical mobile-satellite off-route (OR) service: An aeronautical mobile-satellite service intended for communications, including those relating to flight coordination, primarily outside nautical and international civil air routes [RR 1982].

Aeronautical mobile-satellite route (R) service: An aeronautical mobile-satellite service reserved for communications relating to safety and regularity of flights primarily along national or international civil air routes. [RR 1982].

Aeronautical mobile-satellite service: A mobile-satellite service in which mobile Earth stations are located onboard aircraft; survival craft stations and emergency position-indicating radiobeacon stations may also participate in this service [RR 1982].

Aeronautical mobile service: A mobile service between aeronautical stations and aircraft stations, or between aircraft stations, in which survival craft stations may participate; emergency position-indicating radiobeacon stations may also participate in this service on designated distress and emergency frequencies [RR 1982].

Aeronautical radionavigation-satellite service: A radionavigation-satellite service in which Earth stations are located onboard aircraft [RR 1982].

Aeronautical radionavigation service: A radio-navigation service intended for the benefit and for the safe operation of aircraft [RR 1982].

Aeronautical station: A land station in the aeronautical mobile service. Note: in certain instances, an aeronautical station may be located, for example, onboard ship or on a platform at sea [RR 1982].

Aircraft Earth station: A mobile Earth station in the aeronautical mobile-satellite service located onboard an aircraft [RR 1982].

Aircraft station: A mobile station in the aeronautical mobile service, other than a survival craft station, located onboard an aircraft [RR 1982].

Allocation (of a frequency band): Entry in the Table of Frequency Allocations of a given frequency band for the purpose of its use by one or more terrestrial or space radiocommunication services or the radio astronomy

[7] The terms *permissible interference* and *accepted interference* are used in the coordination of frequency assignments between administrations.

service under specified conditions. This term shall also be applied to the frequency band concerned [RR 1982].

Allotment (of a radio frequency or radio frequency channel): Entry of a designated frequency channel in an agreed plan, adopted by a competent conference, for use by one or more administrations for a terrestrial or space radiocommunication service in one or more identified countries or geographical area and under specified conditions [RR 1982].

Altitude of the apogee or perigee: The altitude of the apogee or perigee above a specified reference surface serving to represent the surface of the Earth [RR 1982].

Amateur-satellite service: A radiocommunication service using space stations on Earth satellites for the same purposes as those of the amateur service [RR 1982].

Amateur service: A radiocommunication service for the purpose of self-training, intercommunication and technical investigations carried out by amateurs, that is, by duly authorized persons interested in radio technique solely with a personal aim and without pecuniary interest [RR 1982].

Amateur station: A station in the amateur service [RR 1982].

Assigned frequency: The center of the frequency band assigned to a station [RR 1982].

Assigned frequency band: The frequency band within which the emission of a station is authorized; the width of the band equals the necessary bandwidth plus twice the absolute value of the frequency tolerance. Where space stations are concerned, the assigned frequency band includes twice the maximum Doppler shift that may occur in relation to any point of the Earth's surface [RR 1982].

Assignment (of a radio frequency or radio frequency channel): Authorization given by an administration for a radio station to use a radio frequency or radio frequency channel under specified conditions [RR 1982].

Base Earth station: An Earth station in the fixed-satellite service or, in some cases, in the land mobile-satellite service, located at a specified fixed point or within a specified area on land to provide a feeder link for the land mobile-satellite service [RR 1982].

Base station: A land station in the land mobile service [RR 1982].

Broadcasting-satellite service: A radiocommunication service in which signals transmitted or retransmitted by space stations are intended for direct reception by the general public. Note: in the broadcasting-satellite service, the term *direct reception* shall encompass both individual reception and community reception [RR 1982].

Broadcasting service: A radiocommunication service in which the transmissions are intended for direct reception by the general public. This service may include sound transmissions, television transmissions or other types of transmission [CONV 1973].

Broadcasting station: A station in the broadcasting service [RR 1982].

Carrier power (of a radio transmitter): The average power supplied to the antenna transmission line by a transmitter during one radio frequency cycle taken under the condition of no modulation [RR 1982].

Characteristic frequency: A frequency that can be easily identified and measured in a given emission. Note: a carrier frequency may, for example, be designated as the characteristic frequency [RR 1982].

Class of emission: The set of characteristics of an emission, designated by standard symbols, for example, type of modulation, modulating signal, type of information to be transmitted, and also if appropriate, any additional signal characteristics [RR 1982].

Coast Earth station: An Earth station in the fixed-satellite service or, in some cases, in the maritime mobile-satellite service, located at a specified fixed point on land to provide a feeder link for the maritime mobile-satellite service [RR 1982].

Coast station: A land station in the maritime mobile service [RR 1982].

Community reception (in the broadcasting-satellite service): The reception of emissions from a space station in the broadcasting-satellite service by receiving equipment, which in some cases may be complex and have antennae larger than those for individual reception, and intended for use: (1) by a group of the general public at one location; or (2) through a distribution system covering a limited area [RR 1982].

Coordinated universal time (UTC): Time scale, based on the second (S.I.), as defined and recommended by the CCIR,[8] and maintained by the Bureau International de l'Heure (BIH). Note: for most practical purposes associated with the Radio Regulations, UTC is equivalent to mean solar time at the prime meridian (0° longitude), formerly expressed in GMT [RR 1982].

Coordination area: The area associated with an Earth station outside of which a terrestrial station sharing the same frequency band neither causes nor is subject to interfering emissions greater than a permissible level [RR 1982].

Coordination contour: The line enclosing the coordination area [RR 1982].

Coordination distance: Distance on a given azimuth from an Earth station beyond which a terrestrial causes nor is subject to interfering emissions greater than a permissible level [RR 1982].

Deep space: Space at distance from the Earth equal to or greater than 2×10^6 km [RR 1982].

Direct sequence systems: A direct sequence system is a spread spectrum system in which the incoming information is usually digitized, if it is not already in a binary format, and modulo 2 added to a higher speed code sequence. The combined information and code are then used to modulate a RF carrier. Since the high-speed code sequence dominates the modulating function, it is the direct cause of the wide spreading of the transmitted signal.

Duplex operation: Operating method in which transmission is possible simultaneously in both directions of a telecommunication channel[9] [RR 1982].

Earth exploration-satellite service: A radiocommunication service between Earth stations and one or more space stations, which may include links between space stations in which:

1. Information relating to the characteristics of the Earth and its natural phenomena is obtained from active sensors or passive sensors on earth satellites
2. Similar information is collected from air-borne or Earth-based platforms
3. Such information may be distributed to Earth stations within the system concerned
4. Platform interrogation may be included

[Note: this service may also include feeder links necessary for its operation [RR 1982].

Earth station: A station located either on the Earth's surface or within the major portion of Earth's atmosphere and intended for communication: (1) with one or more space stations; or (2) with one or more stations of the same kind by means of one or more reflecting satellites or other objects in space [RR 1982].

Effective radiated power (erp) (in a given direction): The product of the power supplied to the antenna and its gain relative to a half-wave dipole in a given direction [RR 1982].

Emergency position-indicating radiobeacon station: A station in the mobile service the emissions of which are intended to facilitate search and rescue operations [RR 1982].

Emission: Radiation produced, or the production of radiation, by a radio transmitting station. Note: for example, the energy radiated by the local oscillator of a radio receiver would not be an emission but a radiation [RR 1982].

Equivalent isotropically radiated power (eirp): The product of the power supplied to the antenna and the antenna gain in a given direction relative to an isotropic antenna [RR 1982].

Equivalent monopole radiated power (emrp) (in a given direction): The product of the power supplied to the antenna and its gain relative to a short vertical antenna in a given direction [RR 1982].

Equivalent satellite link noise temperature: The noise temperature referred to the output of the receiving antenna of the Earth station corresponding to the radio-frequency noise power, which produces the total observed noise at the output of the satellite link excluding the noise due to interference coming from satellite links using other satellites and from terrestrial systems [RR 1982].

Experimental station: A station utilizing radio waves in experiments with a view to the development of science or technique. Note: this definition does not include amateur stations [RR 1982].

Facsimile: A form of telegraphy for the transmission of fixed images, with or without half-tones, with a view to

[8]The full definition is contained in International Radio Consultive Committee (CCRI) Recommendation 460–2.

[9]In general, duplex operation and semi-duplex operation require two frequencies in radiocommunication; simplex operation may use either one or two.

their reproduction in a permanent form. Note: in this definition the term telegraphy has the same general meaning as defined in the Convention [RR 1982].

Feeder link: A radio link from an Earth station at a given location to a space station, or vice versa, conveying information for a space radiocommunication service other than for the fixed-satellite service. The given location may be at a specified fixed point, or at any fixed point within specified areas [RR 1982].

Fixed-satellite service: A radiocommunication service between Earth stations at given positions, when one or more satellites are used; the given position may be a specified fixed point or any fixed point within specified areas; in some cases this service includes satellite-to-satellite links, which may also be operated in the inter-satellite service; the fixed-satellite service may also include feeder links for other space radiocommunication services [RR 1982].

Fixed service: A radiocommunication service between specified fixed points [RR 1982].

Fixed station: A station in the fixed service [RR 1982].

Frequency hopping systems: A frequency hopping system is a spread spectrum system in which the carrier is modulated with the coded information in a conventional manner causing a conventional spreading of the RF energy about the carrier frequency. However, the frequency of the carrier is not fixed but changes at fixed intervals under the direction of a pseudorandom coded sequence. The wide RF bandwidth needed by such a system is not required by a spreading of the RF energy about the carrier but rather to accommodate the range of frequencies to which the carrier frequency can hop.

Frequency-shift telegraphy: Telegraphy by frequency modulation in which the telegraph signal shifts the frequency of the carrier between predetermined values [RR 1982].

Frequency tolerance: The maximum permissible departure by the center frequency of the frequency band occupied by an emission from the assigned frequency or, by the characteristic frequency of an emission from the reference frequency. Note: the frequency tolerance is expressed in parts in 10^6 or in hertz [RR 1982].

Full carrier single-sideband emission: A single-sideband emission without suppression of the carrier [RR 1982].

Gain of an antenna: The ratio, usually expressed in decibels, of the power required at the input of a loss free reference antenna to the power supplied to the input of the given antenna to produce, in a given direction, the same field strength or the same power flux-density at the same distance. When not specified otherwise, the gain refers to the direction of maximum radiation. The gain may be considered for a specified polarization. Note: depending on the choice of the reference antenna a distinction is made between:

1. Absolute or isotropic gain (*Gi*), when the reference antenna is an isotropic antenna isolated in space
2. Gain relative to a half-wave dipole (*Gd*), when the reference antenna is a half-wave dipole isolated in space whose equatorial plane contains the given direction
3. Gain relative to a short vertical antenna (*Gv*), when the reference antenna is a linear conductor, much shorter than one-quarter of the wavelength, normal to the surface of a perfectly conducting plane, which contains the given direction [RR 1982].

General purpose mobile service: A mobile service that includes all mobile communications uses including those within the aeronautical mobile, land mobile, or the maritime mobile services.

Geostationary satellite: A geosynchronous satellite whose circular and direct orbit lies in the plane of the Earth's equator and which, thus, remains fixed relative to the Earth; by extension, a satellite which remains approximately fixed relative to the Earth [RR 1982].

Geostationary satellite orbit: The orbit in which a satellite must be placed to be a geostationary satellite [RR 1982].

Geosynchronous satellite: An Earth satellite whose period of revolution is equal to the period of rotation of the Earth about its axis [RR 1982].

Harmful interference: [10] Interference that endangers the functioning of a radionavigation service or of other safety services or seriously degrades, obstructs, or repeatedly interrupts a radiocommunication service operating in accordance with these (international) Radio Regulations [RR 1982].

[10]See Resolution 68 of the *Radio Regulations*.

Hybrid spread spectrum systems: Hybrid spread spectrum systems are those that use combinations of two or more types of direct sequence, frequency hopping, time hopping and pulsed FM modulation in order to achieve their wide occupied bandwidths.

Inclination of an orbit (of an Earth satellite): The angle determined by the plane containing the orbit and the plane of the Earth's equator [RR 1982].

Individual reception (in the broadcasting-satellite service): The reception of emissions from a space station in the broadcasting-satellite service by simple domestic installations and, in particular, those possessing small antennas [RR 1982].

Industrial, scientific and medical (ISM) (of radio frequency energy) applications: Operation of equipment or appliances designed to generate and use locally radio-frequency energy for industrial, scientific, medical, domestic, or similar purposes, excluding applications in the field of telecommunications [RR 1982].

Instrument landing system (ILS): A radionavigation system that provides aircraft with horizontal and vertical guidance just before and during landing and, at certain fixed points, indicates the distance to the reference point of landing [RR 1982].

Instrument landing system glide path: A system of vertical guidance embodied in the instrument landing system that indicates the vertical deviation of the aircraft from its optimum path of descent [RR 1982].

Instrument landing system localizer: A system of horizontal guidance embodied in the instrument landing system that indicates the horizontal deviation of the aircraft from its optimum path of descent along the axis of the runway [RR 1982].

Interference: The effect of unwanted energy due to one or a combination of emissions, radiations, or inductions upon reception in a radiocommunication system, manifested by any performance degradation, misinterpretation, or loss of information, which could be extracted in the absence of such unwanted energy [RR 1982].

Intersatellite service: A radiocommunication service providing links between artificial Earth satellites [RR 1982].

Ionospheric scatter: The propagation of radio waves by scattering as a result of irregularities or discontinuities in the ionization of the ionosphere [RR 1982].

Land Earth station: An Earth station in the fixed-satellite service or, in some cases, in the mobile-satellite service, located at a specified fixed point or within a specified area on land to provide a feeder link for the mobile-satellite service [RR 1982].

Land mobile Earth station: A mobile Earth station in the land mobile-satellite service capable of surface movement within the geographical limits of a country or continent [RR 1982].

Land mobile-satellite service: A mobile-satellite service in which mobile Earth stations are located on land [RR 1982].

Land mobile service: A mobile service between base stations and land mobile stations, or between land mobile stations [RR 1982].

Land mobile station: A mobile station in the land mobile service capable of surface movement within the geographical limits of a country or continent.

Land station: A station in the mobile service not intended to be used while in motion [RR 1982].

Left-hand (or anticlockwise) polarized wave: An elliptically or circularly-polarized wave, in fixed plane, normal to the direction of propagation, while looking in the direction of propagation, rotates with time in a left-hand or anticlockwise direction [RR 1982].

Line A: Begins at Aberdeen, Washington, running by great circle arc to the intersection of 48° N, 120° W, then along parallel 48° N, to the intersection of 95° W, then by great circle arc through the southernmost point of Duluth, Minnesota, then by great circle arc to 45° N, 85° W, then southward along meridian 85° W, to its intersection with parallel 41° N, then along parallel 41° N, to its intersection with meridian 82° W, then by great circle arc through the southernmost point of Bangor, Maine, then by great circle arc through the southernmost point of Searsport, Maine, at which point it terminates [FCC].

Line B: Begins at Tofino, British Columbia, running by great circle arc to the intersection of 50° N, 125° W, then along parallel 50° N, to the intersection of 90° W, then by great circle arc to the intersection of 45° N, 79°30′ W, then by great circle arc through the northernmost point of Drummondville, Quebec (lat.

45° 52′ N., long. 72° 30′ W), then by great circle arc to 48° 30′ N, 70° W, then by great circle arc through the northernmost point of Compbellton, New Brunswick, then by great circle arc through the northernmost point of Liverpool, Nova Scotia, at which point it terminates [FCC].

Line C: Begins at the intersection of 70° N, 144° W, then by great circle arc to the intersection of 60° N, 143° W, then by great circle arc so as to include all of the Alaskan Panhandle [FCC].

Line D: Begins at the intersection of 70° N, 138° W, then by great circle arc to the intersection of 61° 20′ N, 139° W (Burwash Landing), then by great circle arc to the intersection of 60° 45′ N, 135° W, then by great circle arc to the intersection of 56° N, 128° W, then south along 128° meridian to lat. 55° N, then by great circle arc to the intersection of 54° N, 130° W, then by great circle arc to Port Clements, then to the Pacific Ocean where it ends [FCC].

Maritime mobile-satellite service: A mobile-satellite service in which mobile Earth stations are located onboard ships; survival craft stations and emergency position-indicating radiobeacon stations may also participate in this service [RR 1982].

Maritime mobile service: A mobile service between coast stations and ship stations, or between ship stations, or between associated onboard communication stations; survival craft stations and emergency position-indicating radiobeacon stations may also participate in this service [RR 1982].

Maritime radionavigation-satellite service: A radionavigation-satellite service in which Earth stations are located onboard ships [RR 1982].

Maritime radionavigation service: A radionavigation service intended for the benefit and for the safe operation of ships [RR 1982].

Marker beacon: A transmitter in the aeronautical radionavigation service that radiates vertically a distinctive pattern for providing position information to aircraft [RR 1982].

Mean power (of a radio transmitter): The average power supplied to the antenna transmission line by a transmitter during an interval of time sufficiently long compared with the lowest frequency encountered in the modulation taken under normal operating conditions [RR 1982].

Meteorological aids service: A radiocommunication service used for meteorological, including hydrological, observation and exploration [RR 1982].

Meteorological-satellite service: An Earth exploration-satellite service for meteorological purposes [RR 1982].

Mobile Earth station: An Earth station in the mobile-satellite service intended to be used while in motion or during halts at unspecified points [RR 1982].

Mobile-satellite service: A radiocommunication service (1) between mobile Earth stations and one or more space stations, or between space stations used by this service; or (2) between mobile Earth stations by means of one or more space stations. Note: this service may also include feeder links necessary for its operation [RR 1982].

Mobile service: A radiocommunication service between mobile and land stations, or between mobile stations [CONV 1973].

Mobile station: A station in the mobile service intended to be used while in motion or during halts at unspecified points [RR 1982].

Multisatellite link: A radio link between a transmitting Earth station and a receiving Earth station through two or more satellites, without any intermediate Earth station. Note: a multisatellite link comprises one uplink, one or more satellite-to-satellite links, and one downlink [RR 1982].

Necessary bandwidth: For a given class of emission, the width of the frequency band that is just sufficient to ensure the transmission of information at the rate and with the quality required under specified conditions [RR 1982].

Occupied bandwidth: The width of a frequency band such that, below the lower and above the upper frequency limits, the mean powers emitted are each equal to a specified percentage beta/2 of the total mean power of a given emission. Note: unless otherwise specified by the CCIR for the appropriate class of emission, the value of beta/2 should be taken as 0.5% [RR 1982].

Onboard communication station: A low-powered mobile station in the maritime mobile service intended for use for internal communications onboard a ship, or between a ship and its lifeboats and life-rafts during lifeboat drills or operations, or for communication within a group of vessels being towed or pushed, as well as for line handling and mooring instructions [RR 1982].

Orbit: The path, relative to a specified frame of reference, described by the center of mass of a satellite or other object in space subjected primarily to natural forces, mainly the force of gravity [RR 1982].

Out-of-band emission: Emission on a frequency or frequencies immediately outside the necessary bandwidth, which results from the modulation process, but excluding spurious emissions [RR 1982].

Passive sensor: A measuring instrument in the Earth exploration-satellite service or in the space research service by means of which information is obtained by reception of radio waves of natural origin [RR 1982].

Peak envelope power (of a radio transmitter): The average power supplied to the antenna transmission line by a transmitter during one radio frequency cycle at the crest of the modulation envelope taken under normal operating conditions [RR 1982].

Period (of a satellite): The time elapsing between two consecutive passages of a satellite through a characteristic point on its orbit [RR 1982].

Permissible interference: Observed or predicted interference, which complies with quantitative interference and sharing criteria contained in these [international] Radio Regulations or in CCIR Recommendations or in special agreements as provided for in these regulations [RR 1982].

Port operations service: A maritime mobile service in or near a port, between coast stations and ship stations, or between ship stations, in which messages are restricted to those relating to the operational handling, the movement and the safety of ships and, in emergency, to the safety of persons. Note: messages that are of a public correspondence nature shall be excluded from this service [RR 1982].

Port station: A coast station in the port operations service [RR 1982].

Power: Whenever the power of a radio transmitter, etc., is referred to it shall be expressed in one of the following forms, according to the class of emission, using the arbitrary symbols indicated:

1. Peak envelope power (PX or pX)
2. Mean power (PY or pY)
3. Carrier power (PZ or pZ)

Note: for different classes of emission, the relationships between peak envelope power, mean power and carrier power, under the conditions of normal operation and of no modulation, are contained in CCIR Recommendations, which may be used as a guide. Also note for use in formulas, the symbol p denotes power expressed in watts and the symbol p denotes power expressed in decibels relative to the reference level [RR 1982].

Primary radar: A radiodetermination system based on the comparison of reference signals with radio signals reflected from the position to be determined [RR 1982].

Protection ratio: The minimum value of the wanted-to-unwanted signal ratio, usually expressed in decibels, at the receiver input determined under specified conditions such that a specified reception quality of the wanted signal is achieved at the receiver output [RR 1982].

Pseudorandom sequence: A sequence of binary data that has some of the characteristics of a random sequence but also has some characteristics that are not random. It resembles a true random sequence in that the one bits and zero bits of the sequence are distributed randomly throughout every length N of the sequence and the total numbers of the one and zero bits in that length are approximately equal. It is not a true random sequence, however, because it consists of a fixed number (or length) of coded bits, which repeats itself exactly whenever that length is exceeded, and because it is generated by a fixed algorithm from some fixed initial state.

Public correspondence: Any telecommunication that the offices and stations must, by reason of their being at the disposal of the public, accept for transmission [CONV 1973].

Pulsed FM systems: A pulsed FM system is a spread spectrum system in which a RF carrier is modulated with a fixed period and fixed duty cycle sequence. At the beginning of each transmitted pulse, the carrier frequency is frequency modulated causing an additional spreading of the carrier. The pattern of the frequency modulation will depend on the spreading function that is chosen. In some systems the spreading function is a linear FM chirp sweep, sweeping either up or down in frequency.

Radar: A radiodetermination system based on the comparison of reference signals with radio signals reflected, or retransmitted, from the position to be determined [RR 1982].

Radar beacon (RACON): A transmitter–receiver associated with a fixed navigational mark, which, when triggered by a radar, automatically returns a distinctive signal that can appear on the display of the triggering radar, providing range, bearing and identification information [RR 1982].

Radiation: The outward flow of energy from any source in the form of radio waves [RR 1982].

Radio: A general term applied to the use of radio waves [CONV 1973].

Radio altimeter: Radionavigation equipment, onboard an aircraft or spacecraft or the spacecraft above the Earth's surface or another surface [RR 1982].

Radio astronomy: Astronomy based on the reception of radio waves of cosmic origin [RR 1982].

Radio astronomy service: A service involving the use of radio astronomy [RR 1982].

Radio astronomy station: A station in the radio astronomy service [RR 1982].

Radiobeacon station: A station in the radionavigation service the emissions of which are intended to enable a mobile station to determine its bearing or direction in relation to radiobeacon station [RR 1982].

Radiocommunication: Telecommunication by means of radio waves [CONV 1973].

Radiocommunication service: A service as defined in this section involving the transmission, emission and/or reception of radio waves for specific telecommunication purposes. Note: in these [international] Radio Regulations, unless otherwise stated, any radiocommunication service relates to terrestrial radiocommunication [RR 1982].

Radiodetermination: The determination of the position, velocity, and/or other characteristics of an object, or the obtaining of information relating to these parameters, by means of the propagation properties of radio waves [RR 1982].

Radiodetermination-satellite service: A radiocommunication service for the purpose of radiodetermination involving the use of one or more space stations. This service may also include feeder links necessary for its own operation [RR 1982].

Radiodetermination service: A radiocommunication service for the purpose of radiodetermination [RR 1982].

Radiodetermination station: A station in the radiodetermination service [RR 1982].

Radio direction-finding: Radio-determination using the reception of radio waves for the purpose of determining the direction of a station or object [RR 1982].

Radio direction-finding station: A radiodetermination station using radio direction-finding [RR 1982].

Radiolocation: Radiodetermination used for purposes other than those of radionavigation [RR 1982].

Radiolocation land station: A station in the radiolocation service not intended to be used while in motion [RR 1982].

Radiolocation mobile station: A station in the radiolocation service intended to be used while in motion or during halts at unspecified points [RR 1982].

Radiolocation service: A radiodetermination service for the purpose of radiolocation [RR 1982].

Radionavigation: Radiodetermination used for the purposes of navigation, including obstruction warning.

Radionavigation land station: A station in the radionavigation service not intended to be used while in motion [RR 1982].

Radionavigation mobile station: A station in the radionavigation service intended to be used while in motion or during halts at unspecified points [RR 1982].

Radionavigation-satellite service: A radiodetermination-satellite service used for the purpose of radionavigation. This service may also include feeder links necessary for its operation [RR 1982].

Radionavigation service: A radiodetermination service for the purpose of radionavigation [RR 1982].

Radiosonde: An automatic radio transmitter in the meteorological aids service usually carried on an aircraft, free balloon, kite, or parachute, and which transmits meteorological data [RR 1982].

Radiotelegram: A telegram, originating in or intended for a mobile station or a mobile Earth station transmitted on all or part of its route over the radiocommunication channels of the mobile service or of the mobile-satellite service [RR 1982].

Radiotelemetry: Telemetry by means of radio waves [RR 1982].

Radiotelephone call: A telephone call, originating in or intended for a mobile station or a mobile Earth station, transmitted on all or part of its route over the radiocommunication channels of the mobile service or of the mobile-satellite service [RR 1982].

Radiotelex call: A telex call, originating in or intended for a mobile station or a mobile Earth station, transmitted on all or part of its route over the radiocommunication channels of the mobile service or the mobile-satellite service [RR 1982].

Radio waves or hertzian waves: Electromagnetic waves of frequencies arbitrarily lower than 3000 GHz, propagated in space without artificial guide [RR 1982].

Reduced carrier single-sideband emission: A single-sideband emission in which the degree of carrier suppression enables the carrier to be reconstituted and to be used for demodulation [RR 1982].

Reference frequency: A frequency having a fixed and specified position with respect to the assigned frequency. The displacement of this frequency with respect to the assigned frequency has the same absolute value and sign that the displacement of the characteristic frequency has with respect to the center of the frequency band occupied by the emission [RR 1982].

Reflecting satellite: A satellite intended to reflect radiocommunication signals [RR 1982].

Right-hand (or clockwise) polarized wave: An elliptically or circularly-polarized wave, in which the electric field vector, observed in any fixed plane, normal to the direction of propagation, while looking in the direction of propagation, rotates with time in a right-hand or clockwise direction [RR 1982].

Safety service: Any radiocommunication service used permanently or temporarily for the safe-guarding of human life and property [CONV 1973].

Satellite: A body that revolves around another body of preponderant mass and that has a motion primarily and permanently determined by the force of attraction of that other body [RR 1982].

Satellite link: A radio link between a transmitting Earth station and a receiving Earth station through one satellite. A satellite link comprises one uplink and one downlink [RR 1982].

Satellite network: A satellite system or a part of a satellite system, consisting of only one satellite and the cooperating Earth stations [RR 1982].

Satellite system: A space system using one or more artificial Earth satellites [RR 1982].

Secondary radar: A radiodetermination system based on the comparison of reference signals with radio signals retransmitted from the position to be determined [RR 1982].

Semi-duplex operation: A method that is simplex operation at one end of the circuit and duplex operation at the other. See footnote 9 [RR 1982].

Ship Earth station: A mobile Earth station in the maritime mobile-satellite service located onboard ship [RR 1982].

Ship movement service: A safety service in the maritime mobile service other than a port operations service, between coast stations and ship stations, or between ship stations, in which messages are restricted to those relating to the movement of ships. Messages of a public correspondence nature shall be excluded from this service [RR 1982].

Ship's emergency transmitter: A ship's transmitter to be used exclusively on a distress frequency for distress, urgency, or safety purposes [RR 1982].

Ship station: A mobile station in the maritime mobile service located onboard a vessel which is not permanently moored, other than a survival craft station [RR 1982].

Simplex operation: Operating method in which transmission is made possible alternatively in each direction of a telecommunication channel, for example, by means of manual control. See footnote 9 [RR 1982].

Single-sideband emission: An amplitude modulated emission with one sideband only [RR 1982].

Spacecraft: A man-made vehicle, which is intended to go beyond the major portion of the Earth's atmosphere [RR 1982].

Space operation service: A radiocommunication service concerned exclusively with the operation of spacecraft, in particular space tracking, space telemetry, and space telecommand. Note: these functions will normally be provided within the service in which the space station is operating [RR 1982].

Space radiocommunication: Any radiocommunication involving the use of one or more space stations or the use of one or more reflecting satellites or other objects in space [RR 1982].

Space research service: A radiocommunication service in which spacecraft or other objects in space are used for scientific or technological research purposes [RR 1982].

Space station: A station located on an object that is beyond, is intended to go beyond, or has been beyond, the major portion of the Earth's atmosphere [RR 1982].

Space system: Any group of cooperating Earth stations and/or space stations employing space radiocommunication for specific purposes [RR 1982].

Space telecommand: The use of radiocommunication for the transmission of signals to a space station to initiate, modify or terminate functions of equipment on a space object, including the space station [RR 1982].

Space telemetry: The use of telemetry for transmission for a space station of results of measurements made in a spacecraft, including those relating to the functioning of the spacecraft [RR 1982].

Space tracking: Determination of the orbit, velocity, or instanteneous position of an object in space by means of radiodetermination, excluding primary radar, for the purpose of following the movement of the object [RR 1982].

Special service: A radiocommunication service, not otherwise defined in this section, carried on exclusively for specific needs of general utility, and not open to public correspondence [RR 1982].

Spread spectrum systems: A spread spectrum system is an information bearing communications system in which: (1) information is conveyed by modulation of a carrier by some conventional means, (2) the bandwidth is deliberately widened by means of a spreading function over that which would be needed to transmit the information alone. (In some spread spectrum systems, a portion of the information being conveyed by the system may be contained in the spreading function.)

Spurious emission: Emission on a frequency or frequencies that are outside the necessary bandwidth and the level of which may be reduced without affecting the corresponding transmission of information. Spurious emissions include harmonic emissions, parasitic emissions, intermodulation products and frequency conversion products, but exclude out-of-band emissions [RR 1982].

Standard frequency and time signal-satellite service: A radiocommunication service using space stations on Earth satellites for the same purposes as those of the standard frequency and time signal service. Note: this service may also include feeder links necessary for its operation [RR 1982].

Standard frequency and time signal service: A radiocommunication service for scientific, technical, and other purposes, providing the transmission of specified frequencies, time signals, or both, of stated high precision, intended for general reception [RR 1982].

Standard frequency and time signal station: A station in the standard frequency and time signal service [RR 1982].

Station: One or more transmitters or receivers or a combination of transmitters and receivers, including the accessory equipment, necessary at one location for carrying on a radiocommunication service, or the radio astronomy service. Note: each station shall be classified by the service in which it operates permanently or temporarily [RR 1982].

Suppressed carrier single-sideband emission: A single-sideband emission in which the carrier is virtually suppressed and not intended to be used for demodulation [RR 1982].

Survival craft station: A mobile station in the maritime mobile service or the aeronautical mobile service intended solely for survival purposes and located on any lifeboat, life-raft, or other survival equipment [RR 1982].

Telecommand: The use of telecommunication for the transmission of signals to initiate, modify, or terminate functions of equipment at a distance [RR 1982].

Telecommunication: Any transmission, emission or reception of signs, signals, writing, images, and sounds or intelligence of any nature by wire, radio, optical or other electromagnetic systems [CONV 1973].

Telegram: Written matter intended to be transmitted by telegraphy for delivery to the addressee. This term also includes radiotelegrams unless otherwise specified. Note: in this definition the term telegraphy has the same general meaning as defined in the Convention [CONV 1973].

Telegraphy: A form of telecommunication that is concerned in any process providing transmission and reproduction at a distance of documentary matter, such as written or printed matter or fixed images, or the reproduction at a distance of any kind of information in such a form. For the purposes of the [international] Radio Regulations, unless otherwise specified therein, telegraphy shall mean a form of telecommunication for the transmission of written matter by the use of a signal code. See footnote 10 [RR 1982].

Telemetry: The use of telecommunication for automatical indicating or recording measurements at a distance from the measuring instrument [RR 1982].

Telephony: A form of telecommunication set up for the transmission of speech or, in some cases, other sounds. See footnote 10 [RR 1982].

Television: A form of telecommunication for the transmission of transient images of fixed or moving objects [RR 1982].

Terrestrial radiocommunication: Any radiocommunication other than space radiocommunication or radio astronomy [RR 1982].

Terrestrial station: A station effecting terrestrial radiocommunication. Note: in these [international] Radio Regulations, unless otherwise stated, any station is a terrestrial station [RR 1982].

Time hopping systems: A time hopping system is a spread spectrum system in which the period and duty cycle of a pulsed RF carrier are varied in a pseudorandom manner under the control of a coded sequence. Time hopping is often used effectively with frequency hopping to form a hybrid time-division, multiple access (TDMA) spread spectrum system.

Transponder: A transmitter–receiver facility the function of which is to transmit signals automatically when the proper interrogation is received [FCC].

Tropospheric scatter: The propagation of radio waves by scattering as a result of irregularities or discontinuities in the physical properties of the troposphere [RR 1982].

Unwanted emissions: Consist of spurious emissions and out-of-band emissions [RR 1982].

References

Code of Federal Regulations: 49 FR 2373, Jan. 19, 1964, as amended at 49 FR 44101, Nov. 2, 1964; 49 FR 2368, June 19, 1984, as amended at 50 FR 25239, June 18, 1985; 51 FR 37399, Oct. 22, 1986; 52 FR 7417, Mar. 11, 1987; 54 FR 49980, Dec. 4, 1990; 55 FR 28761, July 13, 1990; 56 FR 42703, Aug. 29, 1991.

CONV. 1973. International Telecommunication Conference, Malaga-Torremolinos.

FCC. 1934. Federal Communications Commission.

RR. 1982. Radio Regulations, International Telecommunications Union, Geneva , Switzerland.

16

International Standards and Constants[1]

16.1 International System of Units (SI)

The International System of units (SI) was adopted by the 11th General Conference on Weights and Measures (CGPM) in 1960. It is a coherent system of units built from seven *SI base units*, one for each of the seven dimensionally independent base quantities: they are the meter, kilogram, second, ampere, kelvin, mole, and candela, for the dimensions length, mass, time, electric current, thermodynamic temperature, amount of substance, and luminous intensity, respectively. The definitions of the SI base units are given subsequently. The *SI derived units* are expressed as products of powers of the base units, analogous to the corresponding relations between physical quantities but with numerical factors equal to unity.

In the International System there is only one SI unit for each physical quantity. This is either the appropriate SI base unit itself or the appropriate SI derived unit. However, any of the approved decimal prefixes, called *SI prefixes*, may be used to construct decimal multiples or submultiples of SI units.

It is recommended that only SI units be used in science and technology (with SI prefixes where appropriate). Where there are special reasons for making an exception to this rule, it is recommended always to define the units used in terms of SI units. This section is based on information supplied by IUPAC.

Definitions of SI Base Units

Meter: The meter is the length of path traveled by light in vacuum during a time interval of 1/299 792 458 of a second (17th CGPM, 1983).

Kilogram: The kilogram is the unit of mass; it is equal to the mass of the international prototype of the kilogram (3rd CGPM, 1901).

Second: The second is the duration of 9 192 631 770 periods of the radiation corresponding the transition between the two hyperfine levels of the ground state of the cesium-133 atom (13th CGPM, 1967).

Ampere: The ampere is that constant current which, if maintained in two straight parallel conductors of infinite length, of negligible circular cross section, and placed 1 m apart in vacuum, would produce between these conductors a force equal to 2×10^{-7} newton per meter of length (9th CGPM, 1948).

Kelvin: The kelvin, unit of thermodynamic temperature, is the fraction 1/273.16 of the thermodynamic temperature of the triple point of water (13th CGPM, 1967).

[1]Material herein was reprinted from the following sources:

Lide, D.R., ed. 1992. *CRC Handbook of Chemistry and Physics*, 76th ed. CRC Press, Boca Raton, FL: International System of Units (SI), symbols and terminology for physical and chemical quantities, classification of electromagnetic radiation.

Zwillinger, D., ed. 1996. *CRC Standard Mathematical Tables and Formulae*, 30th ed. CRC Press, Boca Raton, FL: Greek alphabet, physical constants.

0-8493-0050-9/00/$0.00+$.50

Mole: The mole is the amount of substance of a system that contains as many elementary entities as there are atoms in 0.012 kilogram of carbon-12. When the mole is used, the elementary entities must be specified and may be atoms, molecules, ions, electrons, or other particles, or specified groups of such particles (14th CGPM, 1971).

Examples of the use of the mole:

1 mol of H_2 contains about 6.022×10^{23} H_2 molecules, or 12.044×10^{23} H atoms
1 mol of HgCl has a mass of 236.04 g
1 mol of Hg_2Cl_2 has a mass of 472.08 g
1 mol of Hg_2^{2+} has a mass of 401.18 g and a charge of 192.97 kC
1 mol of $Fe_{0.91}S$ has a mass of 82.88 g
1 mol of e^- has a mass of 548.60 μg and a charge of −96.49 kC
1 mol of photons whose frequency is 10^{14} Hz has energy of about 39.90 kJ

Candela: The candela is the luminous intensity, in a given direction, of a source that emits monochromatic radiation of frequency 540×10^{12} hertz and that has a radiant intensity in that direction of (1/683) watt per steradian (16th CGPM, 1979).

Names and Symbols for the SI Base Units

Physical quantity	Name of SI unit	Symbol for SI unit
length	meter	m
mass	kilogram	kg
time	second	s
electric current	ampere	A
thermodynamic temperature	kelvin	K
amount of substance	mole	mol
luminous intensity	candela	cd

SI Derived Units with Special Names and Symbols

Physical quantity	Name of SI unit	Symbol for SI unit	Expression in terms of SI base units
frequency[a]	hertz	Hz	s^{-1}
force	newton	N	$m\ kg\ s^{-2}$
pressure, stress	pascal	Pa	$N\ m^{-2} = m^{-1}\ kg\ s^{-2}$
energy, work, heat	joule	J	$N\ m = m^2\ kg\ s^{-2}$
power, radiant flux	watt	W	$J\ s^{-1} = m^2\ kg\ s^{-3}$
electric charge	coulomb	C	A s
electric potential, electromotive force	volt	V	$J\ C^{-1} = m^2\ kg\ s^{-3}\ A^{-1}$
electric resistance	ohm	Ω	$V\ A^{-1} = m^2\ kg\ s^{-3}\ A^{-2}$
electric conductance	siemens	S	$\Omega^{-1} = m^{-2}\ kg^{-1}\ s^3\ A^2$
electric capacitance	farad	F	$C\ V^{-1} = m^{-2}\ kg^{-1}\ s^4\ A^2$
magnetic flux density	tesla	T	$V\ s\ m^{-2} = kg\ s^{-2}\ A^{-1}$
magnetic flux	weber	Wb	$V\ s = m^2\ kg\ s^{-2}\ A^{-1}$
inductance	henry	H	$V\ A^{-1}\ s = m^2\ kg\ s^{-2}\ A^{-2}$
Celsius temperature[b]	degree Celsius	°C	K
luminous flux	lumen	lm	cd sr
illuminance	lux	lx	$cd\ sr\ m^{-2}$
activity (radioactive)	becquerel	Bq	s^{-1}
absorbed dose (of radiation)	gray	Gy	$J\ kg^{-1} = m^2 s^{-2}$
dose equivalent (dose equivalent index)	sievert	Sv	$J\ kg^{-1} = m^2\ s^{-2}$
plane angle	radian	rad	$1 = m\ m^{-1}$
solid angle	steradian	sr	$1 = m^2\ m^{-2}$

[a]For radial (circular) frequency and for angular velocity the unit rad s^{-1}, or simply s^{-1}, should be used, and this may not be simplified to Hz. The unit Hz should be used only for frequency in the sense of cycles per second.

[b]The Celsius temperature θ is defined by the equation:

$$\theta/^{\circ}C = T/K - 273.15$$

The SI unit of Celsius temperature interval is the degree Celsius, °C, which is equal to the kelvin, K. °C should be treated as a single symbol, with no space between the ° sign and the letter C. (The symbol °K and the symbol °, should no longer be used.)

Units in Use Together with the SI

These units are not part of the SI, but it is recognized that they will continue to be used in appropriate contexts. SI prefixes may be attached to some of these units, such as milliliter, ml; millibar, mbar; megaelectronvolt, MeV; kilotonne, ktonne.

Physical quantity	Name of unit	Symbol for unit	Value in SI units
time	minute	min	60 s
time	hour	h	3600 s
time	day	d	86 400 s
plane angle	degree	°	$(\pi/180)$ rad
plane angle	minute	′	$(\pi/10\,800)$ rad
plane angle	second	″	$(\pi/648\,000)$ rad
length	ångstrom[a]	Å	10^{-10} m
area	barn	b	10^{-28} m^2
volume	litre	l, L	dm^3 = 10^{-3}m^3
mass	tonne	t	Mg = 10^3 kg
pressure	bar[a]	bar	10^5 Pa = 10^5 N m^{-2}
energy	electronvolt[b]	eV (= e × V)	$\approx 1.60218 \times 10^{-19}$ J
mass	unified atomic mass unit[b,c]	u (= $m_a(^{12}\text{C})/12$)	$\approx 1.66054 \times 10^{-27}$ kg

[a]The ångstrom and the bar are approved by CIPM for temporary use with SI units, until CIPM makes a further recommendation. However, they should not be introduced where they are not used at present.

[b]The values of these units in terms of the corresponding SI units are not exact, since they depend on the values of the physical constants e (for the electronvolt) and N_A (for the unified atomic mass unit), which are determined by experiment.

[c]The unified atomic mass unit is also sometimes called the dalton, with symbol Da, although the name and symbol have not been approved by CGPM.

Greek Alphabet

	Greek letter	Greek name	English equivalent		Greek letter	Greek name	English equivalent
Α	α	Alpha	a	Ν	ν	Nu	n
Β	β	Beta	b	Ξ	ξ	Xi	x
Γ	γ	Gamma	g	Ο	ο	Omicron	ŏ
Δ	δ	Delta	d	Π	π	Pi	p
Ε	ϵ	Epsilon	ĕ	Ρ	ρ	Rho	r
Ζ	ζ	Zeta	z	Σ	σ ς	Sigma	s
Η	η	Eta	ē	Τ	τ	Tau	t
Θ	θ ϑ	Theta	th	Υ	υ	Upsilon	u
Ι	ι	Iota	i	Φ	ϕ φ	Phi	ph
Κ	κ	Kappa	k	Χ	χ	Chi	ch
Λ	λ	Lambda	l	Ψ	ψ	Psi	ps
Μ	μ	Mu	m	Ω	ω	Omega	ō

16.2 Physical Constants

General

Equatorial radius of the Earth = 6378.388 km = 3963.34 miles (statute).

Polar radius of the Earth, 6356.912 km = 3949.99 miles (statute).

1 degree of latitude at 40° = 69 miles.

1 international nautical mile = 1.15078 miles (statute) = 1852 m = 6076.115 ft.

Mean density of the Earth = 5.522 g/cm^3 = 344.7 1b/ft^3

Constant of gravitation $(6.673 \pm 0.003) \times 10^{-8}/\text{cm}^3\text{gm}^{-1}\text{s}^{-2}$.

Acceleration due to gravity at sea level, latitude 45° = 980.6194 cm/s^2 = 32.1726 ft/s^2.

Length of seconds pendulum at sea level, latitude 45° = 99.3575 cm = 39.1171 in.
1 knot (international) = 101.269 ft/min = 1.6878 ft/s = 1.1508 miles (statute)/h.
1 micron = 10^{-4} cm.
1 ångstrom = 10^{-8} cm.
Mass of hydrogen atom = $(1.67339 \pm 0.0031) \times 10^{-24}$ g.
Density of mercury at 0°C = 13.5955 g/ml.
Density of water at 3.98°C = 1.000000 g/ml.
Density, maximum, of water, at 3.98°C = 0.999973 g/cm^3.
Density of dry air at 0°C, 760 mm = 1.2929 g/1.
Velocity of sound in dry air at 0°C = 331.36 m/s −1087.1 ft/s.
Velocity of light in vacuum = $(2.997925 \pm 0.000002) \times 10^{10}$ cm/s.
Heat of fusion of water 0°C = 79.71 cal/g.
Heat of vaporization of water 100°C = 539.55 cal/g.
Electrochemical equivalent of silver 0.001118 g/s international amp.
Absolute wavelength of red cadmium light in air at 15°C, 760 mm pressure = 6438.4696 Å.
Wavelength of orange-red line of krypton 86 = 6057.802 Å.

π Constants

π = 3.14159	26535	89793	23846	26433	83279	50288	41971	69399	37511
$1/\pi$ = 0.31830	98861	83790	67153	77675	26745	02872	40689	19291	48091
π^2 = 9.8690	44010	89358	61883	44909	99876	15113	53136	99407	24079
$\log_e \pi$ = 1.14472	98858	49400	17414	34273	51353	05871	16472	94812	91531
$\log_{10} \pi$ = 0.49714	98726	94133	85435	12682	88290	89887	36516	78324	38044
$\log_{10} \sqrt{2\pi}$ = 0.39908	99341	79057	52478	25035	91507	69595	02099	34102	92128

Constants Involving e

e = 2.71828	18284	59045	23536	02874	71352	66249	77572	47093	69996
$1/e$ = 0.36787	94411	71442	32159	55237	70161	46086	74458	11131	03177
e^2 = 7.38905	60989	30650	22723	04274	60575	00781	31803	15570	55185
$M = \log_{10} e$ = 0.43429	44819	03251	82765	11289	18916	60508	22943	97005	80367
$1/M = \log_e 10$ = 2.30258	50929	94045	68401	79914	54684	36420	76011	01488	62877
$\log_{10} M$ = 9.63778	43113	00536	78912	29674	98645	−10			

Numerical Constants

$\sqrt{2}$ = 1.41421	35623	73095	04880	16887	24209	69807	85696	71875	37695
$\sqrt[3]{2}$ = 1.25992	10498	94873	16476	72106	07278	22835	05702	51464	70151
$\log_e 2$ = 0.69314	71805	59945	30941	72321	21458	17656	80755	00134	36026
$\log_{10} 2$ = 0.30102	99956	63981	19521	37388	94724	49302	67881	89881	46211
$\sqrt{3}$ = 1.73205	08075	68877	29352	74463	41505	87236	69428	05253	81039
$\sqrt[3]{3}$ = 1.44224	95703	07408	38232	16383	10780	10958	83918	69253	49935
$\log_e 3$ = 1.09861	22886	68109	69139	52452	36922	52570	46474	90557	82275
$\log_{10} 3$ = 0.47712	12547	19662	43729	50279	03255	11530	92001	28864	19070

Symbols and Terminology for Physical and Chemical Quantities: Classical Mechanics

Name	Symbol	Definition	SI unit
mass	m		kg
reduced mass	μ	$\mu = m_1 m_2/(m_1 + m_2)$	kg
density, mass density	ρ	$\rho = m/V$	kg m^{-3}
relative density	d	$d = \rho/\rho^\theta$	1
surface density	ρ_A, ρ_S	$\rho_A = m/A$	kg m^{-2}
specific volume	v	$v = V/M = 1/\rho$	m^3 kg^{-1}
momentum	p	$p = mv$	kg ms^{-1}

Symbols and Terminology for Physical and Chemical Quantities: Classical Mechanics (*Continued*)

Name	Symbol	Definition	SI unit
angular momentum, action	$\boldsymbol{L}$	$\boldsymbol{L} = \boldsymbol{r} \times \boldsymbol{p}$	J s
moment of inertia	I, J	$I = \sum m_i r_i^2$	kg m^2
force	$\boldsymbol{F}$	$\boldsymbol{F} = d\boldsymbol{p}/dt = m\boldsymbol{a}$	N
torque, moment of a force	$\boldsymbol{T}, (\boldsymbol{M})$	$\boldsymbol{T} = \boldsymbol{r} \times \boldsymbol{F}$	N m
energy	E		J
potential energy	$E_p, V, \boldsymbol{\Phi}$	$E_p = -\int \boldsymbol{F} \cdot d\boldsymbol{s}$	J
kinetic energy	E_k, T, K	$E_k = (1/2)mv^2$	J
work	W, w	$W = \int \boldsymbol{F} \cdot d\boldsymbol{s}$	J
Hamilton function	H	$H(q, p) = T(q, p) + V(q)$	J
Lagrange function	L	$L(q, \dot{q}) = T(q, \dot{q}) - V(q)$	J
pressure	p, P	$p = F/A$	Pa, N m^{-2}
surface tension	γ, σ	$\gamma = dW/dA$	N m^{-1}, J m^{-2}
weight	$G, (W, P)$	$G = mg$	N
gravitational constant	G	$F = Gm_1m_2/r^2$	N m^2 kg^{-2}
normal stress	σ	$\sigma = F/A$	Pa
shear stress	τ	$\tau = F/A$	Pa
linear strain, relative elongation	ε, e	$\varepsilon - \Delta l/l$	1
modulus of elasticity, Young's modulus	E	$E = \sigma/\varepsilon$	Pa
shear strain	γ	$\gamma = \Delta x/d$	1
shear modulus	G	$G = \tau/\gamma$	Pa
volume strain, bulk strain	θ	$\theta = \Delta V/V_0$	1
bulk modulus,	K	$K = -V_0(dp/dV)$	Pa
compression modulus	η, μ	$\tau_{x,z} = \eta(dv_x/dz)$	Pa s
viscosity, dynamic viscosity			
fluidity	ϕ	$\phi = 1/\eta$	m kg^{-1}s
kinematic viscosity	v	$v = \eta/\rho$	m^2 s^{-1}
friction coefficient	$\mu, (f)$	$F_{\text{frict}} = \mu F_{\text{norm}}$	1
power	P	$P = dW/dt$	W
sound energy flux	P, P_a	$P = dE/dt$	W
acoustic factors			
reflection factor	ρ	$\rho = P_r/P_0$	1
acoustic absorption factor	$\alpha_a, (\alpha)$	$\alpha_a = 1 - \rho$	1
transmission factor	τ	$\tau = P_{tr}/P_0$	1
dissipation factor	δ	$\delta = \alpha_a - \tau$	1

Symbols and Terminology for Physical and Chemical Quantities: Electricity and Magnetism

Name	Symbol	Definition	SI unit
quantity of electricity, electric charge	Q		C
charge density	ρ	$\rho = Q/V$	C m^{-3}
surface charge density	σ	$\sigma = Q/A$	C m^{-2}
electric potential	V, ϕ	$V = dW/dQ$	V, J C^{-1}
electric potential difference	$U, \Delta V, \Delta\phi$	$U = V_2 - V_1$	V
electromotive force	E	$E = \int (\boldsymbol{F}/Q) \cdot d\boldsymbol{s}$	V
electric field strength	$\boldsymbol{E}$	$\boldsymbol{E} = \boldsymbol{F}/Q = -\text{grad}\, V$	V m^{-1}
electric flux	$\boldsymbol{\Psi}$	$\boldsymbol{\Psi} = \int \boldsymbol{D} \cdot d\boldsymbol{A}$	C
electric displacement	$\boldsymbol{D}$	$\boldsymbol{D} = \varepsilon \boldsymbol{E}$	C m^{-2}
capacitance	C	$C = Q/U$	F, C V^{-1}
permittivity	ε	$\boldsymbol{D} = \varepsilon \boldsymbol{E}$	F m^{-1}
permittivity of vacuum	ε_0	$\varepsilon_0 = \mu_0^{-1} c_0^{-2}$	F m^{-1}
relative permittivity	ε_r	$\varepsilon_r = \varepsilon/\varepsilon_0$	1

Symbols and Terminology for Physical and Chemical Quantities: Electricity and Magnetism (*Continued*)

Name	Symbol	Definition	SI unit
dielectric polarization (dipole moment per volume)	$\boldsymbol{P}$	$\boldsymbol{P} = \boldsymbol{D} - \varepsilon_0 \boldsymbol{E}$	$\mathrm{C\,m^{-2}}$
electric susceptibility	χ_e	$\chi_e = \varepsilon_r - 1$	1
electric dipole moment	$\boldsymbol{p}, \mu$	$\boldsymbol{p} = Q\boldsymbol{r}$	C m
electric current	I	$I = dQ/dt$	A
electric current density	$\boldsymbol{j}, \boldsymbol{J}$	$I = \int \boldsymbol{j} \cdot d\boldsymbol{A}$	$\mathrm{A\,m^{-2}}$
magnetic flux density, magnetic induction	$\boldsymbol{B}$	$\boldsymbol{F} = Q\boldsymbol{v} \times \boldsymbol{B}$	T
magnetic flux	$\boldsymbol{\Phi}$	$\boldsymbol{\Phi} = \int \boldsymbol{B} \cdot d\boldsymbol{A}$	Wb
magnetic field strength	$\boldsymbol{H}$	$\boldsymbol{B} = \mu \boldsymbol{H}$	$\mathrm{A\,M^{-1}}$
permeability	μ	$\boldsymbol{B} = \mu \boldsymbol{H}$	$\mathrm{N\,A^{-2}}$, $\mathrm{H\,m^{-1}}$
permeability of vacuum	μ_0		$\mathrm{H\,m^{-1}}$
relative permeability	μ_r	$\mu_r = \mu/\mu_0$	1
magnetization (magnetic dipole moment per volume)	$\boldsymbol{M}$	$M = \boldsymbol{B}/\mu_0 - \boldsymbol{H}$	$\mathrm{A\,m^{-1}}$
magnetic susceptibility	$\chi, \kappa, (\chi_m)$	$\chi = \mu_r - 1$	1
molar magnetic susceptibility	χ_m	$\chi_m = V_m \chi$	$\mathrm{m^3\,mol^{-1}}$
magnetic dipole moment	$\boldsymbol{m}, \mu$	$E_p = -\boldsymbol{m} \cdot \boldsymbol{B}$	$\mathrm{A\,m^2}$, $\mathrm{J\,T^{-1}}$
electrical resistance	R	$R = U/I$	Ω
conductance	G	$G = 1/R$	S
loss angle	δ	$\delta = (\pi/2) + \phi_I - \phi_U$	1, rad
reactance	X	$X = (U/I)\sin\delta$	Ω
impedance (complex impedance)	Z	$Z = R + iX$	Ω
admittance (complex admittance)	Y	$Y = 1/Z$	S
susceptance	B	$Y = G + iB$	S
resistivity	ρ	$\rho = E/j$	Ω m
conductivity	κ, γ, σ	$\kappa = 1/\rho$	$\mathrm{S\,m^{-1}}$
self-inductance	L	$E = -L(dI/dt)$	H
mutual inductance	M, L_{12}	$E_1 = L_{12}(dI_2/dt)$	H
magnetic vector potential	$\boldsymbol{A}$	$\boldsymbol{B} = \boldsymbol{\nabla} \times \boldsymbol{A}$	$\mathrm{Wb\,m^{-1}}$
Poynting vector	$\boldsymbol{S}$	$\boldsymbol{S} = \boldsymbol{E} \times \boldsymbol{H}$	$\mathrm{W\,m^{-2}}$

Symbols and Terminology for Physical and Chemical Quantities: Electromagnetic Radiation

Name	Symbol	Definition	SI unit
wavelength	λ		m
speed of light in vacuum	c_0		$\mathrm{m\,s^{-1}}$
in a medium	c	$c = c_0/n$	$\mathrm{m\,s^{-1}}$
wavenumber in vacuum	$\tilde{v}$	$\tilde{v} = v/c_0 = 1/n\lambda$	$\mathrm{m^{-1}}$
wavenumber (in a medium)	σ	$\sigma = 1/\lambda$	$\mathrm{m^{-1}}$
frequency	v	$v = c/\lambda$	Hz
circular frequency, pulsatance	ω	$\omega = 2\pi v$	$\mathrm{s^{-1}}$, $\mathrm{rad\,s^{-1}}$
refractive index	n	$n = c_0/c$	1
Planck constant	h		J s
Planck constant/2π	$\hbar$	$\hbar = h/2\pi$	J s
radiant energy	Q, W		J
radiant energy density	ρ, w	$\rho = Q/V$	$\mathrm{J\,m^{-3}}$
spectral radiant energy density			
in terms of frequency	ρ_v, w_v	$\rho_v = d\rho/dv$	$\mathrm{J\,m^{-3}Hz^{-1}}$
in terms of wavenumber	$\rho_{\tilde{v}}, w_{\tilde{v}}$	$\rho_{\tilde{v}} = d\rho/d\tilde{v}$	$\mathrm{J\,m^{-2}}$
in terms of wavelength	ρ_λ, w_λ	$\rho_\lambda = d\rho/d\lambda$	$\mathrm{J\,m^{-4}}$

Symbols and Terminology for Physical and Chemical Quantities: Electromagnetic Radiation (*Continued*)

Name	Symbol	Definition	SI unit
Einstein transition probabilities			
spontaneous emission	A_{nm}	$dN_n/dt = -A_{nm}N_n$	s^{-1}
stimulated emission	B_{nm}	$dN_n/dt = -\rho_{\tilde{v}}(\tilde{v}_{nm}) \times B_{nm}N_n$	$s\ kg^{-1}$
stimulated absorption	B_{mn}	$dN_n/dt = \rho_{\tilde{v}}(\tilde{v}_{nm})B_{mn}N_m$	$s\ kg^{-1}$
radiant power, radiant energy per time	Φ, P	$\Phi = dQ/dt$	W
radiant intensity	I	$I = d\Phi/d\Omega$	$W\ sr^{-1}$
radiant exitance (emitted radiant flux)	M	$M = d\Phi/dA_{\text{source}}$	$W\ m^{-2}$
irradiance (radiant flux received)	$E, (I)$	$E = d\Phi/dA$	$W\ m^{-2}$
emittance	ε	$\varepsilon = M/M_{\text{bb}}$	1
Stefan-Boltzman constant	σ	$M_{\text{bb}} = \sigma T^4$	$W\ m^{-2}\ K^{-4}$
first radiation constant	c_1	$c_1 = 2\pi hc_0^2$	$W\ m^2$
second radiation constant	c_2	$c_2 = hc_0/k$	K m
transmittance, transmission factor	τ, T	$\tau = \Phi_{\text{tr}}/\Phi_0$	1
absorptance, absorption factor	α	$\alpha = \Phi_{\text{abs}}/\Phi_0$	1
reflectance, reflection factor	ρ	$\rho = \Phi_{\text{refl}}/\Phi_0$	1
(decadic) absorbance	A	$A = \lg(1 - \alpha_i)$	1
napierian absorbance	B	$B = \ln(1 - \alpha_i)$	1
absorption coefficient			
(linear) decadic	a, K	$a = A/l$	m^{-1}
(linear) napierian	α	$\alpha = B/l$	m^{-1}
molar (decadic)	ε	$\varepsilon = a/c = A/cl$	$m^2\ mol^{-1}$
molar napierian	κ	$\kappa = \alpha/c = B/cl$	$m^2\ mol^{-1}$
absorption index	k	$k = \alpha/4\pi\tilde{v}$	1
complex refractive index	$\hat{n}$	$\hat{n} = n + ik$	1
molar refraction	R, R_m	$R = \dfrac{n^2 - 1}{n^2 + 2}V_m$	$m^3\ mol^{-1}$
angle of optical rotation	α		1, rad

Symbols and Terminology for Physical and Chemical Quantities: Solid State

Name	Symbol	Definition	SI unit
lattice vector	$\boldsymbol{R}, \boldsymbol{R}_0$		m
fundamental translation vectors for the crystal lattice	$\boldsymbol{a}_1; \boldsymbol{a}_2; \boldsymbol{a}_3$, $\boldsymbol{a}; \boldsymbol{b}; \boldsymbol{c}$	$R = n_1\boldsymbol{a}_1 + n_2\boldsymbol{a}_2 + n_3\boldsymbol{a}_3$	m
(circular) reciprocal lattice vector	$\boldsymbol{G}$	$\boldsymbol{G} \cdot \boldsymbol{R} = 2\pi m$	m^{-1}
(circular) fundamental translation vectors for the reciprocal lattice	$\boldsymbol{b}_1; \boldsymbol{b}_2; \boldsymbol{b}_3$, $\boldsymbol{a}^*; \boldsymbol{b}^*; \boldsymbol{c}^*$	$\boldsymbol{a}_i \cdot \boldsymbol{b}_k = 2\pi\delta_{ik}$	m^{-1}
lattice plane spacing	d		m
Bragg angle	θ	$n\lambda = 2d \sin\theta$	1, rad
order of reflection	n		1
order parameters			
short range	σ		1
long range	s		1
Burgers vector	$\boldsymbol{b}$		m
particle position vector	$\boldsymbol{r}, \boldsymbol{R}_j$		m
equilibrium position vector of an ion	$\boldsymbol{R}_0$		m

Symbols and Terminology for Physical and Chemical Quantities: Solid State (*Continued*)

Name	Symbol	Definition	SI unit
equilibrium position vector of an ion	$\boldsymbol{R}_0$		m
displacement vector of an ion	$\boldsymbol{u}$	$\boldsymbol{u} = \boldsymbol{R} - \boldsymbol{R}_0$	m
Debye–Waller factor	B, D		1
Debye circular wavenumber	q_D		m^{-1}
Debye circular frequency	ω_D		s^{-1}
Grüneisen parameter	γ, Γ	$\gamma = \alpha V/\kappa C_V$	1
Madelung constant	$\alpha, \mathcal{M}$	$E_{\text{coul}} = \dfrac{\alpha N_A z_+ z_- e^2}{4\pi\varepsilon_0 R_0}$	1
density of states	N_E	$N_E = dN(E)/dE$	$J^{-1}\,m^{-3}$
(spectral) density of vibrational modes	N_ω, g	$N_\omega = dN(\omega)/d\omega$	$s\,m^{-3}$
resistivity tensor	ρ_{ik}	$E = \rho \cdot j$	$\Omega\,m$
conductivity tensor	σ_{ik}	$\sigma = \rho^{-1}$	$S\,m^{-1}$
thermal conductivity tensor	λ_{ik}	$J_q = -\lambda \cdot \text{grad}\, T$	$W\,m^{-1}\,K^{-1}$
residual resistivity	ρ_R		$\Omega\,m$
relaxation time	τ	$\tau = l/v_F$	s
Lorenz coefficient	L	$L = \lambda/\sigma T$	$V^2\,K^{-2}$
Hall coefficient	A_H, R_H	$E = \rho \cdot j + R_H(\boldsymbol{B} \times \boldsymbol{j})$	$m^3\,C^{-1}$
thermoelectric force	E		V
Peltier coefficient	Π		V
Thomson coefficient	$\mu, (\tau)$		$V\,K^{-1}$
work function	Φ	$\Phi = E_\infty - E_F$	J
number density, number concentration	$n, (p)$		m^{-3}
gap energy	E_g		J
donor ionization energy	E_d		J
acceptor ionization energy	E_a		J
Fermi energy	E_F, ε_F		J
circular wave vector, propagation vector	$\boldsymbol{k}, \boldsymbol{q}$	$k = 2\pi/\lambda$	m^{-1}
Bloch function	$u_k(\boldsymbol{r})$	$\psi(\boldsymbol{r}) = u_k(\boldsymbol{r})\exp(i\boldsymbol{k} \cdot \boldsymbol{r})$	$m^{-3/2}$
charge density of electrons	ρ	$\rho(\boldsymbol{r}) = -e\psi^*(\boldsymbol{r})\psi(\boldsymbol{r})$	$C\,m^{-3}$
effective mass	m^*		kg
mobility	μ	$\mu = v_{\text{drift}}/E$	$m^2\,V^{-1}\,s^{-1}$
mobility ratio	b	$b = \mu_n/\mu_p$	1
diffusion coefficient	D	$dN/dt = -DA(dn/dx)$	$m^2\,s^{-1}$
diffusion length	L	$L = \sqrt{D\tau}$	m
characteristic (Weiss) temperature	ϕ, ϕ_W		K
Curie temperature	T_C		K
Néel temperature	T_N		K

Letter Designations of Microwave Bands

Frequency, GHz	Wavelength, cm	Wavenumber, cm^{-1}	Band
1–2	30–15	0.033–0.067	L-band
1–4	15–7.5	0.067–0.133	S-band
4–8	7.5–3.7	0.133–0.267	C-band
8–12	3.7–2.5	0.267–0.4	X-band
12–18	2.5–1.7	0.4–0.6	Ku-band
18–27	1.7–1.1	0.6–0.9	K-band
27–40	1.1–0.75	0.9–1.33	Ka-band

17

Conversion Factors[1]

Jerry C. Whitaker
Editor-in-Chief

17.1 Introduction

Engineers often find it necessary to convert from one unit of measurement to another. The following table of conversion factors provides a convenient method of accomplishing the task. In Chapter 153, certain conversion factors were listed, grouped by function. In the following table, a more complete listing of conversion factors is presented, grouped in alphabetical order.

To Convert	Into	Multiply by
	A	
abcoulomb	statcoulombs	2.998×10^{10}
acre	square chain (Gunters)	10
acre	rods	160
acre	square links (Gunters)	1×10^{5}
acre	hectare or square hectometer	0.4047
acres	square feet	43,560.0
acres	square meters	4,047
acres	square miles	1.562×10^{-3}
acres	square yards	4,840
acre-feet	cubic feet	43,560.0
acre-feet	gallons	3.259×10^{5}
amperes per square centimeter	ampere per square inch	6.452
amperes per square centimeter	ampere per square meter	10^{4}
amperes per square inch	ampere per square centimeter	0.1550
amperes per square inch	ampere per square meter	1,550.0
amperes per square meter	ampere per square centimeter	10^{-4}
amperes per square meter	ampere per square inch	6.452×10^{-4}
ampere-hours	coulombs	3,600.0
ampere-hours	faradays	0.03731
ampere-turns	gilberts	1.257
ampere-turns per centimeter	ampere-turns per inch	2.540
ampere-turns per centimeter	ampere-turns per meter	100.0
ampere-turns per centimeter	gilberts per centimeter	1.257
ampere-turns per inch	ampere-turns per centimeter	0.3937
ampere-turns per inch	ampere-turns per meter	39.37
ampere-turns per inch	gilberts per centimeter	0.4950

[1]Information contained in this chapter was adapted from Whitaker, J.C. 1991. *Maintaining Electronic Systems.* CRC Press, Boca Raton, FL.

0-8493-0050-9/00/$0.00+$.50

To Convert	Into	Multiply by
ampere-turns per meter	ampere-turns per centimeter	0.01
ampere-turns per meter	ampere-turns per inch	0.0254
ampere-turns per meter	gilberts per centimeter	0.01257
Angstrom unit	inch	3937×10^{-9}
Angstrom unit	meter	1×10^{-10}
Angstrom unit	micron or mu(μ)	1×10^{-4}
are	acre (U.S.)	.02471
ares	square yards	119.60
ares	acres	0.02471
ares	square meters	100.0
astronomical unit	kilometers	1.495×10^{8}
atmospheres	ton per square inch	0.007348
atmospheres	centimeter of mercury	76.0
atmospheres	foot of water (at 4°C)	33.90
atmospheres	inch of mercury (at 0°C)	29.92
atmospheres	kilogram per square centimeter	1.0333
atmospheres	kilogram per square meter	10,332
atmospheres	pounds per square inch	14.70
atmospheres	tons per square foot	1.058
	B	
barrels (U.S., dry)	cubic inch	7056
barrels (U.S., dry)	quarts (dry)	105.0
barrels (U.S., liquid)	gallons	31.5
barrels (oil)	gallons (oil)	42.0
bars	atmospheres	0.9869
bars	dyne per square centimeter	10^{4}
bars	kilogram per square meter	1.020×10^{4}
bars	pound per square foot	2,089
bars	pound per square inch	14.50
Baryl	dyne per square centimeter	1.000
bolt (U.S., cloth)	meter	36.576
British thermal unit	liter-atmosphere	10.409
British thermal unit	erg	1.0550×10^{10}
British thermal unit	foot-pound	778.3
British thermal unit	gram-calorie	252.0
British thermal unit	horsepower-hour	3.931×10^{-4}
British thermal unit	joule	1,054.8
British thermal unit	kilogram-calorie	0.2520
British thermal unit	kilogram-meter	107.5
British thermal unit	kilowatt-hour	2.928×10^{-4}
British thermal unit per hour	foot-pound per second	0.2162
British thermal unit per hour	gram-calorie per second	0.0700
British thermal unit per hour	horsepower-hour	3.929×10^{-4}
British thermal unit per hour	watts	0.2931
British thermal unit per minute	foot-pound per second	12.96
British thermal unit per minute	horsepower	0.02356
British thermal unit per minute	kilowatts	0.01757
British thermal unit per minute	watts	17.57
British thermal unit per square foot per minute	watts per square inch	0.1221
bucket (British, dry)	cubic centimeter	1.818×10^{4}
bushels	cubic foot	1.2445
bushels	cubic inch	2,150.4
bushels	cubic meter	0.03524
bushels	liters	35.24
bushels	pecks	4.0
bushels	pints (dry)	64.0
bushels	quarts (dry)	32.0

To Convert	Into	Multiply by
	C	
calories, gram (mean)	British thermal unit (mean)	3.9685×10^{-3}
candle per square centimeter	lamberts	3.142
candle per square inch	lamberts	.4870
centares (centiares)	square meters	1.0
centigrade	Fahrenheit	$(C° \times 9/5) + 32$
centigrams	grams	0.01
centiliter	ounce fluid (U.S.)	0.3382
centiliter	cubic inch	0.6103
centiliter	drams	2.705
centiliters	liters	0.01
centimeters	feet	3.281×10^{-2}
centimeters	inches	0.3937
centimeters	kilometers	10^{-5}
centimeters	meters	0.01
centimeters	miles	6.214×10^{-6}
centimeters	millimeter	10.0
centimeters	mils	393.7
centimeters	yards	1.094×10^{-2}
centimeter-dynes	centimeter-grams	1.020×10^{-3}
centimeter-dynes	meter-kilogram	1.020×10^{-8}
centimeter-dynes	pound-feet	7.376×10^{-8}
centimeter-grams	centimeter-dynes	980.7
centimeter-grams	meter-kilogram	10^{-5}
centimeter-grams	pound-feet	7.233×10^{-5}
centimeters of mercury	atmospheres	0.01316
centimeters of mercury	feet of water	0.4461
centimeters of mercury	kilogram per square meter	136.0
centimeters of mercury	pounds per square foot	27.85
centimeters of mercury	pounds per square inch	0.1934
centimeters per second	feet per minute	1.9686
centimeters per second	feet per second	0.03281
centimeters per second	kilometer per hour	0.036
centimeters per second	knot	0.1943
centimeters per second	meter per minute	0.6
centimeters per second	mile per hour	0.02237
centimeters per second	mile per minute	3.728×10^{-4}
centimeters per second per second	feet per second per second	0.03281
centimeters per second per second	kilometer per hour per second	0.036
centimeters per second per second	meters per second per second	0.01
centimeters per second per second	miles per hour per second	0.02237
chain	inches	792.00
chain	meters	20.12
chain (surveyors' or Gunter's)	yards	22.00
circular mils	square centimeter	5.067×10^{-6}
circular mils	square mils	0.7854
circular mils	square inches	7.854×10^{-7}
circumference	radians	6.283
cord	cord feet	8
cord feet	cubic feet	16
coulomb	statcoulombs	2.998×10^{9}
coulombs	faradays	1.036×10^{-5}
coulombs per square centimeter	coulombs per square inch	64.52
coulombs per square centimeter	coulombs per square meter	10^{4}
coulombs per square inch	coulombs per square centimeter	0.1550
coulombs per square inch	coulombs per square meter	1,550
coulombs per square meter	coulombs per square centimeter	10^{-4}
coulombs per square meter	coulombs per square inch	6.452×10^{-4}
cubic centimeters	cubic feet	3.531×10^{-5}

To Convert	Into	Multiply by
cubic centimeters	cubic inches	0.06102
cubic centimeters	cubic meters	10^{-6}
cubic centimeters	cubic yards	1.308×10^{-6}
cubic centimeters	gallons (U.S. liquid)	2.642×10^{-4}
cubic centimeters	liters	0.001
cubic centimeters	pints (U.S. liquid)	2.113×10^{-3}
cubic centimeters	quarts (U.S. liquid)	1.057×10^{-3}
cubic feet	bushels (dry)	0.8036
cubic feet	cubic centimeter	28,320.0
cubic feet	cubic inches	1,728.0
cubic feet	cubic meters	0.02832
cubic feet	cubic yards	0.03704
cubic feet	gallons (U.S. liquid)	7.48052
cubic feet	liters	28.32
cubic feet	pints (U.S. liquid)	59.84
cubic feet	quarts (U.S. liquid)	29.92
cubic feet per minute	cubic centimeter per second	472.0
cubic feet per minute	gallons per second	0.1247
cubic feet per minute	liters per second	0.4720
cubic feet per minute	pounds of water per minute	62.43
cubic feet per second	million galllons per day	0.646317
cubic feet per second	gallons per minute	448.831
cubic inches	cubic centimeters	16.39
cubic inches	cubic feet	5.787×10^{-4}
cubic inches	cubic meter	1.639×10^{-5}
cubic inches	cubic yards	2.143×10^{-5}
cubic inches	gallons	4.329×10^{-3}
cubic inches	liters	0.01639
cubic inches	mil-feet	1.061×10^{5}
cubic inches	pints (U.S. liquid)	0.03463
cubic inches	quarts (U.S. liquid)	0.01732
cubic meters	bushels (dry)	28.38
cubic meters	cubic centimeter	10^{6}
cubic meters	cubic feet	35.31
cubic meters	cubic inches	61,023.0
cubic meters	cubic yards	1.308
cubic meters	gallons (U.S. liquid)	264.2
cubic meters	liters	1,000.0
cubic meters	pints (U.S. liquid)	2,113.0
cubic meters	quarts (U.S. liquid)	1,057
cubic yards	cubic centimeter	7.646×10^{5}
cubic yards	cubic feet	27.0
cubic yards	cubic inches	46,656.0
cubic yards	cubic meters	0.7646
cubic yards	gallons (U.S. liquid)	202.0
cubic yards	liters	764.6
cubic yards	pints (U.S. liquid)	1,615.9
cubic yards	quarts (U.S. liquid)	807.9
cubic yards per minute	cubic feet per second	0.45
cubic yards per minute	gallons per second	3.367
cubic yards per minute	liters per second	12.74
	D	
Dalton	gram	1.650×10^{-24}
days	seconds	86,400.0
decigrams	grams	0.1
deciliters	liters	0.1
decimeters	meters	0.1
degrees (angle)	quadrants	0.01111
degrees (angle)	radians	0.01745

To Convert	Into	Multiply by
degrees (angle)	seconds	3,600.0
degrees per second	radians per second	0.01745
degrees per second	revolutions per minute	0.1667
degrees per second	revolutions per second	2.778×10^{-3}
dekagrams	grams	10.0
dekaliters	liters	10.0
dekameters	meters	10.0
drams (apothecaries' or troy)	ounces (avoirdupois)	0.1371429
drams (apothecaries' or troy)	ounces (troy)	0.125
drams (U.S., fluid or apothecaries')	cubic centimeter	3.6967
drams	grams	1.7718
drams	grains	27.3437
drams	ounces	0.0625
dyne per centimeter	erg per square millimeter	0.01
dyne per square centimeter	atmospheres	9.869×10^{-7}
dyne per square centimeter	inch of mercury at 0°C	2.953×10^{-5}
dyne per square centimeter	inch of water at 4°C	4.015×10^{-4}
dynes	grams	1.020×10^{-3}
dynes	joules per centimeter	10^{-7}
dynes	joules per meter (newtons)	10^{-5}
dynes	kilograms	1.020×10^{-6}
dynes	poundals	7.233×10^{-5}
dynes	pounds	2.248×10^{-6}
dynes per square centimeter	bars	10^{-6}
	E	
ell	centimeter	114.30
ell	inches	45
em, pica	inch	0.167
em, pica	centimeter	0.4233
erg per second	dyne–centimeter per second	1.000
ergs	British thermal unit	9.480×10^{-11}
ergs	dyne-centimeters	1.0
ergs	foot-pounds	7.367×10^{-8}
ergs	gram-calories	0.2389×10^{-7}
ergs	gram-centimeter	1.020×10^{-3}
ergs	horsepower-hour	3.7250×10^{-14}
ergs	joules	10^{-7}
ergs	kilogram-calories	2.389×10^{-11}
ergs	kilogram-meters	1.020×10^{-8}
ergs	kilowatt-hour	0.2778×10^{-13}
ergs	watt-hours	0.2778×10^{-10}
ergs per second	British thermal unit per minute	$5,688 \times 10^{-9}$
ergs per second	foot-pound per minute	4.427×10^{-6}
ergs per second	foot-pound per second	7.3756×10^{-8}
ergs per second	horsepower	1.341×10^{-10}
ergs per second	kilogram-calories per minute	1.433×10^{-9}
ergs per second	kilowatts	10^{-10}
	F	
farads	microfarads	10^{6}
faraday per second	ampere (absolute)	9.6500×10^{4}
faradays	ampere-hours	26.80
faradays	coulombs	9.649×10^{4}
fathom	meter	1.828804
fathoms	feet	6.0
feet	centimeters	30.48
feet	kilometers	3.048×10^{-4}
feet	meters	0.3048
feet	miles (nautical)	1.645×10^{-4}
feet	miles (statute)	1.894×10^{-4}

To Convert	Into	Multiply by
feet	millimeters	304.8
feet	mils	1.2×10^4
feet of water	atmospheres	0.02950
feet of water	inch of mercury	0.8826
feet of water	kilogram per square centimeter	0.03048
feet of water	kilogram per square meter	304.8
feet of water	pounds per square foot	62.43
feet of water	pounds per square inch	0.4335
feet per minute	centimeter per second	0.5080
feet per minute	feet per second	0.01667
feet per minute	kilometer per hour	0.01829
feet per minute	meters per minute	0.3048
feet per minute	miles per hour	0.01136
feet per second	centimeter per second	30.48
feet per second	kilometer per hour	1.097
feet per second	knots	0.5921
feet per second	meters per minute	18.29
feet per second	miles per hour	0.6818
feet per second	miles per minute	0.01136
feet per second per second	centimeter per second per second	30.48
feet per second per second	kilometer per hour per second	1.097
feet per second per second	meters per second per second	0.3048
feet per second per second	miles per hour per second	0.6818
feet per 100 feet	per centigrade	1.0
foot-candle	Lumen per square meter	10.764
foot-pounds	British thermal unit	1.286×10^{-3}
foot-pounds	ergs	1.356×10^7
foot-pounds	gram-calories	0.3238
foot-pounds	horsepower per hour	5.050×10^{-7}
foot-pounds	joules	1.356
foot-pounds	kilogram-calories	3.24×10^{-4}
foot-pounds	kilogram-meters	0.1383
foot-pounds	kilowatt-hour	3.766×10^{-7}
foot-pounds per minute	British thermal unit per minute	1.286×10^{-3}
foot-pounds per minute	foot-pounds per second	0.01667
foot-pounds per minute	horsepower	3.030×10^{-5}
foot-pounds per minute	kilogram-calories per minute	3.24×10^{-4}
foot-pounds per minute	kilowatts	2.260×10^{-5}
foot-pounds per second	British thermal unit per hour	4.6263
foot-pounds per second	British thermal unit per minute	0.07717
foot-pounds per second	horsepower	1.818×10^{-3}
foot-pounds per second	kilogram-calories per minute	0.01945
foot-pounds per second	kilowatts	1.356×10^{-3}
furlongs	miles (U.S.)	0.125
furlongs	rods	40.0
furlongs	feet	660.0
	G	
gallons	cubic centimeter	3,785.0
gallons	cubic feet	0.1337
gallons	cubic inches	231.0
gallons	cubic meters	3.785×10^{-3}
gallons	cubic yards	4.951×10^{-3}
gallons	liters	3.785
gallons (liquid British Imperial)	gallons (U.S. liquid)	1.20095
gallons (U.S.)	gallons (Imperial)	0.83267
gallons of water	pounds of water	8.3453
gallons per minute	cubic foot per second	2.228×10^{-3}
gallons per minute	liters per second	0.06308
gallons per minute	cubic foot per hour	8.0208

To Convert	Into	Multiply by
gausses	lines per square inch	6.452
gausses	webers per square centimeter	10^{-8}
gausses	webers per square inch	6.452×10^{-8}
gausses	webers per square meter	10^{-4}
gilberts	ampere-turns	0.7958
gilberts per centimeter	ampere-turns per centimeter	0.7958
gilberts per centimeter	ampere-turns per inch	2.021
gilberts per centimeter	ampere-turns per meter	79.58
gills (British)	cubic centimeter	142.07
gills	liters	0.1183
gills	pints (liquid)	0.25
grade	radian	0.01571
grains	drams (avoirdupois)	0.03657143
grains (troy)	grains (avoirdupois)	1.0
grains (troy)	grams	0.06480
grains (troy)	ounces (avoirdupois)	2.0833×10^{-3}
grains (troy)	pennyweight (troy)	0.04167
grains/U.S. gallon	parts per million	17.118
grains/U.S. gallon	pounds per million gallon	142.86
grains/Imp. gallon	parts per million	14.286
grams	dynes	980.7
grams	grains	15.43
grams	joules per centimeter	9.807×10^{-5}
grams	joules per meter (newtons)	9.807×10^{-3}
grams	kilograms	0.001
grams	milligrams	1,000
grams	ounces (avoirdupois)	0.03527
grams	ounces (troy)	0.03215
grams	poundals	0.07093
grams	pounds	2.205×10^{-3}
grams per centimeter	pounds per inch	5.600×10^{-3}
grams per cubic centimeter	pounds per cubic foot	62.43
grams per cubic centimeter	pounds per cubic inch	0.03613
grams per cubic centimeter	pounds per mil-foot	3.405×10^{-7}
grams per liter	grains per gallon	58.417
grams per liter	pounds per 1000 gallon	8.345
grams per liter	pounds per cubic foot	0.062427
grams per liter	parts per million	1,000.0
grams per square centimeter	pounds per square foot	2.0481
gram-calories	British thermal unit	3.9683×10^{-3}
gram-calories	ergs	4.1868×10^{7}
gram-calories	foot-pounds	3.0880
gram-calories	horsepower-hour	1.5596×10^{-6}
gram-calories	kilowatt-hour	1.1630×10^{-6}
gram-calories	watt-hour	1.1630×10^{-3}
gram-calories per second	British thermal unit per hour	14.286
gram-centimeters	British thermal unit	9.297×10^{-8}
gram-centimeters	ergs	980.7
gram-centimeters	joules	9.807×10^{-5}
gram-centimeters	kilogram-calorie	2.343×10^{-8}
gram-centimeters	kilogram-meters	10^{-5}
	H	
hand	centimeter	10.16
hectares	acres	2.471
hectares	square feet	1.076×10^{5}
hectograms	grams	100.0
hectoliters	liters	100.0
hectometers	meters	100.0
hectowatts	watts	100.0

To Convert	Into	Multiply by
henries	millihenries	1,000.0
hogsheads (British)	cubic foot	10.114
hogsheads (U.S.)	cubic foot	8.42184
hogsheads (U.S.)	gallons (U.S.)	63
horsepower	British thermal unit per minute	42.44
horsepower	foot-pound per minute	33,000
horsepower	foot-pound per second	550.0
horsepower (metric) (542.5 foot-pound per second)	horsepower (550 foot-pound per second)	0.9863
horsepower (550 foot-pound per second)	horsepower (metric) (542.5 foot-pound per second)	1.014
horsepower	kilogram-calories per minute	10.68
horsepower	kilowatts	0.7457
horsepower	watts	745.7
horsepower (boiler)	British thermal unit per hour	33.479
horsepower (boiler)	kilowatts	9.803
horsepower-hour	British thermal unit	2,547
horsepower-hour	ergs	2.6845×10^{13}
horsepower-hour	foot-pound	1.98×10^{6}
horsepower-hour	gram-calories	641,190
horsepower-hour	joules	2.684×10^{6}
horsepower-hour	kilogram-calories	641.1
horsepower-hour	kilogram-meters	2.737×10^{5}
horsepower-hour	kilowatt-hour	0.7457
hours	days	4.167×10^{-2}
hours	weeks	5.952×10^{-3}
hundredweights (long)	pounds	112
hundredweights (long)	tons (long)	0.05
hundredweights (short)	ounces (avoirdupois)	1,600
hundredweights (short)	pounds	100
hundredweights (short)	tons (metric)	0.0453592
hundredweights (short)	tons (long)	0.0446429
	I	
inches	centimeters	2.540
inches	meters	2.540×10^{-2}
inches	miles	1.578×10^{-5}
inches	millimeters	25.40
inches	mils	1,000.0
inches	yards	2.778×10^{-2}
inches of mercury	atmospheres	0.03342
inches of mercury	feet of water	1.133
inches of mercury	kilogram per square centimeter	0.03453
inches of mercury	kilogram per square meter	345.3
inches of mercury	pounds per square foot	70.73
inches of mercury	pounds per square inch	0.4912
inches of water (at 4°C)	atmospheres	2.458×10^{-3}
inches of water (at 4°C)	inches of mercury	0.07355
inches of water (at 4°C)	kilogram per square centimeter	2.540×10^{-3}
inches of water (at 4°C)	ounces per square inch	0.5781
inches of water (at 4°C)	pounds per square foot	5.204
inches of water (at 4°C)	pounds per square inch	0.03613
International ampere	ampere (absolute)	.9998
International volt	volts (absolute)	1.0003
International volt	joules (absolute)	1.593×10^{-19}
International volt	joules	9.654×10^{4}
	J	
joules	British thermal unit	9.480×10^{-4}
joules	ergs	10^{7}

To Convert	Into	Multiply by
joules	foot-pounds	0.7376
joules	kilogram-calories	2.389×10^{-4}
joules	kilogram-meters	0.1020
joules	watt-hour	2.778×10^{-4}
joules per centimeter	grams	1.020×10^{4}
joules per centimeter	dynes	10^{7}
joules per centimeter	joules per meter (newtons)	100.0
joules per centimeter	poundals	723.3
joules per centimeter	pounds	22.48
	K	
kilograms	dynes	980,665
kilograms	grams	1,000.0
kilograms	joules per centimeter	0.09807
kilograms	joules per meter (newtons)	9.807
kilograms	poundals	70.93
kilograms	pounds	2.205
kilograms	tons (long)	9.842×10^{-4}
kilograms	tons (short)	1.102×10^{-3}
kilograms per cubic meter	grams per cubic centimeter	0.001
kilograms per cubic meter	pounds per cubic foot	0.06243
kilograms per cubic meter	pounds per cubic inch	3.613×10^{-5}
kilograms per cubic meter	pounds per mil-foot	3.405×10^{-10}
kilograms per meter	pounds per foot	0.6720
Kilogram per square centimeter	dynes	980,665
kilograms per square centimeter	atmospheres	0.9678
kilograms per square centimeter	feet of water	32.81
kilograms per square centimeter	inches of mercury	28.96
kilograms per square centimeter	pounds per square foot	2,048
kilograms per square centimeter	pounds per square inch	14.22
kilograms per square meter	atmospheres	9.678×10^{-5}
kilograms per square meter	bars	98.07×10^{-6}
kilograms per square meter	feet of water	3.281×10^{-3}
kilograms per square meter	inches of mercury	2.896×10^{-3}
kilograms per square meter	pounds per square foot	0.2048
kilograms per square meter	pounds per square inch	1.422×10^{-3}
kilograms per square millimeter	kilogram per square meter	10^{6}
kilogram-calories	British thermal unit	3.968
kilogram-calories	foot-pounds	3,088
kilogram-calories	horsepower-hour	1.560×10^{-3}
kilogram-calories	joules	4,186
kilogram-calories	kilogram-meters	426.9
kilogram-calories	kilojoules	4.186
kilogram-calories	kilowatt-hour	1.163×10^{-3}
kilogram meters	British thermal unit	9.294×10^{-3}
kilogram meters	ergs	9.804×10^{7}
kilogram meters	foot-pounds	7.233
kilogram meters	joules	9.804
kilogram meters	kilogram-calories	2.342×10^{-3}
kilogram meters	kilowatt-hour	2.723×10^{-6}
kilolines	maxwells	1,000.0
kiloliters	liters	1,000.0
kilometers	centimeters	10^{5}
kilometers	feet	3,281
kilometers	inches	3.937×10^{4}
kilometers	meters	1,000.0
kilometers	miles	0.6214
kilometers	millimeters	10^{4}
kilometers	yards	1,094
kilometers per hour	centimeter per second	27.78

To Convert	Into	Multiply by
kilometers per hour	feet per minute	54.68
kilometers per hour	feet per second	0.9113
kilometers per hour	knots	0.5396
kilometers per hour	meters per minute	16.67
kilometers per hour	miles per hour	0.6214
kilometers per second	centimeter per second per second	27.78
kilometers per second	feet per second per second	0.9113
kilometers per second	meters per second per second	0.2778
kilometers per second	miles per hour per second	0.6214
kilowatts	British thermal unit per minute	56.92
kilowatts	foot-pounds per minute	4.426×10^4
kilowatts	foot-pounds per second	737.6
kilowatts	horsepower	1.341
kilowatts	kilogram-calories per minute	14.34
kilowatts	watts	1,000.0
kilowatt-hour	British thermal unit	3,413
kilowatt-hour	ergs	3.600×10^{13}
kilowatt-hour	foot-pound	2.655×10^6
kilowatt-hour	gram-calories	859,850
kilowatt-hour	horsepower-hour	1.341
kilowatt-hour	joules	3.6×10^6
kilowatt-hour	kilogram-calories	860.5
kilowatt-hour	kilogram-meters	3.671×10^5
kilowatt-hour	pounds of water evaporated from and at 212°F	8.53
kilowatt-hours	pounds of water raised from 62° to 212°F	22.75
knots	feet per hour	6,080
knots	kilometers per hour	1.8532
knots	nautical miles per hour	1.0
knots	statute miles per hour	1.151
knots	yards per hour	2,027
knots	feet per second	1.689
	L	
league	miles (approximate)	3.0
Light year	miles	5.9×10^{12}
Light year	kilometers	9.4637×10^{12}
lines per square centimeter	gausses	1.0
lines per square inch	gausses	0.1550
lines per square inch	webers per square centimeter	1.550×10^{-9}
lines per square inch	webers per square inch	10^{-8}
lines per square inch	webers per square meter	1.550×10^{-5}
links (engineer's)	inches	12.0
links (surveyor's)	inches	7.92
liters	bushels (U.S. dry)	0.02838
liters	cubic centimeter	1,000.0
liters	cubic feet	0.03531
liters	cubic inches	61.02
liters	cubic meters	0.001
liters	cubic yards	1.308×10^{-3}
liters	gallons (U.S. liquid)	0.2642
liters	pints (U.S. liquid)	2.113
liters	quarts (U.S. liquid)	1.057
liters per minute	cubic foot per second	5.886×10^{-4}
liters per minute	gallons per second	4.403×10^{-3}
Lumen	spherical candle power	.07958
Lumen	watt	.001496
lumens per square foot	foot-candles	1.0
lumen per square foot	lumen per square meter	10.76
lux	foot-candles	0.0929

To Convert	Into	Multiply by
	M	
maxwells	kilolines	0.001
maxwells	webers	10^{-8}
megalines	maxwells	10^{6}
megohms	microhms	10^{12}
megohms	ohms	10^{6}
meters	centimeters	100.0
meters	feet	3.281
meters	inches	39.37
meters	kilometers	0.001
meters	miles (nautical)	5.396×10^{-4}
meters	miles (statute)	6.214×10^{-4}
meters	millimeters	1,000.0
meters	yards	1.094
meters	varas	1.179
meters per minute	centimeter per second	1,667
meters per minute	feet per minute	3.281
meters per minute	feet per second	0.05468
meters per minute	kilometer per hour	0.06
meters per minute	knots	0.03238
meters per minute	miles per hour	0.03728
meters per second	feet per minute	196.8
meters per second	feet per second	3.281
meters per second	kilometers per hour	3.6
meters per second	kilometers per minute	0.06
meters per second	miles per hour	2.237
meters per second	miles per minute	0.03728
meters per second per second	centimeter per second per second	100.0
meters per second per second	foot per second per second	3.281
meters per second per second	kilometer per hour per second	3.6
meters per second per second	miles per hour per second	2.237
meter-kilograms	centimeter-dynes	9.807×10^{7}
meter-kilograms	centimeter-grams	10^{5}
meter-kilograms	pound-feet	7.233
microfarad	farads	10^{-6}
micrograms	grams	10^{-6}
microhms	megohms	10^{-12}
microhms	ohms	10^{-6}
microliters	liters	10^{-6}
Microns	meters	1×10^{-6}
miles (nautical)	feet	6,080.27
miles (nautical)	kilometers	1.853
miles (nautical)	meters	1,853
miles (nautical)	miles (statute)	1.1516
miles (nautical)	yards	2,027
miles (statute)	centimeters	1.609×10^{5}
miles (statute)	feet	5,280
miles (statute)	inches	6.336×10^{4}
miles (statute)	kilometers	1.609
miles (statute)	meters	1,609
miles (statute)	miles (nautical)	0.8684
miles (statute)	yards	1,760
miles per hour	centimeter per second	44.70
miles per hour	feet per minute	88
miles per hour	feet per second	1.467
miles per hour	kilometer per hour	1.609
miles per hour	kilometer per minute	0.02682
miles per hour	knots	0.8684
miles per hour	meters per minute	26.82

To Convert	Into	Multiply by
miles per hour	miles per minute	0.1667
miles per hour per second	centimeter per second per second	44.70
miles per hour per second	feet per second per second	1.467
miles per hour per second	kilometer per hour per second	1.609
miles per hour per second	meters per second per second	0.4470
miles per minute	centimeter per second	2,682
miles per minute	feet per second	88
miles per minute	kilometer per minute	1.609
miles per minute	knots per minute	0.8684
miles per minute	miles per hour	60
mil-feet	cubic inches	9.425×10^{-6}
milliers	kilograms	1,000
millimicrons	meters	1×10^{-9}
milligrams	grains	0.01543236
milligrams	grams	0.001
milligrams per liter	parts per million	1.0
millihenries	henries	0.001
milliliters	liters	0.001
millimeters	centimeters	0.1
millimeters	feet	3.281×10^{-3}
millimeters	inches	0.03937
millimeters	kilometers	10^{-6}
millimeters	meters	0.001
millimeters	miles	6.214×10^{-7}
millimeters	mils	39.37
millimeters	yards	1.094×10^{-3}
million gallons per day	cubic foot per second	1.54723
mils	centimeters	2.540×10^{-3}
mils	feet	8.333×10^{-5}
mils	inches	0.001
mils	kilometers	2.540×10^{-8}
mils	yards	2.778×10^{-5}
miner's inches	cubic foot per minute	1.5
minims (British)	cubic centimeter	0.059192
minims (U.S., fluid)	cubic centimeter	0.061612
minutes (angles)	degrees	0.01667
minutes (angles)	quadrants	1.852×10^{-4}
minutes (angles)	radians	2.909×10^{-4}
minutes (angles)	seconds	60.0
myriagrams	kilograms	10.0
myriameters	kilometers	10.0
myriawatts	kilowatts	10.0
	N	
nepers	decibels	8.686
newton	dynes	1 × 105
	O	
ohm (International)	ohm (absolute)	1.0005
ohms	megohms	10^{-6}
ohms	microhms	10^{6}
ounces	drams	16.0
ounces	grains	437.5
ounces	grams	28.349527
ounces	pounds	0.0625
ounces	ounces (troy)	0.9115
ounces	tons (long)	2.790×10^{-5}
ounces	tons (metric)	2.835×10^{-5}
ounces (fluid)	cu. inches	1.805
ounces (fluid)	liters	0.02957
ounces (troy)	grains	480.0
ounces (troy)	grams	31.103481

To Convert	Into	Multiply by
ounces (troy)	ounces (avoirdupois)	1.09714
ounces (troy)	pennyweights (troy)	20.0
ounces (troy)	pounds (troy)	0.08333
Ounce per square inch	dynes per square centimeter	4,309
ounces per square inch	pounds per square inch	0.0625
	P	
parsec	miles	19×10^{12}
parsec	kilometers	3.084×10^{13}
parts per million	grains per U.S. gallon	0.0584
parts per million	grains per Imperial gallon	0.07016
parts per million	pounds per million gallon	8.345
pecks (British)	cubic inches	554.6
pecks (British)	liters	9.091901
pecks (U.S.)	bushels	0.25
pecks (U.S.)	cubic inches	537.605
pecks (U.S.)	liters	8.809582
pecks (U.S.)	quarts (dry)	8
pennyweights (troy)	grains	24.0
pennyweights (troy)	ounces (troy)	0.05
pennyweights (troy)	grams	1.55517
pennyweights (troy)	pounds (troy)	4.1667×10^{-3}
pints (dry)	cubic inches	33.60
pints (liquid)	cubic centimeter	473.2
pints (liquid)	cubic feet	0.01671
pints (liquid)	cubic inches	28.87
pints (liquid)	cubic meters	4.732×10^{-4}
pints (liquid)	cubic yards	6.189×10^{-4}
pints (liquid)	gallons	0.125
pints (liquid)	liters	0.4732
pints (liquid)	quarts (liquid)	0.5
Planck's quantum	erg-second	6.624×10^{-27}
Poise	Gram per centimeter second	1.00
Pounds (avoirdupois)	ounces (troy)	14.5833
poundals	dynes	13,826
poundals	grams	14.10
poundals	joules per centimeter	1.383×10^{-3}
poundals	joules per meter (newtons)	0.1383
poundals	kilograms	0.01410
poundals	pounds	0.03108
pounds	drams	256
pounds	dynes	44.4823×10^{4}
pounds	grains	7,000
pounds	grams	453.5924
pounds	joules per centimeter	0.04448
pounds	joules per meter (newtons)	4.448
pounds	kilograms	0.4536
pounds	ounces	16.0
pounds	ounces (troy)	14.5833
pounds	poundals	32.17
pounds	pounds (troy)	1.21528
pounds	tons (short)	0.0005
pounds (troy)	grains	5,760
pounds (troy)	grams	373.24177
pounds (troy)	ounces (avoirdupois)	13.1657
pounds (troy)	ounces (troy)	12.0
pounds (troy)	pennyweights (troy)	240.0
pounds (troy)	pounds (avoirdupois)	0.822857
pounds (troy)	tons (long)	3.6735×10^{-4}
pounds (troy)	tons (metric)	3.7324×10^{-4}
pounds (troy)	tons (short)	4.1143×10^{-4}

To Convert	Into	Multiply by
pounds of water	cubic foot	0.01602
pounds of water	cubic inches	27.68
pounds of water	gallons	0.1198
pounds of water per minute	cubic foot per second	2.670×10^{-4}
pound-feet	centimeter-dynes	1.356×10^{7}
pound-feet	centimeter-grams	13,825
pound-feet	meter-kilogram	0.1383
pounds per cubic foot	grams per cubic centimeter	0.01602
pounds per cubic foot	kilogram per cubic meter	16.02
pounds per cubic foot	pounds per cubic inch	5.787×10^{-4}
pounds per cubic foot	pounds per mil-foot	5.456×10^{-9}
pounds per cubic inch	gram per cubic centimeter	27.68
pounds per cubic inch	kilogram per cubic meter	2.768×10^{4}
pounds per cubic inch	pounds per cubic foot	1,728
pounds per cubic inch	pounds per mil-foot	9.425×10^{-6}
pounds per foot	kilogram per meter	1.488
pounds per inch	gram per centimeter	178.6
pounds per mil-foot	gram per cubic centimeter	2.306×10^{6}
pounds per square foot	atmospheres	4.725×10^{-4}
pounds per square foot	feet of water	0.01602
pounds per square foot	inches of mercury	0.01414
pounds per square foot	kilogram per square meter	4.882
pounds per square foot	pounds per square inch	6.944×10^{-3}
pounds per square inch	atmospheres	0.06804
pounds per square inch	feet of water	2.307
pounds per square inch	inches of mercury	2.036
pounds per square inch	kilogram per square meter	703.1
pounds per square inch	pounds per square foot	144.0
	Q	
quadrants (angle)	degrees	90.0
quadrants (angle)	minutes	5,400.0
quadrants (angle)	radians	1.571
quadrants (angle)	seconds	3.24×10^{5}
quarts (dry)	cubic inches	67.20
quarts (liquids)	cubic centimeter	946.4
quarts (liquids)	cubic feet	0.03342
quarts (liquids)	cubic inches	57.75
quarts (liquids)	cubic meters	9.464×10^{-4}
quarts (liquids)	cubic yards	1.238×10^{-3}
quarts (liquids)	gallons	0.25
quarts (liquids)	liters	0.9463
	R	
radians	degrees	57.30
radians	minutes	3,438
radians	quadrants	0.6366
radians	seconds	2.063×10^{5}
radians per second	degrees per second	57.30
radians per second	revolutions per minute	9.549
radians per second	revolutions per second	0.1592
radians per second per second	revolution per minute per minute	573.0
radians per second per second	revolution per minute per second	9.549
radians per second per second	revolution per second per second	0.1592
revolutions	degrees	360.0
revolutions	quadrants	4.0
revolutions	radians	6.283
revolutions per minute	degrees per second	6.0
revolutions per minute	radians per second	0.1047
revolutions per minute	revolution per second	0.01667
revolutions per minute per minute	radians per second per second	1.745×10^{-3}
revolutions per minute per minute	revolution per minute per second	0.01667

To Convert	Into	Multiply by
revolutions per minute per minute	revolution per second per second	2.778×10^{-4}
revolutions per second	degrees per second	360.0
revolutions per second	radians per second	6.283
revolutions per second	revolution per minute	60.0
revolutions per second per second	radians per second per second	6.283
revolutions per second per second	revolution per minute per minute	3,600.0
revolutions per second per second	revolution per minute per second	60.0
rod	chain (Gunters)	.25
rod	meters	5.029
rods (surveyors' measure)	yards	5.5
rods	feet	16.5
	S	
scruples	grains	20
seconds (angle)	degrees	2.778×10^{-4}
seconds (angle)	minutes	0.01667
seconds (angle)	quadrants	3.087×10^{-6}
seconds (angle)	radians	4.848×10^{-6}
slug	kilogram	14.59
slug	pounds	32.17
sphere	steradians	12.57
square centimeters	circular mils	1.973×10^{5}
square centimeters	square feet	1.076×10^{-3}
square centimeters	square inches	0.1550
square centimeters	square meters	0.0001
square centimeters	square miles	3.861×10^{-11}
square centimeters	square millimeters	100.0
square centimeters	square yards	1.196×10^{-4}
square feet	acres	2.296×10^{-5}
square feet	circular mils	1.833×10^{8}
square feet	square centimeter	929.0
square feet	square inches	144.0
square feet	square meters	0.09290
square feet	square miles	3.587×10^{-8}
square feet	square millimeters	9.290×10^{4}
square feet	square yards	0.1111
square inches	circular mils	1.273×10^{6}
square inches	square centimeter	6.452
square inches	square feet	6.944×10^{-3}
square inches	square millimeters	645.2
square inches	square mils	10^{6}
square inches	square yards	7.716×10^{-4}
square kilometers	acres	247.1
square kilometers	square centimeter	10^{10}
square kilometers	square foot	10.76×10^{6}
square kilometers	square inches	1.550×10^{9}
square kilometers	square meters	10^{6}
square kilometers	square miles	0.3861
square kilometers	square yards	1.196×10^{6}
square meters	acres	2.471×10^{-4}
square meters	square centimeter	10^{4}
square meters	square feet	10.76
square meters	square inches	1,550
square meters	square miles	3.861×10^{-7}
square meters	square millimeters	10^{6}
square meters	square yards	1.196
square miles	acres	640.0
square miles	square feet	27.88×10^{6}
square miles	square kilometer	2.590
square miles	square meters	2.590×10^{6}
square miles	square yards	3.098×10^{6}

To Convert	Into	Multiply by
square millimeters	circular mils	1,973
square millimeters	square centimeter	0.01
square millimeters	square feet	1.076×10^{-5}
square millimeters	square inches	1.550×10^{-3}
square mils	circular mils	1.273
square mils	square centimeter	6.452×10^{-6}
square mils	square inches	10^{-6}
square yards	acres	2.066×10^{-4}
square yards	square centimeter	8,361
square yards	square feet	9.0
square yards	square inches	1,296
square yards	square meters	0.8361
square yards	square miles	3.228×10^{-7}
square yards	square millimeters	8.361×10^{5}
	T	
temperature (°C) + 273	absolute temperature (°C)	1.0
temperature (°C) + 17.78	temperature (°F)	1.8
temperature(°F) + 460	absolute temperature (°F)	1.0
temperature (°F) − 32	temperature (°C)	5/9
tons (long)	kilograms	1,016
tons (long)	pounds	2,240
tons (long)	tons (short)	1.120
tons (metric)	kilograms	1,000
tons (metric)	pounds	2,205
tons (short)	kilograms	907.1848
tons (short)	ounces	32,000
tons (short)	ounces (troy)	29,166.66
tons (short)	pounds	2,000
tons (short)	pounds (troy)	2,430.56
tons (short)	tons (long)	0.89287
tons (short)	tons (metric)	0.9078
tons (short) per square foot	kilogram per square meter	9,765
tons (short) per square foot	pounds per square inch	2,000
tons of water per 24 hour	pounds of water per hour	83.333
tons of water per 24 hour	gallons per minute	0.16643
tons of water per 24 hour	cubic foot per hour	1.3349
	V	
volt/inch	volt per centimeter	.39370
volt (absolute)	statvolts	.003336
	W	
watts	British thermal unit per hour	3.4129
watts	British thermal unit per minute	0.05688
watts	ergs per second	107
watts	foot-pound per minute	44.27
watts	foot-pound per second	0.7378
watts	horsepower	1.341×10^{-3}
watts	horsepower (metric)	1.360×10^{-3}
watts	kilogram-calories per minute	0.01433
watts	kilowatts	0.001
watts (absolute)	British thermal unit (mean) per minute	0.056884
watts (absolute)	joules per second	1
watt-hours	British thermal unit	3.413
watt-hours	ergs	3.60×10^{10}
watt-hours	foot-pounds	2,656
watt-hours	gram-calories	859.85
watt-hours	horsepower-hour	1.341×10^{-3}
watt-hours	kilogram-calories	0.8605
watt-hours	kilogram-meters	367.2
watt-hours	kilowatt-hour	0.001

To Convert	Into	Multiply by
watt (International)	watt (absolute)	1.0002
webers	maxwells	10^8
webers	kilolines	10^5
webers per square inch	gausses	1.550×10^7
webers per square inch	lines per square inch	10^8
webers per square inch	webers per square centimeter	0.1550
webers per square inch	webers per square meter	1,550
webers per square meter	gausses	10^4
webers per square meter	lines per square inch	6.452×10^4
webers per square meter	webers per square centimeter	10^{-4}
webers per square meter	webers per square inch	6.452×10^{-4}
	Y	
yards	centimeters	91.44
yards	kilometers	9.144×10^{-4}
yards	meters	0.9144
yards	miles (nautical)	4.934×10^{-4}
yards	miles (statute)	5.682×10^{-4}
yards	millimeters	914.4

17.2 Conversion Constants and Multipliers [2]

Recommended Decimal Multiples and Submultiples

Multiples and Submultiples	Prefixes	Symbols	Multiples and Submultiples	Prefixes	Symbols
10^{18}	exa	E	10^{-1}	deci	d
10^{15}	peta	P	10^{-2}	centi	c
10^{12}	tera	T	10^{-3}	milli	m
10^9	giga	G	10^{-6}	micro	μ (Greek mu)
10^6	mega	M	10^{-9}	nano	n
10^3	kilo	k	10^{-12}	pico	p
10^2	hecto	h	10^{-15}	femto	f
10	deca	da	10^{-18}	atto	a

Conversion Factors—Metric to English

To Convert	Into	Multiply By
Inches	Centimeters	0.393700787
Feet	Meters	3.280839895
Yards	Meters	1.093613298
Miles	Kilometers	0.6213711922
Ounces	Grams	$3.527396195 \times 10^{-2}$
Pounds	Kilograms	2.204622622
Gallons (U.S. Liquid)	Liters	0.2641720524
Fluid ounces	Milliliters (cc)	$3.381402270 \times 10^{-2}$
Square inches	Square centimeters	0.1550003100
Square feet	Square meters	10.76391042
Square yards	Square meters	1.195990046
Cubic inches	Milliliters (cc)	$6.102374409 \times 10^{-2}$
Cubic feet	Cubic meters	35.31466672
Cubic yards	Cubic meters	1.307950619

[2]Zwillinger, D., ed. 1996. *CRC Standard Mathematical Tables and Formulae*, 30th ed. CRC Press, Boca Raton, FL: Greek alphabet, conversion constants and multipliers (recommended decimal multiples and submultiples, metric to English, English to metric, general, temperature factors).

Conversion Factors—English to Metric [3]

To Convert	Into	Multiply By
Microns	Mils	**25.4**
Centimeters	Inches	**2.54**
Meters	Feet	**0.3048**
Meters	Yards	**0.9144**
Kilometers	Miles	**1.609344**
Grams	Ounces	28.34952313
Kilograms	Pounds	**0.45359237**
Liters	Gallons (U.S. Liquid)	**3.785411784**
Millimeters (cc)	Fluid ounces	29.57352956
Square centimeters	Square inches	**6.4516**
Square meters	Square feet	**0.09290304**
Square meters	Square yards	**0.83612736**
Milliliters (cc)	Cubic inches	**16.387064**
Cubic meters	Cubic feet	$2.831684659 \times 10^{-2}$
Cubic meters	Cubic yards	0.764554858

Conversion Factors—General [3]

To Convert	Into	Multiply By
Atmospheres	Feet of water @ 4°C	2.950×10^{-2}
Atmospheres	Inches of mercury @ 0°C	3.342×10^{-2}
Atmospheres	Pounds per square inch	6.804×10^{-2}
BTU	Foot-pounds	1.285×10^{-3}
BTU	Joules	9.480×10^{-4}
Cubic feet	Cords	**128**
Degree (angle)	Radians	57.2958
Ergs	Foot-pounds	1.356×10^{7}
Feet	Miles	**5280**
Feet of water @ 4°C	Atmospheres	33.90
Foot-pounds	Horsepower-hours	1.98×10^{6}
Foot-pounds	Kilowatt-hours	2.655×10^{6}
Foot-pounds per min	Horsepower	3.3×10^{4}
Horsepower	Foot-pounds per sec	1.818×10^{-3}
Inches of mercury @ 0°C	Pounds per square inch	2.036
Joules	BTU	1054.8
Joules	Foot-pounds	1.35582
Kilowatts	BTU per min	1.758×10^{-2}
Kilowatts	Foot-pounds per min	2.26×10^{-5}
Kilowatts	Horsepower	0.745712
Knots	Miles per hour	0.86897624
Miles	Feet	1.894×10^{-4}
Nautical miles	miles	0.86897624
Radians	Degrees	1.745×10^{-2}
Squares feet	Acres	**43560**
Watts	BTU per min	17.5796

Temperature Factors

°F = 9/5(°C) + 32

Fahrenheit temperature = 1.8(temperature in kelvins) − 459.67

°C = 5/9[(°F) − 32)]

Celsius temperature = temperature in kelvin − 273.15

Fahrenheit temperature = 1.8 (Celsius temperature) + 32

[3]Boldface numbers are exact; others are given to ten significant figures where so indicated by the multiplier factor.

Conversion of Temperatures

From	To	
°Celsius	°Fahrenheit	$t_F = (t_C \times 1.8) + 32$
	Kelvin	$T_K = t_C + 273.15$
	°Rankine	$T_R = (t_C + 273.15) \times 18$
°Fahrenheit	°Celsius	$t_C = \frac{t_F - 32}{1.8}$
	Kelvin	$T_K = \dfrac{t_F - 32}{1.8} + 273.15$
	°Rankine	$T_R = t_F + 459.67$
Kelvin	°Celsius	$t_C = T_K - 273.15$
	°Rankine	$T_R = T_K \times 1.8$
°Rankine	°Fahrenheit	$t_F = T_R - 459.67$
	Kelvin	$T_K = \dfrac{T_R}{1.8}$

18

General Mathematical Tables[1]

W.F. Ames
Georgia Institute of Technology

George Cain
Georgia Institute of Technology

[1]The material in this chapter was previously published by CRC Press in *The Engineering Handbook*, pp. 2037–2079, 2187–2192, and 2196–2202, 1996.

0-8493-0050-9/00/$0.00+$.50

18.1 Introduction to Mathematics Chapter

Mathematics has been defined as "the logic of drawing unambiguous conclusions from arbitrary assumptions." The assumptions do not come from the mathematics but arise from the engineering or scientific discipline. Mathematical models of real world phenomena have been remarkably useful. The simplest of these express various physical *laws* in mathematical form. For example, Ohms law ($V = IR$), Newton's second law ($F = ma$) and the ideal gas law ($pv = nRT$) provide starting points for many theories. More complicated models employ difference, differential, or integral equations. Whatever the model, the conclusions drawn from it are unambiguous!

In this mathematics chapter we provide for an extensive review of algebra and geometry in the first section. One feature is the collection of common geometric figures together with their areas, volumes, and other data.

The second section reviews the fundamentals of trigonometry and presents the myriad of identities, which are very useful but very easy to forget. The third section presents tables of various series for a number of useful functions, followed by the theory of how these series are calculated.

The next section gives the elements of the differential calculus, discusses the calculation and theory of maxima and minima, and contains a table of derivatives. The fifth section provides a summary of the integral calculus including formulas for calculating such fundamental physical ideas as arclength, area, volume, work, centroids, etc.

Special functions constitute Sec. 18.7. Here are the Bessel functions and various basic polynomials. A special feature found here is the table of three-dimensional quadratic figures.

Because linear algebra and vector calculus are so basic Secs. 18. 8 and 9 are devoted to their summaries. The last section presents the various Fourier transforms and their properties in table form.

Throughout the chapter various references are provided.

18.2 Elementary Algebra and Geometry

Fundamental Properties (Real Numbers)

$a + b = b + a$	Commutative law for addition
$(a + b) + c = a + (b + c)$	Associative law for addition
$a + 0 = 0 + a$	Identity law for addition
$a + (-a) = (-a) + a = 0$	Inverse law for addition
$a(bc) = (ab)c$	Associative law for multiplication
$a(\frac{1}{a}) = (\frac{1}{a})a = 1,\ a \neq 0$	Inverse law for multiplication
$(a)(1) = (1)(a) = a$	Identity law for multiplication
$ab = ba$	Commutative law for multiplication
$a(b + c) = ab + ac$	Distributive law

Division by zero is not defined.

Exponents

For integers m and n,

$$a^n a^m = a^{n+m}$$

$$a^n / a^m = a^{n-m}$$

$$(a^n)^m = a^{nm}$$

$$(ab)^m = a^m b^m$$

$$(a/b)^m = a^m / b^m$$

Fractional Exponents

$$a^{p/q} = (a^{1/q})^p$$

where $a^{1/q}$ is the positive qth root of a if $a > 0$ and the negative qth root of a if a is negative and q is odd. Accordingly, the five rules of exponents just given (for integers) are also valid if m and n are fractions, provided a and b are positive.

Irrational Exponents

If an exponent is irrational (e.g., $\sqrt{2}$), the quantity, such as $a^{\sqrt{2}}$, is the limit of the sequence $a^{1.4}, a^{1.41}, a^{1.414}, \ldots$.

Operations with Zero

$$0^m = 0 \quad a^0 = 1$$

Logarithms

If x, y, and b are positive and $b \neq 1$,

$$\log_b(xy) = \log_b x + \log_b y$$

$$\log_b(x/y) = \log_b x - \log_b y$$

$$\log_b x^p = p \log_b x$$

$$\log_b(1/x) = -\log_b x$$

$$\log_b b = 1$$

$$\log_b 1 = 0 \qquad \text{Note: } b^{\log_b x} = x$$

Change of Base ($a \neq 1$)

$$\log_b x = \log_a x \log_b a$$

Factorials

The factorial of a positive integer n is the product of all of the positive integers less than or equal to the integer n and is denoted $n!$. Thus,

$$n! = 1 \cdot 2 \cdot 3 \cdot \cdots \cdot n$$

Factorial 0 is defined: $0! = 1$.

Stirling's Approximation

$$\lim_{n \to \infty} (n/e)^n \sqrt{2\pi n} = n!$$

Binomial Theorem

For positive integer n

$$(x+y)^n = x^n + nx^{n-1}y + \frac{n(n-1)}{2!}x^{n-2}y^2 + \frac{n(n-1)(n-2)}{3!}x^{n-3}y^3 + \cdots + nxy^{n-1} + y^n$$

Factors and Expansion

$$(a+b)^2 = a^2 + 2ab + b^2$$
$$(a-b)^2 = a^2 - 2ab + b^2$$
$$(a+b)^3 = a^3 + 3a^2b + 3ab^2 + b^3$$
$$(a-b)^3 = a^3 - 3a^2b + 3ab^2 - b^3$$
$$(a^2 - b^2) = (a-b)(a+b)$$
$$(a^3 - b^3) = (a-b)(a^2 + ab + b^2)$$
$$(a^3 + b^3) = (a+b)(a^2 - ab + b^2)$$

Progression

An *arithmetic progression* is a sequence in which the difference between any term and the preceding term is a constant (d):

$$a, a+d, a+2d, \ldots, a+(n-1)d$$

If the last term is denoted $l[= a + (n-1)d]$, then the sum is

$$s = \frac{n}{2}(a+l)$$

A *geometric progression* is a sequence in which the ratio of any term is a constant r. Thus, for n terms,

$$a, ar, ar^2, \ldots, ar^{n-1}$$

The sum is

$$S = \frac{a - ar^n}{1-r}$$

Complex Numbers

A complex number is an ordered pair of real numbers (a, b).

Equality:

$$(a, b) = (c, d) \quad \text{if and only if } a = c \text{ and } b = d$$

Addition:

$$(a, b) + (c, d) = (a+c, b+d)$$

Multiplication:

$$(a, b)(c, d) = (ac - bd, ad + bc)$$

The first element (a, b) is called the *real* part, the second the *imaginary* part. An alternative notation for (a, b) is $a + bi$, where $i^2 = (-1, 0)$, and $i = (0, 1)$ or $0 + 1i$ is written for this complex number as a convenience. With this understanding, i behaves as a number, that is, $(2-3i)(4+i) = 8 - 12i + 2i - 3i^2 = 11 - 10i$. The conjugate of $a + bi$ is $a - bi$, and the product of a complex number and its conjugate is $a^2 + b^2$. Thus, *quotients* are computed by multiplying numerator and denominator by the conjugate of the denominator, illustrated as follows:

$$\frac{2+3i}{4+2i} = \frac{(4-2i)(2+3i)}{(4-2i)(4+2i)} = \frac{14+8i}{20} = \frac{7+4i}{10}$$

Polar Form

The complex number $x + iy$ may be represented by a plane vector with components x and y:

$$x + iy = r(\cos\theta + i\sin\theta)$$

(See Fig. 18.1.) Then, given two complex numbers $z_1 = r_1(\cos\theta_1 + i\sin\theta_1)$, and $z_2 = r_2(\cos\theta_2 + i\sin\theta_2)$, the product and quotient are as follows.

Product:

$$z_1 z_2 = r_1 r_2[\cos(\theta_1 + \theta_2) + i\sin(\theta_1 + \theta_2)]$$

Quotient:

$$z_1/z_2 = (r_1/r_2)[\cos(\theta_1 - \theta_2) + i\sin(\theta_1 - \theta_2)]$$

Powers:

$$z^n = [r\cos\theta + i\sin\theta]^n = r^n[\cos n\theta + i\sin n\theta]$$

Roots:

$$\begin{aligned} z^{1/n} &= [r\cos\theta + i\sin\theta]^{1/n} \\ &= r^{1/n}\left[\cos\frac{\theta + k\cdot 360}{n} + i\sin\frac{\theta + k\cdot 360}{n}\right] \\ &\quad k = 0, 1, 2, \ldots, n-1 \end{aligned}$$

Permutations

A permutation is an ordered arrangement (sequence) of all or part of a set of objects. The number of permutations of n objects taken r at a time is

$$\begin{aligned} p(n,r) &= n(n-1)(n-2)\cdots(n-r+1) \\ &= \frac{n!}{(n-r)!} \end{aligned}$$

A permutation of positive integers is *even* or *odd* if the total number of inversions is an even integer or an odd integer, respectively. Inversions are counted relative to each integer j in the permutation by counting the number of integers that follow j and are less than j. These are summed to give the total number of inversions. For example, the permutation 4132 has four inversions: three relative to 4 and one relative to 3. This permutation is therefore even.

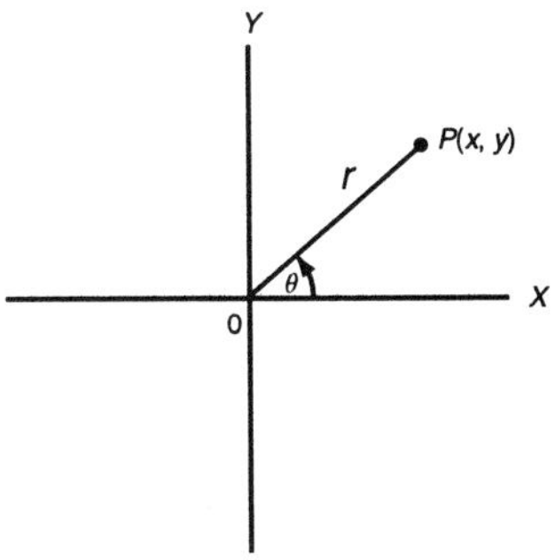

FIGURE 18.1 Polar form of complex number.

Combinations

A combination is a selection of one or more objects from among a set of objects regardless of order. The number of combinations of n different objects taken r at a time is

$$C(n,r) = \frac{P(n,r)}{r!} = \frac{n!}{r!(n-r)!}$$

Algebraic Equations

Quadratic

If $ax^2 + bx + c = 0$, and $a \neq 0$, then roots are

$$x = \frac{-b \pm \sqrt{b^2 - 4ac}}{2a}$$

Cubic

To solve $x^3 + bx^2 + cx + d = 0$, let $x = y - b/3$. Then the *reduced cubic* is obtained,

$$y^3 + py + q = 0$$

where $p = c - (1/3)b^2$ and $q = d - (1/3)bc + (2/27)b^3$. Solutions of the original cubic are then in terms of the reduced cubic roots y_1, y_2, y_3,

$$x_1 = y_1 - (1/3)b \qquad x_2 = y_2 - (1/3)b \qquad x_3 = y_3 - (1/3)b$$

The three roots of the reduced cubic are

$$\begin{aligned} y_1 &= (A)^{1/3} + (B)^{1/3} \\ y_2 &= W(A)^{1/3} + W^2(B)^{1/3} \\ y_3 &= W^2(A)^{1/3} + W(B)^{1/3} \end{aligned}$$

where

$$\begin{aligned} A &= -\frac{1}{2}q + \sqrt{(1/27)p^3 + \frac{1}{4}q^2} \\ B &= -\frac{1}{2}q - \sqrt{(1/27)p^3 + \frac{1}{4}q^2} \\ W &= \frac{-1 + i\sqrt{3}}{2}, \quad W^2 = \frac{-1 - i\sqrt{3}}{2} \end{aligned}$$

When $(1/27)p^3 + (1/4)q^2$ is negative, A is complex; in this case A should be expressed in trigonometric form: $A = r(\cos\theta + i\sin\theta)$ where θ is a first or second quadrant angle, as q is negative or positive. The three roots of the reduced cubic are

$$\begin{aligned} y_1 &= 2(r)^{1/3}\cos(\theta/3) \\ y_2 &= 2(r)^{1/3}\cos\left(\frac{\theta}{3} + 120^\circ\right) \\ y_3 &= 2(r)^{1/3}\cos\left(\frac{\theta}{3} + 240^\circ\right) \end{aligned}$$

Geometry

Figures 18.2–18.12 are a collection of common geometric figures. Area (A), volume (V), and other measurable features are indicated.

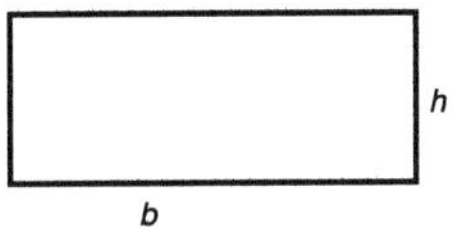

FIGURE 18.2 Rectangle: $A = bh$.

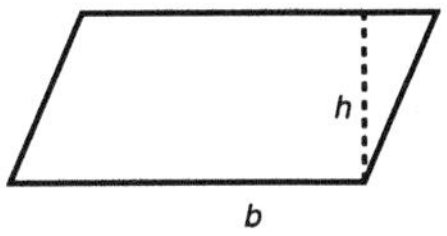

FIGURE 18.3 Parallelogram: $A = bh$.

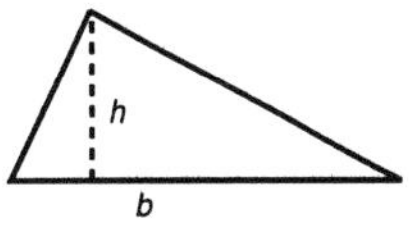

FIGURE 18.4 Triangle: $A = \frac{1}{2}bh$.

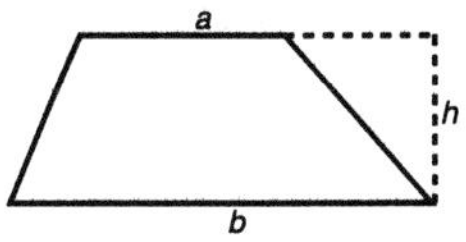

FIGURE 18.5 Trapezoid: $A = \frac{1}{2}(a + b)h$

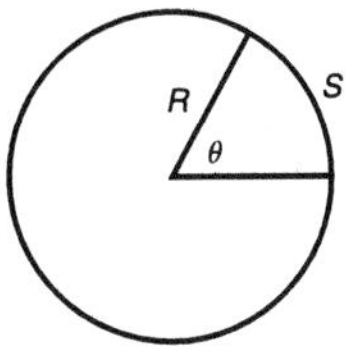

FIGURE 18.6 Circle: $A = \pi R^2$; circumference $= 2\pi R$; arc length $S = R\theta$ (θ in radians).

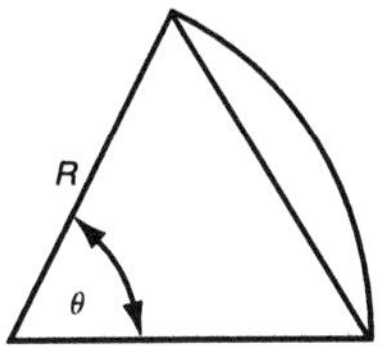

FIGURE 18.7 Sector of a circle: $A_{sector} = \frac{1}{2}R^2\theta$; $A_{segment} = \frac{1}{2}R^2(\theta - \sin\theta)$.

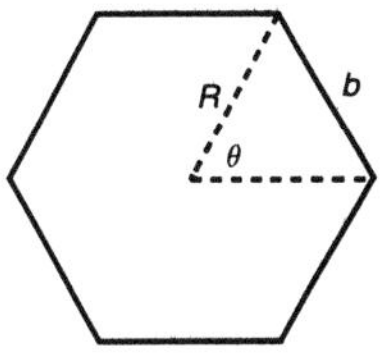

FIGURE 18.8 Regular polygon of n sides: $A = (n/4)b^2 ctn(\pi/n)$; $R = (b/2)csc(\pi/n)$.

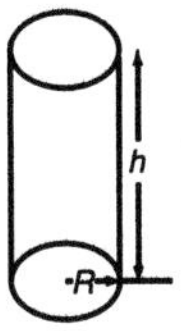

FIGURE 18.9 Right circular cylinder: $V = \pi R^2 h$; lateral surface area $= 2\pi Rh$.

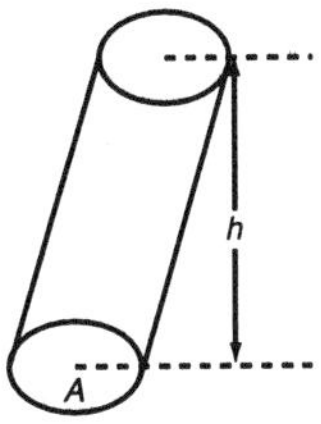

FIGURE 18.10 Cylinder (or prism) with parallel bases: $V = Ah$.

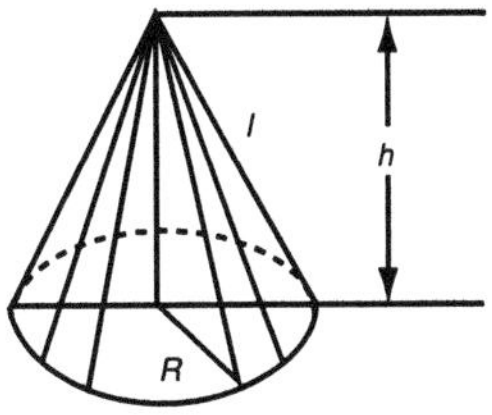

FIGURE 18.11 Right circular cone: $V = \frac{1}{3}\pi R^2 h$; lateral surface area $= \pi Rl = \pi R\sqrt{R^2 + h^2}$.

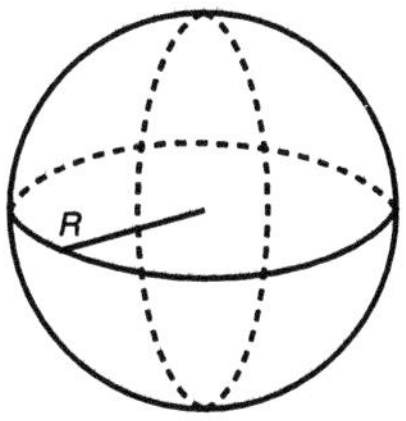

FIGURE 18.12 Sphere: $V = \frac{4}{3}\pi R^3$; surface area $= 4\pi R^2$.

18.3 Trigonometry

Triangles

In any triangle (in a plane) with sides a, b, and c and corresponding opposite angles A, B, and C,

$$\frac{a}{\sin A} = \frac{b}{\sin B} = \frac{c}{\sin C} \qquad \text{(Law of sines)}$$

$$a^2 = b^2 + c^2 - 2cb\cos A \qquad \text{(Law of cosines)}$$

$$\frac{a+b}{a-b} = \frac{\tan\frac{1}{2}(A+B)}{\tan\frac{1}{2}(A-B)} \qquad \text{(Law of Tangents)}$$

$$\sin\frac{1}{2}A = \sqrt{\frac{(s-b)(s-c)}{bc}} \qquad \text{where } s = \frac{1}{2}(a+b+c)$$

$$\cos\frac{1}{2}A = \sqrt{\frac{s(s-a)}{bc}}$$

$$\tan \frac{1}{2}A = \sqrt{\frac{(s-b)(s-c)}{s(s-a)}}$$

$$\text{area} = \frac{1}{2}bc \sin A$$
$$= \sqrt{s(s-a)(s-b)(s-c)}$$

If the vertices have coordinates (x_1, y_1), (x_2, y_2), (x_3, y_3), the area is the *absolute value* of the expression

$$\frac{1}{2}\begin{vmatrix} x_1 & y_1 & 1 \\ x_2 & y_2 & 1 \\ x_3 & y_3 & 1 \end{vmatrix}$$

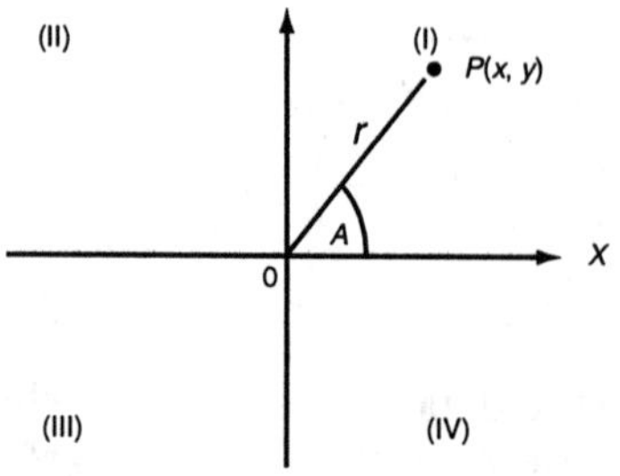

FIGURE 18.13 The trigonometric point: angle A is taken to be positive when the rotation is counterclockwise and negative when the rotation is clockwise. The plane is divided into quadrants as shown.

Trigonometric Functions of an Angle

With reference to Fig. 18.13, $P(x, y)$ is a point in any one of the four quadrants and A is an angle whose initial side is coincident with the positive x axis and whose terminal side contains the point $P(x, y)$. The distance from the origin $P(x, y)$ is denoted by r and is positive. The trigonometric functions of the angle A are defined as:

$$\begin{aligned}
\sin A &= \text{sine } A &&= y/r \\
\cos A &= \text{cosine } A &&= x/r \\
\tan A &= \text{tangent } A &&= y/x \\
\text{ctn } A &= \text{cotangent } A &&= x/y \\
\sec A &= \text{secant } A &&= r/x \\
\csc A &= \text{cosecant } A &&= r/y
\end{aligned}$$

Angles are measured in degrees or radians; $180° = \pi$ radians; 1 radian $= 180/\pi°$.

The trigonometric functions of 0, 30, and 45°, and integer multiples of these are directly computed.

	0°	30°	45°	60°	90°	120°	135°	150°	180°
sin	0	$\frac{1}{2}$	$\frac{\sqrt{2}}{2}$	$\frac{\sqrt{3}}{2}$	1	$\frac{\sqrt{3}}{2}$	$\frac{\sqrt{2}}{2}$	$\frac{1}{2}$	0
cos	1	$\frac{\sqrt{3}}{2}$	$\frac{\sqrt{2}}{2}$	$\frac{1}{2}$	0	$-\frac{1}{2}$	$-\frac{\sqrt{2}}{2}$	$-\frac{\sqrt{3}}{2}$	-1
tan	0	$\frac{\sqrt{3}}{3}$	1	$\sqrt{3}$	∞	$-\sqrt{3}$	-1	$\frac{\sqrt{3}}{3}$	0
ctn	∞	$\sqrt{3}$	1	$\frac{\sqrt{3}}{3}$	0	$-\frac{\sqrt{3}}{3}$	-1	$-\sqrt{3}$	∞
sec	1	$\frac{2\sqrt{3}}{3}$	$\sqrt{2}$	2	∞	-2	$-\sqrt{2}$	$-\frac{2\sqrt{3}}{3}$	-1
csc	∞	2	$\sqrt{2}$	$\frac{2\sqrt{3}}{3}$	1	$\frac{2\sqrt{3}}{3}$	$\sqrt{2}$	2	∞

Trigonometric Identities

$$\sin A = \frac{1}{\csc A}$$

$$\cos A = \frac{1}{\sec A}$$

$$\tan A = \frac{1}{\text{ctn A}} = \frac{\sin A}{\cos A}$$

$$\csc A = \frac{1}{\sin A}$$

$$\sec A = \frac{1}{\cos A}$$

$$\text{ctn}\, A = \frac{1}{\tan A} = \frac{\cos A}{\sin A}$$

$$\sin^2 A + \cos^2 A = 1$$

$$1 + \tan^2 A = \sec^2 A$$

$$1 + \text{ctn}^2 A = \csc^2 A$$

$$\sin(A \pm B) = \sin A \cos B \pm \cos A \sin B$$

$$\cos(A \pm B) = \cos A \cos B \mp \sin A \sin B$$

$$\tan(A \pm B) = \frac{\tan A \pm \tan B}{1 \mp \tan A \tan B}$$

$$\sin 2A = 2 \sin A \cos A$$

$$\sin 3A = 3 \sin A - 4 \sin^3 A$$

$$\sin nA = 2 \sin(n-1)A \cos A - \sin(n-2)A$$

$$\cos 2A = 2 \cos^2 A - 1 = 1 - 2 \sin^2 A$$

$$\cos 3A = 4 \cos^3 A - 3 \cos A$$

$$\cos nA = 2 \cos(n-1)A \cos A - \cos(n-2)A$$

$$\sin A + \sin B = 2 \sin \frac{1}{2}(A+B) \cos \frac{1}{2}(A-B)$$

$$\sin A - \sin B = 2 \cos \frac{1}{2}(A+B) \sin \frac{1}{2}(A-B)$$

$$\cos A + \cos B = 2 \cos \frac{1}{2}(A+B) \cos \frac{1}{2}(A-B)$$

$$\cos A - \cos B = -2 \sin \frac{1}{2}(A+B) \sin \frac{1}{2}(A-B)$$

$$\tan A \pm \tan B = \frac{\sin(A \pm B)}{\cos A \cos B}$$

$$\text{ctn}\, A \pm \text{ctn}\, B = \pm\frac{\sin(A \pm B)}{\sin A \sin B}$$

$$\sin A \sin B = \frac{1}{2}\cos(A-B) - \frac{1}{2}\cos(A+B)$$

$$\cos A \cos B = \frac{1}{2}\cos(A-B) + \frac{1}{2}\cos(A+B)$$

$$\sin A \cos B = \frac{1}{2}\sin(A+B) + \frac{1}{2}\sin(A-B)$$

$$\sin \frac{A}{2} = \pm\sqrt{\frac{1-\cos A}{2}}$$

$$\cos \frac{A}{2} = \pm\sqrt{\frac{1+\cos A}{2}}$$

$$\tan\frac{A}{2} = \frac{1-\cos A}{\sin A} = \frac{\sin A}{1+\cos A} = \pm\sqrt{\frac{1-\cos A}{1+\cos A}}$$

$$\sin^2 A = \frac{1}{2}(1-\cos 2A)$$

$$\cos^2 A = \frac{1}{2}(1+\cos 2A)$$

$$\sin^3 A = \frac{1}{4}(3\sin A - \sin 3A)$$

$$\cos^3 A = \frac{1}{4}(\cos 3A + 3\cos A)$$

$$\sin ix = \frac{1}{2}i(e^x - e^{-x}) = i\sinh x$$

$$\cos ix = \frac{1}{2}(e^x + e^{-x}) = \cosh x$$

$$\tan ix = i\frac{(e^x - e^{-x})}{e^x + e^{-x}} = i\tanh x$$

$$e^{x+iy} = e^x(\cos y + i\sin y)$$

$$(\cos x \pm i\sin x)^n = \cos nx \pm i\sin nx$$

Inverse Trigonometric Functions

The inverse trigonometric functions are multiple valued, and this should be taken into account in the use of the following formulas:

$$\begin{aligned}
\sin^{-1} x &= \cos^{-1}\sqrt{1-x^2} \\
&= \tan^{-1}\frac{x}{\sqrt{1-x^2}} = \text{ctn}^{-1}\frac{\sqrt{1-x^2}}{x} \\
&= \sec^{-1}\frac{1}{\sqrt{1-x^2}} = \csc^{-1}\frac{1}{x} \\
&= -\sin^{-1}(-x)
\end{aligned}$$

$$\begin{aligned}
\cos^{-1} x &= \sin^{-1}\sqrt{1-x^2} \\
&= \tan^{-1}\frac{\sqrt{1-x^2}}{x} = \text{ctn}^{-1}\frac{x}{\sqrt{1-x^2}} \\
&= \sec^{-1}\frac{1}{x} = \csc^{-1}\frac{1}{\sqrt{1-x^2}} \\
&= \pi - \cos^{-1}(-x)
\end{aligned}$$

$$\begin{aligned}
\tan^{-1} x &= \text{ctn}^{-1}\frac{1}{x} \\
&= \sin^{-1}\frac{x}{\sqrt{1+x^2}} = \cos^{-1}\frac{1}{\sqrt{1+x^2}} \\
&= \sec^{-1}\sqrt{1+x^2} = \csc^{-1}\frac{\sqrt{1+x^2}}{x} \\
&= -\tan^{-1}(-x)
\end{aligned}$$

18.4 Series

Bernoulli and Euler Numbers

A set of numbers, $B_1, B_3, \ldots, B_{2n-1}$ (Bernoulli numbers) and $B_2, B_4, \ldots, B_{2n}$ (Euler numbers) appear in the series expansions of many functions. A partial listing follows; these are computed from the following equations:

$$B_{2n} - \frac{2n(2n-1)}{2!}B_{2n-2} + \frac{2n(2n-1)(2n-2)(2n-3)}{4!}B_{2n-4} - \cdots + (-1)^n = 0$$

and

$$\frac{2^{2n}(2^{2n}-1)}{2n}B_{2n-1} = (2n-1)B_{2n-2} - \frac{(2n-1)(2n-2)(2n-3)}{3!}B_{2n-4} + \cdots + (-1)^{n-1}$$

$B_1 = 1/6$	$B_2 = 1$
$B_3 = 1/30$	$B_4 = 5$
$B_5 = 1/42$	$B_6 = 61$
$B_7 = 1/30$	$B_8 = 1385$
$B_9 = 5/66$	$B_{10} = 50521$
$B_{11} = 691/2730$	$B_{12} = 2\,702\,765$
$B_{13} = 7/6$	$B_{14} = 199\,360\,981$
$\vdots$	$\vdots$

Series of Functions

In the following, the interval of convergence is indicated; otherwise it is all x. Logarithms are to the base e. Bernoulli and Euler numbers (B_{2n-1} and B_{2n}) appear in certain expressions.

$$(a+x)^n = a^n + na^{n-1}x + \frac{n(n-1)}{2!}a^{n-2}x^2 + \frac{n(n-1)(n-2)}{3!}a^{n-3}x^3 + \cdots + \frac{n!}{(n-j)!j!}a^{n-j}x^j + \cdots \qquad [x^2 < a^2]$$

$$(a-bx)^{-1} = \frac{1}{a}\left[1 + \frac{bx}{a} + \frac{b^2x^2}{a^2} + \frac{b^3x^3}{a^3} + \cdots\right] \qquad [b^2x^2 < a^2]$$

$$(1 \pm x)^n = 1 \pm nx + \frac{n(n-1)}{2!}x^2 \pm \frac{n(n-1)(n-2)x^3}{3!} + \cdots \qquad [x^2 < 1]$$

$$(1 \pm x)^{-n} = 1 \mp nx + \frac{n(n+1)}{2!}x^2 \mp \frac{n(n+1)(n+2)}{3!}x^3 + \cdots \qquad [x^2 < 1]$$

$$(1 \pm x)^{\frac{1}{2}} = 1 \pm \frac{1}{2}x - \frac{1}{2\cdot 4}x^2 \pm \frac{1\cdot 3}{2\cdot 4\cdot 6}x^3 - \frac{1\cdot 3\cdot 5}{2\cdot 4\cdot 6\cdot 8}x^4 \pm \cdots \qquad [x^2 < 1]$$

$$(1 \pm x)^{-\frac{1}{2}} = 1 \mp \frac{1}{2}x + \frac{1\cdot 3}{2\cdot 4}x^2 \mp \frac{1\cdot 3\cdot 5}{2\cdot 4\cdot 6}x^3 + \frac{1\cdot 3\cdot 5\cdot 7}{2\cdot 4\cdot 6\cdot 8}x^4 \pm \cdots \qquad [x^2 < 1]$$

$$(1 \pm x^2)^{\frac{1}{2}} = 1 \pm \frac{1}{2}x^2 - \frac{x^4}{2\cdot 4} \pm \frac{1\cdot 3}{2\cdot 4\cdot 6}x^6 - \frac{1\cdot 3\cdot 5}{2\cdot 4\cdot 6\cdot 8}x^8 \pm \cdots \qquad [x^2 < 1]$$

$$(1 \pm x)^{-1} = 1 \mp x + x^2 \mp x^3 + x^4 \mp x^5 + \cdots \qquad [x^2 < 1]$$

$$(1 \pm x)^{-2} = 1 \mp 2x + 3x^2 \mp 4x^3 + 5x^4 \mp \cdots \qquad [x^2 < 1]$$

$$e^x = 1 + x + \frac{x^2}{2!} + \frac{x^3}{3!} + \frac{x^4}{4!} + \cdots$$

$$e^{-x^2} = 1 - x^2 + \frac{x^4}{2!} - \frac{x^6}{3!} + \frac{x^8}{4!} - \cdots$$

$$a^x = 1 + x\log a + \frac{(x\log a)^2}{2!} + \frac{(x\log a)^3}{3!} + \cdots$$

$$\log x = (x-1) - \frac{1}{2}(x-1)^2 + \frac{1}{3}(x-1)^3 - \cdots \qquad [0 < x < 2]$$

$$\log x = \frac{x-1}{x} + \frac{1}{2}\left(\frac{x-1}{x}\right)^2 + \frac{1}{3}\left(\frac{x-1}{x}\right)^3 + \cdots \qquad \left[x > \frac{1}{2}\right]$$

$$\log x = 2\left[\left(\frac{x-1}{x+1}\right) + \frac{1}{3}\left(\frac{x-1}{x+1}\right)^3 + \frac{1}{5}\left(\frac{x-1}{x+1}\right)^5 + \cdots\right] \qquad [x > 0]$$

$$\log(1+x) = x - \frac{1}{2}x^2 + \frac{1}{3}x^3 - \frac{1}{4}x^4 + \cdots \qquad [x^2 < 1]$$

$$\log\left(\frac{1+x}{1-x}\right) = 2\left[x + \frac{1}{3}x^3 + \frac{1}{5}x^5 + \frac{1}{7}x^7 + \cdots\right] \qquad [x^2 < 1]$$

$$\log\left(\frac{x+1}{x-1}\right) = 2\left[\frac{1}{x} + \frac{1}{3}\left(\frac{1}{x}\right)^3 + \frac{1}{5}\left(\frac{1}{x}\right)^5 + \cdots\right] \qquad [x^2 < 1]$$

$$\sin x = x - \frac{x^3}{3!} + \frac{x^5}{5!} - \frac{x^7}{7!} + \cdots$$

$$\cos x = 1 - \frac{x^2}{2!} + \frac{x^4}{4!} - \frac{x^6}{6!} + \cdots$$

$$\tan x = x + \frac{x^3}{3} + \frac{2x^5}{15} + \frac{17x^7}{315} + \cdots + \frac{2^{2n}(2^{2n}-1)B_{2n-1}x^{2n-1}}{2n!} \qquad \left[x^2 < \frac{\pi^2}{4}\right]$$

$$\operatorname{ctn} x = \frac{1}{x} - \frac{x}{3} - \frac{x^3}{45} - \frac{2x^5}{945} - \cdots - \frac{B_{2n-1}(2x)^{2n}}{(2n)!x} - \cdots \qquad [x^2 < \pi^2]$$

$$\sec x = 1 + \frac{x^2}{2!} + \frac{5x^4}{4!} + \frac{61x^6}{6!} + \cdots + \frac{B_{2n}x^{2n}}{(2n)!} + \cdots \qquad \left[x^2 < \frac{\pi^2}{4}\right]$$

$$\csc x = \frac{1}{x} + \frac{x}{3!} + \frac{7x^3}{3\cdot 5!} + \frac{31x^5}{3\cdot 7!} + \cdots + \frac{2(2^{2n+1}-1)}{(2n+2)!}B_{2n+1}x^{2n+1} + \cdots \qquad [x^2 < \pi^2]$$

$$\sin^{-1} x = x + \frac{x^3}{6} + \frac{(1\cdot 3)x^5}{(2\cdot 4)5} + \frac{(1\cdot 3\cdot 5)x^7}{(2\cdot 4\cdot 6)7} + \cdots \qquad [x^2 < 1]$$

$$\tan^{-1} x = x - \frac{1}{3}x^3 + \frac{1}{5}x^5 - \frac{1}{7}x^7 + \cdots \qquad [x^2 < 1]$$

$$\sec^{-1} x = \frac{\pi}{2} - \frac{1}{x} - \frac{1}{6x^3} - \frac{1\cdot 3}{(2\cdot 4)5x^5} - \frac{1\cdot 3\cdot 5}{(2\cdot 4\cdot 6)7x^7} - \cdots \qquad [x^2 < 1]$$

$$\sinh x = x + \frac{x^3}{3!} + \frac{x^5}{5!} + \frac{x^7}{7!} + \cdots$$

$$\cosh x = 1 + \frac{x^2}{2!} + \frac{x^4}{4!} + \frac{x^6}{6!} + \frac{x^8}{8!} + \cdots$$

$$\tanh x = (2^2 - 1)2^2 B_1 \frac{x}{2!} - (2^4 - 1)2^4 B_3 \frac{x^3}{4!} + (2^6 - 1)2^6 B_5 \frac{x^5}{6!} - \cdots \qquad \left[x^2 < \frac{\pi^2}{4}\right]$$

$$\operatorname{ctnh} x = \frac{1}{x}\left(1 + \frac{2^2 B_1 x^2}{2!} - \frac{2^4 B_3 x^4}{4!} + \frac{2^6 B_5 x^6}{6!} - \cdots\right) \qquad [x^2 < \pi^2]$$

$$\operatorname{sech} x = 1 - \frac{B_2 x^2}{2!} + \frac{B_4 x^4}{4!} - \frac{B_6 x^6}{6!} + \cdots \qquad \left[x^2 < \frac{\pi^2}{4}\right]$$

$$\operatorname{csch} x = \frac{1}{x} - (2 - 1)2B_1 \frac{x}{2!} + (2^3 - 1)2B_3 \frac{x^3}{4!} - \cdots \qquad [x^2 < \pi^2]$$

$$\sinh^{-1} x = x - \frac{1}{2}\frac{x^3}{3} + \frac{1 \cdot 3}{2 \cdot 4}\frac{x^5}{5} - \frac{1 \cdot 3 \cdot 5}{2 \cdot 4 \cdot 6}\frac{x^7}{7} + \cdots \qquad [x^2 < 1]$$

$$\tanh^{-1} x = x + \frac{x^3}{3} + \frac{x^5}{5} + \frac{x^7}{7} + \cdots \qquad [x^2 < 1]$$

$$\operatorname{ctnh}^{-1} x = \frac{1}{x} + \frac{1}{3x^3} + \frac{1}{5x^5} + \cdots \qquad [x^2 > 1]$$

$$\operatorname{csch}^{-1} x = \frac{1}{x} - \frac{1}{2 \cdot 3x^3} + \frac{1 \cdot 3}{2 \cdot 4 \cdot 5x^5} - \frac{1 \cdot 3 \cdot 5}{2 \cdot 4 \cdot 6 \cdot 7x^7} + \cdots \qquad [x^2 > 1]$$

$$\int_0^x e^{-t^2} dt = x - \frac{1}{3}x^3 + \frac{x^5}{5 \cdot 2!} - \frac{x^7}{7 \cdot 3!} + \cdots$$

Error Function

The following function, known as the error function, erf x, arises frequently in applications:

$$\operatorname{erf} x = \frac{2}{\sqrt{\pi}} \int_0^x e^{-t^2} dt$$

The integral cannot be represented in terms of a finite number of elementary functions; therefore, values of erf x have been compiled in tables. The following is the series for erf x:

$$\operatorname{erf} x = \frac{2}{\sqrt{\pi}}\left[x - \frac{x^3}{3} + \frac{x^5}{5 \cdot 2!} - \frac{x^7}{7 \cdot 3!} + \cdots\right]$$

There is a close relation between this function and the area under the standard normal curve. For evaluation it is convenient to use z instead of x; then erf z may be evaluated from the area $F(z)$ by use of the relation

$$\operatorname{erf} z = 2F(\sqrt{2}z)$$

Example

$$\operatorname{erf}(0.5) = 2F[(1.414)(0.5)] = 2F(0.707)$$

By interpolation, $F(0.707) = 0.260$; thus, erf $(0.5) = 0.520$.

Series Expansion

The expression in parentheses following certain series indicates the region of convergence. If not otherwise indicated, it is understood that the series converges for all finite values of x.

Binomial

$$(x + y)^n = x^n + nx^{n-1}y + \frac{n(n-1)}{2!}x^{n-2}y^2 + \frac{n(n-1)(n-2)}{3!}x^{n-3}y^3 + \cdots \qquad (y^2 < x^2)$$

$$(1 \pm x)^n = 1 \pm nx + \frac{n(n-1)x^2}{2!} \pm \frac{n(n-1)(n-2)x^3}{3!} + \cdots \qquad (x^2 < 1)$$

$$(1 \pm x)^{-n} = 1 \mp nx + \frac{n(n+1)x^2}{2!} \mp \frac{n(n+1)(n+2)x^3}{3!} + \cdots \qquad (x^2 < 1)$$

$$(1 \pm x)^{-1} = 1 \mp x + x^2 \mp x^3 + x^4 \mp x^5 + \cdots \qquad (x^2 < 1)$$

$$(1 \pm x)^{-2} = 1 \mp 2x + 3x^2 \mp 4x^3 + 5x^4 \mp 6x^5 + \cdots \qquad (x^2 < 1)$$

Reversion of Series

Let a series be represented by

$$y = a_1 x + a_2 x^2 + a_3 x^3 + a_4 x^4 + a_5 x^5 + a_6 x^6 + \cdots \qquad (a_1 \neq 0)$$

To find the coefficients of the series

$$x = A_1 y + A_2 y^2 + A_3 y^3 + A_4 y^4 + \cdots$$

$$A_1 = \frac{1}{a_1} \qquad A_2 = -\frac{a_2}{a_1^3} \qquad A_3 = \frac{1}{a_1^5}\left(2a_2^2 - a_1 a_3\right)$$

$$A_4 = \frac{1}{a_1^7}\left(5a_1 a_2 a_3 - a_1^2 a_4 - 5a_2^3\right)$$

$$A_5 = \frac{1}{a_1^9}\left(6a_1^2 a_2 a_4 + 3a_1^2 a_3^2 + 14a_2^4 - a_1^3 a_5 - 21 a_1 a_2^2 a_3\right)$$

$$A_6 = \frac{1}{a_1^{11}}\left(7a_1^3 a_2 a_5 + 7a_1^3 a_3 a_4 + 84 a_1 a_2^3 a_3 - a_1^4 a_6 - 28 a_1^2 a_2^2 a_4 - 28 a_1^2 a_2 a_3^2 - 42 a_2^5\right)$$

$$A_7 = \frac{1}{a_1^{13}}\left(8a_1^4 a_2 a_6 + 8a_1^4 a_3 a_5 + 4a_1^4 a_4^2 + 120 a_1^2 a_2^3 a_4 + 180 a_1^2 a_2^2 a_3^2 + 132 a_2^6 - a_1^5 a_7 - 36 a_1^3 a_2^2 a_5 - 72 a_1^3 a_2 a_3 a_4 - 12 a_1^3 a_3^3 - 330 a_1 a_2^4 a_3\right)$$

Taylor

1.

$$f(x) = f(a) + (x-a)f'(a) + \frac{(x-a)^2}{2!}f''(a) + \frac{(x-a)^3}{3!}f'''(a) + \cdots + \frac{(x-a)^n}{n!}f^{(n)}(a) + \cdots$$

2. (Increment form)

$$f(x+h) = f(x) + hf'(x) + \frac{h^2}{2!}f''(x) + \frac{h^3}{3!}f'''(x) + \cdots$$
$$= f(h) + xf'(h) + \frac{x^2}{2!}f''(h) + \frac{x^3}{3!}f'''(h) + \cdots$$

3. If $f(x)$ is a function possessing derivatives of all orders throughout the interval $a \le x \le b$, then there is a value X, with $a < X < b$, such that

$$f(b) = f(a) + (b-a)f'(a) + \frac{(b-a)^2}{2!}f''(a) + \cdots + \frac{(b-a)^{n-1}}{(n-1)!}f^{(n-1)}(a) + \frac{(b-a)^n}{n!}f^{(n)}(X)$$

$$f(a+h) = f(a) + hf'(a) + \frac{h^2}{2!}f''(a) + \cdots + \frac{h^{n-1}}{(n-1)!}f^{(n-1)}(a)$$
$$+\frac{h^n}{n!}f^{(n)}(a+\theta h), \quad b = a+h, \quad 0<\theta<1$$

or

$$f(x) = f(a) + (x-a)f'(a) + \frac{(x-a)^2}{2!}f''(a) + \cdots + (x-a)^{n-1}\frac{f^{(n-1)}(a)}{(n-1)!} + R_n$$

where

$$R_n = \frac{f^{(n)}[a+\theta\cdot(x-a)]}{n!}(x-a)^n, \quad 0<\theta<1$$

The preceding forms are known as Taylor's series with the remainder term.

4. Taylor's series for a function of two variables: If

$$\left(h\frac{\partial}{\partial x} + k\frac{\partial}{\partial y}\right)f(x,y) = h\frac{\partial f(x,y)}{\partial x} + k\frac{\partial f(x,y)}{\partial y};$$
$$\left(h\frac{\partial}{\partial x} + k\frac{\partial}{\partial y}\right)^2 f(x,y) = h^2\frac{\partial^2 f(x,y)}{\partial x^2} + 2hk\frac{\partial^2 f(x,y)}{\partial x\,\partial y} + k^2\frac{\partial^2 f(x,y)}{\partial y^2}$$

and so forth, and if

$$\left(h\frac{\partial}{\partial x} + k\frac{\partial}{\partial y}\right)^n f(x,y)\Big|_{\substack{x=a\\y=b}}$$

where the bar and subscripts mean that after differentiation we are to replace x by a and y by b,

$$f(a+h, b+k) = f(a,b) + \left(h\frac{\partial}{\partial x} + k\frac{\partial}{\partial y}\right)f(x,y)\Big|_{\substack{x=a\\y=b}} + \cdots$$
$$+\frac{1}{n!}\left(h\frac{\partial}{\partial x} + k\frac{\partial}{\partial y}\right)^n f(x,y)\Big|_{\substack{x=a\\y=b}} + \cdots$$

MacLaurin

$$f(x) = f(0) + xf'(0) + \frac{x^2}{2!}f''(0) + \frac{x^3}{3!}f'''(0) + \cdots + x^{n-1}\frac{f^{n-1}(0)}{(n-1)!} + R_n$$

where

$$R_n = \frac{x^n f^{(n)}(\theta x)}{n!}, 0<\theta<1$$

Exponential

$$e = 1 + \frac{1}{1!} + \frac{1}{2!} + \frac{1}{3!} + \frac{1}{4!} + \cdots$$
$$e^x = 1 + x + \frac{x^2}{2!} + \frac{x^3}{3!} + \frac{x^4}{4!} + \cdots \qquad \text{(all real values of } x\text{)}$$
$$a^x = 1 + x\log_e a + \frac{(x\log_e a)^2}{2!} + \frac{(x\log_e a)^3}{3!} + \cdots$$
$$e^x = e^a\left[1 + (x-a) + \frac{(x-a)^2}{2!} + \frac{(x-a)^3}{3!} + \cdots\right]$$

Logarithmic

$$\log_e x = \frac{x-1}{x} + \frac{1}{2}\left(\frac{x-1}{x}\right)^2 + \frac{1}{3}\left(\frac{x-1}{x}\right)^3 + \cdots \qquad \left(x > \frac{1}{2}\right)$$

$$\log_e x = (x-1) - \frac{1}{2}(x-1)^2 + \frac{1}{3}(x-1)^3 - \cdots \qquad (2 \geq x > 0)$$

$$\log_e x = 2\left[\frac{x-1}{x+1} + \frac{1}{3}\left(\frac{x-1}{x+1}\right)^3 \frac{1}{5}\left(\frac{x-1}{x+1}\right)^5 + \cdots\right] \qquad (x > 0)$$

$$\log_e(1+x) = x - \frac{1}{2}x^2 + \frac{1}{3}x^3 - \frac{1}{4}x^4 + \cdots \qquad (-1 < x \leq 1)$$

$$\log_e(n+1) - \log_e(n-1) = 2\left[\frac{1}{n} + \frac{1}{3n^3} + \frac{1}{5n^5} + \cdots\right]$$

$$\log_e(a+x) = \log_e a + 2\left[\frac{x}{2a+x} + \frac{1}{3}\left(\frac{x}{2a+x}\right)^3 + \frac{1}{5}\left(\frac{x}{2a+x}\right)^5 + \cdots\right]$$

$$(a > 0, -a < x < +\infty)$$

$$\log_e \frac{1+x}{1-x} = 2\left[x + \frac{x^3}{3} + \frac{x^5}{5} + \cdots + \frac{x^{2n-1}}{2n-1} + \cdots\right] \qquad (-1 < x < 1)$$

$$\log_e x = \log_e a + \frac{(x-a)}{a} - \frac{(x-a)^2}{2a^2} + \frac{(x-a)^3}{3a^3} - \cdots \qquad (0 < x \leq 2a)$$

Trigonometric

$$\sin x = x - \frac{x^3}{3!} + \frac{x^5}{5!} - \frac{x^7}{7!} + \cdots \qquad (\text{all real values of } x)$$

$$\cos x = 1 - \frac{x^2}{2!} + \frac{x^4}{4!} - \frac{x^6}{6!} + \cdots \qquad (\text{all real values of } x)$$

$$\tan x = x + \frac{x^3}{3} + \frac{2x^5}{15} + \frac{17x^7}{315} + \frac{62x^9}{2835} + \cdots$$

$$+ \frac{(-1)^{n-1}2^{2n}(2^{2n}-1)B_{2n}}{(2n)!}x^{2n-1} + \cdots$$

($x^2 < \pi^2/4$, and B_n represents the nth Bernoulli number)

$$\cot x = \frac{1}{x} - \frac{x}{3} - \frac{x^2}{45} - \frac{2x^5}{945} - \frac{x^7}{4725} - \cdots$$

$$- \frac{(-1)^{n+1}2^{2n}}{(2n)!}B_{2n}x^{2n-1} + \cdots$$

($x^2 < \pi^2$, and B_n represents the nth Bernoulli number)

18.5 Differential Calculus

Notation

For the following equations, the symbols $f(x)$, $g(x)$, and so forth represent functions of x. The value of a function $f(x)$ at $x = a$ is denoted $f(a)$. For the function $y = f(x)$ the derivative of y with respect to x is denoted by one of the following:

$$\frac{dy}{dx}, \quad f'(x), \quad D_x y, \quad y'$$

Higher derivatives are as follows:

$$\frac{d^2y}{dx^2} = \frac{d}{dx}\left(\frac{dy}{dx}\right) = \frac{d}{dx}f'(x) = f''(x)$$
$$\frac{d^3y}{dx^3} = \frac{d}{dx}\left(\frac{d^2y}{dx^2}\right) = \frac{d}{dx}f''(x) = f'''(x)$$
$$\vdots$$

and values of these at $x = a$ are denoted $f''(a)$, $f'''(a)$, and so on (see Table 18.1, Table of Derivatives).

Slope of a Curve

The tangent line at point $P(x, y)$ of the curve $y = f(x)$ has a slope $f'(x)$ provided that $f'(x)$ exists at P. The slope at P is defined to be that of the tangent line at P. The tangent line at $P(x_1, y_1)$ is given by

$$y - y_1 = f'(x_1)(x - x_1)$$

TABLE 18.1 Table of Derivatives*

1. $\frac{d}{dx}(a) = 0$
2. $\frac{d}{dx}(x) = 1$
3. $\frac{d}{dx}(au) = a\frac{du}{dx}$
4. $\frac{d}{dx}(u+v) = \frac{du}{dx} + \frac{dv}{dx}$
5. $\frac{d}{dx}(uv) = u\frac{dv}{dx} + v\frac{du}{dx}$
6. $\frac{d}{dx}\frac{u}{v} = \frac{v\frac{du}{dx} - u\frac{dv}{dx}}{v^2}$
7. $\frac{d}{dx}(u^n) = nu^{n-1}\frac{du}{dx}$
8. $\frac{d}{dx}e^u = e^u\frac{du}{dx}$
9. $\frac{d}{dx}a^u = (\log_e a)a^u\frac{du}{dx}$
10. $\frac{d}{dx}\log_e u = \frac{1}{u}\frac{du}{dx}$
11. $\frac{d}{dx}\log_a u = (\log_a e)\frac{1}{u}\frac{du}{dx}$
12. $\frac{d}{dx}u^v = vu^{v-1}\frac{du}{dx} + u^v(\log_e u)\frac{dv}{dx}$
13. $\frac{d}{dx}\sin u = \cos u\frac{du}{dx}$
14. $\frac{d}{dx}\cos u = -\sin u\frac{du}{dx}$
15. $\frac{d}{dx}\tan u = \sec^2 u\frac{du}{dx}$
16. $\frac{d}{dx}\text{ctn}\, u = -\csc^2 u\frac{du}{dx}$
17. $\frac{d}{dx}\sec u = \sec u \tan u\frac{du}{dx}$
18. $\frac{d}{dx}\csc u = -\csc u\ \text{ctn}\, u\frac{du}{dx}$
19. $\frac{d}{dx}\sin^{-1} u = \frac{1}{\sqrt{1-u^2}}\frac{du}{dx}, \quad \left(-\frac{1}{2}\pi \le \sin^{-1} u \le \frac{1}{2}\pi\right)$
20. $\frac{d}{dx}\cos^{-1} u = \frac{-1}{\sqrt{1-u^2}}\frac{du}{dx}, \quad (0 \le \cos^{-1} u \le \pi)$
21. $\frac{d}{dx}\tan^{-1} u = \frac{1}{1+u^2}\frac{du}{dx}$
22. $\frac{d}{dx}\text{ctn}^{-1} u = \frac{-1}{1+u^2}\frac{du}{dx}$
23. $\frac{d}{dx}\sec^{-1} u = \frac{1}{u\sqrt{u^2-1}}\frac{du}{dx}, \quad \left(-\pi \le \sec^{-1} u < -\frac{1}{2}\pi;\ 0 \le \sec^{-1} u < \frac{1}{2}\pi\right)$
24. $\frac{d}{dx}\csc^{-1} u = \frac{-1}{u\sqrt{u^2-1}}\frac{du}{dx}, \quad \left(-\pi \le \csc^{-1} u \le -\frac{1}{2}\pi;\ 0 < \csc^{-1} u \le \frac{1}{2}\pi\right)$
25. $\frac{d}{dx}\sinh u = \cosh u\frac{du}{dx}$
26. $\frac{d}{dx}\cosh u = \sinh u\frac{du}{dx}$
27. $\frac{d}{dx}\tanh u = \text{sech}^2 u\frac{du}{dx}$
28. $\frac{d}{dx}\text{ctnh}\, u = -\text{csch}^2 u\frac{du}{dx}$
29. $\frac{d}{dx}\text{sech}\, u = -\text{sech}\, u \tanh u\frac{du}{dx}$
30. $\frac{d}{dx}\text{csch}\, u = -\text{csch}\, u\ \text{ctnh}\, u\frac{du}{dx}$
31. $\frac{d}{dx}\sinh^{-1} u = \frac{1}{\sqrt{u^2+1}}\frac{du}{dx}$
32. $\frac{d}{dx}\cosh^{-1} u = \frac{1}{\sqrt{u^2-1}}\frac{du}{dx}$
33. $\frac{d}{dx}\tanh^{-1} u = \frac{1}{1-u^2}\frac{du}{dx}$
34. $\frac{d}{dx}\text{ctnh}^{-1} u = \frac{-1}{u^2-1}\frac{du}{dx}$
35. $\frac{d}{dx}\text{sech}^{-1} u = \frac{-1}{u\sqrt{1-u^2}}\frac{du}{dx}$
36. $\frac{d}{dx}\text{csch}^{-1} u = \frac{1}{u\sqrt{u^2+1}}\frac{du}{dx}$

*In this table, a and n are constants, e is the base of the natural logarithms, and u and v denote functions of x.

The *normal line* to the curve at $P(x_1, y_1)$ has slope $-1/f'(x_1)$ and thus obeys the equation

$$y - y_1 = \left[-1/f'(x_1)\right](x - x_1)$$

(The slope of a vertical line is not defined.)

Angle of Intersection of Two Curves

Two curves, $y = f_1(x)$ and $y = f_2(x)$, that intersect at a point $P(X, Y)$ where derivatives $f_1'(X)$, $f_2'(X)$ exist have an angle (α) of intersection given by

$$\tan\alpha = \frac{f_2'(X) - f_1'(X)}{1 + f_2'(X) \cdot f_1'(X)}$$

If $\tan\alpha > 0$, then α is the acute angle; if $\tan\alpha < 0$, then α is the obtuse angle.

Radius of Curvature

The radius of curvature R of the curve $y = f(x)$ at the point $P(x, y)$ is

$$R = \frac{\{1 + [f'(x)]^2\}^{3/2}}{f''(x)}$$

In polar coordinates (θ, r) the corresponding formula is

$$R = \frac{\left[r^2 + \left(\dfrac{dr}{d\theta}\right)^2\right]^{3/2}}{r^2 + 2\left(\dfrac{dr}{d\theta}\right)^2 - r\dfrac{d^2r}{d\theta^2}}$$

The *curvature* K is $1/R$.

Relative Maxima and Minima

The function f has a relative maximum at $x = a$ if $f(a) \geq f(a + c)$ for all values of c (positive or negative) that are sufficiently near zero. The function f has a relative minimum at $x = b$ if $f(b) \leq f(b + c)$ for all values of c that are sufficiently close to zero. If the function f is defined on the closed interval $x_1 \leq x \leq x_2$ and has a relative maximum or minimum at $x = a$, where $x_1 < a < x_2$, and if the derivative $f'(x)$ exists at $x = a$, then $f'(a) = 0$. It is noteworthy that a relative maximum or minimum may occur at a point where the derivative does not exist. Further, the derivative may vanish at a point that is neither a maximum nor a minimum for the function. Values of x for which $f'(x) = 0$ are called *critical values*. To determine whether a critical value of x, say x_c, is a relative maximum or minimum for the function at x_c, one may use the second derivative test:

1. If $f''(x_c)$ is positive, $f(x_c)$ is a minimum.
2. If $f''(x_c)$ is negative, $f(x_c)$ is a maximum.
3. If $f''(x_c)$ is zero, no conclusion may be made.

The sign of the derivative as x advances through x_c may also be used as a test. If $f'(x)$ changes from positive to zero to negative, then a maximum occurs at x_c, whereas a change in $f'(x)$ from negative to zero to positive indicates a minimum. If $f'(x)$ does not change sign as x advances through x_c, then the point is neither a maximum nor a minimum.

Points of Inflection of a Curve

The sign of the second derivative of f indicates whether the graph of $y = f(x)$ is concave upward or concave downward:

$$f''(x) > 0\text{: concave upward}$$

$$f''(x) < 0\text{: concave downward}$$

A point of the curve at which the direction of concavity changes is called a point of inflection (Fig. 18.14). Such a point may occur where $f''(x) = 0$ or where $f''(x)$ becomes infinite. More precisely, if the function $y = f(x)$ and its first derivative $y' = f'(x)$ are continuous in the interval $a \le x \le b$, and if $y'' = f''(x)$ exists in $a < x < b$, then the graph of $y = f(x)$ for $a < x < b$ is concave upward if $f''(x)$ is positive and concave downward if $f''(x)$ is negative.

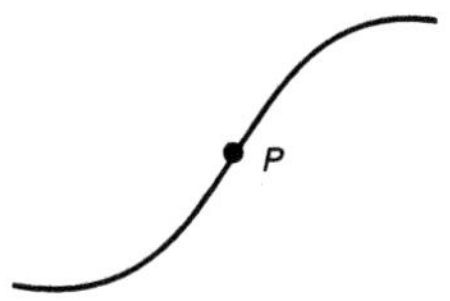

FIGURE 18.14 Point of inflection.

Taylor's Formula

If f is function that is continuous on an interval that contains a and x, and if its first $(n+1)$ derivatives are continuous on this interval, then

$$f(x) = f(a) + f'(a)(x-a) + \frac{f''(a)}{2!}(x-a)^2 + \frac{f'''(a)}{3!}(x-a)^3 + \cdots + \frac{f^{(n)}(a)}{n!}(x-a)^n + R$$

where R is called the *remainder*. There are various common forms of the remainder.

Lagrange's form:

$$R = f^{(n+1)}(\beta)\cdot\frac{(x-a)^{n+1}}{(n+1)!},\quad \beta \text{ between } a \text{ and } x$$

Cauchy's form:

$$R = f^{(n+1)}(\beta)\cdot\frac{(x-B)^n(x-a)}{n!},\quad \beta \text{ between } a \text{ and } x$$

Integral form:

$$R = \int_a^x \frac{(x-t)^n}{n!} f^{(n+1)}(t)dt$$

Indeterminant Forms

If $f(x)$ and $g(x)$ are continuous in an interval that includes $x = a$, and if $f(a) = 0$ and $g(a) = 0$, the limit $\lim_{x\to a}[f(x)/g(x)]$ takes the form 0/0, called an *indeterminant form*. *L'Hôpital's rule* is

$$\lim_{x\to a}\frac{f(x)}{g(x)} = \lim_{x\to a}\frac{f'(x)}{g'(x)}$$

Similarly, it may be shown that if $f(x) \to \infty$ and $g(x) \to \infty$ as $x \to a$, then

$$\lim_{x\to a}\frac{f(x)}{g(x)} = \lim_{x\to a}\frac{f'(x)}{g'(x)}$$

(This holds for $x \to \infty$.)

Examples

$$\lim_{x\to 0}\frac{\sin x}{x} = \lim_{x\to 0}\frac{\cos x}{1} = 1$$

$$\lim_{x\to\infty}\frac{x^2}{e^x} = \lim_{x\to\infty}\frac{2x}{e^x} = \lim_{x\to\infty}\frac{2}{e^x} = 0$$

Numerical Methods

1. *Newton's method* for approximating roots of the equation $f(x) = 0$: A first estimate x_1 of the root is made; then, provided that $f'(x_1) \neq 0$, a better approximation is x_2,

$$x_2 = x_1 - \frac{f(x_1)}{f'(x_1)}$$

The process may be repeated to yield a third approximation, x_3, to the root

$$x_3 = x_2 - \frac{f(x_2)}{f'(x_2)}$$

provided $f'(x_2)$ exists. The process may be repeated. (In certain rare cases the process will not converge.)

2. *Trapezoidal rule for areas* (Fig. 18.15): For the function $y = f(x)$ defined on the interval (a, b) and positive there, take n equal subintervals of width $\Delta x = (b - a)/n$. The area bounded by the curve between $x = a$ and $x = b$ [or definite integral of $f(x)$] is approximately the sum of trapezoidal areas, or

$$A \sim \left(\frac{1}{2}y_0 + y_1 + y_2 + \cdots + y_{n-1} + \frac{1}{2}y_n\right)(\Delta x)^2$$

Estimation of the error (E) is possible if the second derivative can be obtained:

$$E = \frac{b-a}{12}f''(c)(\Delta x)^2$$

where c is some number between a and b.

Functions of Two Variables

For the function of two variables, denoted $z = f(x, y)$, if y is held constant, say at $y = y_1$, then the resulting function is a function of x only. Similarly, x may be held constant at x_1, to give the resulting function of y.

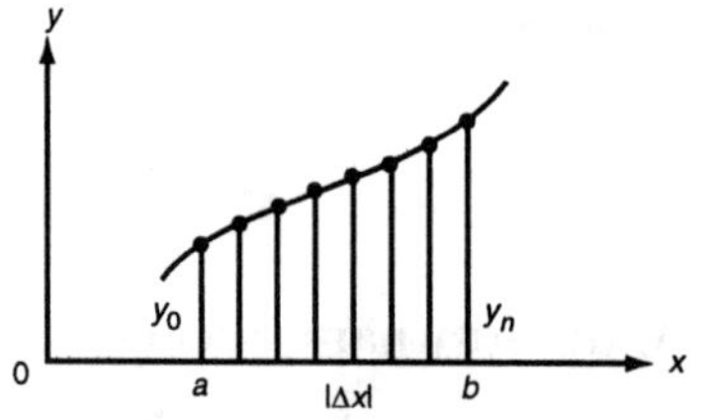

FIGURE 18.15 Trapezoidal rule for area.

The Gas Laws

A familiar example is afforded by the ideal gas law relating the pressure p, the volume V, and the absolute temperature T of an ideal gas:

$$pV = nRT$$

where n is the number of moles and R is the gas constant per mole, 8.31 ($J \cdot K^{-1} \cdot mole^{-1}$). By rearrangement, any one of the three variables may be expressed as a function of the other two. Further, either one of these two may be held constant. If T is held constant, then we get the form known as Boyle's law:

$$p = kV^{-1} \qquad \text{(Boyle's law)}$$

where we have denoted nRT by the constant k and, of course, $V > 0$. If the pressure remains constant, we have Charles' law:

$$V = bT \qquad \text{(Charles' law)}$$

where the constant b denotes nR/p. Similarly, volume may be kept constant:

$$p = aT$$

where now the constant, denoted a, is nR/V.

Partial Derivatives

The physical example afforded by the ideal gas law permits clear interpretations of processes in which one of the variables is held constant. More generally, we may consider a function $z = f(x, y)$ defined over some region of the xy plane in which we hold one of the two coordinates, say y, constant. If the resulting function of x is differentiable at a point (x, y), we denote this derivative by one of the notations.

$$f_x, \quad \frac{\partial f}{\partial x}$$

called the *partial derivative with respect to x*. Similarly, if x is held constant and the resulting function of y is differentiable, we get the *partial derivative with respect to y*, denoted by one of the following:

$$f_y, \quad \frac{\partial f}{\partial y}$$

Example

Given $z = x^4y^3 - y \sin x + 4y$, then

$$\frac{\partial z}{\partial x} = 4(xy)^3 - y \cos x$$

$$\frac{\partial z}{\partial y} = 3x^4y^2 - \sin x + 4$$

In Table 18.1, a and n are constants, e is the base of the natural logarithms, and u and v denote functions of x.

Additional Relations with Derivatives

$$\frac{d}{dt}\int_a^t f(x)\,dx = f(t) \qquad \frac{d}{dt}\int_t^a f(x)\,dx = -f(t)$$

If $x = f(y)$, then

$$\frac{dy}{dx} = \frac{1}{\frac{dx}{dy}}$$

If $y = f(u)$ and $u = g(x)$, then

$$\frac{dy}{dx} = \frac{dy}{du} \cdot \frac{du}{dx} \qquad \text{(chain rule)}$$

If $x = f(t)$ and $y = g(t)$, then

$$\frac{dy}{dx} = \frac{g'(t)}{f'(t)}, \qquad \text{and} \qquad \frac{d^2}{dx^2} = \frac{f'(t)g''(t) - g'(t)f''(t)}{[f'(t)]^3}$$

(*Note*: Exponent in denominator is 3.)

18.6 Integral Calculus

Indefinite Integral

If $F(x)$ is differentiable for all values of x in the interval (a, b) and satisfies the equation $dy/dx = f(x)$, then $F(x)$ is an integral of $f(x)$ with respect to x. The notation is $F(x) = \int f(x)\,dx$ or, in differential form, $dF(x) = f(x)\,dx$.

For any function $F(x)$ that is an integral of $f(x)$, it follows that $F(x) + C$ is also an integral. We thus write

$$\int f(x)dx = F(x) + C$$

Definite Integral

Let $f(x)$ be defined on the interval $[a, b]$ which is partitioned by points $x_1, x_2, \ldots, x_j, \ldots, x_{n-1}$ between $a = x_0$ and $b = x_n$. The jth interval has length $\Delta x_j = x_j - x_{j-1}$, which may vary with j. The sum $\sum_{j=1}^{n} f(v_j)\Delta x_j$, where v_j is arbitrarily chosen in the jth subinterval, depends on the numbers $x_0, \ldots, x_n$ and the choice of the v as well as f; but if such sums approach a common value as all Δx approach zero, then this value is the definite integral of f over the interval (a, b) and is denoted $\int_a^b f(x)\,dx$. The *fundamental theorem of integral calculus* states that

$$\int_a^b f(x)\,dx = F(b) - F(a),$$

where F is any continuous indefinite integral of f in the interval (a, b).

Properties

$$\int_a^b [f_1(x) + f_2(x) + \cdots + f_j(x)]\,dx = \int_a^b f_1(x)\,dx + \int_a^b f_2(x)\,dx + \cdots + \int_a^b f_j(x)\,dx$$

$$\int_a^b cf(x)\,dx = c\int_a^b f(x)\,dx, \qquad \text{if } c \text{ is a constant}$$

$$\int_a^b f(x)\,dx = -\int_b^a f(x)\,dx$$

$$\int_a^b f(x)\,dx = \int_a^c f(x)\,dx + \int_c^b f(x)\,dx$$

Common Applications of the Definite Integral

Area (Rectangular Coordinates)

Given the function $y = f(x)$ such that $y > 0$ for all x between a and b, the area bounded by the curve $y = f(x)$, the x axis, and the vertical lines $x = a$ and $x = b$ is

$$A = \int_a^b f(x)\,dx$$

Length of Arc (Rectangular Coordinates)

Given the smooth curve $f(x, y) = 0$ from point (x_1, y_1) to point (x_2, y_2), the length between these points is

$$L = \int_{x_1}^{x_2} \sqrt{1 + \left(\frac{dy}{dx}\right)^2}\,dx$$

$$L = \int_{y_1}^{y_2} \sqrt{1 + \left(\frac{dx}{dy}\right)^2}\, dy$$

Mean Value of a Function

The mean value of a function $f(x)$ continuous on $[a, b]$ is

$$\frac{1}{(b-a)} \int_a^b f(x)\, dx$$

Area (Polar Coordinates)

Given the curve $r = f(\theta)$, continuous and nonnegative for $\theta_1 \le \theta \le \theta_2$, the area enclosed by this curve and the radial lines $\theta = \theta_1$ and $\theta = \theta_2$ is given by

$$A = \int_{\theta_1}^{\theta_2} \frac{1}{2}[f(\theta)]^2 d\theta$$

Length of Arc (Polar Coordinates)

Given the curve $r = f(\theta)$ with continuous derivative $f'(\theta)$ on $\theta_1 \le \theta \le \theta_2$, the length of arc from $\theta = \theta_1$ to $\theta = \theta_2$ is

$$L = \int_{\theta_1}^{\theta_2} \sqrt{[f(\theta)]^2 + [f'(\theta)]^2} d\theta$$

Volume of Revolution

Given a function $y = f(x)$ continuous and nonnegative on the interval (a, b), when the region bounded by $f(x)$ between a and b is revolved about the x axis, the volume of revolution is

$$V = \pi \int_a^b [f(x)]^2\, dx$$

Surface Area of Revolution (Revolution about the x axis, between *a* and *b*)

If the portion of the curve $y = f(x)$ between $x = a$ and $x = b$ is revolved about the x axis, the area A of the surface generated is given by the following:

$$A = \int_a^b 2\pi f(x)\{1 + [f'(x)]^2\}^{\frac{1}{2}}\, dx$$

Work

If a variable force $f(x)$ is applied to an object in the direction of motion along the x axis between $x = a$ and $x = b$, the work done is

$$W = \int_a^b f(x)\, dx$$

Cylindrical and Spherical Coordinates

1. Cylindrical coordinates (Fig. 18.16):

$$x = r\cos\theta$$

$$y = r\sin\theta$$

Element of volume $dV = r\, dr\, d\theta\, dz$.

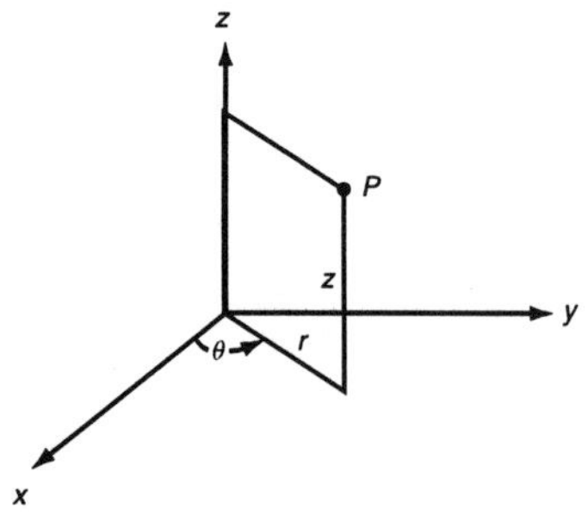

FIGURE 18.16 Cylindrical coordinates.

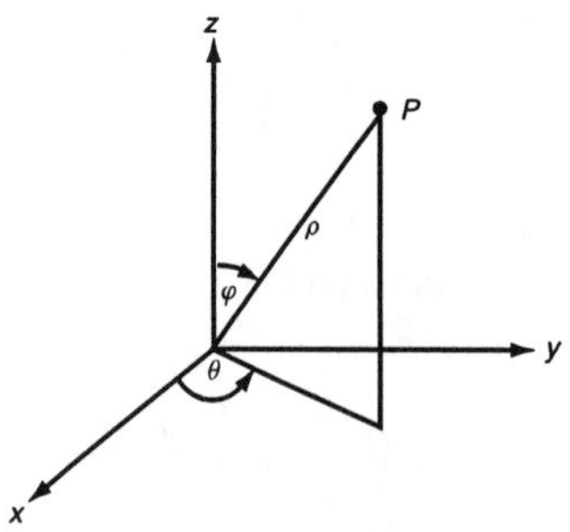

FIGURE 18.17 Spherical coordinates.

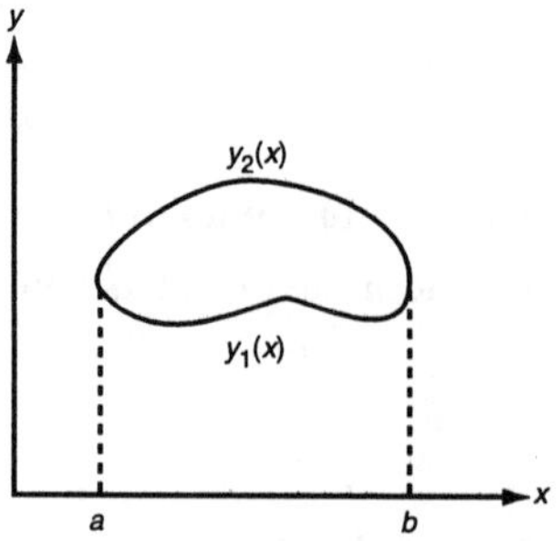

FIGURE 18.18 Region R bounded by $y_2(x)$ and $y_1(x)$.

2. Spherical coordinates (Fig. 18.17):

$$x = \rho \sin\phi \cos\theta$$
$$y = \rho \sin\phi \sin\theta$$
$$z = \rho \cos\phi$$

Element of volume $dV = \rho^2 \sin\phi \, d\rho \, d\phi \, d\theta$.

Double Integration

The evaluation of a double integral of $f(x, y)$ over a plane region R,

$$\iint_R f(x, y)\, dA$$

is practically accomplished by iterated (repeated) integration. For example, suppose that a vertical straight line meets the boundary of R in at most two points so that there is an upper boundary, $y = y_2(x)$, and a lower boundary, $y = y_1(x)$. Also, it is assumed that these functions are continuous from a to b (see Fig. 18.18). Then

$$\iint_R f(x, y)\, dA = \int_a^b \left(\int_{y_1(x)}^{y_2(y)} f(x, y)\, dy \right) dx$$

If R has left-hand boundary $x = x_1(y)$ and a right-hand boundary $x = x_2(y)$, which are continuous from c to d (the extreme values of y in R), then

$$\iint_R f(x, y)\, dA = \int_c^d \left(\int_{x_1(y)}^{x_2(y)} f(x, y)\, dx \right) dy$$

Such integrations are sometimes more convenient in polar coordinates, $x = r\cos\theta$, $y = r\sin\theta$, $dA = r\, dr\, d\theta$.

Surface Area and Volume by Double Integration

For the surface given by $z = f(x, y)$, which projects onto the closed region R of the xy plane, one may calculate the volume V bounded above by the surface and below by R, and the surface area S by the following:

$$V = \iint_R z\, dA = \iint_R f(x, y)\, dx\, dy$$

$$S = \iint_R [1 + (\partial z/\partial x)^2 + (\partial z/\partial y)^2]^{\frac{1}{2}}\, dx\, dy$$

[In polar coordinates, (r, θ), we replace dA by $r\, dr\, d\theta$.]

Centroid

The centroid of a region R of the xy plane is a point (x', y') where

$$x' = \frac{1}{A}\iint_R x\,dA, \quad y' = \frac{1}{A}\iint_R y\,dA$$

and A is the area of the region.

Example

For the circular sector of angle 2α and radius R, the area A is αR^2; the integral needed for x', expressed in polar coordinates, is

$$\iint x\,dA = \int_{-\alpha}^{\alpha}\int_0^R (r\cos\theta) r\,dr\,d\theta = \left[\frac{R^3}{3}\sin\theta\right]_{-\alpha}^{+\alpha} = \frac{2}{3}R^3\sin\alpha$$

and thus,

$$x' = \frac{\frac{2}{3}R^3\sin\alpha}{\alpha R^2} = \frac{2}{3}R\frac{\sin\alpha}{\alpha}$$

Centroids of some common regions are shown in Table 18.2

TABLE 18.2 Centroids

	Area	x'	y'
Rectangle 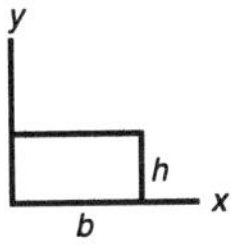	bh	$b/2$	$h/2$
Isosceles triangle 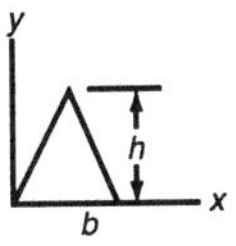 ($y' = h/3$ for any triangle of altitude h.)	$bh/2$	$b/2$	$h/3$
Semicircle 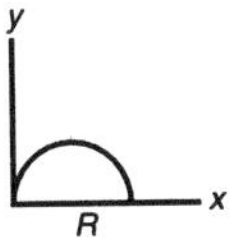	$\pi R^2/2$	R	$4R/3\pi$
Quarter circle 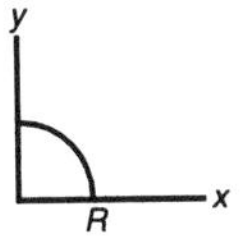	$\pi R^2/4$	$4R/3\pi$	$4R/3\pi$
Circular sector 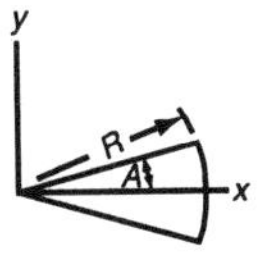	R^2A	$2R\sin A/3A$	0

18.7 Special Functions

Hyperbolic Functions

$$\sinh x = \frac{e^x - e^{-x}}{2} \qquad \operatorname{csch} x = \frac{1}{\sinh x}$$

$$\cosh x = \frac{e^x + e^{-x}}{2} \qquad \operatorname{sech} x = \frac{1}{\cosh x}$$

$$\tanh x = \frac{e^x - e^{-x}}{e^x + e^{-x}} \qquad \operatorname{ctnh} x = \frac{1}{\tanh x}$$

$$\sinh(-x) = -\sinh x \qquad \operatorname{ctnh}(-x) = -\operatorname{ctnh} x$$

$$\cosh(-x) = \cosh x \qquad \operatorname{sech}(-x) = \operatorname{sech} x$$

$$\tanh(-x) = -\tanh x \qquad \operatorname{csch}(-x) = -\operatorname{csch} x$$

$$\tanh x = \frac{\sinh x}{\cosh x} \qquad \operatorname{ctnh} x = \frac{\cosh x}{\sinh x}$$

$$\cosh^2 x - \sinh^2 x = 1 \qquad \cosh^2 x = \frac{1}{2}(\cosh 2x + 1)$$

$$\sinh^2 x = \frac{1}{2}(\cosh 2x - 1) \qquad \operatorname{ctnh}^2 x - \operatorname{csch}^2 x = 1$$

$$\operatorname{csch}^2 x - \operatorname{sech}^2 x = \operatorname{csch}^2 x \operatorname{sech}^2 x \qquad \tanh^2 x + \operatorname{sech}^2 x = 1$$

$$\sinh(x + y) = \sinh x \cosh y + \cosh x \sinh y$$

$$\cosh(x + y) = \cosh x \cosh y + \sinh x \sinh y$$

$$\sinh(x - y) = \sinh x \cosh y - \cosh x \sinh y$$

$$\cosh(x - y) = \cosh x \cosh y - \sinh x \sinh y$$

$$\tanh(x + y) = \frac{\tanh x + \tanh y}{1 + \tanh x \tanh y}$$

$$\tanh(x - y) = \frac{\tanh x - \tanh y}{1 - \tanh x \tanh y}$$

Bessel Functions

Bessel functions, also called cylindrical functions, arise in many physical problems as solutions of the differential equation

$$x^2 y'' + xy' + (x^2 - n^2)y = 0$$

which is known as Bessel's equation. Certain solutions, known as *Bessel functions of the first kind of order n*, are given by

$$J_n(x) = \sum_{k=0}^{\infty} \frac{(-1)^k}{k!\Gamma(n + k + 1)} \left(\frac{x}{2}\right)^{n+2k}$$

$$J_{-n}(x) = \sum_{k=0}^{\infty} \frac{(-1)^k}{k!\Gamma(-n + k + 1)} \left(\frac{x}{2}\right)^{-n+2k}$$

see Fig. 18.19a.

In the preceding it is noteworthy that the gamma function must be defined for the negative argument q : $\Gamma(q) = \Gamma(q+1)/q$, provided that q is not a negative integer. When q is a negative integer, $1/\Gamma(q)$ is defined to be zero. The functions $J_{-n}(x)$ and $J_n(x)$ are solutions of Bessel's equation for all real n. It is seen, for $n = 1, 2, 3, \ldots$, that

$$J_{-n}(x) = (-1)^n J_n(x)$$

and, therefore, these are not independent; hence, a linear combination of these is not a general solution. When, however, n is not a positive integer, a negative integer, or zero, the linear combination with arbitrary constants c_1 and c_2,

$$y = c_1 J_n(x) + c_2 J_{-n}(x)$$

is the general solution of the Bessel differential equation.

The zero-order function is especially important as it arises in the solution of the heat equation (for a long cylinder):

$$J_0(x) = 1 - \frac{x^2}{2^2} + \frac{x^4}{2^2 4^2} - \frac{x^6}{2^2 4^2 6^2} + \cdots$$

whereas the following relations show a connection to the trigonometric functions:

$$J_{1/2}(x) = \left[\frac{2}{\pi x}\right]^{1/2} \sin x$$

$$J_{-1/2}(x) = \left[\frac{2}{\pi x}\right]^{1/2} \cos x$$

The following recursion formula gives $J_{n+1}(x)$ for any order in terms of lower order functions:

$$\frac{2n}{x} J_n(x) = J_{n-1}(x) + J_{n+1}(x)$$

Bessel Functions of the Second Kind, $Y_n(x)$ (Also Called *Neumann Functions* or *Weber Functions*) (Fig. 18.19b)

Domain: $[x > 0]$

Recurrence relation:

$$Y_{n+1}(x) = \frac{2n}{x} Y_n(x) - Y_{n-1}(x), \quad n = 0, 1, 2, \ldots$$

Symmetry: $Y_{-n}(x) = (-1)^n Y_n(x)$

Legendre Polynomials

If Laplace's equation, $\nabla^2 V = 0$, is expressed in spherical coordinates, it is

$$r^2 \sin\theta \frac{\delta^2 V}{\delta r^2} + 2r \sin\theta \frac{\delta V}{\delta r} + \sin\theta \frac{\delta^2 V}{\delta \theta^2} + \cos\theta \frac{\delta V}{\delta \theta} + \frac{1}{\sin\theta} \frac{\delta^2 V}{\delta \phi^2} = 0$$

and any of its solutions, $V(r, \theta, \phi)$, are known as *spherical harmonics*. The solution as a product

$$V(r, \theta, \phi) = R(r)\Theta(\theta)$$

which is independent of ϕ, leads to

$$\sin^2\theta\Theta'' + \sin\theta\cos\theta\Theta' + [n(n+1)\sin^2\theta]\Theta = 0$$

Rearrangement and substitution of $x = \cos\theta$ leads to

$$(1 - x^2)\frac{d^2\Theta}{dx^2} - 2x\frac{d\Theta}{dx} + n(n+1)\Theta = 0$$

known as *Legendre's equation*. Important special cases are those in which n is zero or a positive integer, and, for

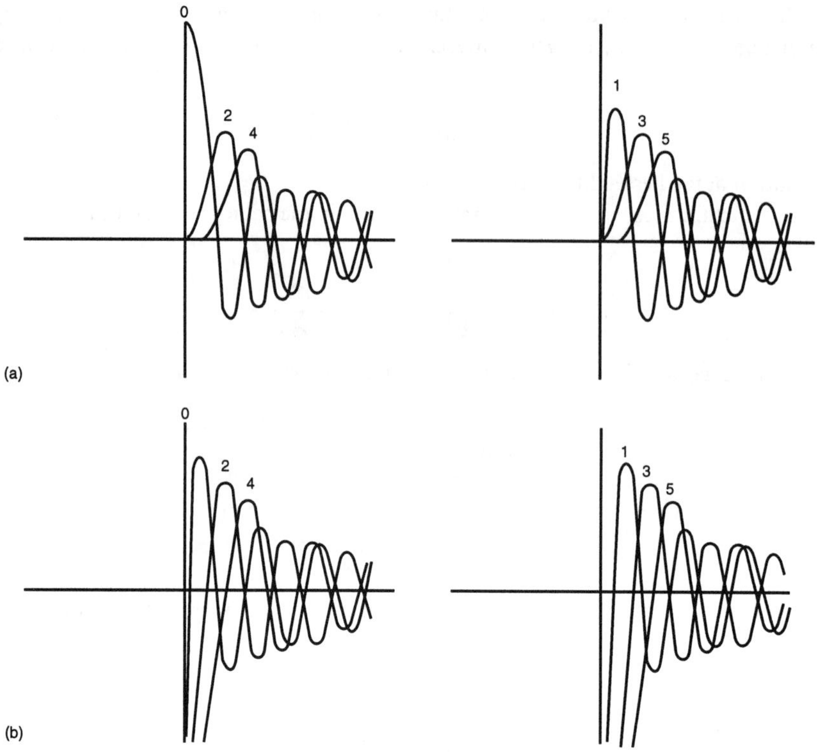

FIGURE 18.19 Bessel Functions: (a) of the first kind, (b) of the second kind.

such cases, Legendre's equation is satisfied by polynomials called Legendre polynomials, $P_n(x)$. A short list of Legendre polynomials, expressed in terms of x and $\cos\theta$, is given next. These are given by the following general formula:

$$P_n(x) = \sum_{j=0}^{L} \frac{(-1)^j (2n-2j)!}{2^n j!(n-j)!(n-2j)!} x^{n-2j}$$

where $L = n/2$ if n is even and $L = (n-1)/2$ if n is odd,

$$P_0(x) = 1$$

$$P_1(x) = x$$

$$P_2(x) = \frac{1}{2}(3x^2 - 1)$$

$$P_3(x) = \frac{1}{2}(5x^3 - 3x)$$

$$P_4(x) = \frac{1}{8}(35x^4 - 30x^2 + 3)$$

$$P_5(x) = \frac{1}{8}(63x^5 - 70x^3 + 15x)$$

$$P_0(\cos\theta) = 1$$

$$P_1(\cos\theta) = \cos\theta$$

$$P_2(\cos\theta) = \frac{1}{4}(3\cos 2\theta + 1)$$

$$P_3(\cos\theta) = \frac{1}{8}(5\cos 3\theta + 3\cos\theta)$$

$$P_4(\cos\theta) = \frac{1}{64}(35\cos 4\theta + 20\cos 2\theta + 9)$$

Additional Legendre polynomials may be determined from the *recursion formula*

$$(n+1)P_{n+1}(x) - (2n+1)xP_n(x) + nP_{n-1}(x) = 0 \quad (n = 1, 2, \ldots)$$

or the *Rodrigues formula*

$$P_n(x) = \frac{1}{2^n n!}\frac{d^n}{dx^n}(x^2-1)^n$$

Laguerre Polynomials

Laguerre polynomials, denoted $L_n(x)$, are solutions of the differential equation

$$xy'' + (1-x)y' + ny = 0$$

and are given by

$$L_n(x) = \sum_{j=0}^{n} \frac{(-1)^j}{j!} C_{(n,j)} x^j \quad (n = 0, 1, 2, \ldots)$$

Thus,

$$L_0(x) = 1$$

$$L_1(x) = 1 - x$$

$$L_2(x) = 1 - 2x + \frac{1}{2}x^2$$

$$L_3(x) = 1 - 3x + \frac{3}{2}x^2 - \frac{1}{6}x^3$$

Additional Laguerre polynomials may be obtained from the recursion formula

$$(n+1)L_{n+1}(x) - (2n+1-x)L_n(x) + nL_{n-1}(x) = 0$$

Hermite Polynomials

The Hermite polynomials, denoted $H_n(x)$, are given by

$$H_0 = 1, \quad H_n(x) = (-1)^n e^{x^2}\frac{d^n e^{-x^2}}{dx^n}, \qquad (n = 1, 2, \ldots)$$

and are solutions of the differential equation

$$y'' - 2xy' + 2ny = 0 \qquad (n = 0, 1, 2, \ldots)$$

The first few Hermite polynomials are

$$H_0 = 1 \qquad H_1(x) = 2x$$

$$H_2(x) = 4x^2 - 2 \qquad H_3(x) = 8x^3 - 12x$$

$$H_4(x) = 16x^4 - 48x^2 + 12$$

Additional Hermite polynomials may be obtained from the relation

$$H_{n+1}(x) = 2xH_n(x) - H_n'(x)$$

where prime denotes differentiation with respect to x.

Orthogonality

A set of functions $\{f_n(x)\}(n = 1, 2, \ldots)$ is orthogonal in an interval (a, b) with respect to a given weight function $w(x)$ if

$$\int_a^b w(x) f_m(x) f_n(x) dx = 0 \qquad \text{when } m \neq n$$

The following polynomials are orthogonal on the given interval for the given $w(x)$.

Legendre polynomials:

$$P_n(x)\; w(x) = 1$$
$$a = -1, \quad b = 1$$

Leguerre polynomials:

$$L_n(x)\; w(x) = \exp(-x)$$
$$a = 0, \quad b = \infty$$

Hermite polynomials:

$$H_n(x) w(x) = \exp(-x^2)$$
$$a = -\infty, \quad b = \infty$$

The Bessel functions of *order* n, $J_n(\lambda_1 x)$, $J_n(\lambda_2 x)$, $\ldots$, are orthogonal with respect to $w(x) = x$ over the interval $(0, c)$ provided that the λ_i are the positive roots of $J_n(\lambda c) = 0$,

$$\int_0^c x J_n(\lambda_j x) J_n(\lambda_k x)\, dx = 0 \quad (j \neq k)$$

where n is fixed and $n \geq 0$.

Functions with $x^2/a^2 \pm y^2/b^2$

Elliptic Paraboloid (Fig. 18.20)

$$z = c(x^2/a^2 + y^2/b^2)$$
$$x^2/a^2 + y^2/b^2 - z/c = 0$$

Hyperbolic Paraboloid (Commonly Called *Saddle*) (Fig. 18.21):

$$z = c(x^2/a^2 - y^2/b^2)$$
$$x^2/a^2 - y^2/b^2 - z/c = 0$$

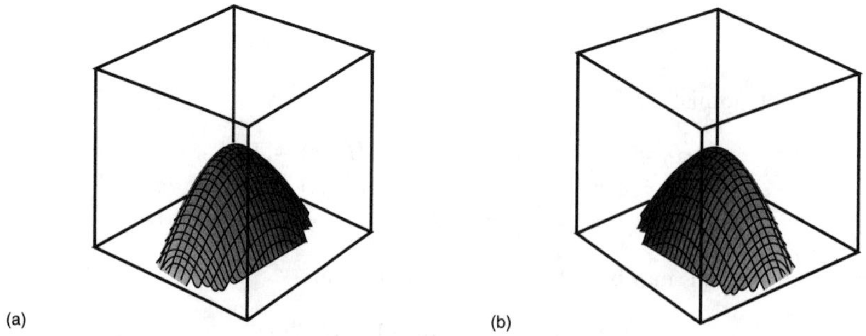

FIGURE 18.20 Elliptic paraboloid: (a) $a = 0.5, b = 1.0, c = -1.0$, viewpoint $= (5, -6, 4)$; (b) $a = 1.0, b = 1.0, c = -2.0$, viewpoint $= (5, -6, 4)$.

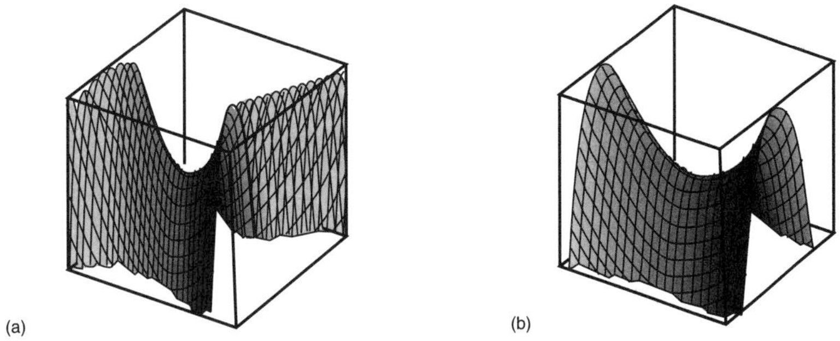

FIGURE 18.21 Hyperbolic paraboloid: (a) $a = 0.50, b = 0.5, c = 1.0$, viewpoint $= (4, -6, 4)$; (b) $a = 1.00, b = 0.5, c = 1.0$, viewpoint $= (4, -6.4)$.

Elliptic Cylinder (Fig. 18.22):

$$1 = x^2/a^2 + y^2/b^2$$

$$x^2/a^2 + y^2/b^2 - 1 = 0$$

Hyperbolic Cylinder (Fig. 18.23)

$$1 = x^2/a^2 - y^2/b^2$$

$$x^2/a^2 - y^2/b^2 - 1 = 0$$

Functions with $(x^2/a^2 + y^2/b^2 \pm c^2)^{1/2}$

Sphere (Fig. 18.24)

$$z = (1 - x^2 - y^2)^{1/2}$$

$$x^2 + y^2 + z^2 - 1 = 0$$

Ellipsoid (Fig. 18.25)

$$z = c(1 - x^2/a^2 - y^2/b^2)^{1/2}$$

$$x^2/a^2 + y^2/b^2 + z^2/c^2 - 1 = 0$$

Special cases:

$a = b > c$ gives oblate spheroid

$a = b < c$ gives prolate spheroid

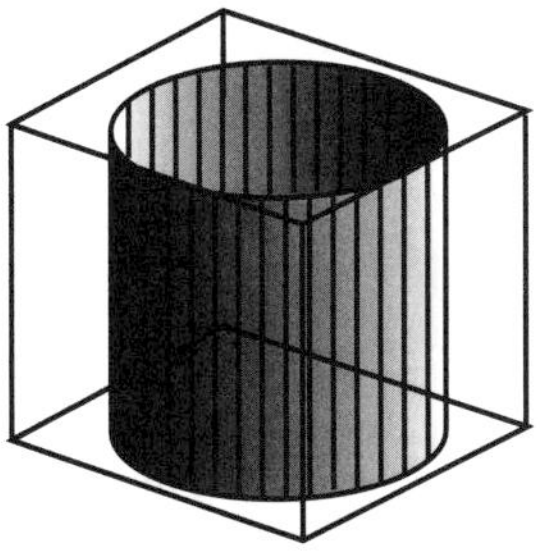

FIGURE 18.22 Elliptic cylinder. $a = 1.0, b = 1.0$, viewpoint $= (4, -5, 2)$.

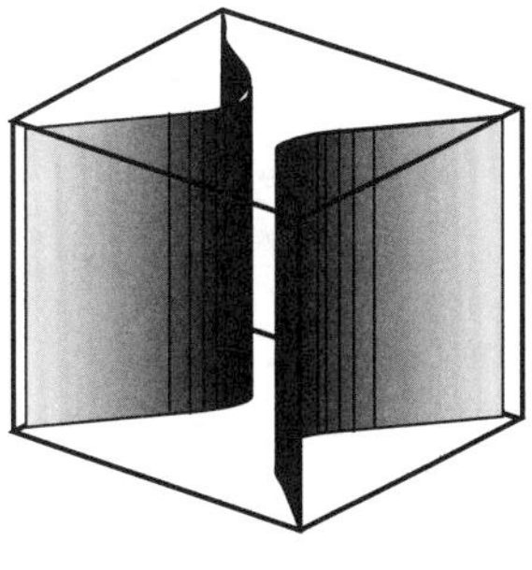

FIGURE 18.23 Hyperbolic cylinder. $a = 1.0, b = 1.0$, viewpoint $= (4, -6, 3)$.

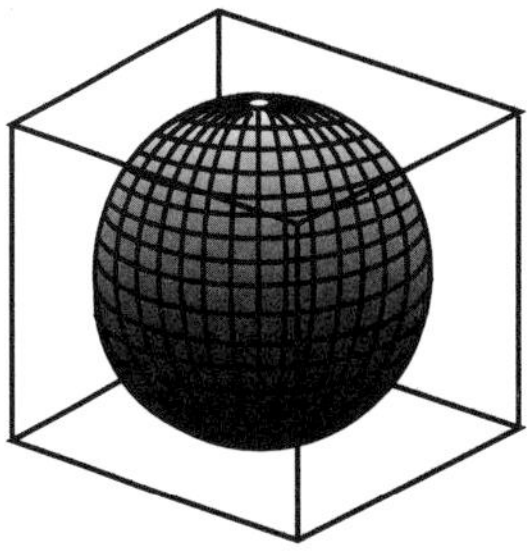

FIGURE 18.24 Sphere: viewpoint $= (4, -5, 2)$.

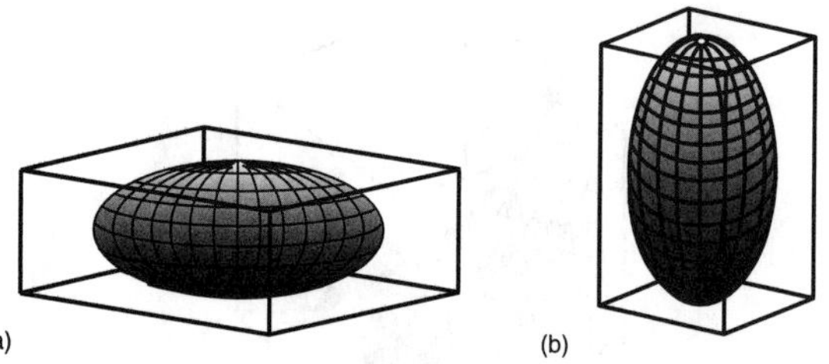

(a) (b)

FIGURE 18.25 Ellipsoid: (a) $a = 1.00, b = 1.00, c = 0.5$, viewpoint = (4, −5, 2) ; (b) $a = 0.50, b = 0.50, c = 1.0$, viewpoint = (4, −5, 2).

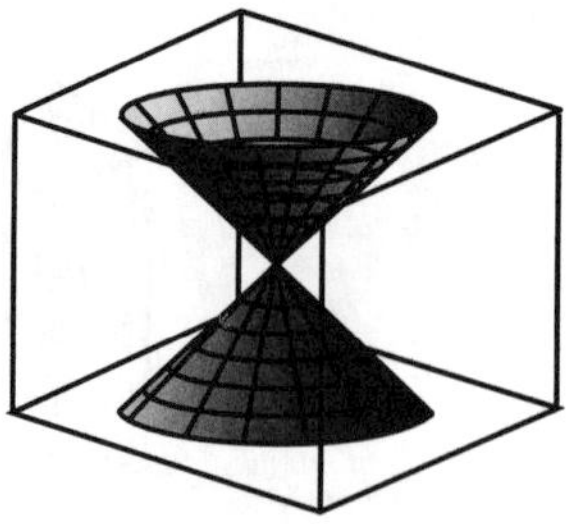

FIGURE 18.26 Cone: viewpoint = (4, −5, 2).

Cone (Fig. 18.26)

$$z = (x^2 + y^2)^{1/2}$$

$$x^2 + y^2 - z^2 = 0$$

Elliptic Cone (Circular Cone if *a* = *b*) (Fig. 18.27)

$$z = c(x^2/a^2 + y^2/b^2)^{1/2}$$

$$x^2/a^2 + y^2/b^2 - z^2/c^2 = 0$$

Hyperboloid of One Sheet (Fig. 18.28)

$$z = c(x^2/a^2 + y^2/b^2 - 1)^{1/2}$$

$$x^2/a^2 + y^2/b^2 - z^2/c^2 - 1 = 0$$

Hyperboloid of Two Sheets (Fig. 18.29)

$$z = c(x^2/a^2 + y^2/b^2 + 1)^{1/2}$$

$$x^2/a^2 + y^2/b^2 - z^2/c^2 + 1 = 0$$

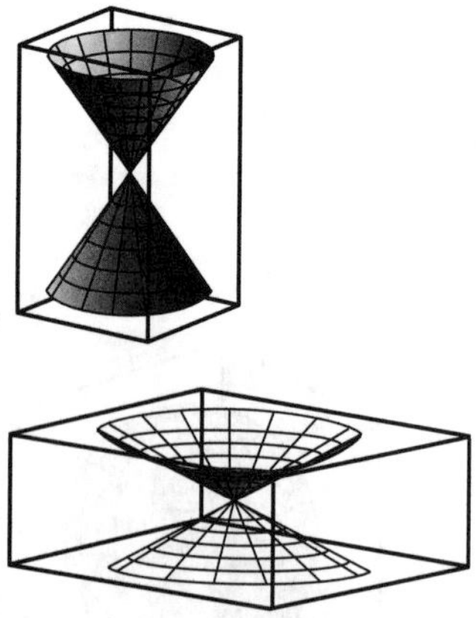

(a) (b)

FIGURE 18.27 Elliptic cone: (a) $a = 0.5, b = 0.5, c = 1.00$, viewpoint = (4, −5, 2); (b) $a = 1.0, b = 1.0, c = 0.50$, viewpoint = (4, −5, 2).

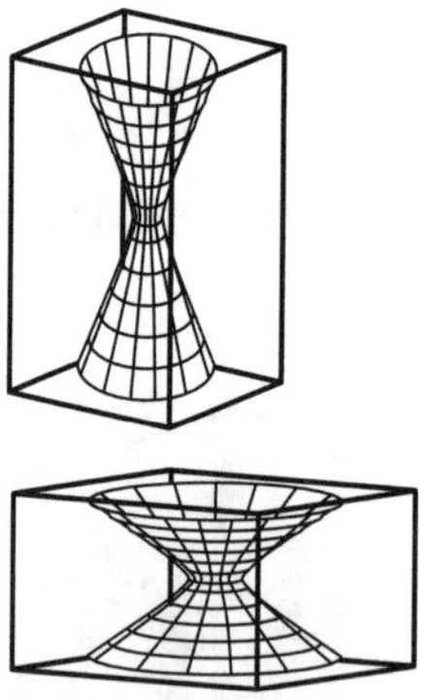

(a) (b)

FIGURE 18.28 Hyperboloid of one sheet: (a) $a = 0.1, b = 0.1, c = 0.2, \pm z = c\sqrt{15}$, viewpoint = (4, −5, 2); (b) $a = 0.2, b = 0.2, c = 0.2, \pm z = c\sqrt{15}$, viewpoint = (4, −5, 2).

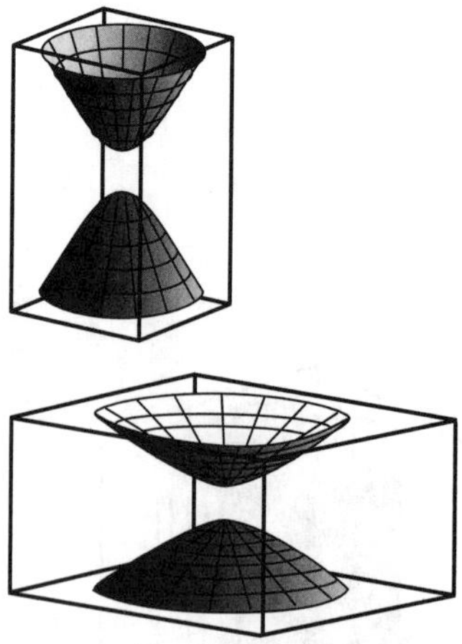

(a) (b)

FIGURE 18.29 Hyperboloid of two sheets: (a) $a = 0.125, b = 0.125, c = 0.2, \pm z = c\sqrt{17}$, viewpoint = (4, −5, 2); (b) $a = 0.25, b = 0.25, c = 0.2, \pm z = c\sqrt{17}$, viewpoint = (4, −5, 2).

18.8 Basic Definitions: Linear Algebra Matrices

A *matrix* A is a rectangular array of numbers (real or complex):

$$A = \begin{bmatrix} a_{11} & a_{12} & \cdots & a_{1m} \\ a_{21} & a_{22} & \cdots & a_{2m} \\ \vdots & & & \\ a_{n1} & a_{n2} & \cdots & a_{nm} \end{bmatrix}$$

The *size* of the matrix is said to be $n \times m$. The $1 \times m$ matrices $[a_{i1}\ a_{i2}\ \cdots\ a_{im}]$ are called *rows* of $\boldsymbol{A}$, and the $n \times 1$ matrices

$$\begin{bmatrix} a_{1j} \\ a_{2j} \\ \vdots \\ a_{nj} \end{bmatrix}$$

are called *columns* of $\boldsymbol{A}$. An $n \times m$ matrix thus consists of n rows and m columns; a_{ij} denotes the *element*, or *entry*, of $\boldsymbol{A}$ in the ith row and jth column. A matrix consisting of just one row is called a *row vector*, whereas a matrix of just one column is called a *column vector*. The elements of a vector are frequently called *components* of the vector. When the size of the matrix is clear from the context, we sometimes write $\boldsymbol{A} = (a_{ij})$.

A matrix with the same number of rows as columns is a *square* matrix, and the number of rows and columns is the *order* of the matrix. The diagonal of an $n \times n$ square matrix $\boldsymbol{A}$ from a_{11} to a_{nn} is called the *main*, or *principal, diagonal.* The word *diagonal* with no modifier usually means the main diagonal. The *transpose* of a matrix $\boldsymbol{A}$ is the matrix that results from interchanging the rows and columns of $\boldsymbol{A}$. It is usually denoted by $\boldsymbol{A}^T$. A matrix $\boldsymbol{A}$ such that $\boldsymbol{A} = \boldsymbol{A}^T$ is said to be *symmetric*. The *conjugate transpose* of $\boldsymbol{A}$ is the matrix that results from replacing each element of $\boldsymbol{A}^T$ by its complex conjugate, and is usually denoted by $\boldsymbol{A}^H$. A matrix such that $\boldsymbol{A} = \boldsymbol{A}^H$ is said to be *Hermitian.*

A square matrix $\boldsymbol{A} = (a_{ij})$ is *lower triangular* if $a_{ij} = 0$ for $j > i$ and is *upper triangular* if $a_{ij} = 0$ for $j < i$. A matrix that is both upper and lower triangular is a *diagonal* matrix. The $n \times n$ *identity matrix* is the $n \times n$ diagonal matrix in which each element of the main diagonal is 1. It is traditionally denoted $\boldsymbol{I}_n$, or simply $\boldsymbol{I}$ when the order is clear from the context.

Algebra of Matrices

The sum and difference of two matrices $\boldsymbol{A}$ and $\boldsymbol{B}$ are defined whenever $\boldsymbol{A}$ and $\boldsymbol{B}$ have the same size. In that case $\boldsymbol{C} = \boldsymbol{A} \pm \boldsymbol{B}$ is defined by $\boldsymbol{C} = (c_{ij}) = (a_{ij} \pm b_{ij})$. The product $t\boldsymbol{A}$ of a scalar t (real or complex number) and a matrix $\boldsymbol{A}$ is defined by $t\boldsymbol{A} = (ta_{ij})$. If $\boldsymbol{A}$ is an $n \times m$ matrix and $\boldsymbol{B}$ is an $m \times p$ matrix, the product $\boldsymbol{C} = \boldsymbol{AB}$ is defined to be the $n \times p$ matrix $\boldsymbol{C} = (c_{ij})$ given by $c_{ij} = \sum_{k=1}^{m} a_{ik}b_{kj}$. Note that the product of an $n \times m$ matrix and an $m \times p$ matrix is an $n \times p$ matrix, and the product is defined only when the number of column of the first factor is the same as the number of rows of the second factor. Matrix multiplication is, in general, associative: $\boldsymbol{A}(\boldsymbol{BC}) = (\boldsymbol{AB})$C. It also distributes over addition (and subtraction):

$$\boldsymbol{A}(\boldsymbol{B} + \boldsymbol{C}) = \boldsymbol{AB} + \boldsymbol{AC} \qquad \text{and} \qquad (\boldsymbol{A} + \boldsymbol{B})\boldsymbol{C} = \boldsymbol{AC} + \boldsymbol{BC}$$

It is, however, not in general true that $\boldsymbol{AB} = \boldsymbol{BA}$, even in case both products are defined. It is clear that $(\boldsymbol{A} + \boldsymbol{B})^T = \boldsymbol{A}^T + \boldsymbol{B}^T$ and $(\boldsymbol{A} + \boldsymbol{B})^H = \boldsymbol{A}^H + \boldsymbol{B}^H$. It is also true, but not so obvious perhaps, that $(\boldsymbol{AB})^T = \boldsymbol{B}^T\boldsymbol{A}^T$ and $(\boldsymbol{AB})^H = \boldsymbol{B}^H\boldsymbol{A}^H$.

The $n \times n$ identity matrix $\boldsymbol{I}$ has the property that $\boldsymbol{IA} = \boldsymbol{AI} = \boldsymbol{A}$ for every $n \times n$ matrix $\boldsymbol{A}$. If $\boldsymbol{A}$ is square, and if there is a matrix $\boldsymbol{B}$ such that $\boldsymbol{AB} = \boldsymbol{BA} = \boldsymbol{I}$, then $\boldsymbol{B}$ is called the *inverse* of A and is denoted $\boldsymbol{A}^{-1}$. This terminology and notation are justified by the fact that a matrix can have at most one inverse. A matrix having an inverse is said to be *invertible*, or *nonsingular*, whereas a matrix not having an inverse is said to be *noninvertible*, or *singular*. The product of two invertible matrices is invertible and, in fact, $(\boldsymbol{AB})^{-1} = \boldsymbol{B}^{-1}\boldsymbol{A}^{-1}$. The sum of two invertible matrices is, obviously, not necessarily invertible.

Systems of Equations

The system of n linear equations in m unknowns

$$\begin{aligned} a_{11}x_1 + a_{12}x_2 + a_{13}x_3 + \cdots + a_{1m}x_m &= b_1 \\ a_{21}x_1 + a_{22}x_2 + a_{23}x_3 + \cdots + a_{2m}x_m &= b_2 \\ &\vdots \\ a_{n1}x_1 + a_{n2}x_2 + a_{n3}x_3 + \cdots + a_{nm}x_m &= b_n \end{aligned}$$

may be written $\boldsymbol{Ax} = \boldsymbol{b}$, where $\boldsymbol{A} = (a_{ij})$, $x = [x_1 \quad x_2 \quad \cdots \quad x_m]^T$ and $\boldsymbol{b} = [b_1 \quad b_2 \quad \cdots \quad b_n]^T$. Thus, $\boldsymbol{A}$ is an $n \times m$ matrix, and $\boldsymbol{x}$ and $\boldsymbol{b}$ are column vectors of the appropriate sizes.

The matrix $\boldsymbol{A}$ is called the *coefficient matrix* of the system. Let us first suppose the coefficient matrix is square; that is, there are an equal number of equations and unknowns. If $\boldsymbol{A}$ is upper triangular, it is quite easy to find all solutions of the system. The ith equation will contain only the unknowns $x_i, x_{i+1}, \ldots, x_n$, and one simply solves the equations in reverse order: the last equation is solved for x_n; the result is substituted into the $(n-1)$st equation, which is then solved for x_{n-1}; these values of x_n and x_{n-1} are substituted in the $(n-2)$th equation, which is solved for x_{n-2}, and so on. This procedure is known as *back substitution*.

The strategy for solving an arbitrary system is to find an upper-triangular system equivalent with it and solve this upper-triangular system using back substitution. First, suppose the element $a_{11} \neq 0$. We may rearrange the equations to ensure this, unless, of course the first column of $\boldsymbol{A}$ is all 0s. In this case proceed to the next step, to be described later. For each $i \geq 2$ let $m_{i1} = a_{i1}/a_{11}$. Now replace the ith equation by the result of multiplying the first equation by m_{i1} and subtracting the new equation from the ith equation. Thus,

$$a_{i1}x_1 + a_{i2}x_2 + a_{i3}x_3 + \cdots + a_{im}x_m = b_i$$

is replaced by

$$0 \cdot x_1 + (a_{i2} + m_{i1}a_{12})x_2 + (a_{i3} + m_{i1}a_{13})x_3 + \cdots + (a_{im} + m_{i1}a_{1m})x_m = b_i + m_{i1}b_1$$

After this is done for all $i = 2, 3, \ldots, n$, there results the equivalent system

$$\begin{aligned} a_{11}x_1 + a_{12}x_2 + a_{13}x_3 + \cdots + a_{1n}x_n &= b_1 \\ 0 \cdot x_1 + a'_{22}x_2 + a'_{23}x_3 + \cdots + a'_{2n}x_n &= b'_2 \\ 0 \cdot x_1 + a'_{32}x_2 + a'_{33}x_3 + \cdots + a'_{3n}x_n &= b'_3 \\ &\vdots \\ 0 \cdot x_1 + a'_{n2}x_2 + a'_{n3}x_3 + \cdots + a'_{nn}x_n &= b'_n \end{aligned}$$

in which all entries in the first column below a_{11} are 0. (Note that if all entries in the first column were 0 to begin with, then $a_{11} = 0$ also.) This procedure is now repeated for the $(n-1) \times (n-1)$ system

$$\begin{aligned} a'_{22}x_2 + a'_{23}x_3 + \cdots + a'_{2n}x_n &= b'_2 \\ a'_{32}x_2 + a'_{33}x_3 + \cdots + a'_{3n}x_n &= b'_3 \\ &\vdots \\ a'_{n2}x_2 + a'_{n3}x_3 + \cdots + a'_{nn}x_n &= b'_n \end{aligned}$$

to obtain an equivalent system in which all entries of the coefficient matrix below a'_{22} are 0. Continuing, we obtain an upper-triangular system $\boldsymbol{Ux} = \boldsymbol{c}$ equivalent with the original system. This procedure is known as *Gaussian elimination*. The numbers m_{ij} are known as the *multipliers*.

Essentially the same procedure may be used in case the coefficient matrix is not square. If the coefficient matrix is not square, we may make it square by appending either rows or columns of 0s as needed. Appending rows of 0s and appending 0s to make $\boldsymbol{b}$ have the appropriate size is equivalent to appending equations $0 = 0$ to the system. Clearly the new system has precisely the same solutions as the original system. Appending columns of 0s and adjusting the size of $\boldsymbol{x}$ appropriately yields a new system with additional unknowns, each appearing only with coefficient 0, thus not affecting the solutions of the original system. In either case we may assume the coefficient matrix is square, and apply the Gauss elimination procedure.

Suppose the matrix $\boldsymbol{A}$ is invertible. Then if there were no row interchanges in carrying out the above Gauss elimination procedure, we have the *LU factorization* of the matrix $\boldsymbol{A}$:

$$\boldsymbol{A} = \boldsymbol{LU}$$

where $\boldsymbol{U}$ is the upper-triangular matrix produced by elimination and $\boldsymbol{L}$ is the lower-triangular matrix given by

$$\boldsymbol{L} = \begin{bmatrix} 1 & 0 & \cdots & \cdots 0 \\ m_{21} & 1 & 0 & \cdots 0 \\ \vdots & & \ddots & \\ m_{n1} & m_{n2} & \cdots & 1 \end{bmatrix}$$

A *permutation* $\boldsymbol{P}_{ij}$ matrix is an $n \times n$ matrix such that $\boldsymbol{P}_{ij}\boldsymbol{A}$ is the matrix that results from exchanging row i and j of the matrix $\boldsymbol{A}$. The matrix $\boldsymbol{P}_{ij}$ is the matrix that results from exchanging row i and j of the identity matrix. A product $\boldsymbol{P}$ of such matrices $\boldsymbol{P}_{ij}$ is called a *permutation* matrix. If row interchanges are required in the Gauss elimination procedure, then we have the factorization

$$\boldsymbol{PA} = \boldsymbol{LU}$$

where $\boldsymbol{P}$ is the permutation matrix giving the required row exchanges.

Vector Spaces

The collection of all column vectors with n real components is *Euclidean n-space*, and is denoted $\boldsymbol{R}^n$. The collection of column vectors with n complex components is denoted $\boldsymbol{C}^n$. We shall use *vector space* to mean either $\boldsymbol{R}^n$ or $\boldsymbol{C}^n$. In discussing the space $\boldsymbol{R}^n$, the word *scalar* will mean a real number, and in discussing the space $\boldsymbol{C}^n$, it will mean a complex number. A subset $\boldsymbol{S}$ of a vector space is a *subspace* such that if $\boldsymbol{u}$ and $\boldsymbol{v}$ are vectors in $\boldsymbol{S}$, and if c is any scalar, then $\boldsymbol{u} + \boldsymbol{v}$ and $c\boldsymbol{u}$ are in $\boldsymbol{S}$. We shall sometimes use the word *space* to mean a subspace. If $B = \{\boldsymbol{v}_1, \boldsymbol{v}_2, \ldots, \boldsymbol{v}_k\}$ is a collection of vectors in a vector space, then the set $\boldsymbol{S}$ consisting of all vectors $c_1\boldsymbol{v}_1 + c_2\boldsymbol{v}_2 + \cdots + c_m\boldsymbol{v}_m$ for all scalars $c_1, c_2, \ldots, c_m$ is a subspace, called the *span* of B. A collection $\{\boldsymbol{v}_1, \boldsymbol{v}_2, \ldots, \boldsymbol{v}_m\}$ of vectors $c_1\boldsymbol{v}_1 + c_2\boldsymbol{v}_2 + \cdots + c_m\boldsymbol{v}_m\}$ is a *linear combination* of B. If $\boldsymbol{S}$ is a subspace and $B = \{v_1, v_2, \ldots, v_m\}$ is a subset of $\boldsymbol{S}$ such that $\boldsymbol{S}$ is the space of B, then B is said to *span* $\boldsymbol{S}$.

A collection $\{\boldsymbol{v}_1, \boldsymbol{v}_2, \ldots, \boldsymbol{v}_m\}$ of n vectors is *linearly dependent* if there exist scalars $c_1, c_2, \ldots, c_m$, not all zero, such that $c_1\boldsymbol{v}_1 + c_2\boldsymbol{v}_2 + \cdots + c_m\boldsymbol{v}_m = 0$. A collection of vectors that is not linearly dependent is said to be *linearly independent*. The modifier *linearly* is frequently omitted, and we speak simply of dependent and independent collections. A linearly independent collection of vectors in a space $\boldsymbol{S}$ that spans $\boldsymbol{S}$ is a *basis* of $\boldsymbol{S}$. Every basis of a space $\boldsymbol{S}$ contains the same number of vectors; this number is the *dimension* of $\boldsymbol{S}$. The dimension of the space consisting of only the zero vector is 0. The collection $B = \{\boldsymbol{e}_1, \boldsymbol{e}_2, \ldots, \boldsymbol{e}_n\}$, where $\boldsymbol{e}_1 = [1, 0, 0, \ldots, 0]^T$, $\boldsymbol{e}_2 = [0, 1, 0, \ldots, 0]^T$, and so forth ($\boldsymbol{e}_i$ has 1 as its ith component and zero for all other components) is a basis for the spaces $\boldsymbol{R}^n$ and $\boldsymbol{C}^n$. This is the *standard basis* for these spaces. The dimension of these spaces is thus n. In a space $\boldsymbol{S}$ of dimension n, no collection of fewer than n vectors can span $\boldsymbol{S}$, and no collection of more than n vectors in $\boldsymbol{S}$ can be independent.

Rank and Nullity

The *column space* of an $n \times m$ matrix A is the subspace of R^n or C^n spanned by the columns of A. The *row space* is the subspace of R^m or C^m spanned by the rows of A. Note that for any vector $x = [x_1 x_2 \cdots x_m]^T$,

$$Ax = x_1 \begin{bmatrix} a_{11} \\ a_{21} \\ \vdots \\ a_{n1} \end{bmatrix} + x_2 \begin{bmatrix} a_{12} \\ a_{22} \\ \vdots \\ a_{n2} \end{bmatrix} + \cdots + x_m \begin{bmatrix} a_{1m} \\ a_{2m} \\ \vdots \\ a_{nm} \end{bmatrix}$$

so that the column space is the collection of all vectors Ax, and thus the system $Ax = b$ has a solution if and only if b is a member of the column space of A.

The dimension of the column space is the *rank* of A. The row space has the same dimension as the column space. The set of all solutions of the system $Ax = 0$ is a subspace called the *null space* of A, and the dimension of this null space is the *nullity* of A. A fundamental result in matrix theory is the fact that, for an $n \times m$ matrix A,

$$\text{rank } A + \text{nullity } A = m$$

The difference of any two solutions of the linear system $Ax = b$ is a member of the null space of A. Thus this system has at most one solution if and only if the nullity of A is zero. If the system is square (that is, if A is $n \times n$), then there will be a solution for every right-hand side b if and only if the collection of columns of A is linearly independent, which is the same as saying the rank of A is n. In this case the nullity must be zero. Thus, for any b, the square system $Ax = b$ has exactly one solution if and only if rank $A = n$. In other words the $n \times n$ matrix A is invertible if and only if rank $A = n$.

Orthogonality and Length

The *inner product* of two vectors x and y is the scalar $x^H y$. The *length*, or *norm*, $\|x\|$, of the vector x is given by $\|x\| = \sqrt{x^H x}$. A *unit vector* is a vector of norm 1. Two vectors x and y are *orthogonal* if $x^H y = 0$. A collection of vectors $\{v_1, v_2, \ldots, v_m\}$ in a space S is said to be an *orthonormal* collection if $v_i^H v_j = 0$ for $i \neq j$ and $v_i^H v_i = 1$. An orthonormal collection is necessarily linearly independent. If S is a subspace (or R^n or C^n) spanned by the orthonormal collection $\{v_1, v_2, \ldots, v_m\}$, then the *projection* of a vector x onto S is the vector

$$\text{proj}\,(x; S) = (x^H v_1)v_1 + (x^H v_2)v_2 + \cdots + (x^H v_m)v_m$$

The projection of x onto S minimizes the function $f(y) = \|x - y\|^2$ for $y \in S$. In other words the projection of x onto S is the vector in S that is closest to x.

If b is a vector and A is an $n \times m$ matrix, then a vector x minimizes $\|b - Ax\|^2$ if and only if it is a solution of $A^H Ax = A^H b$. This system of equations is called the *system of normal equations* for the least-squares problem of minimizing $\|b - Ax\|^2$.

If A is an $n \times m$ matrix and rank $A = k$, then there is an $n \times k$ matrix Q whose columns form an orthonormal basis for the column space of A and a $k \times m$ upper-triangular matrix R of rank k such that

$$A = QR$$

This is called the *QR factorization* of A. It now follows that x minimizes $\|b - Ax\|^2$ if and only if it is a solution of the upper-triangular system $Rx = Q^H b$.

If $\{w_1, w_2, \ldots, w_m\}$ is a basis for a space S, the following procedure produces an orthonormal basis $\{v_1, v_2, \ldots, v_m\}$ for S:

- Set $v_1 = w_1/\|w_1\|$
- Let $\tilde{v}_2 = w_2 - \text{proj}(w_2; S_1)$, where S_1 is the span of $\{v_1\}$; set $v_2 = v_2/\|\tilde{v}_2\|$
- Next, let $\tilde{v}_3 = w_3 - \text{proj}(w_3; s_2)$, where S_2 is the span of $\{v_1, v_2\}$; set $v_3 = \tilde{v}_3/\|\tilde{v}_3\|$.

And so on: $\tilde{v}_i = w_i - \text{proj}(w_i; S_{i-1})$, where S_{i-1} is the span of $\{v_1, v_2, \ldots, v_{i-1}\}$ set $v_i = \tilde{v}_i/\|\tilde{v}_i\|$ This is the *Gram–Schmidt procedure.*

If the collection of columns of a square matrix is an orthonormal collection, the matrix is called a *unitary matrix.* In case the matrix is a real matrix, it is usually called an *orthogonal matrix.* A unitary matrix U is invertible, and $U^{-1} = U^H$. (In the real case an orthogonal matrix Q is invertible, and $Q^{-1} = Q^T$.)

Determinants

The *determinant* of a square matrix is defined inductively. First, suppose the determinant $\det A$ has been defined for all square matrices of order $< n$. Then

$$\det A = a_{11}C_{11} + a_{12}C_{12} + \cdots + a_{1n}C_{1n}$$

where the numbers C_{ij} are *cofactors* of the matrix A,

$$C_{ij} = (-1)^{i+j} \det M_{ij}$$

where M_{ij} is the $(n-1) \times (n-1)$ matrix obtained by deleting the ith row and jth column of A. Now $\det A$ is defined to be the only entry of a matrix of order 1. Thus, for a matrix of order 2, we have

$$\det \begin{bmatrix} a & b \\ c & d \end{bmatrix} = ad - bc$$

There are many interesting but not obvious properties of determinants. It is true that

$$\det A = a_{i1}C_{i1} + a_{12}C_{i2} + \cdots + a_{in}C_{in}$$

for any $1 \leq i \leq n$. It is also true that $\det A = \det A^T$, so that we have

$$\det A = a_{1j}C_{1j} + a_{2j}C_{2j} + \cdots + a_{nj}C_{nj}$$

for any $1 \leq j \leq n$.

If A and B are matrices of the same order, then $\det AB = (\det A)(\det B)$, and the determinant of any identity matrix is 1. Perhaps the most important property of the determinant is the fact that a matrix is invertible if and only if its determinant is not zero.

Eigenvalues and Eigenvectors

If A is a square matrix, and $Av = \lambda v$ for a scalar λ and a nonzero v, then λ is an *eigenvalue* of A and v is an *eigenvector* of A that *corresponds* to λ. Any nonzero linear combination of eigenvectors corresponding to the same eigenvalue λ is also an eigenvector corresponding to λ. The collection of all eigenvectors corresponding to a given eigenvalue λ is thus a subspace, called an *eigenspace* of A. A collection of eigenvectors corresponding to different eigenvalues is necessarily linear independent. It follows that a matrix of order n can have at most n distinct eigenvectors. In fact, the eigenvalues of A are the roots of the nth degree polynomial equation

$$\det(A - \lambda I) = 0$$

called the *characteristic equation* of A. (Eigenvalues and eigenvectors are frequently called *characteristic values* and *characteristic vectors.*)

If the nth order matrix A has an independent collection of n eigenvectors, then A is said to have a *full set* of eigenvectors. In this case there is a set of eigenvectors of A that is a basis for R^n or, in the complex case, C^n. In case there are n distinct eigenvalues of A, then, of course, A has a full set of eigenvectors. If there are fewer than

n distinct eigenvalues, then $\boldsymbol{A}$ may or may not have a full set of eigenvectors. If there is a full set of eigenvectors, then

$$\boldsymbol{D} = \boldsymbol{S}^{-1}\boldsymbol{A}\,\boldsymbol{S} \qquad \text{or} \qquad \boldsymbol{A} = \boldsymbol{S}\,\boldsymbol{D}\,\boldsymbol{S}^{-1}$$

where $\boldsymbol{D}$ is a diagonal matrix the eigenvalues of $\boldsymbol{A}$ on the diagonal, and $\boldsymbol{S}$ is a matrix whose columns are the full set of eigenvectors. If $\boldsymbol{A}$ is symmetric, there are n real distinct eigenvalues of $\boldsymbol{A}$ and the corresponding eigenvectors are orthogonal. There is thus an orthonormal collection of eigenvectors that span $\boldsymbol{R}^n$, and we have

$$\boldsymbol{A} = \boldsymbol{Q}\,\boldsymbol{D}\,\boldsymbol{Q}^T \qquad \text{and} \qquad \boldsymbol{D} = \boldsymbol{Q}^T\,\boldsymbol{A}\,\boldsymbol{Q}$$

where $\boldsymbol{Q}$ is a real orthogonal matrix and $\boldsymbol{D}$ is diagonal. For the complex case, if $\boldsymbol{A}$ is Hermitian, we have

$$\boldsymbol{A} = \boldsymbol{U}\,\boldsymbol{D}\,\boldsymbol{U}^H \qquad \text{and} \qquad \boldsymbol{D} = \boldsymbol{U}^H\,\boldsymbol{A}\,\boldsymbol{U}$$

where $\boldsymbol{U}$ is a unitary matrix and $\boldsymbol{D}$ is a *real* diagonal matrix. (A Hermitian matrix also has n distinct real eigenvalues.)

18.9 Basic Definitions: Vector Algebra and Calculus

A vector is a directed line segment, with two vectors being equal if they have the same length and the same direction. More precisely, a *vector* is an equivalence class of directed line segments, where two directed segments are equivalent if they have the same length and the same direction. The *length* of a vector is the common length of its directed segments, and the *angle between* vectors is the angle between any of their segments. The length of a vector $\boldsymbol{u}$ is denoted $|\boldsymbol{u}|$. There is defined a distinguished vector having zero length, which is usually denoted $\boldsymbol{0}$. It is frequently useful to visualize a directed segment as an arrow; we then speak of the nose and the tail of the segment. The *sum* $\boldsymbol{u} + \boldsymbol{v}$ of two vectors $\boldsymbol{u}$ and $\boldsymbol{v}$ is defined by taking directed segments from $\boldsymbol{u}$ and $\boldsymbol{v}$ and placing the tail of the segment representing $\boldsymbol{v}$ at the nose of the segment representing $\boldsymbol{u}$ and defining $\boldsymbol{u} + \boldsymbol{v}$ to be the vector determined by the segment from the tail of the $\boldsymbol{u}$ representative to the nose of the $\boldsymbol{v}$ representative. It is easy to see that $\boldsymbol{u}+\boldsymbol{v}$ is well defined and that $\boldsymbol{u}+\boldsymbol{v} = \boldsymbol{v}+\boldsymbol{u}$. Subtraction is the inverse operation of addition. Thus the *difference* $\boldsymbol{u} - \boldsymbol{v}$ of two vectors is defined to be the vector that when added to $\boldsymbol{v}$ gives $\boldsymbol{u}$. In other words, if we take a segment from $\boldsymbol{u}$ and a segment from $\boldsymbol{v}$ and place their tails together, the difference is the segment from the nose of $\boldsymbol{v}$ to the nose of $\boldsymbol{u}$. The zero vector behaves as one might expect: $\boldsymbol{u} + 0 = \boldsymbol{u}$, and $\boldsymbol{u} - \boldsymbol{u} = 0$. Addition is associative: $\boldsymbol{u} + (\boldsymbol{v} + \boldsymbol{w}) = (\boldsymbol{u} + \boldsymbol{v}) + \boldsymbol{w}$.

To distinguish them from vectors, the real numbers are called *scalars*. The product $t\boldsymbol{u}$ of a scalar t and a vector $\boldsymbol{u}$ is defined to be the vector having length $|t|\ |\boldsymbol{u}|$ and direction the same as $\boldsymbol{u}$ if $t > 0$, the opposite direction if $t < 0$. If $t = 0$, then $t\boldsymbol{u}$ is defined to be the zero vector. Note that $t(\mathbf{u}+\boldsymbol{v}) = t\boldsymbol{u}+t\boldsymbol{v}$, and $(t+s)\boldsymbol{u} = t\boldsymbol{u}+s\boldsymbol{u}$. From this it follows that $\boldsymbol{u} - \boldsymbol{v} = \boldsymbol{u} + (-1)\boldsymbol{v}$.

The *scalar product* $\boldsymbol{u} \cdot \boldsymbol{v}$ of two vectors is $|\boldsymbol{u}|\ |\boldsymbol{v}| \cos\theta$, where θ is the angle between $\boldsymbol{u}$ and $\boldsymbol{v}$. The scalar product is frequently called the *dot product*. The scalar product distributes over addition,

$$\boldsymbol{u} \cdot (\boldsymbol{v} + \boldsymbol{w}) = \boldsymbol{u} \cdot \boldsymbol{v} + \boldsymbol{u} \cdot \boldsymbol{w}$$

and it is clear that $(t\boldsymbol{u}) \cdot \boldsymbol{v} = t(\boldsymbol{u} \cdot \boldsymbol{v})$. The *vector product* $\boldsymbol{u} \times \boldsymbol{v}$ of two vectors is defined to be the vector perpendicular to both $\boldsymbol{u}$ and $\boldsymbol{v}$ and having length $|\boldsymbol{u}|\ |\boldsymbol{v}| \sin\theta$, where θ is the angle between $\boldsymbol{u}$ and $\boldsymbol{v}$. The direction of $\boldsymbol{u} \times \boldsymbol{v}$ is the direction a right-hand threaded bolt advances if the vector $\boldsymbol{u}$ is rotated to $\boldsymbol{v}$. The vector product is frequently called the *cross product*. The vector product is both associative and distributive, but not commutative: $\boldsymbol{u} \times \boldsymbol{v} = -\boldsymbol{v} \times \boldsymbol{u}$.

Coordinate Systems

Suppose we have a right-handed Cartesian coordinate system is space. For each vector $\boldsymbol{u}$, we associate a point in space by placing the tail of a representative of $\boldsymbol{u}$ at the origin and associating with $\boldsymbol{u}$ the point at the nose of the segment. Conversely, associated with each point in space is the vector determined by the directed segment from the origin to that point. There is thus a one-to-one correspondence between the points in space and all vectors. The origin corresponds to the zero vector. The coordinates of the point associated with a vector $\boldsymbol{u}$ are called *coordinates* of $\boldsymbol{u}$. One frequently refers to the vector $\boldsymbol{u}$ and writes $\boldsymbol{u} = (x, y, z)$, which is, strictly speaking, incorrect, because the left side of this equation is a vector and the right side gives the coordinates of a point in space. What is meant is that (x, y, z) are the coordinates of the point associated with $\boldsymbol{u}$ under the correspondence described. In terms of coordinates, for $\boldsymbol{u} = (u_1, u_2, u_3)$ and $\boldsymbol{v} = (v_1, v_2, v_3)$, we have

$$\boldsymbol{u} + \boldsymbol{v} = (u_1 + v_1, u_2 + v_2, u_3 + v_3)$$

$$t\boldsymbol{u} = (tu_1, tu_2, tu_3)$$

$$\boldsymbol{u} \cdot \boldsymbol{v} = u_1v_1 + u_2v_2 + u_3v_3$$

$$\boldsymbol{u} \times \boldsymbol{v} = (u_2v_3 - v_2u_3, u_3v_1 - v_3u_1, u_1v_2 - v_1u_2)$$

The *coordinate vectors* $\boldsymbol{i}$, $\boldsymbol{j}$, and $\boldsymbol{k}$ are the unit vectors $\boldsymbol{i} = (1, 0, 0), \boldsymbol{j} = (0, 1, 0)$, and $k = (0, 0, 1)$. Any vector $\boldsymbol{u} = (u_1, u_2, u_3)$ is thus a linear combination of these coordinate vectors: $\boldsymbol{u} = u_1\boldsymbol{i} + u_2\boldsymbol{j} + u_3\boldsymbol{k}$. A convenient form for the vector product is the formal determinant

$$\boldsymbol{u} \times \boldsymbol{v} = \det \begin{bmatrix} \boldsymbol{i} & \boldsymbol{j} & \boldsymbol{k} \\ u_1 & u_2 & u_3 \\ v_1 & v_2 & v_2 \end{bmatrix}$$

Vector Functions

A *vector function* $\boldsymbol{F}$ *of one variable* is a rule that associates a vector $\boldsymbol{F}$(t) with each real number t is some set, called the *domain* of $\boldsymbol{F}$. The expression $\lim_{t\to t_0} \boldsymbol{F}(t) = \boldsymbol{a}$ means that for any $\varepsilon > 0$, there is a $\delta > 0$ such that $|\boldsymbol{F}(t) - \boldsymbol{a}| < \varepsilon$ whenever $0 < |t - t_0| < \delta$. If $\boldsymbol{F}(t) = [x(t), y(t), z(t)]$ and $\boldsymbol{a} = (a_1, a_2, a_3)$, then $\lim_{t\to t_0} \mathbf{F}(t) = \boldsymbol{a}$ if and only if

$$\lim_{t\to t_0} x(t) = a_1$$

$$\lim_{t\to t_0} y(t) = a_2$$

$$\lim_{t\to t_0} z(t) = a_3$$

A vector function $\boldsymbol{F}$ is *continuous* at t_0 if $\lim_{t\to t_0} \boldsymbol{F}(t) = \boldsymbol{F}(t_0)$. The vector function $\boldsymbol{F}$ is continuous at t_0 if and only if each of the coordinates $x(t)$, $y(t)$, and $z(t)$ is continuous at t_0.

The function $\boldsymbol{F}$ is *differentiable* at t_0 if the limit

$$\lim_{h\to 0} \frac{1}{h}[\boldsymbol{F}(t + h) - \boldsymbol{F}(t)]$$

exists. This limit is called the *derivative* of $\boldsymbol{F}$ at t_0 and is usually written $\boldsymbol{F}'(t_0)$, or $(d\boldsymbol{F}/dt)(t_0)$. The vector function $\boldsymbol{F}$ is differentiable at t_0 if and only if each of its coordinate functions is differentiable at t_0. Moreover, $(d\boldsymbol{F}/dt)(t_0) = [(dx/dt)(t_0), (dy/dt)(t_0), (dz/dt)(t_0)]$. The usual rules for derivatives of real valued functions all hold for vector functions. Thus, if $\boldsymbol{F}$ and $\boldsymbol{G}$ are vector functions and s is a scalar function, then

$$\frac{d}{dt}(\boldsymbol{F}+\boldsymbol{G}) = \frac{d\boldsymbol{F}}{dt}+\frac{d\boldsymbol{G}}{dt}$$
$$\frac{d}{dt}(s\boldsymbol{F}) = s\frac{d\boldsymbol{F}}{dt}+\frac{ds}{dt}\boldsymbol{F}$$
$$\frac{d}{dt}(\boldsymbol{F}\cdot\boldsymbol{G}) = \boldsymbol{F}\cdot\frac{d\boldsymbol{G}}{dt}+\frac{d\boldsymbol{F}}{dt}\cdot\boldsymbol{G}$$
$$\frac{d}{dt}(\boldsymbol{F}\times\boldsymbol{G}) = \boldsymbol{F}\times\frac{d\boldsymbol{G}}{dt}+\frac{d\boldsymbol{F}}{dt}\times\boldsymbol{G}$$

If $\boldsymbol{R}$ is a vector function defined for t is some interval, then, as t varies, with the tail of $\boldsymbol{R}$ at the origin, the nose traces out some object C in space. For nice functions $\boldsymbol{R}$, the object C is a *curve.* If $\boldsymbol{R}(t) = [x(t), y(t), z(t)]$, then the equations

$$x = x(t)$$
$$y = y(t)$$
$$z = z(t)$$

are called *parametric equations* of C. At points where $\boldsymbol{R}$ is differentiable, the derivative $d\boldsymbol{R}/dt$ is a vector *tangent* to the curve. The unit vector $\boldsymbol{T} = (d\boldsymbol{R}/dt)/|d\boldsymbol{R}/dt|$ is called the *unit tangent vector.* If $\boldsymbol{R}$ is differentiable and if the length of the arc of curve described by $\boldsymbol{R}$ between $\boldsymbol{R}(a)$ and $\boldsymbol{R}(t)$ is given by $s(t)$, then

$$\frac{ds}{dt} = \left|\frac{d\boldsymbol{R}}{dt}\right|$$

Thus the length L of the arc from $\boldsymbol{R}(t_0)$ to $\boldsymbol{R}(t_1)$ is

$$L = \int_{t_0}^{t_1}\frac{ds}{dt}dt = \int_{t_0}^{t_1}\left|\frac{d\boldsymbol{R}}{dt}\right|dt$$

The vector $d\boldsymbol{T}/ds = (d\boldsymbol{T}/dt)/(ds/dt)$ is perpendicular to the unit tangent $\boldsymbol{T}$, and the number $\kappa = |d\boldsymbol{T}/ds|$ is the *curvature* of C. The unit vector $\boldsymbol{N} = (1/\kappa)(d\boldsymbol{T}/ds)$ is the *principal normal.* The vector $\boldsymbol{B} = \boldsymbol{T}\times\boldsymbol{N}$ is the *binormal,* and $d\boldsymbol{B}/ds = -\tau\boldsymbol{N}$. The number τ is the *torsion.* Note that C is a plane curve if and only if τ is zero for all t.

A *vector function* $\boldsymbol{F}$ *of two variables* is a rule that assigns a vector $\boldsymbol{F}(s,t)$ to each point (s,t) is some subset of the plane, called the *domain* of $\boldsymbol{F}$. If $\boldsymbol{R}(s,t)$ is defined for all (s,t) in some region D of the plane, then as the point (s,t) varies over D, with its rail at the origin, the nose of $\boldsymbol{R}(s,t)$ traces out an object in space. For a nice function $\boldsymbol{R}$, this object is a *surface,* S. The partial derivatives $(\partial\boldsymbol{R}/\partial s)(s,t)$ and $(\partial\boldsymbol{R}/\partial t)(s,t)$ are tangent to the surface at $\boldsymbol{R}(s,t)$ and the vector $(\partial\boldsymbol{R}/\partial s)\times(\partial\boldsymbol{R}/\partial t)$ is thus *normal* to the surface. Of course, $(\partial\boldsymbol{R}/\partial t)\times(\partial\boldsymbol{R}/\partial s) = -(\partial\boldsymbol{R}/\partial s)\times(\partial\boldsymbol{R}/\partial t)$ is also normal to the surface and points in the direction opposite that of $(\partial\boldsymbol{R}/\partial s)\times(\partial\boldsymbol{R}/\partial t)$. By electing one of these normals, we are choosing an *orientation* of the surface. A surface can be oriented only if it has two sides, and the process of orientation consists of choosing which side is positive and which is negative.

Gradient, Curl, and Divergence

If $f(x,y,z)$ is a scalar field defined in some region D, the *gradient* of f is the vector function

$$\text{grad } f = \frac{\partial f}{\partial x}\boldsymbol{i}+\frac{\partial f}{\partial y}\boldsymbol{j}+\frac{\partial f}{\partial z}\boldsymbol{k}$$

If $\boldsymbol{F}(x, y, z) = F_1(x, y, z)\boldsymbol{i} + F_2(x, y, z)\boldsymbol{j} + F_3(x, y, z)\boldsymbol{k}$ is a vector field defined in some region D, then the *divergence* of $\boldsymbol{F}$ is the scalar function

$$\text{div } \boldsymbol{F} = \frac{\partial F_1}{\partial x} + \frac{\partial F_2}{\partial y} + \frac{\partial F_3}{\partial z}$$

The curl is the vector function

$$\text{curl } \boldsymbol{F} = \left(\frac{\partial F_3}{\partial y} - \frac{\partial F_2}{\partial z}\right)\boldsymbol{i} + \left(\frac{\partial F_1}{\partial z} - \frac{\partial F_3}{\partial x}\right)\boldsymbol{j} + \left(\frac{\partial F_2}{\partial x} - \frac{\partial F_1}{\partial y}\right)\boldsymbol{k}$$

In terms of the vector operator *del*, $\nabla = \boldsymbol{i}(\partial/\partial x) + \boldsymbol{j}(\partial/\partial y) + \boldsymbol{k}(\partial/\partial z)$, we can write

$$\text{grad } f = \nabla f$$

$$\text{div } \boldsymbol{F} = \nabla \cdot \boldsymbol{F}$$

$$\text{curl } \boldsymbol{F} = \nabla \times \boldsymbol{F}$$

The *Laplacian operator* is div (grad) $= \nabla \cdot \nabla = \nabla^2 = (\partial^2/\partial x^2) + (\partial^2/\partial y^2) + (\partial^2/\partial z^2)$.

Integration

Suppose C is a curve from the point (x_0, y_0, z_0) to the point (x_1, y_1, z_1) are is described by the vector function $\boldsymbol{R}(t)$ for $t_0 \le t \le t_1$. If f is a scalar function (sometimes called a *scalar field*) defined on C, then the integral of f over C is

$$\int_C f(x, y, z)\, ds = \int_{t_0}^{t_1} f[\boldsymbol{R}(t)]\left|\frac{d\boldsymbol{R}}{dt}\right| dt$$

If $\boldsymbol{F}$ is a vector function (sometimes called a *vector field*) defined on C, then the integral of $\boldsymbol{F}$ over C is

$$\int_C \boldsymbol{F}(x, y, z) \cdot d\boldsymbol{R} = \int_{t_0}^{t_1} \boldsymbol{F}[\boldsymbol{R}(t)] \cdot \frac{d\boldsymbol{R}}{dt}\, dt$$

These integrals are called *line integrals.*

In case there is a scalar function f such that $\boldsymbol{F} = \text{grad} f$, then the line integral

$$\int_C \boldsymbol{F}(x, y, z) \cdot d\boldsymbol{R} = f[\boldsymbol{R}(t_1)] - f[\boldsymbol{R}(t_0)]$$

The value of the integral thus depends only on the end points of the curve C and not on the curve C itself. The integral is said to be *path independent.* The function f is called a *potential function* for the vector field $\boldsymbol{F}$, and $\boldsymbol{F}$ is said to be a *conservative field.* A vector field $\boldsymbol{F}$ with domain D is conservative if and only if the integral of $\boldsymbol{F}$ around every closed curve in D is zero. If the domain D is simply connected (that is, every closed curve in D can be continuously deformed in D to a point), then $\boldsymbol{F}$ is conservative if and only if curl $\boldsymbol{F} = 0$ in D.

Suppose S is a surface described by $\boldsymbol{R}(s, t)$ for (s, t) in a region D of the plane. If f is a scalar function defined on D, then the integral of f over S is given by

$$\iint_S f(x, y, z)\, dS = \iint_D f[\boldsymbol{R}(s, t)]\left|\frac{\partial \boldsymbol{R}}{\partial s} \times \frac{\partial \boldsymbol{R}}{\partial t}\right| ds\, dt$$

If $\boldsymbol{F}$ is a vector function defined on S, and if an orientation for S is chosen, then the integral of $\boldsymbol{F}$ over S, sometimes called the *flux* of **F** through S, is

$$\iint_S \mathbf{F}(x, y, z) \cdot d\mathbf{S} = \iint_D \mathbf{F}[\mathbf{R}(s,t)] \cdot \left(\frac{\partial \mathbf{R}}{\partial s} \times \frac{\partial \mathbf{R}}{\partial t}\right) ds\, dt$$

Integral Theorems

Suppose F is a vector field with a closed domain D bounded by the surface S oriented so that the normal points out from D. Then the *divergence theorem* states that

$$\iiint_D \text{div } \mathbf{F} dV = \iint_S \mathbf{F} \cdot d\mathbf{S}$$

If S is an orientable surface bounded by a closed curve C, the orientation of the closed curve C is chosen to be consistent with the orientation of the surface S. Then we have *Stokes's theorem:*

$$\iint_S (\text{curl } \mathbf{F}) \cdot d\mathbf{S} = \oint_C \mathbf{F} \cdot d\mathbf{s}$$

18.10 The Fourier Transforms: Overview

For a piecewise continuous function $F(x)$ over a finite interval $0 \le x \le \pi$, the *finite Fourier cosine transform* of $F(x)$ is

$$f_c(n) = \int_0^\pi f(x) \cos nx\, dx \quad (n = 0, 1, 2, \ldots) \tag{18.1}$$

If x ranges over the interval $0 \le x \le L$, the substitution $x' = \pi x/L$ allows the use of this definition also. The inverse transform is written

$$\bar{F}(x) = \frac{1}{\pi} f_c(0) + \frac{2}{\pi} \sum_{n=1}^{\infty} f_c(n) \cos nx \quad (0 < x < \pi) \tag{18.2}$$

where $\bar{F}(x) = [F(x + 0) + F(x - 0)]/2$. We observe that $\bar{F}(x) = F(x)$ at points of continuity. The formula

$$\begin{aligned} f_c^{(2)}(n) &= \int_0^\pi F''(x) \cos nx\, dx \\ &= -n^2 f_c(n) - F'(0) + (-1)^n F'(\pi) \end{aligned} \tag{18.3}$$

makes the finite Fourier cosine transform useful in certain boundary-value problems.

Analogously, the *finite Fourier sine transform* of $F(x)$ is

$$f_s(n) = \int_0^\pi F(x) \sin nx\, dx \quad (n = 1, 2, 3, \ldots) \tag{18.4}$$

and

$$\bar{F}(x) = \frac{2}{\pi} \sum_{n=1}^{\infty} f_s(n) \sin\ nx \quad (0 < x < \pi) \tag{18.5}$$

Corresponding to Eq. (18.3), we have

$$f_s^{(2)}(n) = \int_0^\pi F''(x) \sin nx\, dx \tag{18.6}$$

$$= -n^2 f_s(n) - nF(0) - n(-1)^n F(\pi)$$

Fourier Transforms

If $F(x)$ is defined for $x \geq 0$ and is piecewise continuous over any finite interval, and if

$$\int_0^\infty F(x)\,dx$$

is absolutely convergent, then

$$f_c(\alpha) = \sqrt{\frac{2}{\pi}} \int_0^\infty f(x) \cos(\alpha x)\,dx \tag{18.7}$$

is the *Fourier cosine transform of F(x)*. Furthermore,

$$\bar{F}(x) = \sqrt{\frac{2}{\pi}} \int_0^\infty f_c(\alpha) \cos(\alpha x)\,d\alpha \tag{18.8}$$

If $\lim_{x\to\infty} d^n F/dx^n = 0$, an important property of the Fourier cosine transform,

$$\begin{aligned} f_c^{(2r)}(\alpha) &= \sqrt{\frac{2}{\pi}} \int_0^\infty \left(\frac{d^{2r}F}{dx^{2r}}\right) \cos\,(\alpha x)\,dx \\ &= -\sqrt{\frac{2}{\pi}} \sum_{n=0}^{r-1} (-1)^n a_{2r-2n-1} \alpha^{2n} + (-1)^r \alpha^{2r} f_c(\alpha) \end{aligned} \tag{18.9}$$

where $\lim_{x\to 0} d^r F/dx^r = a_r$, makes it useful in the solution of many problems.

Under the same conditions,

$$f_s(\alpha) = \sqrt{\frac{2}{\pi}} \int_0^\infty F(x) \sin(\alpha x)\,dx \tag{18.10}$$

defines the *Fourier sine transform of F(x)*, and

$$\bar{F}(x) = \sqrt{\frac{2}{\pi}} \int_0^\infty f_s(\alpha) \sin(\alpha x)\,d\alpha \tag{18.11}$$

Corresponding to Eq. (18.9) we have

$$\begin{aligned} f_s^{(2r)}(\alpha) &= \sqrt{\frac{2}{\pi}} \int_0^\infty \frac{d^{2r}F}{dx^{2r}} \sin\,(\alpha x)\,dx \\ &= -\sqrt{\frac{2}{\pi}} \sum_{n=1}^{r} (-1)^n \alpha^{2n-1} a_{2r-2n} + (-1)^{r-1} \alpha^{2r} f_s(\alpha) \end{aligned} \tag{18.12}$$

Similarly, if $F(x)$ is defined for $-\infty < x < \infty$, and if $\int_{-\infty}^{\infty} F(x)dx$ is absolutely convergent, then

$$f(\alpha) = \frac{1}{\sqrt{2\pi}} \int_{-\infty}^{\infty} F(x) e^{i\alpha x}\,dx \tag{18.13}$$

is the *Fourier transform of F(x)*, and

$$\bar{F}(x) = \frac{1}{\sqrt{2\pi}} \int_{-\infty}^{\infty} f(\alpha) e^{-i\alpha x}\,d\alpha \tag{18.14}$$

Also, if

$$\lim_{|x|\to\infty}\left|\frac{d^nF}{dx^n}\right| = 0 \quad (n = 1, 2, \ldots, r-1)$$

then

$$f^{(r)}(\alpha) = \frac{1}{\sqrt{2\pi}}\int_{-\infty}^{\infty} F^{(r)}(x)e^{i\alpha x}\,dx = (-i\alpha)^r f(\alpha) \tag{18.15}$$

TABLE 18.3 Finite Sine Transforms

	$f_s(n)$	$F(x)$
1.	$f_s(n) = \int_0^\pi F(x)\sin nx\,dx \quad (n = 1, 2, 3, \ldots)$	$F(x)$
2.	$(-1)^{n+1} f_s(n)$	$F(\pi - x)$
3.	$\frac{1}{n}$	$\frac{\pi - x}{\pi}$
4.	$\frac{(-1)^{n+1}}{n}$	$\frac{x}{\pi}$
5.	$\frac{1-(-1)^n}{n}$	1
6.	$\frac{2}{n^2}\sin\frac{n\pi}{2}$	$\begin{cases} x & \text{when } 0 < x < \pi/n \\ \pi - x & \text{when } \pi/2 < x < \pi \end{cases}$
7.	$\frac{(-1)^{n+1}}{n^3}$	$\frac{x(\pi^2 - x^2)}{6\pi}$
8.	$\frac{1-(-1)^n}{n^3}$	$\frac{x(\pi - x)}{2}$
9.	$\frac{\pi^2(-1)^{n-1}}{n} - \frac{2[1-(-1)^n]}{n^3}$	x^2
10.	$\pi(-1)^n\left(\frac{6}{n^3} - \frac{\pi^2}{n}\right)$	x^3
11.	$\frac{n}{n^2+c^2}[1-(-1)^n e^{c\pi}]$	e^{cx}
12.	$\frac{n}{n^2+c^2}$	$\frac{\sin h\, c(\pi - x)}{\sin h\, c\pi}$
13.	$\frac{n}{n^2-k^2} \quad (k \neq 0, 1, 2, \ldots)$	$\frac{\sin hk(\pi - x)}{\sin k\pi}$
14.	$\begin{cases} \frac{\pi}{2} & \text{when } n = m \\ 0 & \text{when } n \neq m \end{cases} \quad (m = 1, 2, \ldots)$	$\sin mx$
15.	$\frac{n}{n^2-k^2}[1-(-1)^n\cos k\pi] \quad (k \neq 1, 2, \ldots)$	$\cos kx$
16.	$\begin{cases} \frac{n}{n^2-m^2}[1-(-1)^{n+m}] & \text{when } n \neq m = 1, 2, \ldots \\ 0 & \text{when } n = m \end{cases}$	$\cos mx$
17.	$\frac{n}{(n^2-k^2)^2} \quad (k \neq 0, 1, 2, \ldots)$	$\frac{\pi\sin kx}{2k\sin^2 k\pi} - \frac{x\cos k(\pi - x)}{2k\sin k\pi}$
18.	$\frac{b^n}{n} \quad (\lvert b\rvert \leq 1)$	$\frac{2}{\pi}\arctan\frac{b\sin x}{1-b\cos x}$
19.	$\frac{1-(-1)^n}{n}b^n \quad (\lvert b\rvert \leq 1)$	$\frac{2}{\pi}\arctan\frac{2b\sin x}{1-b^2}$

TABLE 18.4 Finite Cosine Transforms

	$f_c(n)$	$F(x)$
1.	$f_c(n) = \int_0^\pi F(x)\cos nx\,dx \quad (n = 0, 1, 2, \ldots)$	$F(x)$
2.	$(-1)^n f_c(n)$	$F(\pi - x)$
3.	0 when $n = 1, 2, \ldots;\quad f_c(0) = \pi$	1
4.	$\frac{2}{\pi}\sin\frac{n\pi}{2};\quad f_c(0) = 0$	$\begin{cases} 1 & \text{when } 0 < x < \pi/2 \\ -1 & \text{when } \pi/2 < x < \pi \end{cases}$
5.	$-\frac{1-(-1)^n}{n^2};\quad f_c(0) = \frac{\pi^2}{2}$	x
6.	$\frac{(-1)^n}{n^2};\quad f_c(0) = \frac{\pi^2}{6}$	$\frac{x^2}{2\pi}$
7.	$\frac{1}{n^2};\quad f_c(0) = 0$	$\frac{(\pi - x)^2}{2\pi} - \frac{\pi}{6}$
8.	$3\pi^2\frac{(-1)^n}{n^2} - 6\frac{1-(-1)^n}{n^4};\quad f_c(0) = \frac{\pi^4}{4}$	x^3
9.	$\frac{(-1)^n e^c \pi - 1}{n^2 + c^2}$	$\frac{1}{c}e^{cx}$
10.	$\frac{1}{n^2 + c^2}$	$\frac{\cos\text{h}\, c(\pi - x)}{c \sin\text{h}\, c\pi}$
11.	$\frac{k}{n^2 - k^2}[(-1)^n \cos \pi k - 1] \quad (k \neq 0, 1, 2, \ldots)$	$\sin kx$
12.	$\frac{(-1)^{n+m} - 1}{n^2 - m^2}$	$\frac{1}{m}\sin mx$
13.	$\frac{1}{n^2 - k^2} \quad (k \neq 0, 1, 2, \ldots)$	$-\frac{\cos k(\pi - x)}{k \sin k\pi}$
14.	0 when $n = 1, 2, \ldots;$	$\cos mx$

TABLE 18.5 Fourier Sine Transforms

	$F(x)$	$f_s(\alpha)$
1.	$\begin{cases} 1 & (0 < x < a) \\ 0 & (x > a) \end{cases}$	$\sqrt{\frac{2}{\pi}}\left[\frac{1 - \cos\alpha}{\alpha}\right]$
2.	$x^{p-1} \quad (0 < p < 1)$	$\sqrt{\frac{2}{\pi}}\frac{\Gamma(p)}{\alpha^p}\sin\frac{p\pi}{2}$
3.	$\begin{cases} \sin x & (0 < x < a) \\ 0 & (x > a) \end{cases}$	$\frac{1}{\sqrt{2\pi}}\left[\frac{\sin[a(1-\alpha)]}{1-\alpha} - \frac{\sin[a(1+\alpha)]}{1+\alpha}\right]$
4.	e^{-x}	$\sqrt{\frac{2}{\pi}}\left[\frac{\alpha}{1+\alpha^2}\right]$
5.	$xe^{-x^2/2}$	$\alpha e^{-\alpha^2/2}$
6.	$\cos\frac{x^2}{2}$	$\sqrt{2}\left[\sin\frac{\alpha^2}{2}C\left(\frac{\alpha^2}{2}\right) - \cos\frac{\alpha^2}{2}S\left(\frac{\alpha^2}{2}\right)\right]$[a]
7.	$\sin\frac{x^2}{2}$	$\sqrt{2}\left[\cos\frac{\alpha^2}{2}C\left(\frac{\alpha^2}{2}\right) + \sin\frac{\alpha^2}{2}S\left(\frac{\alpha^2}{2}\right)\right]$[a]

[a] $C(y)$ and $S(y)$ are the Fresnel integrals

$$C(y) = \frac{1}{\sqrt{2\pi}}\int_0^y \frac{1}{\sqrt{t}}\cos t\,dt$$

$$S(y) = \frac{1}{\sqrt{2\pi}}\int_0^y \frac{1}{\sqrt{t}}\sin t\,dt$$

TABLE 18.6 Fourier Cosine Transforms

	$F(x)$	$f_c(\alpha)$
1.	$\begin{cases} 1 & (0 < x < a) \\ 0 & (x < a) \end{cases}$	$\sqrt{\frac{2}{\pi}}\frac{\sin a\alpha}{\alpha}$
2.	$x^{p-1} \quad (0 < p < 1)$	$\sqrt{\frac{2}{\pi}}\frac{\Gamma(p)}{\alpha^p}\cos\frac{p\pi}{2}$
3.	$\begin{cases} \cos x & (0 < x < a) \\ 0 & (x > a) \end{cases}$	$\frac{1}{\sqrt{2\pi}}\left[\frac{\sin[a(1-\alpha)]}{1-\alpha} + \frac{\sin[a(1+\alpha)]}{1+\alpha}\right]$
4.	e^{-x}	$\sqrt{\frac{2}{\pi}}\left(\frac{1}{1+\alpha^2}\right)$
5.	$e^{-x^2/2}$	$e^{-\alpha^2/2}$
6.	$\cos\frac{x^2}{2}$	$\cos\left(\frac{\alpha^2}{2} - \frac{\pi}{4}\right)$
7.	$\sin\frac{x^2}{2}$	$\cos\left(\frac{\alpha^2}{2} - \frac{\pi}{4}\right)$

TABLE 18.7 Fourier Transforms

	$F(x)$	$f(\alpha)$
1.	$\frac{\sin ax}{x}$	$\begin{cases} \sqrt{\frac{\pi}{2}} & \lvert\alpha\rvert < a \\ 0 & \lvert\alpha\rvert > a \end{cases}$
2.	$\begin{cases} e^{iwx} & (p < x < q) \\ 0 & (x < p, x > q) \end{cases}$	$\frac{i}{\sqrt{2\pi}}\frac{e^{ip(w+\alpha)} - e^{iq(w+\alpha)}}{(w+\alpha)}$
3.	$\begin{cases} e^{-cx+iwx} & (x > 0) \\ 0 & (x < 0) \end{cases} \quad (c > 0)$	$\frac{i}{\sqrt{2\pi}(w+\alpha+ic)}$
4.	$e^{-px^2} \quad R(p) > 0$	$\frac{1}{\sqrt{2p}}e^{-\alpha^2/4p}$
5.	$\cos px^2$	$\frac{1}{\sqrt{2p}}\cos\left[\frac{\alpha^2}{4p} - \frac{\pi}{4}\right]$
6.	$\sin px^2$	$\frac{1}{\sqrt{2p}}\cos\left[\frac{\alpha^2}{4p} + \frac{\pi}{4}\right]$
7.	$\lvert x\rvert^{-p} \quad (0 < p < 1)$	$\sqrt{\frac{2}{\pi}}\frac{\Gamma(1-p)\sin\frac{p\pi}{2}}{\lvert\alpha\rvert^{(1-p)}}$
8.	$\frac{e^{-a\lvert x\rvert}}{\sqrt{\lvert x\rvert}}$	$\frac{\sqrt{\sqrt{(a^2+\alpha^2)}+a}}{\sqrt{a^2+\alpha^2}}$
9.	$\frac{\cos\text{h}\, ax}{\cos\text{h}\, \pi x} \quad (-\pi < a < \pi)$	$\sqrt{\frac{2}{\pi}}\frac{\cos\frac{a}{2}\cos\text{h}\frac{\alpha}{2}}{\cos\text{h}\alpha + \cos a}$
10.	$\frac{\sin\text{h}\, ax}{\sin\text{h}\, \pi x} \quad (-\pi < a < \pi)$	$\frac{1}{\sqrt{2\pi}}\frac{\sin a}{\cos\text{h}\,\alpha + \cos a}$
11.	$\begin{cases} \frac{1}{\sqrt{a^2-x^2}} & (\lvert x\rvert < a) \\ 0 & (\lvert x\rvert > a) \end{cases}$	$\sqrt{\frac{\pi}{2}}J_0(a\alpha)$
12.	$\frac{\sin[b\sqrt{a^2+x^2}]}{\sqrt{a^2+x^2}}$	$\begin{cases} 0 & (\lvert\alpha\rvert > b) \\ \sqrt{\frac{\pi}{2}}J_0(a\sqrt{b^2-\alpha^2}) & (\lvert\alpha\rvert < b) \end{cases}$

TABLE 18.7 Fourier Transforms (*Continued*)

	$F(x)$	$f(\alpha)$
13.	$\begin{cases} P_n(x) & (\lvert x\rvert < 1) \\ 0 & (\lvert x\rvert > 1) \end{cases}$	$\dfrac{i^n}{\sqrt{\alpha}} J_{n+1/2}(\alpha)$
14.	$\begin{cases} \dfrac{\cos[b\sqrt{a^2 - x^2}]}{\sqrt{a^2 - x^2}} & (\lvert x\rvert < a) \\ 0 & (\lvert x\rvert > a \end{cases}$	$\sqrt{\dfrac{\pi}{2}} J_0(a\sqrt{a^2 + b^2})$
15.	$\begin{cases} \dfrac{\cos h[b\sqrt{a^2 - x^2}]}{\sqrt{a^2 - x^2}} & (\lvert x\rvert < a) \\ 0 & (\lvert x\rvert > a) \end{cases}$	$\sqrt{\dfrac{\pi}{2}} J_0(a\sqrt{a^2 - b^2})$

TABLE 18.8 Functions Among Transforms Tables Entries

Funtion	Definition	Name
$\mathrm{Ei}(x)$	$\int_{-\infty}^{x} \frac{e^v}{v} dv$; or sometimes defined as $-Ei(-x) = \int_x^\infty \frac{e^{-v}}{v} dv$	Exponential integral function
$\mathrm{Si}(x)$	$\int_0^x \frac{\sin v}{v} dv$	Sine integral function
$\mathrm{Ci}(x)$	$\int_\infty^x \frac{\cos v}{v} dv$; or sometimes defined as negative of this integral	Cosine integral function
$\mathrm{erf}(x)$	$\frac{2}{\sqrt{\pi}} \int_0^x e^{-v^2} dv$	Error function
$\mathrm{erfc}(x)$	$1 - \mathrm{erf}(x) = \frac{2}{\pi} \int_x^\infty e^{-v^2} dv$	Complementary function to error function
$L_n(x)$	$\frac{e^x}{n!} \frac{d^n}{dx^n}(x^n e^{-x})$, $n = 0, 1, \ldots$	Laguerre polynomial of degree n

References

Daniel, J.W. and Noble, B. 1988. *Applied Linear Algebra.* Prentice Hall, Englewood Cliffs, NJ.

Davis, H.F. and Snider, A.D. 1991. *Introduction to Vector Analysis,* 6th ed. Wm. C. Brown, Dubuque, IA.

Strang, G. 1993. *Introduction to Linear Algebra.* Wellesley–Cambridge Press, Wellesley, MA.

Wylie, C.R. 1975. *Advanced Engineering Mathematics,* 4th ed. McGraw–Hill, New York.

Further Information

More advanced topics leading into the theory and applications of tensors may be found in J.G. Simmonds, *A Brief on Tensor Analysis* (1982, Springer–Verlag, New York).

19

Communications Terms: Abbreviations[1]

A

A	ampere
Å	Angstrom
A-to-D converter	analog to digital converter
AALU	arithmetic and logical unit
AAMPS	advanced mobile phone system
AC	access control
AC	alternating current

[1]Courtesy of Intertec Publishing, Overland Park, KS.

ACC	automatic color correction
ADC	analog to digital converter
ADPCM	adaptive differential pulse-code modulation
ADR	automatic dialog replacement
AES	Audio Engineering Society
AFC	automatic frequency control
AFCEA	Armed Forces Communications and Electronics Association
AFP	AppleTalk filing protocol
AFRTS	Armed Forces Radio and Television Service
AFV	audio-follow-video
AGC	automatic gain control
AI	artificial intelligence
AIN	advanced intelligent network
ALAP	AppleTalk link access protocol
ALGOL	algorithmic language, algorithmic oriented language
AM	amplitude modulation
AMI	alternate mark inversion
ANI	automatic number identification
ANSI	American National Standards Institute
APD	avalanche photodiode
API	application program interface
APL	average picture level
APPC	advanced program-to-program communications
ARIS	access request information system
ARP	address resolution protocol
ASCII	American standard code for information interchange
ASI	adapter support interface
ASIC	application specific integrated circuit
ASK	amplitude-shift keying
ASR	access service request
ATE	automatic test equipment
ATM	asynchronous transfer mode
ATR	audio tape recorder
ATSC	Advanced Television Systems Committee
ATV	advanced television
AUI	attachment unit interface
AVK	audio/video kernel
AVL	automatic vehicle location
AWG	American wire gauge

B

B-link	bridge link
B8ZS	bipolar with eight zeros substitution
BBC	British Broadcasting Corporation
BCC	background color cancellation
BCC	Bellcore Client Company
BCD	binary coded decimal
BCS	background color suppression
BER	bit error rate
BER	bit error ratio
BETRS	Basic Exchange Telecommunications Radio Service

BEXR	Basic Exchange Radio Service
BF	burst flag
BG	burst gate
BIP-N	bit interleaved parity-N
BISDN	broadband integrated services digital network
BITS	building integrated timing supply
BKGD	background
BLKG	blanking
BNC	bayonet Neill–Concelman
BNZS	bipolar with N zeros substitution
BOC	Bell Operating Company
BORSCHT	battery feed, overvoltage protection, ringing, supervision, coding/decoding, hybrid, and testing
BT	British Telecom
BTSC	Broadcast Television Systems Committee
BVB	black-video-black

C

C/N	carrier-to-noise ratio
CABSC	Canadian Advanced Broadcast Systems Committee
CAD/CAM	computer-aided design/computer-aided manufacture
CAMA	centralized automatic message accounting
CARS	community antenna relay service
CATV	community antenna television
CAV	component analog video
CBD	central business district
CBU	Caribbean Broadcasting Union
CC	calling channel
CCC	clear channel capability
CCD	charge coupled device
CCIR	Comité Consultatif International de Radiocommunications (International Radio Consultative Committee)
CCITT	Comité Consultatif International Télégraphique et Téléphonique (Consultative Committee for International Telephone and Telegraph)
CCS	centum call seconds
CCU	camera control unit
CD	compact disc
CD-ROM	compact disc-read only memory
CDI	compact disc-interactive
CDMA	code division multiple access
CEPT	Conference of European Postal and Telecommunications Administrations
CEV	controlled environmental vault
CG	character generator
CGSA	cellular geographic service area
CIBER	cellular intercarrier billing exchange roamer
CIC	circuit identification code
CIE	Commission Internationale de l'Eclairageclear
CIF	common intermediate format
CIMAP/CC	circuit installation and maintenance assistance package/control center
CIMAP/SSC	circuit installation and maintenance assistance package/special service center
CLONES	common language on-line entry system

CMDS	centralized message data system
CMOS	complementary metal oxide semiconductor
CMR	common mode rejection
CMRR	common mode rejection ratio
CMRS	Cellular Mobile Radiotelephone Service
CO	central office
COMSAT	Communications Satellite Corporation
CPE	customer premise equipment
CPU	central processing unit
CRC	cyclical redundancy check
CRCC	cyclic redundancy check code
CRT	cathode ray tube
CS	composite sync
CSMA	carrier-sense multiple access
CSMA/CD	carrier sense multiple access with collision detection
CSU	channel service unit
CTIA	Cellular Telecommunications Industry Association
CUCRIT	capital utilization criteria
CVGB	cable vault ground bar
CWA	Communications Workers of America
CWCG	copper wire counterpoise ground

D

D-to-A converter	digital to analog converter
D-to-D	digital to digital transfer
D/I	drop and insert
DA	distribution amplifier
DAC	digital to analog converter
DARPA	Defense Advanced Research Projects Agency
DAS	data acquisition system
dB	decibel
dBi	decibels relative to an isotropic antenna
dBk	decibels relative to 1 kilowatt
dBm	decibels relative to 1 milliwatt
dBmv	decibels relative to 1 millivolt
dBrn	decibels above reference noise
DBS	direct-broadcast satellite
dBV	decibels relative to 1 volt
dBW	decibels relative to 1 watt
DC	direct current
DCE	data communications equipment
DCPSK	differentially coherent phase-shift keying
DCT	discrete cosine transform
DDD	direct distance dialing
DDS	digital data system
DES	data encryption standard
DID	direct inward dialing
DILEP	Digital Line Engineering Program
DIP	dual in-line package
DIR/ECT	directory project
DLC	data-link control

DMA	direct memory access
dmW	digital milliwatt
DNS	domain name service
DOD	direct outward dialing
DOMSAT	domestic satellite
DOS	disk operating system
DPCM	differential pulse-code modulation
DPSK	differential phase-shift keying
DRAM	dynamic random access memory
DSB	double sideband (AM)
DSE	data switching exchange
DSK	downstream keyer
DSSC	double-sideband suppressed carrier
DSX	digital signal cross connect
DTE	data terminal equipment
DTL	diode-transistor logic
DTMF	dual tone multifrequency
DVE	digital video effects
DVI	digital video interactive
DVTR	digital videotape recorder
DWG	drilled well ground

E

E-link	extension link
e-mail	electronic mail
EADAS	Engineering and Administrative Data Acquisition System
EAROM	electrically alterable read-only memory
EBCDIC	extended binary-coded decimal interchange code
EBU	European Broadcasting Union
EC	European Community
ECC	error correcting code
ECCS	economic CCS
ECL	emitter-coupled logic
ECS	European communication satellite
EDFA	erbium doped fiber amplifier
EDI	electronic data interchange
EDL	edit decision list
EDTV	extended definition television
EEPROM	electrically erasable programmable read-only memory
EFM	eight to fourteen modulation
EFP	electronic field production
efx	effects
EHF	extremely high frequency
EIA	Electronic Industries Association
EIRP	effective isotropic radiated power
EISA	expanded industry standard architecture
ELF	extremely low frequency
EME	economic modular evaluation
EMF	electromotive force
EMI	electromagnetic interference
EMP	electromagnetic pulse

ENG	electronic news gathering
EPLD	erasable programmable logic device
EPROM	erasable programmable read-only memory
EQ	equalization
ERP	effective radiated power
ESA	European Space Agency
ESD	electrostatic discharge
ESDI	enhanced small device interface
ESI	equivalent step index
ESN	electronic service number
ESPRIT	European Strategic Program for Research in Information Technology
ETM	eight to ten modulation
ETSI	European Telecommunications Standards Institute

F

F-link	fully associated link
FACTS	fully automated collect and third-number service
FCC	Federal Communications Commission
FCD	frame continuity date
FD	frequency distance
FDDI	fiber distributed data interface
FDM	frequency division multiplexing
FDMA	frequency division multiple access
FDR	frequency dependent rejection
FDRL	filed data of regions and LECS
FEC	forward error correction
FEO	foreign exchange office
FET	field effect transistor
FFSK	fast frequency shift keying
FID	field identifier
FIFO	first-in-first-out
FIR filter	finite impulse response filter
FIT	failure in time
FIVE	format independent visual exchange
FM	frequency modulation
FNPA	foreign numbering plan area
FO	fiber optics
FOMS/FUSA	frame operations management system/frame user switch access system
FPGA	field programmable gate array
FPLA	field programmable logic array
FPLF	field programmable logic family
FPLS	field programmable logic sequence
fs	femtosecond
FSK	frequency shift keying
FSS	fixed satellite service
FTA	fault tree analysis
FTAM	file transfer, access, and management
FTB	fade-to-black
fx	effects
FX	foreign exchange
FXS	foreign exchange station

G

G/T	gain-over-noise temperature
GaAs	gallium arsenide
GADS	generic advisory diagnostic system
GBR	green, blue, red
GHz	gigahertz
GIGO	garbage-in-garbage-out
GMT	Greenwich mean time
GND	ground
GOS	grade of service
GOSIP	government open systems interconnection profile
GPI	general purpose interface
GSM	Global System for Mobile Communications

H

H	horizontal
HCS fiber	hard clad silica fiber
HDTV	high definition television
HF	high frequency
HLL	high level language
HVAC	heating, ventilation, and air conditioning
Hz	hertz

I

I/O	input/output
IACC	interaural cross correlation
IBA	Independent Broadcasting Authority
IBG	interblock gap
IC	integrated circuit
IC	interexchange carrier
IDTV	improved definition television
IEC	International Electrotechnical Commission
IEE	Institution of Electrical Engineers
IEEE	Institute of Electrical and Electronics Engineers
IFRB	International Frequency Registration Board
IIR	infinite impulse response
ILD	injection laser diode
ILF	infra low frequency
IM	intensity modulation
IMD	intermodulation distortion
IMTS	improved mobile telephone service
IN	intelligent network
INA	information networking architecture
INA	integrated network access
INFORMS	integrated forecasting management system
Inmarsat	International Maritime Satellite Organization
INREFS	integrated reference system
INTELSAT	International Telecommunications Satellite Consortium
INWATS	inward WATS

IOC	integrated optical circuit
IP	Internet protocol
IPC	interprocess communications
IPX	internetwork packet exchange
IPX/SPX	internetwork packet exchange/sequenced packet exchange
IR	infrared
IRQ	interrupt request
ISA	industry standard architecture
ISD	international subscriber dialing
ISDN	Integrated Services Digital Network
ISO	International Standards Organization
ITC	Independent Television Commission
ITDG	initial-time-delay gap
ITS	International Teleproduction Society
ITU	International Telecommunication Union

J

JF	junction frequency
JPEG	Joint Photographic Experts Group

K

K	Kelvin
k/s	kilobits per second
kHz	kilohertz
kV	kilovolt

L

LAN	local area network
LAP	link access protocol
LAPB	link access protocol—balanced
LAPD	link access protocol on D channel
LAT	local area transport
LATA	local access and transport area
LCD	liquid crystal display
LCRIS	loop cable record inventory system
LEC	light energy converter
LEC	local exchange carrier (company)
LED	light emitting diode
LF	low frequency
LFACS	loop facilities assignment and control system
LIFO	last-in-first-out
LLC	logical link control
LMOS	loop maintenance operations system
LNA	launch numerical aperture
LNA	low noise amplifier
LOMS	loop assignment center operations management system
LOS	line of sight
LP	linearly polarized
LPC	linear predictive coding

LPIE	loop plant improvement evaluator
LPTV	low power television
LRC	longitudinal redundancy check
LSB	least significant bit
LSB	lower sideband
LSI	large-scale integration
LSL	link support layer
LTC	longitudinal time code
LU	logical unit
LU 6.2	logical unit 6.2
LUM	luminance

M

M/E	mix/effects
mA	milliamperes
MAC	multiplexed analog components
MACS	major apparatus and cable system
MAP	manufacturing automation protocol
master SPG	master reference synchronizing pulse generator
MATV	master antenna television
MAU	media access unit
MAU	multistation access unit
MAVEN	mapping and access for valid equipment nomenclature
mb/s	megabits per second
MCA	media control architecture
MCA	microchannel architecture
MCI	media control interface
MDS	multipoint distribution system
MESFET	metal semiconductor field effect transistor
MF	medium frequency
MFJ	modification of final judgment
MFSK	multiple frequency shift keying
MHz	megahertz
MIB	management information base
MICR	magnetic ink character recognition
MIDI	musical instrument digital interface
MIPS	millions of instructions per second
MITI	Ministry of International Trade and Industry
MLID	multiple link interface driver
MNOS	metal, nitride, oxide semiconductor
MOPS	millions of operations per second
MOS	metal-oxide semiconductor
MPCD	minimum perceptible color difference
MPEG	Motion Picture Experts Group
MPT	Ministry of Posts and Telecommunications
ms	millisecond
MSA	metropolitan statistical area
MSB	most significant bit
MSU	medium-scale integration
MTBF	mean time between failures

MTSO	mobile telephone switching office
MTTF	mean time to failure
MTTR	mean time to repair
MUF	maximum usable frequency
MUSA	multiple unit steerable antenna
MUX	multiplex
mV	millivolt
MW	medium wave
mW	milliwatt

N

NAM	negative nonadditive mix
NAM	numeric assignment module
NANP	North American Numbering Plan
NARUC	National Association of Regulatory Utility Commissioners
NASC	Number Administration and Service Center
NBP	name binding protocol
NCP	network control point
NCS	National Communications System
NCTA	National Telephone Cooperative Association
NEBS	network equipment building system
NEC	National Electrical Code
NECA	National Exchange Carriers Association
NEP	noise equivalent power
NF	noise factor
NHK	Nippon Hoso Kyokai
NIC	network interface card
NICAM	near instantaneously companded audio multiplex
NIST	National Institute of Standards and Technology
NLM	network loadable module
NOS	network operating system
NPA	number plan area
NPR	noise power ratio
NRZ	nonreturn-to-zero
ns	nanosecond
NSDB	network and services database
NSEP	National Security Emergency Preparedness
NSIS	network server interface specification
NT1	network termination 1
NTC/C	network configuration
NTIA	National Telecommunications and Information Administration
NTL	National Transcommunications Limited
NTSC	National Television System Committee
NTT	Nippon Telegraph and Telephone

O

OCR	optical character recognition
ODI	open data-link interface
OEM	original equipment manufacturer
ONI	operator number identification

OPASTCO	Organization for the Protection and Advancement of Small Telephone Companies
OPS/INE	operations process system/intelligent network elements
OSI	open systems interconnection
OTDR	optical time domain reflectometer
OTF	optimum traffic frequency
OXO	ovenized crystal oscillator

P

p-p	peak-to-peak
PAD	packet assembler/disassembler
PAL	phase alternate each line
PAL	programmable array logic
PAM	pulse amplitude modulation
PAP	printer access protocol
PBX	private branch exchange
PC board	printed circuit board
PCF	physical control fields
PCM	pulse code modulation
PCN	personal communications network
PCS	personal communications services
PCSA	personal computing system architecture
PDM	pulse duration modulation
PDN	public data network
PDP	plasma display panel
PERT	program evaluation and review technique
PFM	pulse frequency modulation
PGM	program
PICS/DCPR	plug-in inventory control system/detailed continuing property record
PIN	personal identification number
PIN	positive-intrinsic-negative
PLD	programmable logic device
PLL	phase locked loop
PLV	production level video
PM	phase modulation
PM	pulse modulation
PMR	public mobile radio
POSIX	portable operating system interface
POTS	plain old telephone service
PPM	pulse position modulation
PPSN	public packet switched network
PPSS	public packet switched service
PRBS	pseudorandom bit stream
programmable GPI	programmable general purpose interface
PROM	programmable read-only memory
PSDS	public switched digital service
PSK	phase-shift keying
PSTN	public switched telephone network
PTM	pulse time modulation
PTT	post, telephone, and telegraph
PTT	push to talk
PUC	public utilities commission

pulse DA	pulse distribution amplifier
pulse delay DA	pulse delay distribution amplifier
PVC	polyvinylchloride

Q

Q	quality factor
QA	quality assurance
QBE	query by example
QC	quality control
QPSK	quadrature phase-shift keying
QUIL	quad-in-line

R

R-Y	red minus luminance
RAA	rural allocation area
RACE	Research in Advanced Communications in Europe
RAM	random access memory
RAR	read after read
RARP	reverse address resolution protocol
RC	resistor–capacitor
RCC	radio common carrier
REA	Rural Electrification Administration
RF	radio frequency
RFI	radio frequency interference
RGB	red, green, blue
RIFF	resource interchange file format
RISC	reduced instruction set computer
RIT	rate of information transfer
RMAS	Remote Memory Administration System
RMS	root means square
ROM	read-only memory
RPC	remote procedure call
RPG	report program generator
RSA	rural service area
RSC	Reed Solomon code
RSL	received signal level
RTC	real time clock
RTL	resistor-transistor logic
RU	rack unit
RZ	return to zero

S

(S+N)/N	signal-plus-noise to noise ratio
S/N	signal-to-noise ratio
SAP	secondary audio program
SAP	service access point
SAT	supervisory audio tone
SAW	surface acoustic wave
SC	subcarrier

SC/H phase	subcarrier to horizontal phase
SCCP	signaling connection control part
SCP	service control point
SCP/800	service control point/800
SCPC	single-channel-per-carrier
SCSI	small computer systems interface
SDLC	synchronous data link control
SECAM	sequential couleur avec memoire
SEF	source explicit forwarding
SGML	Standard Generalized Markup Language
SHF	super high frequency
SID	sudden ionospheric disturbance
SID	system identification
SINAD	signal-to-noise and distortion
SMB	server message block
SMDS	switched multimegabit data service
SMPTE	Society of Motion Picture and Television Engineers
SMRS	Specialized Mobile Radio Service
SMSA	standard metropolitan statistical area
SMT	surface mount technology
SMTP	simple mail transfer protocol
SNA	systems network architecture
SNET	Southern New England Telecommunications Corporation
SNMP	simple network management protocol
SNR	signal-to-noise ratio
SOAC	service order analysis and control
SOH	start-of-heading
SONET	synchronous optical network
SOS	silicon on sapphire
SPC	stored-program control
SPDT	single-pole double-throw
SPG	sync pulse generator
SPL	sound pressure level
SPP	sequenced packet protocol
SPST	single-pole single-throw
SQL	structured query language
SS6	signaling system 6
SS7	signaling system 7
SSB	single sideband
SSBSC	single-sideband suppressed carrier
SSC	special services center
SSP	service switching point
SST	single-sideband transmission
STA	spanning tree algorithm
STL	studio-transmitter link
STP	signaling transfer point
STSL	synchronous transport signal level
STX	start-of-text
SW	short wave
SWR	standing wave ratio

T

TA	terminal adapter
TAP	test access port
TASI	time-assignment speech interpolation
TBC	time base corrector
TCP	transmission control protocol
TCP/IP	transmission control protocol/internet protocol
TCXO	temperature compensated crystal oscillator
TDD	telecommunications device for the deaf
TDM	time division multiplexing
TDMA	time division multiple access
TE	transverse electric
TEM	transverse electromagnetic
THD	total harmonic distortion
THF	tremendously high frequency
TIE	terminal interface equipment
TM	transverse magnetic
TMDA	time-division multiple access
TND	telephone network for the deaf
TNDS/TK	total network data system/trunking
TOP	technical and office protocols
TRF	tuned radio frequency
TTL	transistor-transistor logic
TVI	television interference
TVRO	television receive-only

U

UART	universal asynchronous receiver/transmitter
UHF	ultrahigh frequency
UPS	uninterruptible power supply
USART	universal synchronous/asynchronous receiver/transmitter
USOA	uniform system of accounts
USOAR	uniform system of accounts rewrite
USOC	uniform service order code
USTA	United States Telephone Association
USTSA	United States Telephone Suppliers Association
UTC	coordinated universal time

V

VA	volt-amperes
VCR	video cassette recorder
VCXO	voltage controlled crystal oscillator
VDRV	variable data rate video
VDT	video display terminal
VFD	vacuum fluorescent display
VFO	variable frequency oscillator
VHF	very high frequency
VIA-D	voice interface access-disabled
VIR	vertical interval reference

VITC	vertical interval time code
VITS	vertical interval test signal
VLF	very low frequency
VLSI	very large-scale integration
VOM	volt-ohm-milliammeter
vox	voice-operated relay
VSAT	very small aperture terminal
VSWR	voltage standing wave ratio
VT	virtual tributary
VTR	videotape recorder
VU	volume unit

W

WAN	wide area network
WARC	World Administrative Radio Conference
WATS	wide area telecommunications service
WDM	wavelength division multiplexing
WF monitor	waveform monitor
WORD	work order record and details
WORM	write once read many
WYSIWYG	what you see is what you get

X

XFMR	transformer
XMTR	transmitter
XOR	exclusive OR
XTALK	crosstalk

Y

YAG	yttrium-aluminum garnet
YIG	yttrium-iron garnet

Z

Z	impedance

Index

0-8493-0050-9/00/$0.00+$.50